MW01346386

Ecosystems: Functions, Sustainability and Management

Ecosystems: Functions, Sustainability and Management

Editor: Sierra Adkins

www.callistoreference.com

Callisto Reference,
118-35 Queens Blvd., Suite 400,
Forest Hills, NY 11375, USA

Visit us on the World Wide Web at:
www.callistoreference.com

ISBN: 978-1-64116-193-0 (Hardback)

Cataloging-in-Publication Data

Ecosystems : functions, sustainability and management / edited by Sierra Adkins.
p. cm.
Includes bibliographical references and index.
ISBN 978-1-64116-193-0
1. Ecosystem management. 2. Sustainability.
3. Biotic communities. I. Adkins, Sierra.

QH75 .E26 2019
333.95--dc23

Table of Contents

Preface

Ecosystems are communities made of biotic and abiotic components, which interact within the system through energy flows and nutrient cycles. Ecosystems are complex as they involve interactions between organisms and between organisms and their environment. The functioning of an ecosystem is affected by its biodiversity. Anthropogenic activities have a significant influence on the environment. Climate change, environmental pollution, biodiversity loss, soil degradation, microplastics pollution, etc. are some of the prominent threats to the ecosystem. Therefore, management of ecosystems is a growing concern. This involves the conservation and restoration of natural resources, in order to meet the need of current and future generations. This book provides significant information of this discipline to help develop a good understanding of ecosystems, their management, functions and sustainability. The topics covered herein deal with the core aspects in this area. Those in search of information to further their knowledge will be greatly assisted by this book.

This book has been the outcome of endless efforts put in by authors and researchers on various issues and topics within the field. The book is a comprehensive collection of significant researches that are addressed in a variety of chapters. It will surely enhance the knowledge of the field among readers across the globe.

It gives us an immense pleasure to thank our researchers and authors for their efforts to submit their piece of writing before the deadlines. Finally in the end, I would like to thank my family and colleagues who have been a great source of inspiration and support.

Editor

1

Small phytoplankton contribution to the standing stocks and the total primary production in the Amundsen Sea

Sang H. Lee[1], Bo Kyung Kim[1], Yu Jeong Lim[1], HuiTae Joo[1], Jae Joong Kang[1], Dabin Lee[1], Jisoo Park[2], Sun-Yong Ha[2], and Sang Hoon Lee[2]

[1]Department of Oceanography, Pusan National University, Geumjeong-gu, Busan 609-735, South Korea
[2]Korea Polar Research Institute, Incheon 406-840, South Korea

Correspondence to: Sang H. Lee (sanglee@pnu.ac.kr)

Abstract. Small phytoplankton are anticipated to be more important in a recently warming and freshening ocean condition. However, little information on the contribution of small phytoplankton to overall phytoplankton production is currently available in the Amundsen Sea. To determine the contributions of small phytoplankton to total biomass and primary production, carbon and nitrogen uptake rates of total and small phytoplankton were obtained from 12 productivity stations in the Amundsen Sea. The daily carbon uptake rates of total phytoplankton averaged in this study were $0.42\,g\,C\,m^{-2}\,d^{-1}$ ($SD = \pm 0.30\,g\,C\,m^{-2}\,d^{-1}$) and $0.84\,g\,C\,m^{-2}\,d^{-1}$ ($SD = \pm 0.18\,g\,C\,m^{-2}\,d^{-1}$) for non-polynya and polynya regions, respectively, whereas the daily total nitrogen (nitrate and ammonium) uptake rates were $0.12\,g\,N\,m^{-2}\,d^{-1}$ ($SD = \pm 0.09\,g\,N\,m^{-2}\,d^{-1}$) and $0.21\,g\,N\,m^{-2}\,d^{-1}$ ($SD = \pm 0.11\,g\,N\,m^{-2}\,d^{-1}$), respectively, for non-polynya and polynya regions, all of which were within the ranges reported previously. Small phytoplankton contributed 26.9 and 27.7 % to the total carbon and nitrogen uptake rates of phytoplankton in this study, respectively, which were relatively higher than the chlorophyll *a* contribution (19.4 %) of small phytoplankton. For a comparison of different regions, the contributions for chlorophyll *a* concentration and primary production of small phytoplankton averaged from all the non-polynya stations were 42.4 and 50.8 %, which were significantly higher than those (7.9 and 14.9 %, respectively) in the polynya region. A strong negative correlation ($r^2 = 0.790$, $p < 0.05$) was found between the contributions of small phytoplankton and the total daily primary production of phytoplankton in this study. This finding implies that daily primary production decreases as small phytoplankton contribution increases, which is mainly due to the lower carbon uptake rate of small phytoplankton than large phytoplankton.

1 Introduction

The Amundsen Sea is located in West Antarctica between the Ross Sea and Bellingshausen Sea (Fig. 1), which is one of the least biologically studied regions in the Southern Ocean. Recently, several international research programs (KOPRI Amundsen project, iSTAR, ASPIRE and DynaLiFe) were launched to improve the understanding of this remote area. Field-measurement data revealed that annual primary production of phytoplankton reaching to $220\,g\,C\,m^{-2}\,yr^{-1}$ in the Amundsen Sea polynya is as high as that of Ross Sea polynya ($200\,g\,C\,m^{-2}\,yr^{-1}$) which was previously known for the highest productivity region in the Southern Ocean (Lee et al., 2012). Given the fact that the chlorophyll *a* concentration averaged from all the chlorophyll *a* measured stations was twice as high as that of the only productivity-measured stations, Lee et al. (2012) argued that the annual production in the Amundsen Sea polynya could be even 2 times higher than that of Ross Sea polynya.

Over the past several decades, a rapid climate change has been detected and subsequently physical changes have occurred in the marine ecosystem in the western Antarctic Peninsula (WAP), which was mainly based on the results from Palmer Antarctic Long-Term Ecological Research project focusing on the north of ~69° S (Ducklow et al., 2007; Montes-Hugo et al., 2009). Recent studies revealed

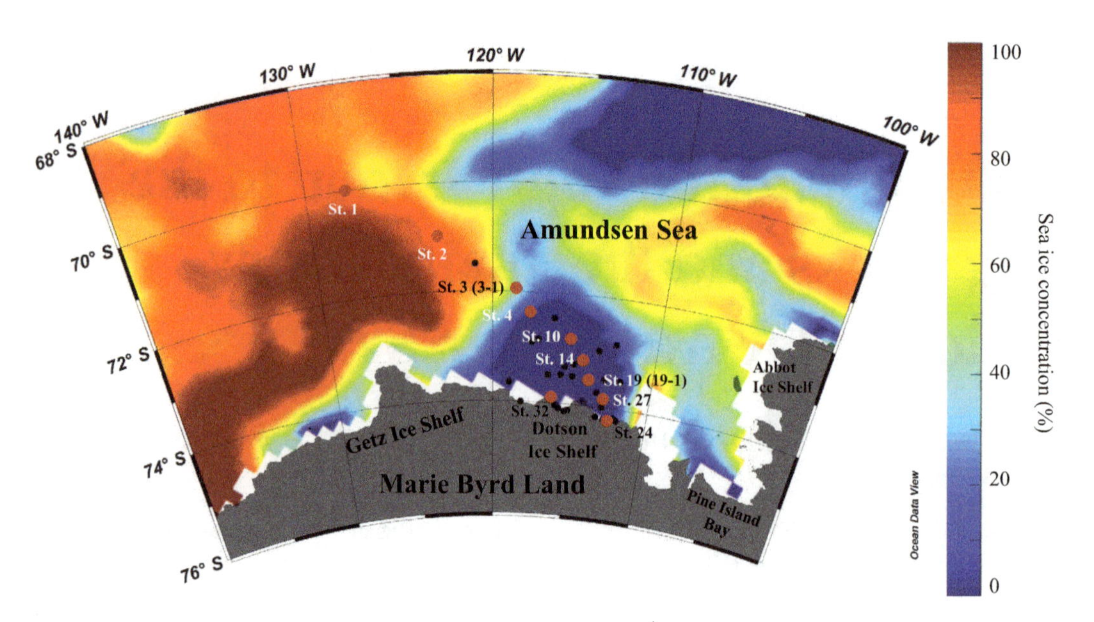

Figure 1. Sampling locations in the Amundsen Sea. Red closed circles represent productivity stations. Sea ice concentration data during the cruise period in 2013 from Nimbus-7 SMMR and DMSP SSM/I-SSMIS Passive Microwave data provided by National Snow and Ice Data Center.

that the Thwaites Glacier in Pine Island Bay is retreating fast and the ice volume loss in the nearby Getz Ice shelf is accelerating (Joughin et al., 2014; Paolo et al., 2015). Shoaling warm Circumpolar Deep Water is believed to be a main cause of the ice sheet mass loss through the ice shelf basal melt underside of the ice shelves (Yager et al., 2012; Schmidtko et al., 2014). Climate change from a cold-dry polar type to a warm-humid sub-Antarctic type has driven subsequent changes in ocean biological productivity along the WAP shelf over the recent 3 decades (Montes-Hugo et al., 2009).

Phytoplankton, as the base of oceanic food webs, can be an indicator for changes in marine ecosystems responding to environmental changes (Moline et al., 2004; Wassman et al., 2011; Arrigo and van Dijken, 2015). For example, a recurrent shift in phytoplankton community structure from large diatoms to relatively small cryptophytes could be tightly associated with changes in glacial meltwater runoff (Moline et al., 2004). To date, little information on the contribution of small phytoplankton to primary production is available in the Antarctic Ocean (Saggiomo et al., 1998), especially in the Amundsen Sea with a rapid melting of ice shelf (Yager et al., 2012; Schmidtko et al., 2014). Thus, our main objective in this study is to determine contributions of small phytoplankton to the overall total biomass and primary production of phytoplankton in the Amundsen Sea for monitoring marine ecosystem responding to environmental condition change.

2 Materials and methods

2.1 Total and size-fractionated chlorophyll *a* concentration

Water samples for total and size-fractionated chlorophyll *a* concentrations of phytoplankton were obtained at the 12 productivity stations in the Amundsen Sea (Fig. 1) during the KOPRI Amundsen cruise from 1 to 15 January 2014 onboard the Korean research icebreaker *Araon*. Based on the sea ice concentration data from the National Snow and Ice Data Center during the cruise period in 2013 (Fig. 1), our study region was further separated into polynya and non-polynya areas for comparison. Four stations (St. 1, St. 2, St. 3 and St. 3-1) among the 12 stations belong to the non-polynya region and the rest of the stations belong to the polynya region. St. 3 and St. 3-1 were on the fringe of the polynya area, experiencing approximately 30 % sea ice cover. Following the definition of polynya as an area of open water within sea ice zone, we grouped them into non-polynya regions. Six different light depths (100, 50, 30, 12, 5 and 1 % penetration of the surface irradiance, PAR) were determined with an LI-COR underwater 4π light sensor. Total chlorophyll *a* concentrations were measured at the six different light depths (100, 50, 30, 12, 5 and 1 % of PAR). For size-fractionated chlorophyll *a* concentrations, water samples were collected at three light depths (100, 30 and 1 %). Water samples (0.3–0.5 L) for total chlorophyll *a* concentrations were filtered using Whatman glass fiber filters (GF/F; 25 mm). For different size-fractionated chlorophyll *a* concentrations water samples (0.7–1 L) were passed sequentially through 20 and 5 µm Nucleopore filters (47 mm) and 0.7 µm GF/F filters (47 mm).

Table 1. Percentage contributions (%) of small phytoplankton to depth-integrated total concentrations of chlorophyll *a*, POC, PON and carbon and nitrogen uptake rates in the Amundsen Sea.

	Chlorophyll *a*	POC	PON	Daily carbon uptake rate	Daily nitrate uptake rate	Daily ammonium uptake rate	Total nitrogen uptake rate
All stations	19.4 ± 26.0	41.1 ± 10.6	41.3 ± 11.5	26.9 ± 29.3	21.5 ± 11.1	38.7 ± 24.9	27.7 ± 14.4
Non-polynya	42.4 ± 37.2	49.5 ± 14.4	50.0 ± 15.1	50.8 ± 42.8	28.2 ± 15.9	52.8 ± 40.5	36.2 ± 23.0
Polynya	7.9 ± 3.5	36.9 ± 4.6	37.0 ± 6.9	14.9 ± 8.4	18.1 ± 6.8	31.6 ± 10.1	23.5 ± 6.0

After the filters were extracted using the method described by Kim et al. (2015), all chlorophyll *a* concentrations were subsequently determined onboard using a Trilogy fluorometer (Turner Designs, USA). The methods and calculations for chlorophyll *a* were based on Parsons et al. (1984).

2.2 Carbon and nitrogen uptake experiments

Water samples were collected for and carbon and nitrogen uptake measurements of phytoplankton. Using a dual stable isotope technique (Lee et al., 2012; Kim et al., 2015), the experiments of carbon and nitrogen uptake rates of phytoplankton were conducted at 12 selected productivity stations including two revisited stations (St. 3-1 and St. 19-1) when on-deck incubations were available during daytime at oceanographic survey stations. Water samples from the six light depths for the uptake experiments were obtained from a CTD-rosette sampler system equipped with twenty-four 10 L Niskin bottles. Water sample from each light depth was transferred into different screened polycarbonate incubation bottles (1 L) which matched each light depth. The bottles were placed in large polycarbonate incubators cooled with running surface seawater on deck under natural light conditions for 4–5 h, after the water samples in the incubation bottles were inoculated with labeled carbon ($NaH^{13}CO_3$) and nitrate ($K^{15}NO_3$) or ammonium ($^{15}NH_4Cl$) substrates. After 4–5 h incubations, the incubated waters were well mixed and distributed into two filtration sets for the carbon and nitrogen uptake rates of total (> 0.7 µm) and small-sized cells (0.7–5 µm). Small-sized cells are generally defined as small phytoplankton in comparison to large diatoms (> 5 µm) (Robineau et al., 1994; references therein). The incubated waters (0.3 L) for total uptake rates were filtered through pre-combusted GF/F filters (25 mm diameter), whereas waters samples (0.5 L) for the uptake rates of small phytoplankton were passed through 5 µm Nuclepore filters (47 mm) to remove large phytoplankton cells (> 5 µm) and then the filtrate was passed through pre-combusted GF/F (25 mm) for the small phytoplankton (Lee et al., 2013). The values for large phytoplankton in this study were obtained from the difference between small and total fractions (Lee et al., 2013). The filters were immediately preserved at −80 °C until mass spectrometric analysis. After acid fuming overnight to remove carbonate, the concentrations of particulate organic carbon (POC) and nitrogen (PON) and the abundance of ^{13}C and ^{15}N were determined by a Finnigan Delta+XL mass spectrometer at the Alaska Stable Isotope Facility, USA. All contribution results of small phytoplankton in this study were estimated from comparison of small phytoplankton to total phytoplankton integral values from 100 to 1 % light depth at each station based on the trapezoidal rule. Daily carbon and nitrogen uptake rates of phytoplankton were based on our hourly uptake rates measured in this study and a 24 h photoperiod per day during the summer period in the Amundsen Sea (Lee et al., 2012).

3 Results

3.1 Chlorophyll *a*, POC and PON contributions of small phytoplankton

The depth-integrated total (large + small phytoplankton) chlorophyll *a* concentration was 11.1–80.3 mg Chl *a* m^{-2} (mean ± SD = 57.4 ± 25.2 mg Chl *a* m^{-2}) in this study (Fig. 2). The contribution of small phytoplankton to the total chlorophyll *a* concentration was 4.9–76.5 % (19.4 ± 26.0 %). Large phytoplankton (> 5 µm) were generally predominant (approximately 80 %) based on different-sized chlorophyll *a* concentrations. For a regional comparison, the average contributions of small phytoplankton to the total chlorophyll *a* concentration were 42.4 % (±37.2 %) and 7.9 % (±3.5 %) for non-polynya and polynya regions, respectively (Table 1). The chlorophyll *a* contribution of small phytoplankton was larger in the non-polynya region than in the polynya region although they were not significantly different (*t* test, $p > 0.05$).

The depth-integrated total POC concentration of phytoplankton showed no large spatial variation ranging from 4.72 to 9.22 mg C m^{-2} (7.40 ± 1.55 mg C m^{-2}) (Fig. 3). In comparison, the depth-integrated total PON concentration of phytoplankton was 0.76–1.74 mg N m^{-2} (1.33 ± 0.32 mg N m^{-2}). The POC contribution of small phytoplankton was 30.7–65.5 % (41.1 ± 10.6 %), whereas the PON contribution was 30.8–67.2 % (41.3 ± 11.5 %) in the Amundsen Sea (Fig. 3). Specifically, the POC and PON contributions of small phytoplankton averaged from all the productivity stations in the polynya region were 36.9 % (±4.6 %) and 37.0 % (±6.9 %), respectively, whereas they were 49.5 % (±14.4 %) and 50.0 % (±15.1 %), respectively, in the non-polynya re-

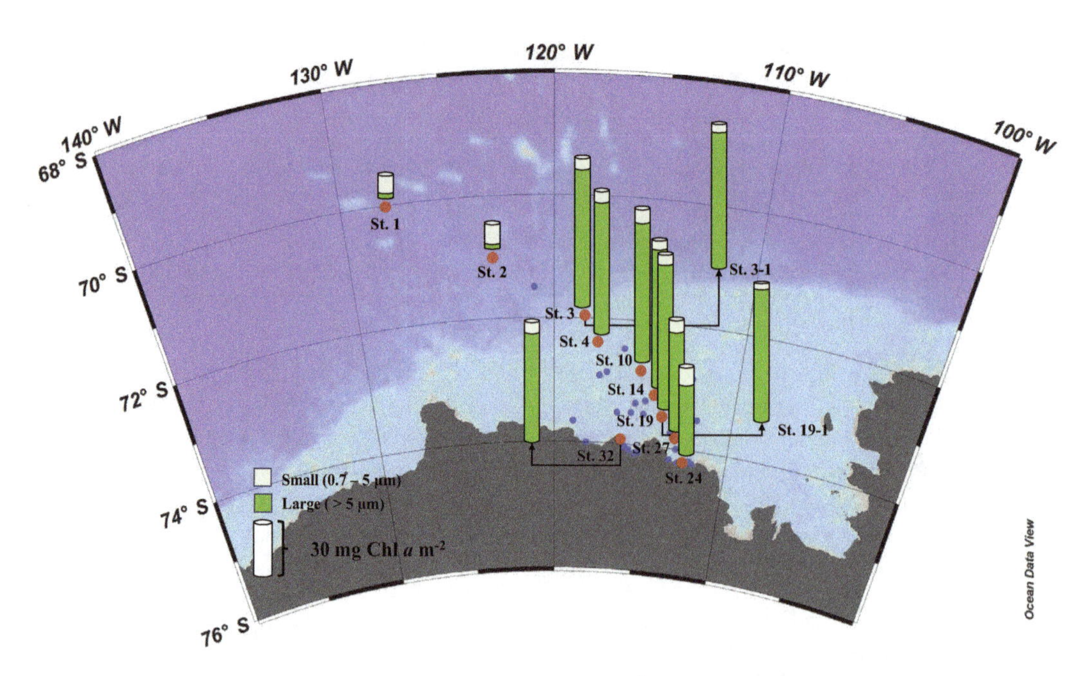

Figure 2. Water-column-integrated chlorophyll *a* concentrations (mg Chl *a* m^{-2}) of small (0.7–5 µm) and large (>5 µm) phytoplankton at the productivity stations in the Amundsen Sea.

gion (Table 1). The POC and PON contributions of small phytoplankton were not statistically different between the polynya and non-polynya regions (t test, $p>0.05$).

3.2 Carbon uptake rate contributions of small phytoplankton

The depth-integrated total daily carbon uptake rate of phytoplankton (large + small phytoplankton) was 150.4–1213.4 mg C m^{-2} d^{-1} (696.5 ± 298.4 mg C m^{-2} d^{-1}) in this study (Fig. 4). In contrast, the rate of small phytoplankton was 58.6–266.4 mg C m^{-2} d^{-1} (124.9±62.4 mg C m^{-2} d^{-1}). Small phytoplankton contributed 26.9 % (±29.3 %) to total daily carbon uptake rate of total phytoplankton.

Specifically, the total daily carbon uptake rate of phytoplankton was 150.4–796.4 mg C m^{-2} d^{-1} (415.0 ± 298.2 mg C m^{-2} d^{-1}) in the non-polynya region, whereas it was 654.8–1213.4 mg C m^{-2} d^{-1} (837.3 ± 184.1 mg C m^{-2} d^{-1}) in the polynya region. The total daily carbon uptake rates of phytoplankton were significantly higher (t test, $p<0.05$) in the polynya region than the non-polynya region. The rate of small phytoplankton was 58.6–193.6 mg C m^{-2} d^{-1} (126.5 ± 55.2 mg C m^{-2} d^{-1}) in the non-polynya region, whereas it was 62.2–266.4 mg C m^{-2} d^{-1} (124.1 ± 69.3 mg C m^{-2} d^{-1}) in the polynya region. The daily carbon uptake rates of small phytoplankton were not significantly different (t test, $p>0.05$) between the polynya and non-polynya stations. The average contributions of small phytoplankton to total daily carbon uptake rates were 50.8 % (±42.8 %) and 14.9 % (±8.4 %), respectively, for the non-polynya and polynya regions (Table 1). The average contributions were largely different between the polynya and non-polynya regions but they were not statistically significant (t test, $p>0.05$).

3.3 Nitrogen uptake rate contributions of small phytoplankton

The depth-integrated total daily nitrate uptake rate of phytoplankton (large + small phytoplankton) was 34.0–174.2 mg N m^{-2} d^{-1} (93.7 ± 43.2 mg N m^{-2} d^{-1}), whereas the rate of small phytoplankton was 6.1–40.9 mg N m^{-2} d^{-1} (19.0 ± 11.3 mg N m^{-2} d^{-1}) in this study (Fig. 5). Small phytoplankton contributed 21.5 % (±11.1 %) to total daily nitrate uptake rates. In comparison, the total daily ammonium uptake rate of phytoplankton was 12.4–173.8 mg N m^{-2} d^{-1} (86.7 ± 75.9 mg N m^{-2} d^{-1}), whereas the rate of small phytoplankton was 9.1–81.1 mg N m^{-2} d^{-1} (25.7 ± 21.1 mg N m^{-2} d^{-1}) in this study (Fig. 6). Small phytoplankton contributed 38.7 % (±24.9 %) to total daily ammonium uptake rates. The contributions of small phytoplankton were significantly higher in ammonium uptake rate than nitrate uptake rate (t test, $p<0.05$).

For different regions, the total daily nitrate uptake rates of phytoplankton were 34.0–142.1 mg N m^{-2} d^{-1} (71.9 ± 48.4 mg N m^{-2} d^{-1}) in the non-polynya region and 44.2–174.2 mg N m^{-2} d^{-1} (104.6 ± 39.0 mg N m^{-2} d^{-1}) in the polynya region, respectively. In comparison, the daily nitrate uptake rates of small phytoplankton were 7.5–26.6 mg N m^{-2} d^{-1} (16.7 ± 7.8 mg N m^{-2} d^{-1}) and 6.1–40.9 mg N m^{-2} d^{-1} (20.1 ± 13.1 mg N m^{-2} d^{-1}), respectively, for the non-polynya and polynya regions.

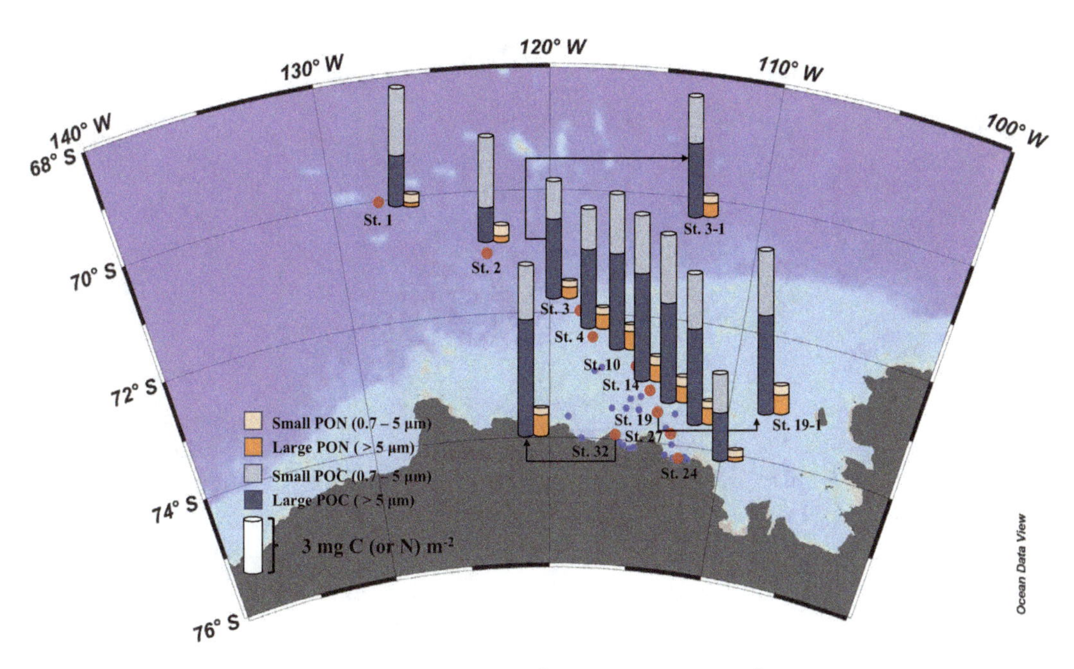

Figure 3. Water-column-integrated concentrations of POC ($mg\,C\,m^{-2}$) and PON ($mg\,N\,m^{-2}$) of small (0.7–5 μm) and large (>5 μm) phytoplankton.

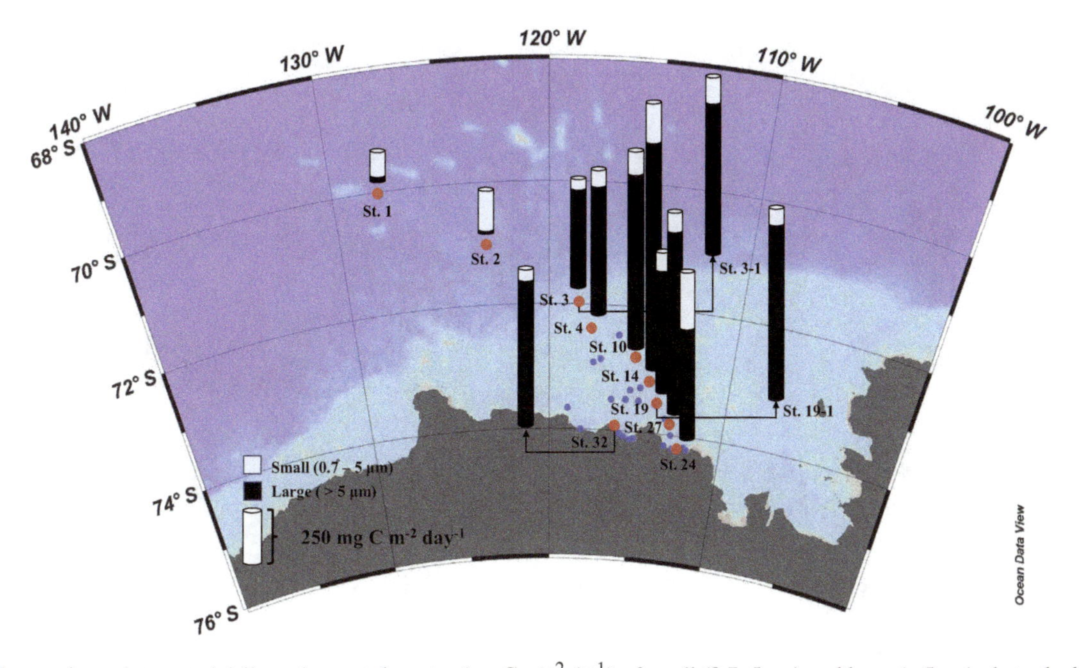

Figure 4. Water-column-integrated daily carbon uptake rates ($mg\,C\,m^{-2}\,d^{-1}$) of small (0.7–5 μm) and large (>5 μm) phytoplankton.

The contributions of small phytoplankton to the total daily nitrate uptake rates were 28.2 % (±15.9 %) in the non-polynya region and 18.1 % (±6.8 %) in the polynya region, respectively (Table 1). The total daily ammonium uptake rates of total phytoplankton were 12.3 and 106.1 $mg\,N\,m^{-2}\,d^{-1}$ ($49.7 \pm 41.2\,mg\,N\,m^{-2}\,d^{-1}$) in the non-polynya region and 18.1–269.3 $mg\,N\,m^{-2}\,d^{-1}$ ($105.2 \pm 84.6\,mg\,N\,m^{-2}\,d^{-1}$) in the polynya region. In comparison, the rates of small phytoplankton were 9.1–22.4 $mg\,N\,m^{-2}\,d^{-1}$ ($15.8 \pm 6.4\,mg\,N\,m^{-2}\,d^{-1}$) in the non-polynya region and 9.9–81.1 $mg\,N\,m^{-2}\,d^{-1}$ ($30.7 \pm 24.5\,mg\,N\,m^{-2}\,d^{-1}$) in the polynya region. Small phytoplankton contributed 52.8 % (±40.5 %) and 31.6 % (±10.1 %) to the total daily ammonium uptake rates in the non-polynya and polynya regions, respectively, which were not significantly different (t test, $p>0.05$).

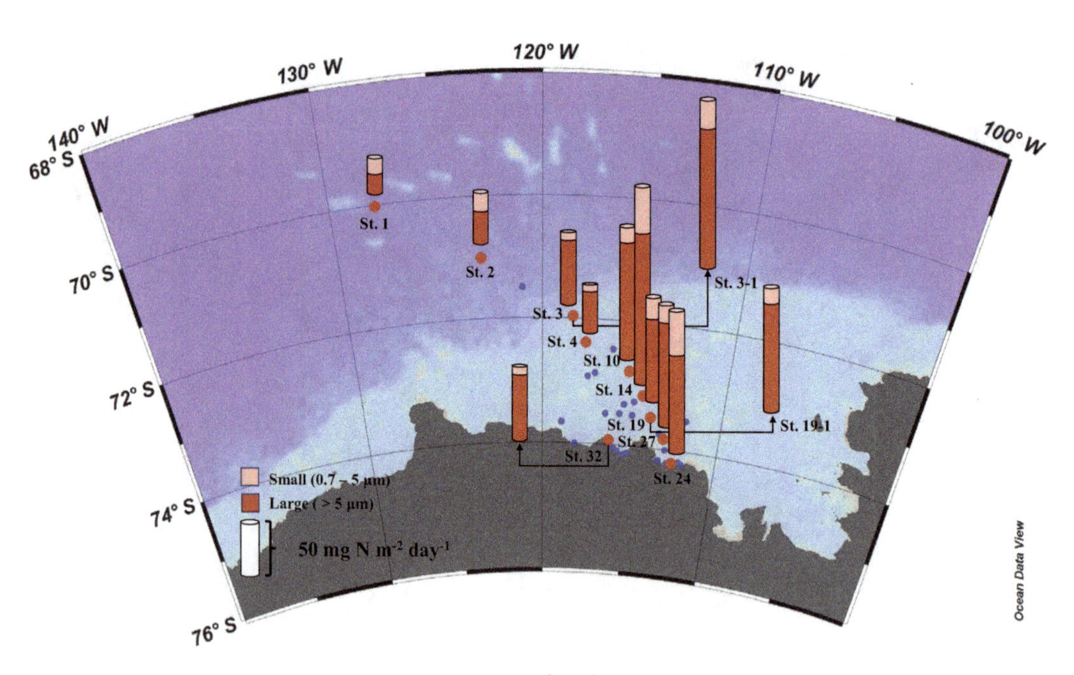

Figure 5. Water-column-integrated daily nitrate uptake rates ($mg\,N\,m^{-2}\,d^{-1}$) of small (0.7–5 µm) and large (> 5 µm) phytoplankton.

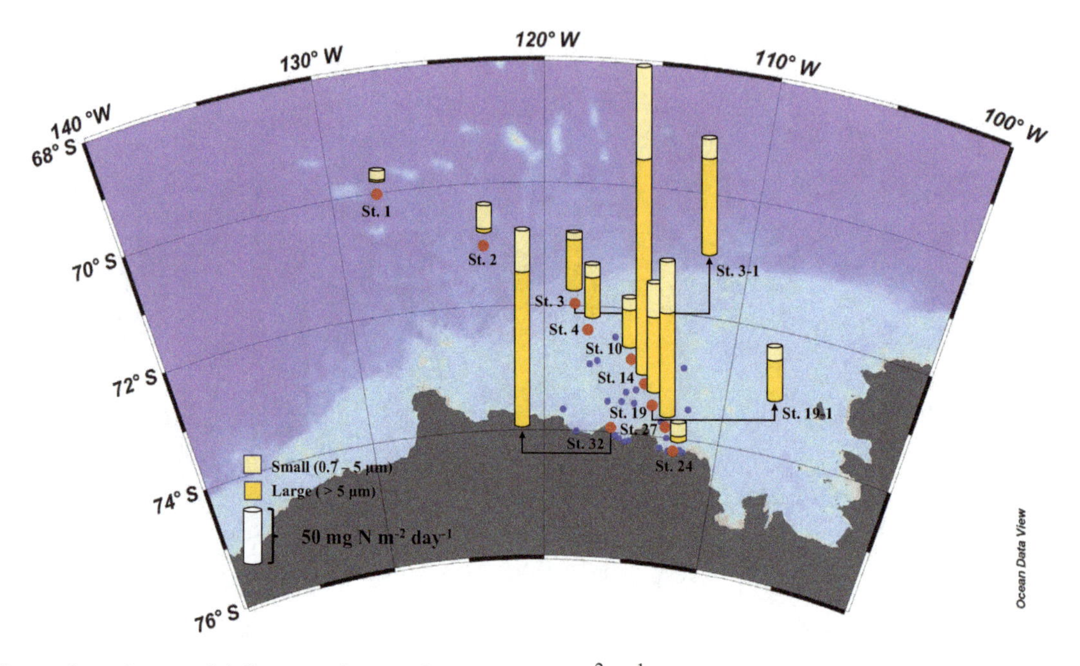

Figure 6. Water-column-integrated daily ammonium uptake rates ($mg\,N\,m^{-2}\,d^{-1}$) of small (0.7–5 µm) and large (> 5 µm) phytoplankton.

The total integral daily nitrogen uptake rate (nitrate + ammonium uptake rates) of phytoplankton was 46.4–443.5 $mg\,N\,m^{-2}\,d^{-1}$ (180.4 ± 106.7 $mg\,N\,m^{-2}\,d^{-1}$) in this study. For the non-polynya and polynya regions, they were 46.4–248.1 $mg\,N\,m^{-2}\,d^{-1}$ (121.6 ± 89.3 $mg\,N\,m^{-2}\,d^{-1}$) and 91.7–443.5 $mg\,N\,m^{-2}\,d^{-1}$ (209.8 ± 107.3 $mg\,N\,m^{-2}\,d^{-1}$), respectively. In comparison, the total integral daily nitrogen uptake rates of small phytoplankton were 16.6–46.6 $mg\,N\,m^{-2}\,d^{-1}$ (32.5 ± 13.2 $mg\,N\,m^{-2}\,d^{-1}$) and 17.6–122.0 $mg\,N\,m^{-2}\,d^{-1}$ (50.8 ± 32.4 $mg\,N\,m^{-2}\,d^{-1}$) for the non-polynya and polynya regions, respectively. Small phytoplankton contributed 36.2 % (±23.0 %) to the total integral daily nitrogen uptake rates in the non-polynya region, whereas they contributed 23.5 % (±6.0 %) for the polynya region (Table 1). The integral daily nitrogen uptake rates and contributions of small phytoplankton were not statistically different between the non-polynya and polynya regions.

4 Discussion and conclusion

The total daily carbon uptake rates of phytoplankton averaged for the non-polynya and polynya regions were $0.42\,\mathrm{g\,C\,m^{-2}\,d^{-1}}$ ($\pm 0.30\,\mathrm{g\,C\,m^{-2}\,d^{-1}}$) and $0.84\,\mathrm{g\,C\,m^{-2}\,d^{-1}}$ ($\pm 0.18\,\mathrm{g\,C\,m^{-2}\,d^{-1}}$), respectively, in this study. According to the previous reports in the Amundsen Sea (Lee et al., 2012; Kim et al., 2015), the total daily carbon uptake rates ranged from 0.2 to $0.12\,\mathrm{g\,C\,m^{-2}\,d^{-1}}$ in the non-polynya region. Our rate ($0.42\,\mathrm{g\,C\,m^{-2}\,d^{-1}}$) in the non-polynya region is somewhat higher than those reported previously but they are not significantly different (t test, $p>0.05$). In comparison, our total daily carbon uptake rate in the polynya region ($0.84\,\mathrm{g\,C\,m^{-2}\,d^{-1}}$) is lower than that ($2.2\,\mathrm{g\,C\,m^{-2}\,d^{-1}}$) of Lee et al. (2012) and higher than that ($0.2\,\mathrm{g\,C\,m^{-2}\,d^{-1}}$) of Kim et al. (2015). The carbon uptake rates of phytoplankton in Lee et al. (2012) and Kim et al. (2015) were measured during the periods 21 December 2010 to 23 January 2011 and 11 February to 14 March 2012, respectively. Our measurements in this study were executed mainly during the period 1–15 January 2014. For the Amundsen polynya region, a large seasonal variation in the total daily carbon uptake rate of phytoplankton was already reported by Kim et al. (2015) and Arrigo et al. (2012) based on field-measured data and satellite-derived approach, respectively. Generally, late December is the time of peak uptake rate in this region (Arrigo et al., 2012). Previous studies reported that the total daily nitrogen uptake rates in non-polynya region were $0.24\,\mathrm{g\,N\,m^{-2}\,d^{-1}}$ during the period 21 December 2010 to 23 January 2011 and $0.04\,\mathrm{g\,N\,m^{-2}\,d^{-1}}$ during the period 11 February to 14 March 2012, whereas the uptake rates in polynya region were $0.93\,\mathrm{g\,N\,m^{-2}\,d^{-1}}$ in 2010/2011 and $0.06\,\mathrm{g\,N\,m^{-2}\,d^{-1}}$ in 2012 in the Amundsen Sea (Lee et al., 2012; Kim et al., 2015). Our total daily nitrogen uptake rates of phytoplankton in non-polynya ($0.12 \pm 0.09\,\mathrm{g\,N\,m^{-2}\,d^{-1}}$) and polynya regions ($0.21 \pm 0.11\,\mathrm{g\,N\,m^{-2}\,d^{-1}}$) were between the ranges of two previous studies (Lee et al., 2012; Kim et al., 2015). Based on the nitrate and ammonium uptake rates in this study, f ratios (nitrate uptake rate/nitrate + ammonium uptake rates) averaged for non-polynya and polynya regions were 0.62 (± 0.08) and 0.54 (± 0.20), respectively. These ratios were also between the ranges of two previous studies. Although they were not significantly different because of a large spatial variation, larger f ratios in non-polynya than in polynya region are consistent with the results of the previous studies (Lee et al., 2012; Kim et al., 2015). At this point, we do not have a solid explanation for that and a further future study is needed for the higher f-ratio mechanism in non-polynya region.

The percentage contributions of small phytoplankton to chlorophyll *a*, POC/PON, daily carbon and nitrogen uptake rates are shown in Table 1. The result of significantly higher chlorophyll *a* contribution than the POC contribution of small phytoplankton is consistent with the result in the Chukchi Sea, Arctic Ocean reported by Lee et al. (2013). They explain that higher POC content per chlorophyll *a* unit of small phytoplankton could have caused the higher POC contribution in their study (Lee et al., 2013). Given C / N ratio (6.6 ± 0.6) and δ^{13}C (-25.9 ± 1.0 ‰) of sample filters attained for POC and PON in this study, our filtered samples are believed to be mainly phytoplankton-originated POC and PON (Kim et al., 2016). Thus, a significant potential overestimated contribution of POC caused by non-phytoplankton materials could be excluded for the higher POC contribution than chlorophyll *a* contribution of small phytoplankton. Therefore, small phytoplankton contributions based on conventional assessments of chlorophyll *a* concentration might lead to an underestimated contribution of small phytoplankton (Lee et al., 2013). In fact, several authors argue that chlorophyll *a* concentration might not be a good index for phytoplankton biomass since it depends largely on environmental factors such as nutrient and light conditions, as well as dominant groups and physiological status of phytoplankton (Desortová, 1981; Behrenfeld et al., 2005; Kruskopf and Flynn, 2006; Behrenfeld and Boss, 2006). However, the effects of non-phytoplankton carbon materials such as extracellular carbon mucilage cannot be completely excluded for the POC contribution as discussed below.

The overall contributions of carbon (26.9 %) and nitrogen (27.7 %) uptake rates of small phytoplankton at all the productivity stations in this study are relatively higher than the chlorophyll *a* contribution of small phytoplankton but they are not statistically different (t test, $p>0.05$). In general, the contribution of daily ammonium uptake rate of small phytoplankton is significantly (t test, $p<0.05$) higher than the contribution of daily nitrate uptake rate of small phytoplankton at all the stations in this study. This is well known for the ammonium preference of small phytoplankton in various regions (Koike et al., 1986; Tremblay et al., 2000; Lee et al., 2008, 2013).

In terms of the contributions in different regions, all the contributions (chlorophyll *a*, POC/PON, carbon and nitrogen uptake rates) of small phytoplankton were higher in the non-polynya region than in the polynya region (Table 1). In addition, the chlorophyll *a* contribution of small phytoplankton (7.9 ± 3.5 %) was significantly (t test, $p<0.05$) lower than the POC contribution (36.9 ± 4.6 %) in the polynya region, whereas they were not statistically different in the non-polynya region (Table 1). This indicates that small phytoplankton contributed more to the total POC than to the chlorophyll *a* concentration in the polynya region. We do not have species compositions of phytoplankton in this study, but previous results reported that *Phaeocystis* spp. are dominant in the Amundsen Sea polynya region (Lee et al., 2012). Generally, *Phaeocystis* spp. release a large amount (up to 46 %) of extracellular carbon mucilage which constitutes their colonial form (Matrai et al., 1995). This non-phytoplankton carbon material without chlorophyll *a* might have caused a high POC contribution of small phytoplankton in the polynya region in this study. In fact, the contribution of the daily carbon

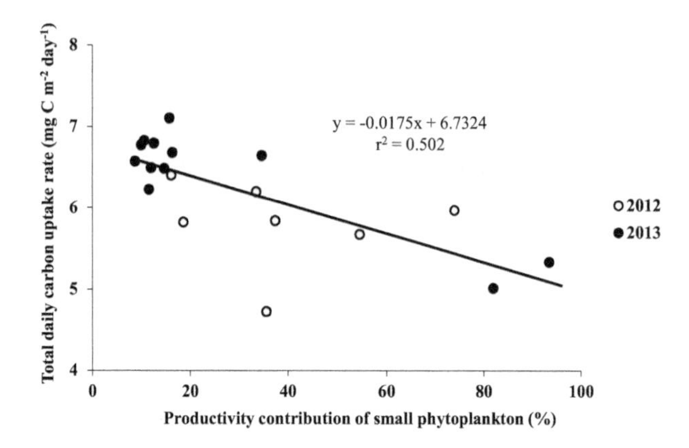

Figure 7. Relationship between productivity contributions (%) of small phytoplankton and the total daily carbon uptake rates ($mg\,C\,m^{-2}\,d^{-1}$) of phytoplankton (large + small). The total daily carbon uptake rates were transformed into natural logs for a linear regression. Red circles represent data obtained in 2013 (this study). Yellow circles representing 2012 data (unpublished) were included for a better regression.

uptake rates of small phytoplankton (14.9 ± 8.4 %) was not as high as the POC contribution (36.9 ± 4.6 %) in the polynya region. The chlorophyll *a* contributions of small phytoplankton were lower than those of the daily carbon uptake rate in this study, which is consistent with the results from polynya and marginal ice zone stations in the Ross Sea, Antarctica, during austral spring and summer (Saggiomo et al., 1998). They reported that the chlorophyll *a* and primary production contributions of pico-phytoplankton (< 2 µm) were 29 and 40 % at polynya stations, whereas the contributions were 17 and 32 % at marginal ice zone stations, respectively. In the polynya region, they found much higher contributions in chlorophyll *a* and primary production of small phytoplankton than those in this study, although their size of the small phytoplankton is somewhat smaller than our size (< 5 µm).

In conclusion, we found a strong negative correlation ($r^2 = 0.502$, $p<0.05$) between the productivity contributions of small phytoplankton and total daily carbon uptake rates of total phytoplankton in the Amundsen Sea (Fig. 7), which indicates that daily primary production decreases as small phytoplankton contribution increases. With respect to food quality of small phytoplankton as a basic food source to herbivores, macromolecular compositions such as proteins, lipids and carbohydrates as photosynthetic end products will be needed for a better understanding of a small-cell-dominant marine ecosystem in response to environmental changes (Lee et al., 2013). According to Kang et al. (2017), small phytoplankton assimilate more food materials and calorific contents per unit of chlorophyll *a* concentration and thus provide more contributions with respect to the energy aspect than do other phytoplankton communities in the East/Japan Sea. However, this change in dominant phytoplankton community from large to small cells will likely cause further alteration in the higher trophic levels because of the prey size available to higher trophic grazers (Moline et al., 2004). Monitoring the contributions of small phytoplankton to total biomass and primary production of the total phytoplankton community is important, as it provides a valuable indicator to sense environmental changes and consequently their potential influence on higher trophic animals in marine ecosystem.

Competing interests. The authors declare that they have no conflict of interest.

Acknowledgements. We thank the captain and crew members of the Korean research icebreaker *Araon*, for their outstanding assistance during the cruise. This research was supported by the Korea Polar Research Institute (KOPRI; PP15020).

Edited by: Gerhard Herndl

References

Arrigo, K. R. and van Dijken, G. L.: Continued increases in Arctic Ocean primary production, Prog. Oceanogr., 136, 60–70, 2015.

Arrigo, K. R., Lowry, K. E., and van Dijken, G. L.: Annual changes in sea ice and phytoplankton in polynyas of the Amundsen Sea, Antarctica, Deep-Sea Res. Pt. II, 71, 5–15, 2012.

Behrenfeld, M. J. and Boss, E.: Beam attenuation and chlorophyll concentration as alternative optical indices of phytoplankton biomass, J. Mar. Res., 64, 431-451, 2006.

Behrenfeld, M. J., Boss, E., Siegel, D. A. and Shea, D. M.: Carbon-based ocean productivity and phytoplankton physiology from space, Global Biogeochem. Cy., 19, GB1006, https://doi.org/10.1029/2004GB002299, 2005.

Desortová, B.: Relationship between chlorophyll-a concentration and phytoplankton biomass in several reservoirs in Czechoslovakia, Int. Rev. ges. Hydrobio., 66, 153–169, 1981.

Ducklow, H. W., Baker, K., Martinson, D. G., Quetin, L. B., Ross, R. M., Smith, R. C., Stammerjohn, S. E., Vernet, M., and Fraser, W.: Marine pelagic ecosystems: the west Antarctic Peninsula, Philos. Trans. R. Soc. Lond. B. Biol. Sci., 362, 67–94, 2007.

Joughin, I., Smith, B. E., and Medley, B.: Marine ice sheet collapse potentially under way for the Thwaites Glacier Basin, West Antarctica, Science, 344, 735–738, 2014.

Kang, J. J., Joo, H. T., Lee, J. H., Lee, J. H., Lee, H. W., Lee, D., Kang, C. K., Yun, M. S., and Lee, S. H.: Composition of biochemical compositions of phytoplankton during spring and fall seasons in the northern East/Japan Sea, Deep-Sea Res. Pt. II, https://doi.org/https://doi.org/10.1016/j.dsr2.2017.06.006, 2017.

Kim, B. K., Joo, H., Song, H. J., Yang, E. J., Lee, S. H., Hahm, D., Rhee, T. S., and Lee, S. H.: Large seasonal variation in phytoplankton production in the Amundsen Sea, Polar Biol., 38, 319–331, 2015.

Kim, B. K., Lee, J. H., Joo, H., Song, H. J., Yang, E. J., Lee, S. H., and Lee, S. H.: Macromolecular compositions of phytoplankton in the Amundsen Sea, Antarctica, Deep-Sea Res. Pt. II, 123, 42–49, 2016.

Koike, I., Holm-Hansen, O., and Biggs, D. C.: Inorganic nitrogen metabolism by Antarctic phytoplankton with special reference to ammonium cycling, Mar. Ecol.-Prog. Ser., 30, 105–116, 1986.

Kruskopf, M. and Flynn, K. J.: Chlorophyll content and fluorescence responses cannot be used to gauge reliably phytoplankton biomass, nutrient status or growth rate, New Phytol., 169, 525–536, 2006.

Lee, S. H., Whitledge, T. E., and Kang, S.: Spring time production of bottom ice algae in the landfast sea ice zone at Barrow, Alaska, J. Exp. Mar. Biol., 367, 204–212, 2008.

Lee, S. H., Kim, B. K., Yun, M. S., Joo, H., Yang, E. J., Kim, Y. N., Shin, H. C., and Lee, S.: Spatial distribution of phytoplankton productivity in the Amundsen Sea, Antarctica, Polar Biol., 35, 1721–1733, 2012.

Lee, S. H., Yun, M. S., Kim, B. K., Joo, H., Kang, S.-J., Kang, C. K., and Whitledge, T. E.: Contribution of small phytoplankton to total primary production in the Chukchi Sea, Cont. Shelf Res., 68, 43–50, 2013.

Matrai, P., Vernet, M., Hood, R., Jennings, A., Brody, E., and Saemundsdóttir, S.: Light-dependence of carbon and sulfur production by polar clones of the genus Phaeocystis, Mar. Biol., 124, 157–167, 1995.

Moline, M. A., Claustre, H., Frazer, T. K., Schofield, O., and Vernet, M.: Alteration of the food web along the Antarctic Peninsula in response to a regional warming trend, Glob. Change Biol., 10, 1973–1980, 2004.

Montes-Hugo, M., Doney, S. C., Ducklow, H. W., Fraser, W., Martinson, D., Stammerjohn, S. E., and Schofield, O.: Recent changes in phytoplankton communities associated with rapid regional climate change along the western Antarctic Peninsula, Science, 323, 1470–1473, 2009.

Paolo, F. S., Fricker, H. A., and Padman, L.: Ice sheets. Volume loss from Antarctic ice shelves is accelerating, Science, 348, 327–331, 2015.

Parsons, T. R., Maita, Y., and Lalli, C. M.: A manual of chemical and biological methods for seawater analysis, Publ. Pergamon Press, Oxford, 1984.

Robineau, B., Legendre, L., Therriault, J.-C., Fortier, L., Rosenberg, G., and Demers, S.: Ultra-algae (<5 μm) in the ice, at the ice-water interface and in the under-ice water column (southeastern Hudson Bay, Canada), 115, 169–180, 1994.

Saggiomo, V., Carrada, G., Mangoni, O., d'Alcala, M. R., and Russo, A.: Spatial and temporal variability of size-fractionated biomass and primary production in the Ross Sea (Antarctica) during austral spring and summer, J. Mar. Syst., 17, 115–127, 1998.

Schmidtko, S., Heywood, K. J., Thompson, A. F., and Aoki, S.: Multidecadal warming of Antarctic waters, Science, 346, 1227–1231, 2014.

Tremblay, J. É., Legendre, L., Klein, B., and Therriault, J.: Size-differential uptake of nitrogen and carbon in a marginal sea (Gulf of St. Lawrence, Canada): significance of diel periodicity and urea uptake, Deep-Sea Res. Pt. II, 47, 489–518, 2000.

Wassmann, P., Duarte, C. M., Agusti, S., and Sejr, M. K.: Footprints of climate change in the Arctic marine ecosystem, Glob. Change Biol., 17, 1235–1249, 2011.

Yager, P. L., Sherrell, L., Stammerjohn, S. E., Alderkamp, A., Schofield, O., Abrahamsen, E. P., Arrigo, K. R., Bertilsson, S., Garay, D., and Guerrero, R.: ASPIRE: the Amundsen Sea Polynya international research expedition, Oceanography, 25, 40–53, 2012.

2

Soil moisture control of sap-flow response to biophysical factors in a desert-shrub species, *Artemisia ordosica*

Tianshan Zha[1,3,*], Duo Qian[2,*], Xin Jia[1,3], Yujie Bai[1], Yun Tian[1], Charles P.-A. Bourque[4], Jingyong Ma[1], Wei Feng[1], Bin Wu[1], and Heli Peltola[5]

[1]Yanchi Research Station, School of Soil and Water Conservation, Beijing Forestry University, Beijing 100083, China
[2]Beijing Vocational College of Agriculture, Beijing 102442, China
[3]Key Laboratory of State Forestry Administration on Soil and Water Conservation, Beijing Forestry University, Beijing, China
[4]Faculty of Forestry and Environmental Management, 28 Dineen Drive, P.O. Box 4400, University of New Brunswick, New Brunswick, E3B5A3, Canada
[5]Faculty of Science and Forestry, School of Forest Sciences, University of Eastern Finland, Joensuu, 80101, Finland
[*]These authors contributed equally to this work.

Correspondence to: Tianshan Zha (tianshanzha@bjfu.edu.cn)

Abstract. The current understanding of acclimation processes in desert-shrub species to drought stress in dryland ecosystems is still incomplete. In this study, we measured sap flow in *Artemisia ordosica* and associated environmental variables throughout the growing seasons of 2013 and 2014 (May–September period of each year) to better understand the environmental controls on the temporal dynamics of sap flow. We found that the occurrence of drought in the dry year of 2013 during the leaf-expansion and leaf-expanded periods caused sap flow per leaf area (J_s) to decline significantly, resulting in transpiration being 34 % lower in 2013 than in 2014. Sap flow per leaf area correlated positively with radiation (R_s), air temperature (T), and water vapor pressure deficit (VPD) when volumetric soil water content (VWC) was greater than $0.10\,m^3\,m^{-3}$. Diurnal J_s was generally ahead of R_s by as much as 6 hours. This time lag, however, decreased with increasing VWC. The relative response of J_s to the environmental variables (i.e., R_s, T, and VPD) varied with VWC, J_s being more strongly controlled by plant-physiological processes during periods of dryness indicated by a low decoupling coefficient and low sensitivity to the environmental variables. According to this study, soil moisture is shown to control sap-flow (and, therefore, plant-transpiration) response in *Artemisia ordosica* to diurnal variations in biophysical factors. This species escaped (acclimated to) water limitations by invoking a water-conservation strategy with the regulation of stomatal conductance and advancement of J_s peaking time, manifesting in a hysteresis effect. The findings of this study add to the knowledge of acclimation processes in desert-shrub species under drought-associated stress. This knowledge is essential in modeling desert-shrub-ecosystem functioning under changing climatic conditions.

1 Introduction

This study provides a significant contribution to the understanding of acclimation processes in desert-shrub species to drought-associated stress in dryland ecosystems.

Due to the low amount of precipitation and high potential evapotranspiration in desert ecosystems, low soil water availability limits both plant water- and gas-exchange and, as a consequence, limits vegetation productivity (Razzaghi et al., 2011). Shrub and semi-shrub species are replacing grass species in arid and semi-arid lands in response to ongoing aridification of the land surface (H. Huang et al., 2011). This progression is predicted to continue under a changing climate (Houghton et al., 1999; Pacala et al., 2001; Asner et al., 2003). Studies have shown that desert shrubs are able to adapt to hot, dry environments as a result of their small plant surface area, thick epidermal hairs, and large root-to-shoot ratios (Eberbach and Burrows, 2006; Forner et al., 2014). Plant traits related to water use are likely to adapt differentially with species and habitat type (Brouillette et al., 2014). Plants may select water-acquisition or water-conservation strategies in response to water limitations (Brouillette et al., 2014). Knowledge of physiological acclimation of changing species to water shortages in deserts, particularly with respect to transpiration, is inadequate and, in the context of plant adaptation to changing climatic conditions, is of immense interest (Jacobsen et al., 2007; H. Huang et al., 2011). Transpiration maintains ecosystem balance through the soil–plant–atmosphere continuum and its magnitude and timing are related to the prevailing biophysical factors (Jarvis, 1976; Jarvis and McNaughton, 1986).

Sap flow can be used to reflect species-specific water consumption by plants (Ewers et al., 2002; Baldocchi, 2005; Naithani et al., 2012). Sap flow can also be used to continuously monitor canopy conductance and its response to environmental variables (Ewers et al., 2007; Naithani et al., 2012). Biotic and abiotic effects on sap flow and transpiration are often interactive and confounded. The decoupling coefficient (Ω) was used to examine the relative contribution of plant control through stomatal regulation of transpiration (Jarvis and McNaughton, 1986). Stomatal regulation becomes stronger as Ω approaches zero. Stomatal conductance (g_s) on the plant scale exerts a large biotic control on transpiration particularly during dry conditions (Jarvis, 1976; Jarvis and McNaughton, 1986). Stomatal conductance couples photosynthesis and transpiration (Cowan and Farquhar, 1977), making this parameter an important component of climate models in quantifying biospheric–atmospheric interactions (Baldocchi et al., 2002).

Studies have shown that xylem hydraulic conductivity was closely correlated with drought resistance (Cochard et al., 2008, 2010; Ennajeh et al., 2008). With increasing aridity, trees can progressively lessen their stomatal conductance, resulting in lower transpiration (McAdam et al., 2016). Generally, desert shrubs can close their stomata, reducing stomatal conductance, and reduce their water consumption when exposed to dehydration stresses. However, differences exist among shrub species in terms of their stomatal response to changes in air and soil moisture deficits (Pacala et al., 2001).

In *Elaeagnus angustifolia*, transpiration is observed to peak at noon, i.e., just before stomatal closure under water-deficit conditions (Liu et al., 2011), peaking earlier than radiation, temperature, and water vapor pressure deficit (VPD). This response lag or hysteresis effect has been widely noticed in dryland species (Du et al., 2011; Naithani et al., 2012), but its function is not completely understood. Transpiration in *Hedysarum scoparium* peaks multiple times during the day. During dry periods of the year, sap flow in *Artemisia ordosica* has been observed to be controlled by volumetric soil water content (VWC) at about a 30 cm depth in the soil (Li et al., 2014). For other shrubs, sap flow has been observed to decrease rapidly when the VWC is lower than the water loss through evapotranspiration (Buzkova et al., 2015). Sap flow in *Caragana korshinskii* and *Hippophae rhamnoides* have been found to increase with increasing rainfall intensity (Jian et al., 2016); whereas in *Haloxylon ammodendron*, it was found to vary in response to rainfall, from an immediate decline after a heavy rainfall to no observable change after a small rainfall event (Zheng and Wang, 2014). Drought-insensitive shrubs have relatively strong stomatal regulation and, therefore, tend to be insensitive to soil water deficits and rainfall, unlike their drought-sensitive counterparts (Du et al., 2011). Support for the relationship between sap flow in desert shrubs and prevailing environmental factors is decidedly variable (McDowell et al., 2013; Sus et al., 2014), potentially varying with plant habitat and species (Liu et al., 2011).

Artemisia ordosica, a shallow-rooted desert shrub, is the dominant species in the Mu Us Desert of northwestern China. It plays an important role in combating desertification and in stabilizing sand dunes (Li et al., 2010). Increases in air temperature, precipitation variability, and associated shorter wet and longer dry periods are expected to ensue under the influence of climate change (Lioubimtseva and Henebry, 2009). However, our understanding of the mechanisms of desert-shrub acclimation during periods of water shortage remains incomplete. Questions that need answers from our research include (1) how do changes in sap flow relate to changes in biotic and abiotic factors, and (2) whether *Artemisia ordosica* selects a strategy of water-conservation or water-acquisition under conditions of drought? To attend to these questions, we continuously measured stem sap flow in *Artemisia ordosica* and associated environmental variables in situ throughout the growing seasons of 2013 and 2014 (May–September period of each year). Our findings lead to insights concerning the main environmental factors affecting transpiration in *Artemisia ordosica*, e.g., optimal temperature, VPD, and VWC. This understanding can lead to improving phytoremediation practices in desert-shrub ecosystems.

2 Materials and methods

2.1 Experimental site

Continuous sap-flow measurements were made at the Yanchi Research Station (37°42′31″ N, 107°13′47″ E; 1530 m above mean sea level), Ningxia, northwestern China. The research station is located between the arid and semi-arid climatic zones along the southern edge of the Mu Us Desert. The sandy soil in the upper 10 cm of the soil profile has a bulk density of $1.54 \pm 0.08\,\mathrm{g\,cm^{-3}}$ (mean ± standard deviation, $n = 16$). Mean annual precipitation in the region is about 287 mm, of which 62 % falls between July and September. Mean annual potential evapotranspiration and air temperature are about 2024 mm and 8.1 °C, respectively, based on meteorological data (1954–2004) from the Yanchi County weather station. Normally, shrub leaf-expansion,

Table 1. Seasonal changes in transpiration (T_r) per month, leaf area index (LAI), and stomatal conductance (g_s) in *Artemisia ordosica* during the growing seasons (May–September period) of 2013 and 2014.

	T_r (mm month^{-1})		LAI (m^2 m^{-2})		g_s (mol m^{-2} s^{-1})	
	2013	2014	2013	2014	2013	2014
May	0.57	1.59	0.02	0.04	0.07	0.18
June	1.03	2.28	0.05	0.06	0.08	0.13
July	3.36	3.46	0.10	0.06	0.09	0.14
August	1.04	2.45	0.08	0.06	0.10	0.08
September	1.23	1.13	0.05	0.04	0.15	0.05

leaf-expanded, and leaf-coloration stages begin in April, June, and September, respectively (Chen et al., 2015).

2.2 Environmental measurements

Shortwave radiation (R_s in W m^{-2}; CMP3, Kipp & Zonen, the Netherland), air temperature (T in °C), wind speed (u in m s^{-1}, 034B, Met One Instruments Inc., USA), and relative humidity (RH in %; HMP155A, Vaisala, Finland) were measured simultaneously near the sap-flow measurement plot. Half-hourly data were recorded by data logger (CR3000 data logger, Campbell Scientific Inc., USA). VWC at a 30 cm depth were measured using three ECH_2O-5TE soil moisture probes (Decagon Devices, USA). In the analysis, we used half-hourly averages of VWC from the three soil moisture probes. VPD (in kPa) was calculated from recorded RH and T.

2.3 Measurements of sap flow, leaf area, and stomatal conductance

The experimental plot (10 × 10 m) was located on the western side of Yanchi Research Station in an *Artemisia ordosica*-dominated area. The mean age of the *Artemisia ordosica* was 10-years old. Maximum monthly mean leaf area index (LAI) for plant specimens with full leaf expansion was about 0.1 m^2 m^{-2} (Table 1). Over 60 % of their roots were distributed in the first 60 cm of the soil complex (Zhao et al., 2010; Jia et al., 2016). Five stems of *Artemisia ordosica* were randomly selected within the plot as replicates for sap-flow measurement. Mean height and sapwood area of sampled shrubs were 84 cm and 0.17 cm^2, respectively. Sampled stems represented the average size of stems in the plot. A heat-balance sensor (Flow32-1K, Dynamax Inc., Houston, USA) was installed at about 15 cm above the ground surface on each of the five stems (Dynamax, 2005). Sap-flow measurements from each stem were taken once per minute. Half-hourly data were recorded by a Campbell CR1000 data logger from 1 May to 30 September, for both 2013 and 2014 (Campbell Scientific, Logan, UT, USA).

Leaf area was estimated for each stem every 7–10 days by sampling about 50–70 leaves from five randomly sampled neighboring shrubs with similar characteristics to the shrubs being used for sap-flow measurements. Leaf area was measured immediately at the station laboratory with a portable leaf-area meter (LI-3000, LI-COR Inc., Lincoln, NE, USA). LAI was measured on a weekly basis from a 4 × 4 grid of 16 quadrats (10 × 10 m each) within a 100 × 100 m plot centered on a flux tower using measurements of sampled leaves and allometric equations (Jia et al., 2014). Stomatal conductance (g_s) was measured in situ for three to four leaves on each of the sampled shrubs with a LI-6400 portable photosynthesis analyzer (LI-COR Inc., Lincoln, NE, USA). The g_s measurements were made every 2 h from 07:00 to 19:00 (all times are local time) every 10 days from May to September 2013 and 2014.

The degree of coupling between the ecosystem surface and the atmospheric boundary layer was estimated with the decoupling coefficient (Ω). The decoupling coefficient varies from zero (i.e., leaf transpiration is mostly controlled by g_s) to one (i.e., leaf transpiration is mostly controlled by radiation). The Ω was calculated as described by Jarvis and McNaughton (1986),

$$\Omega = \frac{\Delta + \gamma}{\Delta + \gamma\left(1 + \frac{g_a}{g_s}\right)}, \tag{1}$$

where Δ is the rate of change of saturation vapor pressure vs. temperature (kPa K^{-1}), γ is the psychrometric constant (kPa K^{-1}), and g_a is the aerodynamic conductance (m s^{-1}; Monteith and Unsworth, 1990),

$$g_a = \left(\frac{u}{u^{*2}} + 6.2u^{*^{-0.67}}\right)^{-1}, \tag{2}$$

where u is the wind speed (m s^{-1}) at 6 m above the ground, and u^* is the friction velocity (m s^{-1}), measured by a nearby eddy covariance system (Jia et al., 2014).

2.4 Data analysis

In our analysis, drought days were defined as those days with daily mean VWC < 0.1 m^3 m^{-3}. This is based on a VWC threshold of 0.1 m^3 m^{-3} for the sap flow per leaf area J_s (Fig. 1), with J_s increasing as VWC increased, saturating at a VWC of 0.1 m^3 m^{-3}, and decreasing as VWC continued

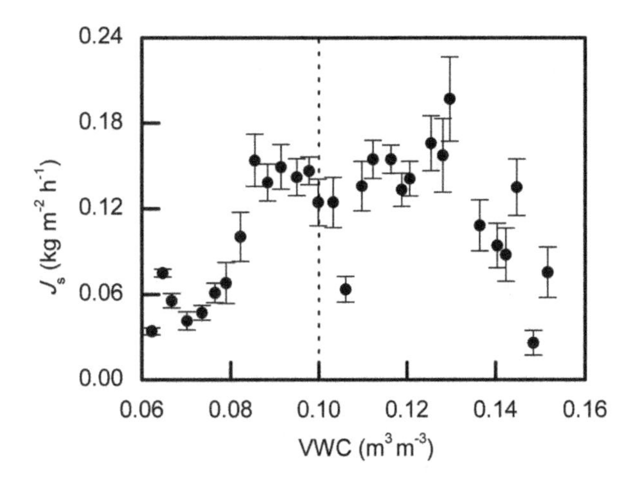

Figure 1. Sap flow per leaf area (J_S) as a function of soil water content (VWC) at 30 cm depth in non-rainy, daytime hours during the mid-growing period from 1 June to 31 August 2013 and 2014. Data points are binned values from pooled data over 2 years at a VWC increment of 0.003 m^3 m^{-3}. The dotted line represents the VWC threshold for J_S.

to increase. The VWC threshold of 0.1 m^3 m^{-3} is equivalent to a relative extractable soil water (REW) of 0.4 for drought conditions (Granier et al., 1999, 2007; Zeppel et al., 2004; Fig. 2d, e). Duration and severity of "drought" were defined based on a VWC threshold and REW of 0.4. REW was calculated with

$$\mathrm{REW} = \frac{\mathrm{VWC} - \mathrm{VWC_{min}}}{\mathrm{VWC_{max}} - \mathrm{VWC_{min}}}, \tag{3}$$

where VWC is the specific daily soil water content (m^3 m^{-3}), and VWC$_{min}$ and VWC$_{max}$ are the minimum and maximum VWC during the measurement period in each year, respectively.

Sap-flow analysis was conducted using mean data from five sensors. Sap flow per leaf area (J_s) was calculated according to

$$J_s = \left(\sum_{i=1}^{n} E_i / A_{li}\right) / n, \tag{4}$$

where J_s is the sap flow per leaf area (kg m^{-2} h^{-1} or kg m^{-2} d^{-1}), E is the measured sap flow of a stem (g h^{-1}), A_l is the leaf area of the sap-flow stem, and n is the number of stems sampled (e.g., $n = 5$).

Transpiration per ground area (T_r) was estimated in this study according to:

$$T_r = \left(\sum_{i=1}^{n} J_s \cdot \mathrm{LAI}\right) / n, \tag{5}$$

where T_r is transpiration per ground area (mm d^{-1}).

Linear and non-linear regressions were used to analyze abiotic control on sap flow. In order to minimize the effects of different phenophases and rainfall, we only used data from the mid-growing season, non-rainy days, and daytime hours from 08:00 to 20:00, i.e., from 1 June to 31 August, with hourly shortwave radiation greater than 10 W m^{-2}. Relations between mean sap flow at specific times over a period of 08:00–20:00 and corresponding environmental factors from 1 June to 31 August were derived from linear regression ($p < 0.05$; Fig. 3). Regression slopes were used as indicators of sap-flow sensitivity (degree of response) to the various environmental variables (see Zha et al., 2013). All statistical analyses were performed with SPSS version 17.0 for Windows software (SPSS Inc., USA). Significance level was set at 0.05.

3 Results

3.1 Seasonal variations in environmental factors and sap flow

The range of daily means (24 h mean) for R_s, T, VPD, and VWC during the 2013 growing season (May–September) were 31.1–364.9 W m^{-2}, 8.8–24.4 °C, 0.05–2.3 kPa, and 0.06–0.17 m^3 m^{-3} (Fig. 2a–d), respectively, annual means being 224.8 W m^{-2}, 17.7 °C, 1.03 kPa, and 0.08 m^3 m^{-3}. The corresponding range of daily means for 2014 were 31.0–369.9 W m^{-2}, 7.1–25.8 °C, 0.08–2.5 kPa, and 0.06–0.16 m^3 m^{-3} (Fig. 2a–d), respectively, annual means being 234.9 W m^{-2}, 17.2 °C, 1.05 kPa, and 0.09 m^3 m^{-3}.

Total precipitation and number of days with rainfall events during the 2013 measurement period (257.2 and 46 days) were about 5.6 and 9.8 % lower than those during 2014 (272.4 mm and 51 days; Fig. 2d), respectively. More irregular rainfall events occurred in 2013 than in 2014, with 45.2 % of rainfall falling in July and 8.8 % in August.

Drought mainly occurred in May, June, and August of 2013 and in May and June of 2014 (shaded sections in Fig. 2d, e). Both years had dry springs. Over a 1-month period of summer drought occurred in 2013.

The range of daily J_s during the growing season was 0.01–4.36 kg m^{-2} d^{-1} in 2013 and 0.01–2.91 kg m^{-2} d^{-1} in 2014 (Fig. 2f), with annual means of 0.89 kg m^{-2} d^{-1} in 2013 and 1.31 kg m^{-2} d^{-1} in 2014. Mean daily J_s over the growing season of 2013 was 32 % lower than that of 2014. Mean daily T_r were 0.05 and 0.07 mm d^{-1} over the growing season of 2013 and 2014 (Fig. 2f), respectively, being 34 % lower in 2013 than in 2014. The total T_r over the growing season (1 May–30 September) of 2013 and 2014 were 7.3 and 10.9 mm, respectively. Seasonal fluctuations in J_s and T_r corresponded with seasonal patterns in VWC (Fig. 2d, f). Daily mean J_s and T$_r$ decreased or remained nearly constant during dry-soil periods (Fig. 2d, f), with the lowest J_s and T_r observed in spring and mid-summer (August) of 2013.

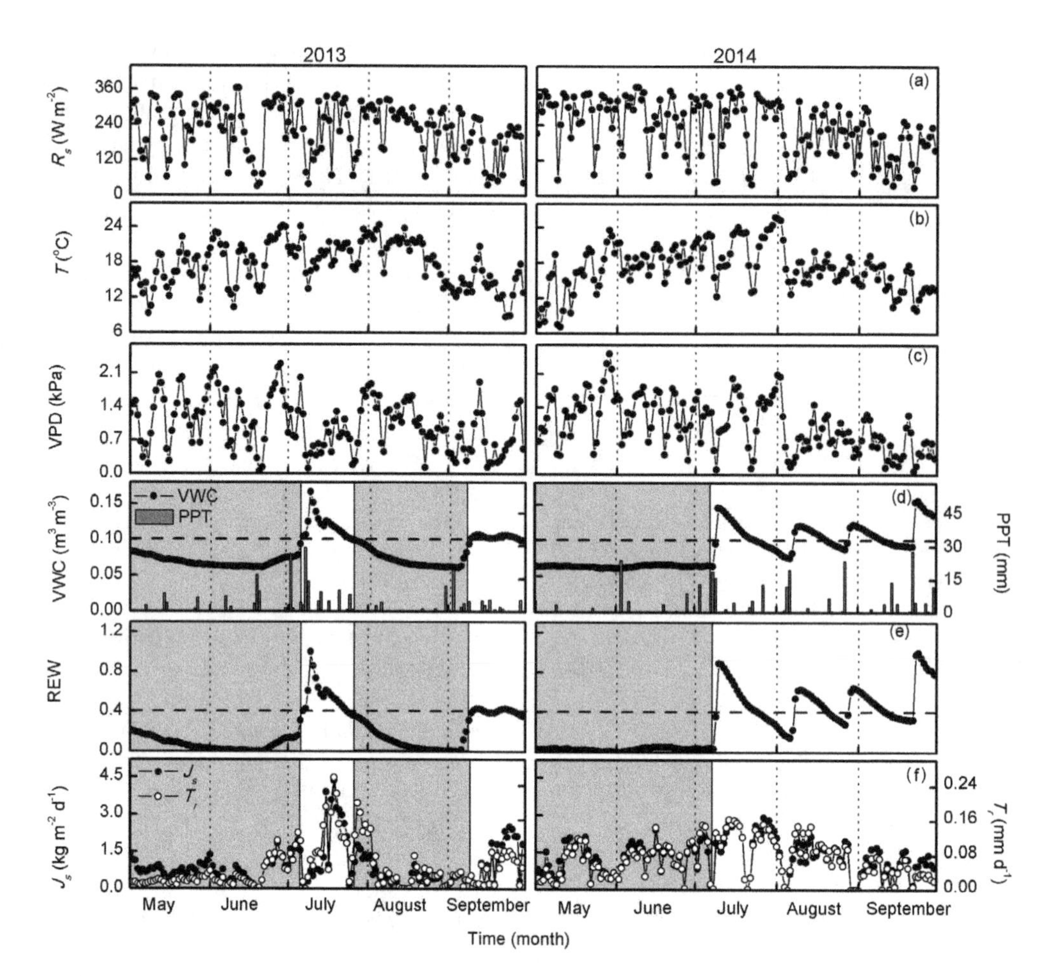

Figure 2. Seasonal changes in daily (24 h) mean shortwave radiation (R_S; **a**), air temperature (T; **b**), water vapor pressure deficit (VPD; **c**), volumetric soil water content (VWC; **d**), relative extractable water (REW; **e**), daily total precipitation (PPT; **d**), and daily sap flow per leaf area (J_S; **f**), and daily transpiration (T_r, mm d^{-1}; **f**) from May to September for both 2013 and 2014. The horizontal dashed lines in **(d, e)** represent VWC and REW thresholds of 0.1 m^3 m^{-3} and 0.4, respectively. Shaded bands indicate periods of drought.

3.2 Sap-flow response to environmental factors

In summer, J_s increased with increasing VWC, R_S, T, and VPD (Figs. 2d, f, and 3). Sap flow increased more rapidly with increases in R_S, T, and VPD under high VWC (i.e., VWC > 0.1 m^3 m^{-3} in both 2013 and 2014; Fig. 4) compared with periods with lower VWC (i.e., VWC < 0.1 m^3 m^{-3} in both 2013 and 2014; Fig. 4). Sap flow was more sensitive to R_S, T, and VPD under high VWC (Fig. 4), which coincided with a steeper regression slope under high VWC conditions.

Sensitivity of J_s to environmental variables (in particular, R_S, T, VPD, and VWC) varied depending on time of day (Fig. 5). Regression slopes for the relations of J_S–R_S, J_S–T, and J_S–VPD were greater in the morning before 11:00, and lower during mid-day and early afternoon (12:00–16:00). In contrast, regression slopes of the relation of J_S–VWC were lower in the morning (Fig. 5), increasing thereafter, peaking at ~13:00, and subsequently decreasing in late afternoon. Regression slopes of the response of J_s to R_S, T, and VPD in 2014 were steeper than those in 2013.

3.3 Diurnal changes and hysteresis between sap flow and environmental factors

Diurnal patterns of J_s were similar in both years (Fig. 6), initiating at 07:00 and increasing thereafter, peaking before noon (12:00), and subsequently decreasing thereafter and remaining near zero from 20:00 to 06:00. Diurnal changes in g_S were similar to J_S, but peaking about 2 and 1 h earlier than J_s in July and August, respectively (Fig. 6).

There were pronounced time lags between J_s and R_S over the 2 years (Fig. 7), J_s peaking earlier than R_S and, thus, earlier than either VPD or T. These time lags differed seasonally. For example, mean time lag between J_s and R_S was 2 h during July, 5 h during May, and 3 h during June, August, and September of 2013. However, the time lags in 2014 were generally shorter than those observed in 2013 (Table 2).

Clockwise hysteresis loops between J_s and R_S during the growing period were observed (Fig. 7). As R_S increased in the morning, J_s increased until it peaked at ~10:00. Sap flow

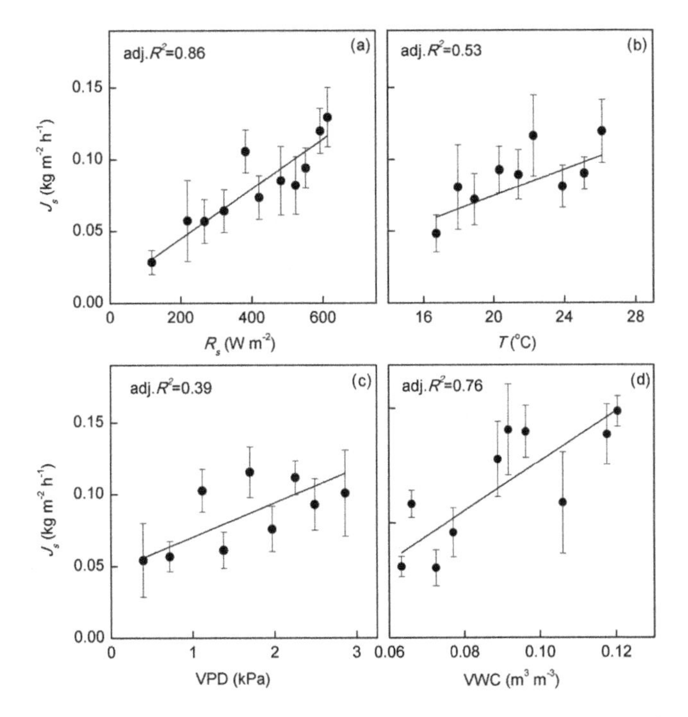

Figure 3. Relationships between sap flow per leaf area (J_S) and environmental factors (shortwave radiation, R_S; air temperature, T; water vapor pressure deficit, VPD; and soil water content at 30 cm depth, VWC) in non-rainy days between 08:00 and 20:00 during the mid-growing season of 1 June to 31 August for 2013 and 2014. Data points are binned values from pooled data over 2 years at increments of $40\,W\,m^{-2}$, 1.2 °C, 0.3 kPa, and $0.005\,m^3\,m^{-3}$ for R_S, T, VPD, and VWC, respectively.

declined with decreasing R_S during the afternoon. Sap flow (J_S) was higher in the morning than in the afternoon.

Diurnal time lag in the relation of J_S–R_S were influenced by VWC (Figs. 8, 9). For example, J_S peaked about 2 h earlier than R_S on days with low VWC (Fig. 8a), 1 h earlier than R_S on days with moderate VWC (Fig. 8b), and at the same time as R_S on days with high VWC (Fig. 8c). Lag hours between J_S and R_S over the growing season were negatively and linearly related to VWC (Fig. 9: lag (h) $= -133.5 \times \mathrm{VWC} + 12.24$, $R^2 = 0.41$). The effect of VWC on time lags between J_S and R_S was smaller in 2014, with evenly distributed rainfall during the growing season, than in 2013, with a pronounced summer drought (Fig. 9). Variables g_S and Ω showed a significantly increasing trend with increasing VWC in 2013 and 2014 (Fig. 10). This trend was more obvious in the dry year of 2013 than in 2014.

4 Discussion and conclusions

4.1 Sap-flow response to environmental factors

Drought tolerance of some plants may be related to lower overall sensitivity of plant physiological attributes to environmental stress and/or stomatal regulation (Y. Huang et al., 2011; Naithani et al., 2012). In this study, steep regression slopes between J_S and the environmental variables (R_S, VPD, and T) in the morning indicated that sap flow was less sensitive to variations in R_S, VPD, and T during the drier and hotter part of the day (Fig. 5). The lower sensitivity combined with lower stomatal conductances led to lower sap flow, and, thus, lower transpiration (water consumption) during hot mid-day summer hours, pointing to a water-conservation strategy in plant acclimation during dry and hot conditions. When R_S peaked during mid-day (13:00–14:00) in summer, there was often insufficient soil water to meet the atmospheric demand, causing g_S to be limited by available soil moisture and making J_S more responsive to VWC at noon, but less responsive to R_S and T. Similarly, sap flow in *Hedysarum mongolicum* and some other shrubs in a nearby region were positively correlated with VWC at noon (Qian et al., 2015). For instance, sap flow in *Picea crassifolia* peaked at noon (12:00 and 14:00) and then decreased, heightening by increasing R_S, T, and VPD, when $R_S < 800\,W\,m^{-2}$, $T < 18.0$ °C, and VPD < 1.4 kPa (Chang et al., 2014); sap flow in *Caragana korshinskii* was significantly lower during the stress period, its conductance decreasing linearly after the wilting point (She et al., 2013). The fact that J_S was less sensitive to meteorological variables when VWC $< 0.10\,m^3\,m^{-3}$, highlights the water-conservation strategy taken by drought-afflicted *Artemisia ordosica*. The positive linear relationship between g_S and VWC in this study further supports this conclusion.

Precipitation, being the most important source of soil moisture and, thus, VWC, affected transpiration directly. Frequent small rainfall events (< 5 mm) are crucially important to the survival and growth of desert plants (Zhao and Liu, 2010). Variations in J_S were clearly associated with the intermittent supply of water to the soil during rainfall events (see Fig. 2d, f). Reduced J_S during rainy days can be largely explained by a reduction in incident R_S and liquid water-induced saturation of the leaf surface, which led to a decrease in leaf turgor and stomatal closure. After each rainfall event, J_S increased quickly when soil moisture was replenished. Schwinning and Sala (2004) have previously shown that VWC contributed the most to the post-rainfall response in plant transpiration at similar sites. The study shows that *Artemisia ordosica* responded differently to wet and dry conditions. In the mid-growing season, high J_S in July was related to rainfall-fed soil moisture, which increased the rate of transpiration. However, dry soil conditions combined with high T and R_S led to a reduction in J_S in August of 2013 (Fig. 2). In some deep-rooting desert shrubs, groundwater may replenish water lost by transpiration (Yin et al., 2014). *Artemisia ordosica* roots are generally distributed in the upper 60 cm of the soil (Zhao et al., 2010), and as a result the plant usually depends on water directly supplied by precipitation. This is because groundwater levels in drylands can often be below the rooting zone of many shrub species, typ-

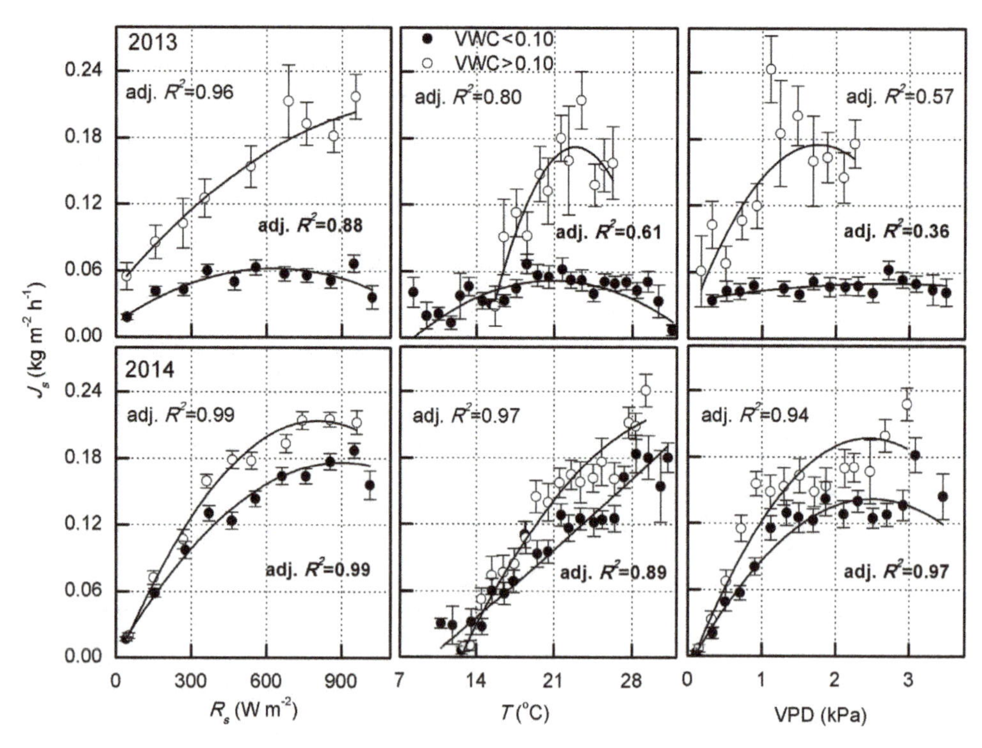

Figure 4. Sap flow per leaf area (J_S) in non-rainy, daytime hours during the mid-growing season of 1 June to 31 August for both 2013 and 2014 as a function of shortwave radiation (R_S), air temperature (T), and vapor pressure deficit (VPD) under high volumetric soil water content (VWC > 0.10 m^3 m^{-3} both in 2013 and 2014) and low VWC (< 0.10 m^3 m^{-3}, 2013 and 2014). J_S is given as binned averages according to R_S, T, and VPD based on increments of 100 W m^{-2}, 1 °C, and 0.2 kPa, respectively. Bars indicate standard error.

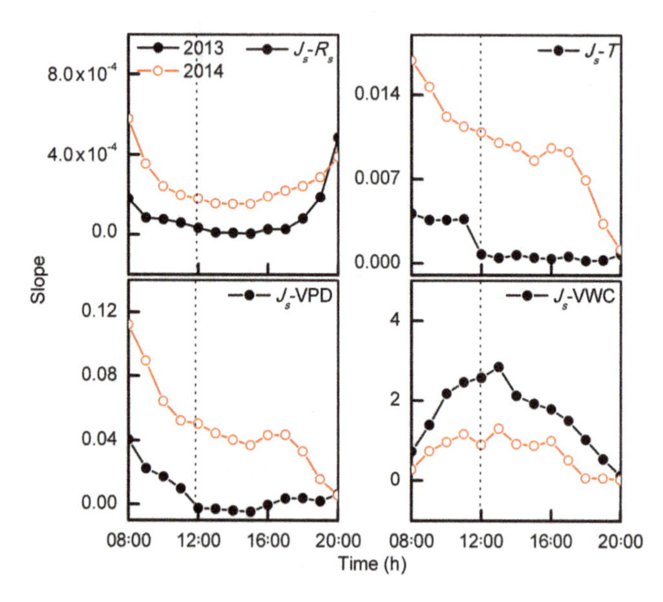

Figure 5. Regression slopes of linear fits between sap flow per leaf area (J_S) in non-rainy days and shortwave radiation (R_S), vapor pressure deficit (VPD), air temperature (T), and volumetric soil water content (VWC) between 08:00 and 20:00 during the mid-growing season of 1 June to 31 August for 2013 and 2014.

ically at depths greater than 10 m as witnessed at our site. Similar findings regarding the role of rainfall and VWC in desert vegetation are reported by Wang et al. (2017).

4.2 Hysteresis between sap flow and environmental factors

Diurnal patterns in J_S corresponded with those of R_S from sunrise until diverging later in the day (Fig. 7), suggesting that R_S was a primary controlling factor of diurnal J_S. As an initial energy source, R_S also can force T and VPD to increase, causing a phase difference in time lags among the relations of J_S–R_S, J_S–T, and J_S–VPD.

The hysteresis effect reflects plant acclimation to water limitations, due to stomatal conductance being inherently dependent on plant hydrodynamics (Matheny et al., 2014). The large g_S in the morning promoted higher rates of transpiration (Figs. 6, 7), while lower g_S in the afternoon reduced transpiration. Therefore, diurnal curves (hysteresis) were mainly caused by a g_S-induced hydraulic process (Fig. 7). The finding that hysteresis varied seasonally, decreasing with increasing VWC, further reflects the acclimation to water limitation causing J_S to peak in advance of the environmental factors. At our site, dry soils accompanied with high VPD in summer, led to a decreased g_S and a more significant control of the stomata on J_S relative to the environmental factors.

Table 2. Mean monthly diurnal cycles of sap flow (J_S) response to shortwave radiation (R_S), air temperature (T), and water vapor pressure deficit (VPD), including lag times (h) as a function of R_S, T, and VPD.

Relationship	May		June		July		August		September	
	2013	2014	2013	2014	2013	2014	2013	2014	2013	2014
J_S–R_S	5	2	3	0	2	1	3	1	3	2
J_S–T	8	6	7	4	4	4	6	5	6	6
J_S–VPD	8	5	7	4	6	4	6	5	6	5

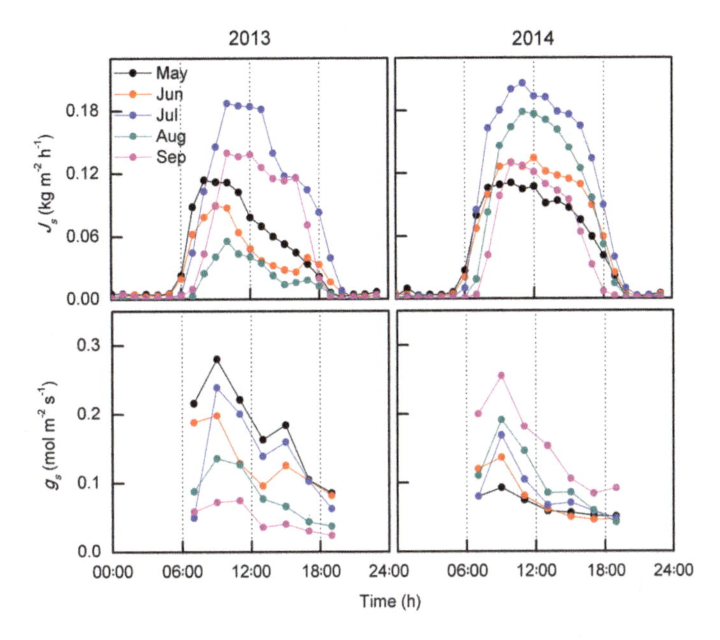

Figure 6. Mean monthly diurnal changes in sap flow per leaf area (J_S) and stomatal conductance (g_S) in *Artemisia ordosica* during the growing season (May–September) for both 2013 and 2014. Each point is given as the mean at specific times during each month.

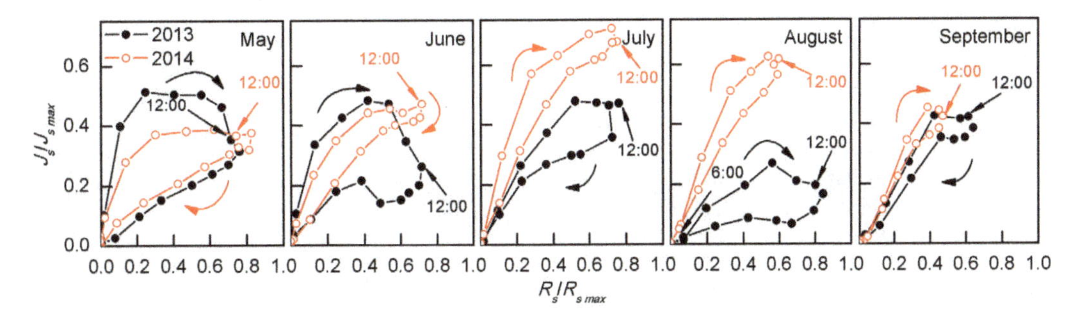

Figure 7. Seasonal variation in hysteresis loops between sap flow per leaf area (J_S) and shortwave radiation (R_S) using normalized plots for both 2013 and 2014. The y axis represents the proportion of maximum J_S (dimensionless), and the x axis represents the proportion of maximum R_S (dimensionless). The curved arrows indicate the clockwise direction of response during the day.

The result that g_S increased with increasing VWC (Fig. 10a), along with the synchronization of J_S and g_S, suggests that J_S is more sensitive to g_S in low VWC and less so to R_S. Due to the incidence of small rainfall events in drylands, soil water supplied by rainfall pulses was largely insufficient to meet the transpiration demand under high mid-day R_S, resulting in clockwise loops. Lower Ω values (< 0.4) at our site also support the idea that g_S has a greater control on transpiration than R_S under situations of water limitation (Fig. 10).

Altogether, stomatal control on the diurnal evolution of J_s by reducing g_S combined with lower sensitivity to meteorological variables during the mid-day dry hours help to reduce water consumption in *Artemisia ordosica*. Seasonally, plant-moderated reductions in g_S and increased hystere-

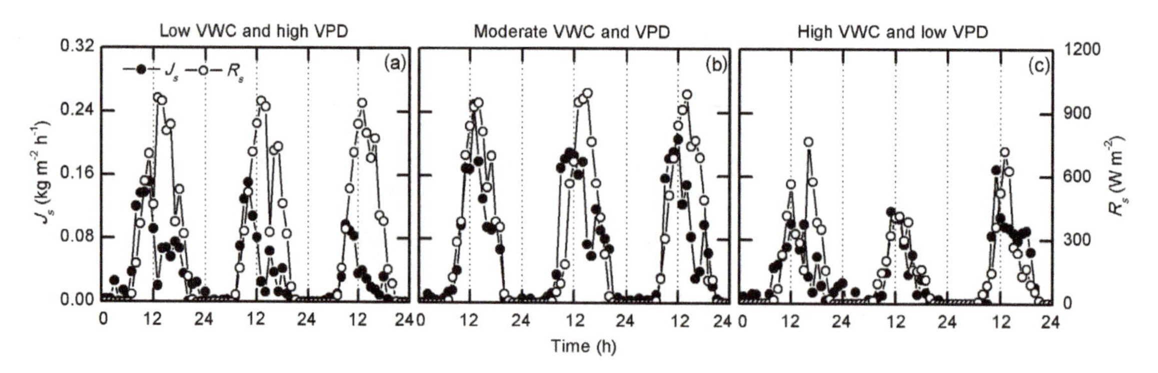

Figure 8. Sap flow per leaf area (J_S) and shortwave radiation (R_S) over three consecutive days in 2013, i.e., **(a)** under low volumetric soil water content (VWC) and high vapor pressure deficit (VPD; DOY 153–155, VWC = 0.064 m^3 m^{-3}, REW = 0.025, VPD = 2.11 kPa), **(b)** moderate VWC and VPD (DOY 212–214, VWC = 0.092 m^3 m^{-3}, REW = 0.292, VPD = 1.72 kPa), and **(c)** high VWC and low VPD (DOY 192–194, VWC = 0.152 m^3 m^{-3}, REW = 0.865, VPD = 0.46 kPa); REW is the relative extractable soil water and DOY is the ordinal day of the year. VWC, REW, and VPD give the 3-day mean value.

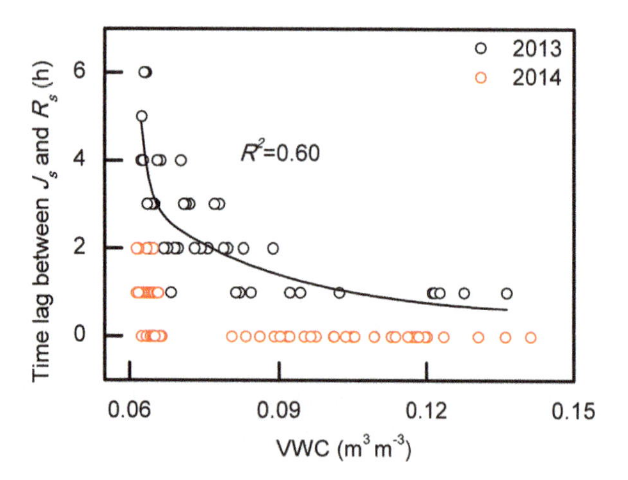

Figure 9. Time lag between sap flow per leaf area (J_S) and short wave radiation (R_S) in relation to volumetric soil water content (VWC). Hourly data in non-rainy days during the mid-growing season of 1 June to 31 August 2013 and 2014. The lag hours were calculated by a cross-correlation analysis using a 3-day moving window with a 1-day time step. Rainy days were excluded. The solid line is based on an exponential regression ($p < 0.05$).

sis, leads to reduced J_S and acclimation to drought conditions. It is suggested here that water limitation invokes a water-conservation strategy in *Artemisia ordosica*. Contrary to our findings, counterclockwise hysteresis has been observed to occur between J_S and R_S in tropical and temperate forests (Meinzer et al., 1997; O'Brien et al., 2004; Zeppel et al., 2004), which is reported to be consistent with the capacitance of the particular soil–plant–atmosphere system being considered. Unlike short-statured vegetation, it usually takes more time for water to move up and expand vascular elements in tree stems during the transition from night to day.

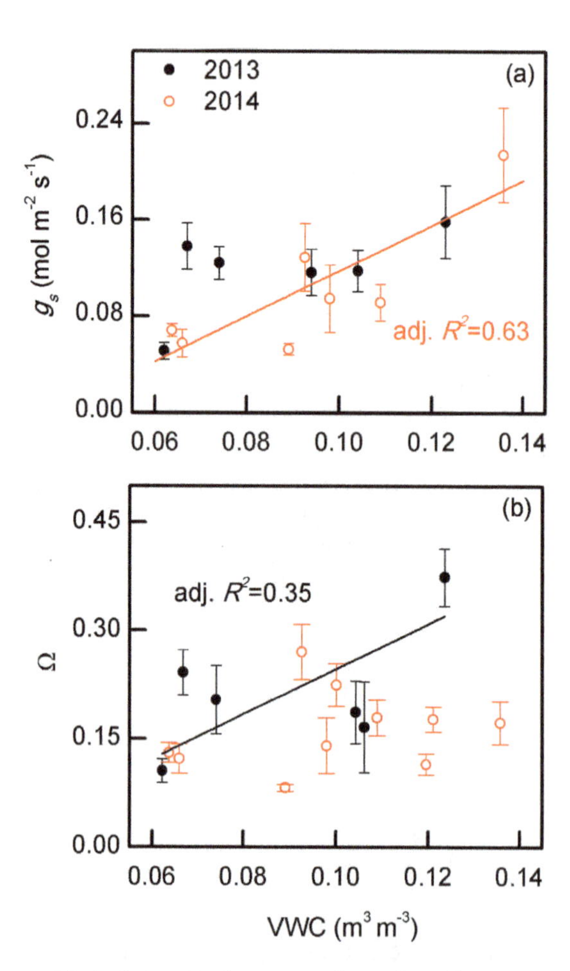

Figure 10. Relationship between volumetric soil water content (VWC) and **(a)** stomatal conductance (g_S) in *Artemisia ordosica*, and **(b)** decoupling coefficient (Ω) for 2013 and 2014. Hourly values are given as binned averages based on a VWC-increment of 0.005 m^3 m^{-3}. Bars indicate standard error. Only statistically significant regressions (with p values < 0.05) are shown.

4.3 Conclusions

The relative influence of R_S, T, and VPD on J_s in *Artemisia ordosica* was modified by soil water, indicating J_s's lessened sensitivity to the environmental variables during dry periods. Sap flow was constrained by soil water deficits, causing J_s to peak several hours prior to the peaking of R_S. Diurnal hysteresis between J_s and R_s varied seasonally and was mainly controlled by hydraulic stresses. Soil moisture controlled sap-flow response in *Artemisia ordosica* to meteorological factors. This species escaped and acclimated to water limitations by invoking a water-conservation strategy through the hysteresis effect and stomatal regulation. Our findings add to our understanding of acclimation in desert-shrub species under stress of dehydration. The information advanced here can assist in modeling desert-shrub-ecosystem functioning under changing climatic conditions.

Author contributions. DQ and TZ contributed equally to the design and implementation of the field experiment, data collection and analysis, and writing the first draft of the manuscript. XJ gave helpful suggestions concerning the analysis of the field data and contributed to the scientific revision and editing of the manuscript. BW contributed to the design of the experiment. CP-AB and HP contributed to the scientific revision and editing of the manuscript. YB, WF, and YT were involved in the implementation of the experiment and in the revision of the manuscript.

Competing interests. The authors declare that they have no conflict of interest.

Acknowledgements. This research was financially supported by grants from the National Natural Science Foundation of China (NSFC no. 31670710, 31670708, 31361130340, 31270755) and the Academy of Finland (project no. 14921). Xin Jia and Wei Feng are also grateful for financial support from the Fundamental Research Funds for the Central Universities (project no. 2015ZCQ-SB-02). This work is related to the Finnish–Chinese collaborative research project EXTREME (2013–2016), between Beijing Forestry University (team led by Tianshan Zha) and the University of Eastern Finland (team led by Heli Peltola), and the US–China Carbon Consortium (USCCC). We thank Ben Wang, Sijing Li, Qiang Yang, and others for their assistance in the field.

Edited by: Paul Stoy

References

Asner, G. P., Archer, S., Hughes, R. F., Ansley, R. J., and Wessman, C. A.: Net changes in regional woody vegetation cover and carbon storage in Texas Drylands, 1937–1999, Glob. Change Biol., 9, 316–335, 2003.

Baldocchi, D. D.: The role of biodiversity on the evaporation of forests, in: Forest diversity and function ecological studies, edited by: Scherer-Lorenzen, M., Körner, C., Schulze, E.-D., Springer, Berlin, Heidelberg, 131–148, 2005.

Baldocchi, D. D., Wilson, K. B., and Gu, L.: How the environment, canopy structure and canopy physiological functioning influence carbon, water and energy fluxes of a temperate broad-leaved deciduous forest – an assessment with the biophysical model CANOAK, Tree Physiol., 22, 1065–1077, 2002.

Brouillette, L. C., Mason, C. M., Shirk, R. Y., and Donovan, L. A.: Adaptive differentiation of traits related to resource use in a desert annual along a resource gradient, New Phytol., 201, 1316–1327, 2014.

Buzkova, R., Acosta, M., Darenova, E., Pokorny, R., and Pavelka, M.: Environmental factors influencing the relationship between stem CO_2 efflux and sap flow, Trees-Struct. Funct., 29, 333-343, 2015.

Chang, X., Zhao, W., and He, Z.: Radial pattern of sap flow and response to microclimate and soil moisture in Qinghai spruce (*Picea crassifolia*) in the upper Heihe River Basin of arid northwestern China, Agr. Forest Meteorol., 187, 14–21, 2014.

Chen, Z. H., Zha, T., Jia, X., Wu, Y., Wu, B., Zhang, Y., Guo, J., Qin, S., Chen, S., and Peltola, H.: Leaf nitrogen is closely coupled to phenophases in a desert shrub ecosystem in China, J. Arid Environ., 122, 124–131, 2015.

Cochard, H., Barigah, S., and Kleinhentz, M.: Is xylem cavitation resistance a relevant criterion for screening drought resistance among Prunus species?, J. Plant Physiol., 165, 976–982, 2008.

Cochard, H., Herbette, S., and Hernandez, E.: The effects of sap ionic composition on xylem vulnerability to cavitation, J. Exp. Bot., 61, 275–285, 2010.

Cowan, I. R. and Farquhar, G. D.: Stomatal function in relation to leaf metabolism and environment, Sym. Soc. Exp. Biol., 31, 471–505, 1977.

Du, S., Wang, Y.-L., Kume, T., Zhang, J.-G., Otsuki, K., Yamanaka, N., and Liu, G.-B.: Sapflow characteristics and climatic responses in three forest species in the semiarid Loess Plateau region of China, Agr. Forest Meteorol., 151, 1–10, 2011.

Dynamax: Dynagage® Installation and Operation Manual, Dynamax, Houston, TX, 2005.

Eberbach, P. L. and Burrows, G. E.: The transpiration response by four topographically distributed *Eucalyptus* species, to rainfall occurring during drought in south eastern Australia, Physiol. Plantarum, 127, 483–493, 2006.

Ennajeh, M., Tounekti, T., and Vadel, A. M.: Water relations and droughtinduced embolism in olive (*Olea europaea*) varieties "Meski" and "Chemlali" during severe drought, Tree Physiol., 28, 971–976, 2008.

Ewers, B. E., Mackay, D. S., Gower, S. T., Ahl, D. E., and Samanta, S. N. B.: Tree species effects on stand transpiration in northern Wisconsin, Water Resour. Res., 38, 1–11, 2002.

Ewers, B. E., Mackay, D. S., and Samanta, S.: Interannual consistency in canopy stomatal conductance control of leaf water potential across seven tree species, Tree Physiol., 27, 11–24, 2007.

Forner, A., Aranda, I., Granier, A., and Valladares, F.: Differential impact of the most extreme drought event over the last half century on growth and sap flow in two coexisting Mediterranean trees, Plant Ecol., 215, 703–719, 2014.

Granier, A., Bréda, N., Biron, P., and Villette, S.: A lumped water balance model to evaluate duration and intensity of drought constraints in forest stands, Ecol. Model., 116, 269–283, 1999.

Granier, A., Reichstein, M., Bréda, N., Janssens, I. A., Falge, E., Ciais, P., and Buchmann, N.: Evidence for soil water control on carbon and water dynamics in European forests during the extremely dry year, Agr. Forest Meteorol., 143, 123–145, 2007.

Houghton, R. A., Hackler, J. L., and Lawrence, K. T.: The U.S. carbon budget: contributions from land-use change, Science, 285, 574–578, 1999.

Huang, H., Gang, W., and NianLai, C.: Advanced studies on adaptation of desert shrubs to environmental stress, Sciences in Cold and Arid Regions, 3, 0455–0462, 2011.

Huang, Y., Li, X., Zhang, Z., He, C., Zhao, P., You, Y., and Mo, L.: Seasonal changes in Cyclobalanopsis glauca transpiration and canopy stomatal conductance and their dependence on subterranean water and climatic factors in rocky karst terrain, J. Hydrol., 402, 135–143, 2011.

Jacobsen, A. L., Agenbag, L., Esler, K. J., Pratt, R. B., Ewers, F. W., and Davis, S. D.: Xylem density, biomechanics and anatomical traits correlate with water stress in 17 evergreen shrub species of the Mediterranean-type climate region of South Africa, J. Ecol., 95, 171–183, 2007.

Jarvis, P. G.: The interpretation of the variations in leaf water potential and stomatal conductance found in canopies in the field, Philos. T. Roy. Soc. B, 273, 593–610, 1976.

Jarvis, P. G. and McNaughton, K. G.: Stornatal Control of Transpiration: Scaling Up from Leaf' to Region, Adv. Ecol. Res., 15, 1–42, 1986.

Jia, X., Zha, T. S., Wu, B., Zhang, Y. Q., Gong, J. N., Qin, S. G., Chen, G. P., Qian, D., Kellomäki, S., and Peltola, H.: Biophysical controls on net ecosystem CO_2 exchange over a semiarid shrubland in northwest China, Biogeosciences, 11, 4679–4693, https://doi.org/10.5194/bg-11-4679-2014, 2014.

Jia, X., Zha, T, Gong, J., Wang, B., Zhang, Y., Wu, B., Qin, S., and Peltola, H.: Carbon and water exchange over a temperate semi-arid shrubland during three years of contrasting precipitation and soil moisture patterns, Agr. Forest Meteorol., 228, 120–129, 2016.

Jian, S. Q., Wu, Z. N., Hu, C. H., and Zhang, X. L.: Sap flow in response to rainfall pulses for two shrub species in the semiarid Chinese Loess Plateau, J. Hydrol. Hydromech., 64, 121–132, 2016.

Li, S. J., Zha, T. S., Qin, S. G., Qian, D., and Jia, X.: Temporal patterns and environmental controls of sap flow in Artemisia ordosica, Chinese Journal of Ecology, 33, 1–7, 2014.

Li, S. L,, Werger, M. A., Zuidema, P., Yu, F., and Dong, M.: Seedlings of the semi-shrub Artemisia ordosica are resistant to moderate wind denudation and sand burial in Mu Us sandland, China, Trees, 24, 515–521, 2010.

Lioubimtseva, E. and Henebry, G. M.: Climate and environmental change in arid Central Asia: Impacts, vulnerability, and adaptations, J. Arid Environ., 73, 963–977, 2009.

Liu, B., Zhao, W., and Jin, B.: The response of sap flow in desert shrubs to environmental variables in an arid region of China, Ecohydrology, 4, 448–457, 2011.

Matheny, A. M., Bohrer, G., Vogel, C. S., Morin, T. H., He, L., Frasson, R. P. de M., Mirfenderesgi, G., Schäfer, K. V. R., Gough, C. M., Ivanov, V. Y., and Curtis, P. S.: Species-specific transpiration responses to intermediate disturbance in a northern hardwood forest, J. Geophys. Res.-Biogeo., 119, 2292–2311, 2014.

McAdam, S. A., Sussmilch, F. C., and Brodribb, T. J.: Stomatal responses to vapour pressure deficit are regulated by high speed gene expression in angiosperms, Plant Cell Environ., 39, 485–491, 2016.

McDowell, N. G., Fisher, R. A., and Xu, C.: Evaluating theories of drought-induced vegetationmortality using a multimodel-experiment framework, New Phytol., 200, 304–321, 2013.

Meinzer, F. C., Andrade, J. L., Goldstein, G., Holbrook, N. M., Cavelier, J., and Jackson, P.: Control of transpiration from the upper canopy of a tropical forest: the role of stomatal, boundary layer and hydraulic architecture components, Plant Cell Environ., 20, 1242–1252, 1997.

Monteith, J. L. and Unsworth, M. H.: Principles of Environmental Physics, Butterworth-Heinemann, Oxford, 1990.

Naithani, K. J., Ewers, B. E., and Pendall, E.: Sap flux-scaled transpiration and stomatal conductance response to soil and atmospheric drought in a semi-arid sagebrush ecosystem, J. Hydrol., 464, 176–185, 2012.

O'Brien, J. J., Oberbauer, S. F., and Clark, D. B.: Whole tree xylem sap flow responses to multiple environmental variables in a wet tropical forest, Plant Cell Environ., 27, 551–567, 2004.

Pacala, S. W., Hurtt, G. C., Baker, D., Peylin, P., Houghton, R. A., Birdsey, R. A., Heath, L., Sundquist, E. T., Stallard, R. F., Ciais, P., Moorcroft, P., Caspersen, J. P., Shevliakova, E., Moore, B., Kohlmaier, G., Holland, E., Gloor, M., Harmon, M. E., Fan, S.-M., Sarmiento, J. L., Goodale, C. L., Schimel, D., and Field, C. B.: Consistent land- and atmosphere-based U.S. carbon sink estimates, Science, 292, 2316–2320, 2001.

Qian, D., Zha, T., Jia, X., Wu, B., Zhang, Y., Bourque, C. P., Qin, S., and Peltola, H.: Adaptive, water-conserving strategies in Hedysarum mongolicum endemic to a desert shrubland ecosystem, Environmental Earth Sciences, 74, 6039–6046, 2015.

Razzaghi, F., Ahmadi, S. H., Adolf, V. I., Jensen, C. R., Jacobsen, S. E., and Andersen, M. N.: Water Relations and Transpiration of Quinoa (*Chenopodium quinoa* Willd.) Under Salinity and Soil Drying, J. Agron. Crop Sci., 197, 348–360, 2011.

Schwinning, S. and Sala, O. E.: Hierarchy of responses to resource pulses in arid and semi-arid ecosystems, Oecologia, 141, 211–220, 2004.

She, D., Xia, Y., Shao, M., Peng, S., and Yu, S.: Transpiration and canopy conductance of Caragana korshinskii trees in response to soil moisture in sand land of China, Agroforest. Syst., 87, 667–678, 2013.

Sus, O., Poyatos, R., Barba, j., Carvalhais, N., Llorens, P., Williams, M., and Vilalta, J. M.: Time variable hydraulic parameters improve the performance of amechanistic stand transpiration model. A case study of MediterraneanScots pine sap flow data assimilation, Agr. Forest Meteorol., 198–199, 168–180, 2014.

Wang, X. P., Schaffer, B. E., Yang, Z., and Rodriguez-Iturbe, I.: Probabilistic model predicts dynamics of vegetation biomass in a desert ecosystem in NW China, P. Natl. Acad. Sci. USA, 114, E4944–E4950, https://doi.org/10.1073/pnas.1703684114, 2017.

Yin, L., Zhou, Y., Huang, J., Wenninger, J., Hou, G., Zhang, E., Wang, X., Dong, J., Zhang, J., and Uhlenbrook, S.: Dynamics of willow tree (*Salix matsudana*) water use and its response to environmental factors in the semi-arid Hailiutu River catchment, Northwest China, Environmental Earth Sciences, 71, 4997–5006, 2014.

Zeppel, M. J. B., Murray, B. R., Barton, C., and Eamus, D.: Seasonal responses of xylem sap velocity to VPD and solar radiation during drought in a stand of native trees in temperate Australia, Funct. Plant Biol., 31, 461–470, 2004.

Zha, T., Li, C., Kellomäki, S., Peltola, H., Wang, K.-Y., and Zhang, Y.: Controls of evapotranspiration and CO^2 fluxes from scots pine by surface conductance and abiotic factors, PloS ONE, 8, e69027, https://doi.org/10.1371/journal.pone.0069027, 2013.

Zhao, W. and Liu, B.: The response of sap flow in shrubs to rainfall pulses in the desert region of China, Agr. Forest Meteorol., 150, 1297–1306, 2010.

Zhao, Y., Yuan, W., Sun, B., Yang, Y., Li, J., Li, J., Cao, B., and Zhong, H.: Root distribution of three desert shrubs and soil moisture in Mu Us sand land, Research of Soil and Water Conservation, 17, 129–133, 2010.

Zheng, C. and Wang, Q.: Water-use response to climate factors at whole tree and branch scale for a dominant desert species in central Asia: Haloxylon ammodendron, Ecohydrology, 7, 56–63, 2014.

3

Accumulation of soil organic C and N in planted forests fostered by tree species mixture

Yan Liu[1], **Pifeng Lei**[1,2], **Wenhua Xiang**[1], **Wende Yan**[1,2], **and Xiaoyong Chen**[2,3]

[1]Faculty of Life Science and Technology, Central South University of Forestry and Technology, Changsha 410004, Hunan, China
[2]National Engineering Laboratory for Applied Technology of Forestry & Ecology in South China, Central South University of Forestry and Technology, Changsha 410004, Hunan, China
[3]Division of Science, College of Arts and Sciences, Governors State University, University Park, Illinois 60484, USA

Correspondence to: Pifeng Lei (pifeng.lei@outlook.com)

Abstract. With the increasing trend of converting monocultures into mixed forests, more and more studies have been carried out to investigate the admixing effects on tree growth and aboveground carbon storage. However, few studies have considered the impact of mixed forests on belowground carbon sequestration, particularly changes in soil carbon and nitrogen stocks as a forest grows. In this study, paired pure *Pinus massoniana* plantations, *Cinnamomum camphora* plantations and mixed *Pinus massoniana–Cinnamomum camphora* plantations at ages of 10, 24 and 45 years were selected to test whether the mixed plantations sequestrate more organic carbon (OC) and nitrogen (N) in soils and whether this admixing effect becomes more pronounced with stand ages. The results showed that tree species identification, composition and stand age significantly affected soil OC and N stocks. The soil OC and N stocks were the highest in mixed *Pinus–Cinnamomum* stands compared to those in counterpart monocultures with the same age in the whole soil profile or specific soil depth layers (0–10, 10–20 and 20–30 cm) for most cases, followed by *Cinnamomum* stands and *Pinus* stands with the lowest. These positive admixing effects were mostly nonadditive. Along the chronosequence, the soil OC stock peaked in the 24-year-old stand and was maintained as relatively stable thereafter. The admixing effects were also the highest at this stage. However, in the topsoil layer, the admixing effects increased with stand ages in terms of soil OC stocks. When comparing mixed *Pinus–Cinnamomum* plantations with corresponding monocultures within the same age, the soil N stock in mixed stands was 8.30, 11.17 and 31.45 % higher than the predicted mean value estimated from counterpart pure species plantations in 10-, 24- and 45-year-old stands, respectively. This suggests that these admixing effects were more pronounced along the chronosequence.

1 Introduction

Soil carbon is more stable than that stored in plants, which makes soil carbon more resistant to disturbance (Cunningham et al., 2015). Organic carbon from forest soils accounts for 70–73 % of global soil organic carbon (Six et al., 2002), and a notable portion of global forests are plantations. Thereby, plantations play an important role in soil carbon sequestration and mitigating the global atmospheric carbon budget (Cunningham et al., 2015). However, considering the problems caused by pure plantations (Peng et al., 2008), nowadays planting mixed forest with different species in one stand is becoming more and more popular in worldwide plantation management (Felton et al., 2016; Oxbrough et al., 2012) for the potential positive effects (Knoke et al., 2007). Mixed forests, compared with monocultures, are generally characterized by the sustainability to resist disturbance, potential for higher yield and better ecological services (Grime, 1998; Knoke et al., 2007). It is generally accepted that mixed forests exert favorable effects over monocultures by two main mechanisms: the complementary effect and the selection effect (Isbell et al., 2009; Grossiord et al., 2014). The former is explained by inducing facilitation or easing interspecific competition and enhancing resource use efficiency by niche partitioning. The latter is explained by increasing the possibilities of including highly productive species to increase the yield. Though a lot of work has been done to study the positive effects of biodiversity on ecological function (Balvanera et al., 2006; Marquard et al., 2009), attention is rarely given to soil organic carbon (Vesterdal et al., 2013) and nitrogen stocks, which are closely related to global climate change. The limited research shows that the soil OC and N stocks in mixed stands do not necessarily exceed those of corresponding pure stands (Berger et al., 2002; Forrester et al., 2012; Wiesmeier et al., 2013; Wang et al., 2013; Cremer et al., 2016). Most previous studies did not consider the relative portion of the component species (Berger et al., 2002; Wang et al., 2013; Cremer et al., 2016), which may result in an underestimation of admixing effects. For example, admixing effects could also exist because of the small portion of higher-production species, which leads to low expected production even if the production of a two-species mixed stand is between that of two corresponding pure stands.

Studies of monocultures have suggested that soil organic carbon stocks in the upper soil layer fluctuate in the early stage of afforestation until reaching a new equilibrium that depends on the rates of litter input and decomposition (Paul et al., 2002; Tremblay et al., 2006; Sartori et al., 2007). The soil nitrogen stock increases with increasing stand age in the upper soil layer because of cumulative biological fixation, litterfall, recycling from deeper mineral soil via root mortality and even atmospheric N deposition (Hume et al., 2016). However, the changes in soil organic carbon and nitrogen stocks at different stand ages are rarely quantified and poorly understood, and this is especially true in the context of comparing pairwise mixed forests with pure forest stands along the chronosequence.

In this study we investigated the soil organic carbon and nitrogen stocks of *Pinus massoniana* and *Cinnamomum camphora* pure stands as well as mixed stands at 10, 24 and 45 years old. All the soil samples were taken from the 0–10, 10–20 and 20–30 cm soil layers in the above-mentioned stands in the Hunan Province in China. We hypothesized that (1) the soil organic carbon and nitrogen stocks under mixed stands are higher than those under corresponding monocultures within the same age in the whole soil profile, and (2) these positive admixing effects become more pronounced along the chronosequence.

2 Materials and methods

2.1 Study site

Three parallel forest stands at different stand ages were selected, comprised of mixed *P. massoniana* and *C. camphora* plantations, pure *P. massoniana* plantations and pure *C. camphora* at 10, 24 and 45 years old. Some *Pinus elliottii* were spotted in between and were treated as *P. massoniana* because of the similarity in growth and biological characteristics (we grouped them as "*Pinus*" thereafter). Three 20 × 20 m plots were established in each of the forest stands, including mixed *Pinus–Cinnamomum* stands and corresponding pure species stands (*Pinus* and *Cinnamomum*) at the age of 24 and 45 years in March 2013 in the Botanical Garden in the Hunan Province, China (28°06′ N, 113°02′ E); this amounted to 18 plots. Due to the lack of young stands in the Botanical Garden, we selected another site located in the Taolin forestry station (28°55′ N, 113°03′ E) in the Hunan Province, China. This site was used as a nursery and abandoned in 2000; it was replanted in 2003. Three plots for each of the three plantation types at 10 years old were set up with same method mentioned above, but with a size of 12 × 12 m constrained by the smaller patches there. In total, our study included 27 plots consisting of mixed *Pinus–Cinnamomum* plantations and corresponding monocultures at ages of 10, 24 and 45 years (Table 1). These two sites are 200 km in distance from each other with similar climates and soil types. The regional climate is typical midrange subtropical monsoonal with a mean annual air temperature of 17.2°, mean annual precipitation of 1422 mm in the Botanical Garden and a mean annual air temperature of 16.9°. Mean annual precipitation is 1353.6 mm in the Taolin forestry station. The soil is well-drained clay-loam red soil developed from slate parent rock and classified as Alliti-Udic Ferrosols according to the World Reference Base for Soil Resources (Institute of Soil Science, Chinese Academy of Science, 2001). The depth of the soil layer is deeper than 80 cm, but the content of the soil humus was not rich and pH values range from 4.12 to 4.86. The forests remained unmanaged and all soil-forming factors had remained constant since forest establishment owning to either the foundation of the Botanical Garden or the young age of the forests. Here we used two sites for this experiment due to the difficulties of finding one field site with all these plantations in gradients of forest ages. The principle of field site selection here was to put three plantation types of the same age in one site. Therefore, we think these two sites are suitable for our experiment since our purpose here was to evaluate the admixing effects on soil C and N stocks by comparing mixed forests and corresponding monocultures at the same age rather than comparing soil C and N in forests at different ages. For more detailed information about the experimental site and soil conditions, refer to Wen et al. (2014).

2.2 Soil sampling and laboratory analyses

Four soil samples were randomly collected in each plot using a metal corer (10 cm in diameter) from three depths: 0–10, 10–20, 20–30 cm. Each was treated as one individual sample, then air-dried and sieved through a 0.25 mm sieve. We collected the soil samples down to 30 cm of depth as the soils in this area were susceptible to environmental varia-

Table 1. Stand characteristics in pure species *Pinus* stands, pure *Cinnamomum* stands and mixed *Pinus–Cinnamomum* stands at age 10, 24 and 45 years (mean ± standard deviation).

Stand	Age	Species	Density ($n\,ha^{-1}$)	Diameter at breast height (cm)	Height (cm)	Basal area ($m^2\,ha^{-1}$)
Pinus stands	10	*Pinus*	2592	9.38 ± 3.26	5.28 ± 3.97	20.06
	24	*Pinus*	2050	14.18 ± 4.34	12.86 ± 6.52	35.37
	45	*Pinus*	600	21.40 ± 5.30	12.47 ± 1.88	22.84
Cinnamomum stands	10	*Cinnamomum*	2708	7.77 ± 2.60	5.99 ± 1.25	14.26
	24	*Cinnamomum*	900	17.02 ± 6.52	13.71 ± 2.74	23.46
	45	*Cinnamomum*	800	21.06 ± 6.73	13.24 ± 2.29	30.63
Mixed *Pinus–Cinnamomum* stands	10	*Pinus*	902	7.64 ± 1.82	4.73 ± 0.82	4.37
		Cinnamomum	1689	8.14 ± 2.81	7.2 ± 0.73	9.83
	24	*Pinus*	267	19.88 ± 5.06	12.35 ± 1.64	7.80
		Cinnamomum	592	15.27 ± 5.92	11.41 ± 3.13	12.45
	45	*Pinus*	250	19.69 ± 4.10	12.37 ± 2.60	7.91
		Cinnamomum	325	20.94 ± 8.54	13.75 ± 2.79	12.91

tions and sensitive to the carbon input by litter and fine roots, which matches our purposes of assessing the admixing effects on soil OC and N stocks over time (Wang et al., 2013; Cremer et al., 2016). Soil organic carbon was determined by the wet combustion method through the oxidization of potassium bichromate (Walkley–Black method) followed by titration with 0.5 N ferrous ammonium sulfate solution by using a diphenylamine indicator. Soil total nitrogen was measured using the semimicro Kjeldahl method digested with a mixture of H_2SO_4, K_2SO_4, $CuSO_4$ and Se (Institute of Soil Science, Chinese Academy of Science).

2.3 Data analysis

Soil OC stock ($t\,ha^{-1}$) and N stock ($t\,ha^{-1}$) at different soil depths in different stands were calculated with the following formula: stock = bulk density • depth • [OC or N concentration]. The effects of the experimental factors, species composition, stand age and soil depth on soil OC, N concentrations and soil OC and N stocks were tested by means of three-way analysis of variance (ANOVA). Differences in soil OC or N concentrations and stocks among mixed *P. massoniana* and *C. camphora* stands, pure *P. massoniana* stands and pure *Cinnamomum* stands within the same age stages and soil layers were analyzed by using one-way ANOVA followed by a Tukey's test. In order to detect whether the admixing effects were additive or nonadditive, an alternative analytical method was used as suggested by Ball et al. (2008). Expected values of OC and N stock in mixed stands at different stand ages in different soil depths were calculated by adjusted values based on basal area in monocultures with the following formula:

$$\text{expected value} = \left(\mathrm{Ba}_{\mathrm{p.mix}}/\mathrm{Ba}_{\mathrm{p.pure}}\right) \cdot \mathrm{Stock}_{\mathrm{p.pure}} + \left(\mathrm{Ba}_{\mathrm{c.mix}}/\mathrm{Ba}_{\mathrm{c.pure}}\right) \cdot \mathrm{Stock}_{\mathrm{c.pure}},$$

where $\mathrm{Ba}_{\mathrm{p.mix}}$ is the basal area of *Pinus* in the mixed stand, $\mathrm{Ba}_{\mathrm{p.pure}}$ is the basal area of *Pinus* in the pure *Pinus* stand, $\mathrm{Ba}_{\mathrm{c.mix}}$ is the basal area of *C. camphora* in the mixed stand, $\mathrm{Ba}_{\mathrm{c.pure}}$ is the basal area of *Cinnamomum* in the *Cinnamomum* pure stand, $\mathrm{Stock}_{\mathrm{p.pure}}$ is the soil OC or N stock under the pure *Pinus* stand and $\mathrm{Stock}_{\mathrm{c.pure}}$ is the soil OC or N stock under the pure *Cinnamomum* stand. All the calculations above are conducted in the same given stand age (10, 24 or 45 years old). These expected values for OC or N stock were then compared with observed values for each individual sample that were measured experimentally in mixed stands with the following formula: (observed − expected)/expected. For soil OC and N stocks at specific soil depths at given stand ages, 95 % confidence intervals (CI) were calculated with the above formula. If the CIs for mixtures did not cross $y = 0$, the admixing effect was considered nonadditive (Ball et al., 2008). Otherwise, we regard the admixing effect as additive. All the statistical analyses were conducted with statistical software from the R project (R 3.0.2; R Development Core Team, 2013).

3 Results

3.1 Soil organic carbon and nitrogen stocks

The three-way ANOVA indicated that the forest stand types, soil depths and their interactions exerted significant influence on the soil OC and N concentrations, while the age effects were not significant for soil OC concentration (Table 2). When compared with parallel forest stands with the

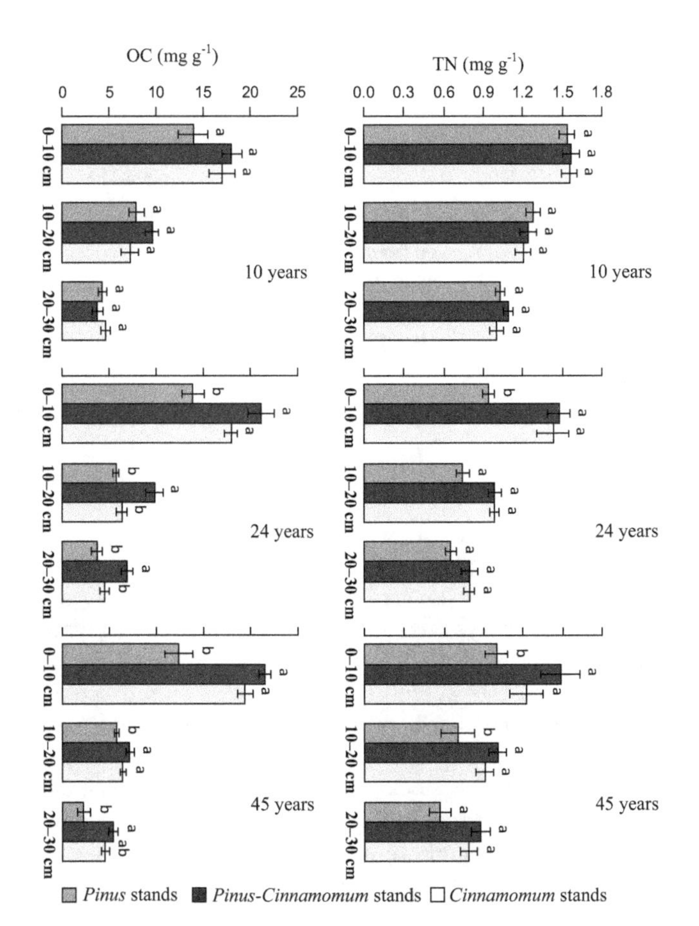

Figure 1. Soil organic carbon (OC) concentration and nitrogen (N) concentration in pure *Pinus*, *Cinnamomum* and mixed *Pinus–Cinnamomum* stands in 0–10, 10–20 and 20–30 cm of soil depth at the age of 10, 24 and 45 years. Error bars indicate standard errors. Different letters indicate significant differences among different stands within the same soil profile and age stage ($p < 0.05$).

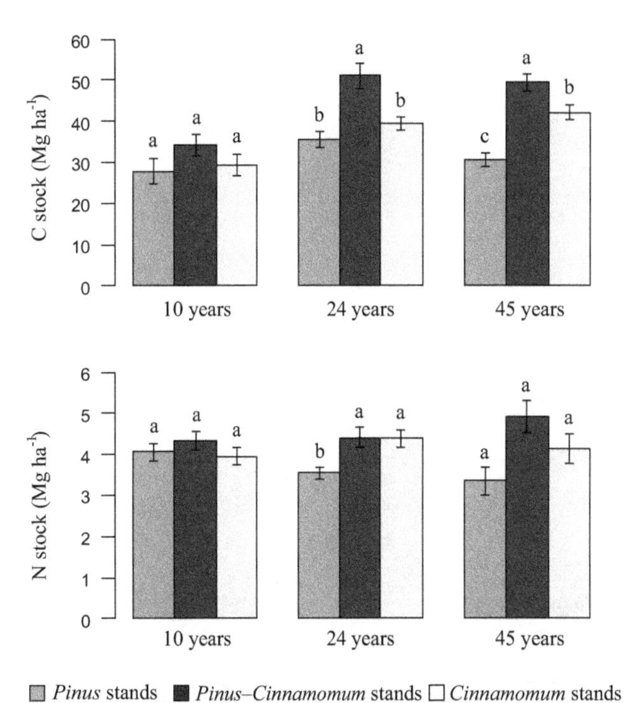

Figure 2. Total soil organic carbon (OC) and nitrogen (N) stocks in pure *Pinus*, *Cinnamomum* and mixed *Pinus–Cinnamomum* stands in 0–30 cm of soil depth at the age of 10, 24 and 45 years. Error bars indicate standard errors. Different letters indicate significant differences among different stands with the same age ($p < 0.05$).

same age, soil OC and N concentrations were the highest in *Pinus–Cinnamomum* mixed stands for almost all cases in the whole soil profile or in the specific soil layers, but significant differences were only detected in 24- and 45-year-old stands ($P < 0.05$) (Fig. 1).

Total soil OC stock was highest in *Pinus–Cinnamomum* mixed stands (44.86 Mg ha^{-1}) compared to corresponding monocultures at the same stand age in all the soil profiles investigated followed by the evergreen broad-leaved *Cinnamomum* stand (36.37 Mg ha^{-1}), and the conifer *Pinus* stand showed the lowest values (31.51 Mg ha^{-1}). Significant differences were detected in 24- and 45-year-old stands but not in 10-year-old stands ($P < 0.05$). Along the chronosequence, soil OC stocks increased with increasing stand age in these three forest types with a mean value of 30.50, 41.96 and 43.85 Mg ha^{-1} in 10-, 24- and 45-year-old stands, respectively (Fig. 2). When we take a closer look at the stratification distribution of the soil OC stocks in these three forest stands at different stand ages, the results show that soil OC stocks decreased significantly with increasing soil depth and a similar pattern: the highest OC stock in mixed stands over monocultures was observed within given soil layers at given stand ages. The over-performance of OC stock in *Pinus–Cinnamomum* mixed stands compared to the counterpart monoculture stands with the same age was mainly attributed to the top 10 cm of soil depth (Fig. 3). In 0–10 cm soil layers, OC stocks under mixed stands showed no significant differences compared with individual stands in the young stands. Those under mixed stands significantly exceeded the *Pinus* individual stands in middle-aged stands and finally exceeded both individual stands in the oldest stands. In the 10–20 cm soil layers, both 10- and 45-year-old stands exerted no significant differences in soil OC stock among three stand types, while in 24-year-old stands the mixture had a significantly higher soil OC stock than the monocultures. In the 20–30 cm soil layers, mixed stands showed similar soil OC stock patterns when compared with corresponding monocultures, and in 24-year-old stands the mixture showed a significantly higher soil OC stock than two pure *Pinus* and *Cinnamomum* stands.

Soil depth and stand type, but not stand age, exerted significant effects on soil N stock in mineral soil (Table 2). Within the same stand, the soil N stocks decreased with soil depth, and among them, the superficial soil layers (0–10 cm) exhibited significantly higher soil N stocks than the other two

Table 2. The effects of plantation stand, stand age and soil depth on soil OC concentration, N concentration, OC stock and N stock. Values shown are ANOVA F values and P values. Bold font indicates significant differences at $p < 0.05$.

Factor	OC concentration		N concentration		OC stock		N stock	
	F value	P value	F value	P value	F value	P value	F value	P value
Stand	24.30	**<0.0001**	11.89	**<0.0001**	36.73	**<0.0001**	15.17	**<0.0001**
Age	0.00	0.9942	60.53	**<0.0001**	35.99	**<0.0001**	0.54	0.5814
Depth	699.57	**<0.0001**	205.33	**<0.0001**	489.87	**<0.0001**	65.62	**<0.0001**
Stand × age	2.07	0.1283	10.50	**<0.0001**	2.87	**0.0236**	3.92	**0.0042**
Stand × depth	8.61	**0.0002**	2.39	0.0937	5.53	**0.0003**	1.15	0.3354
Age × depth	2.00	0.1588	0.05	0.8163	16.79	**<0.0001**	4.18	**0.0027**
Stand × age × depth	0.15	0.8579	0.73	0.4834	1.03	0.4142	0.67	0.7217

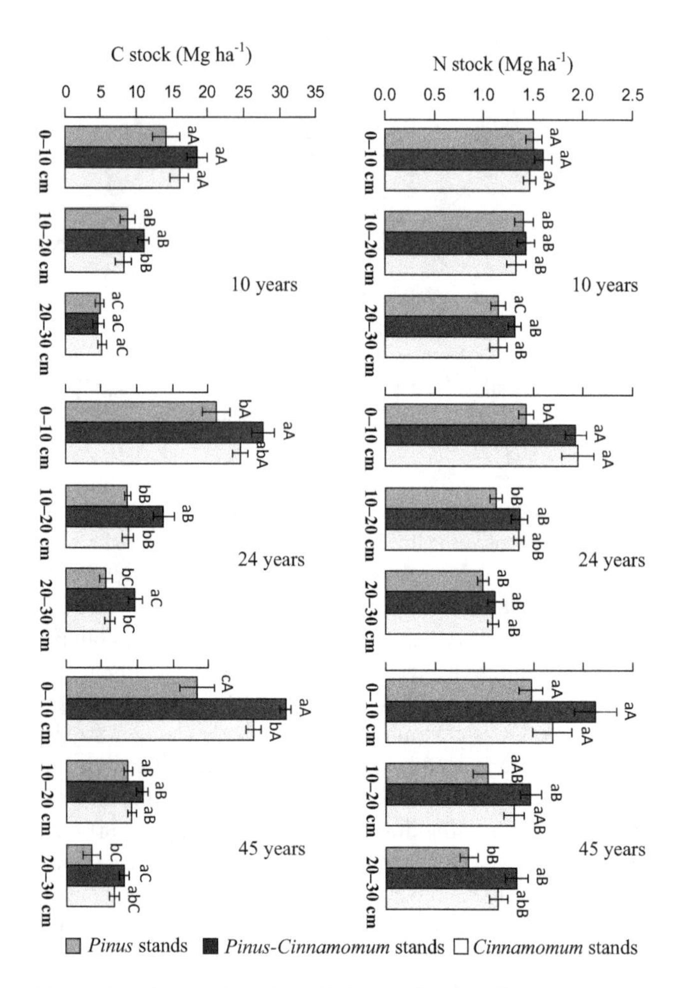

Figure 3. Soil organic carbon (OC) and nitrogen (N) stocks in pure *Pinus*, *Cinnamomum* and mixed *Pinus–Cinnamomum* stands in 0–10, 10–20 and 20–30 cm of soil depth at the age of 10, 24 and 45 years. Error bars indicate standard errors. Different small letters indicate significant differences among different stands within the same soil profile and age stage ($p < 0.05$). Different capital letters indicate significant differences among different soil layers within the same stand at a given stand age ($p < 0.05$).

deeper soil layers (Fig. 3). Only in the 0–10 and 10–20 cm soil layers of 24-year-old stands and the 20–30 cm soil layer of 45-year-old stands did N stock significantly differ among stand types. Pure *Cinnamomum* and *Pinus–Cinnamomum* mixed stands always had similar N stocks, which were higher than *Pinus* stands. The N stock in the topsoil of mixed stands increased along the chronosequence for all stands, except in the *Cinnamomum* stand, which marginally decreased from the 24- to the 45-year-old stand. In the 10–20 cm soil layer of the *Pinus* stand, N stock increased from the 10- to the 24-year-old stand and then stayed stable. All the other soil layers did not exert significant differences along the chronosequence within the stand type (Figs. 2 and 3).

3.2 Admixing effects on soil OC and N stocks

The relative performance of soil OC and N stocks in the *Pinus–Cinnamomum* mixed stands was calculated to determine the additive or nonadditive effects for each stand at different soil layers and different stand ages by comparing observed soil OC or N stock values with expected values based on counterpart monocultures. In our study the mixed planting always exerted positive effects, as all the relative values of soil-sequestrated OC and N stocks were positive in all the soil profiles in the mixed stands. Almost all the CIs did not cross $y = 0$, suggesting that these positive admixing effects were strongly nonadditive, except the relative OC sequestration in the 20–30 cm soil layers of 10-year-old stands. In the 10-year-old stand, the over-performance of soil OC sequestration was mainly attributed to the topsoil and subsoil layers, and these positive effects were stable in the two upper soil layers and then decreased in the deepest layer. In the 24-year-old stand, it increased with soil depth. In the 45-year-old stand, the relative percentage was the highest in the deepest soil layer (20–30 cm). Overall, the percentage of over-performance in mixture was significantly higher in the 24-year-old stand than in the 10-year-old stand, and then it marginally decreased in the 45-year-old stand. For soil N stock, however, the results showed a consistent pattern

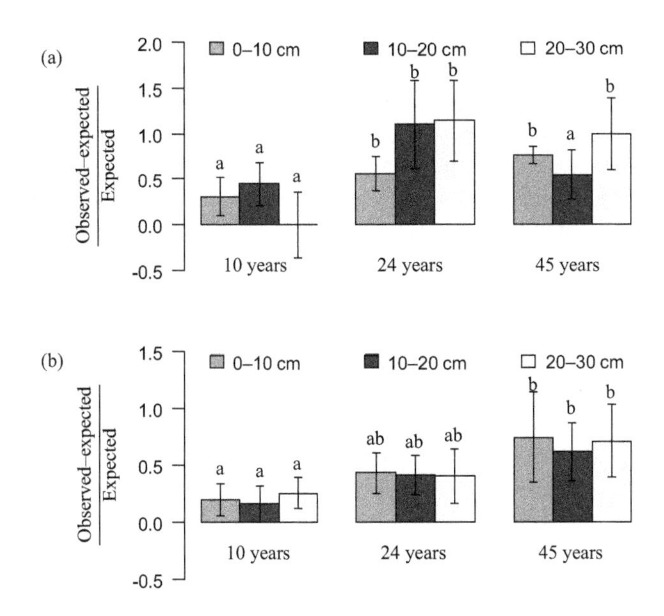

Figure 4. Investigation of additive or nonadditive interactions for soil OC stock **(a)** and N stock **(b)** in *Pinus–Cinnamomum* mixed stands in 0–10, 10–20 and 20–30 cm of soil depth at the age of 10, 24 and 45 years. Observed values were compared to expected values calculated as the average value in monocultures of *Pinus* and *Cinnamomum*. Error bars represent 95 % CI, and mixtures for which the CIs do not cross $y = 0$ are considered to be significantly nonadditive. Different letters indicate significant differences among different stand ages within the same soil depth profile.

that the positive effects increased with increasing stand ages (Fig. 4).

4 Discussion

4.1 Soil organic carbon and nitrogen stocks

The soil organic carbon stocks under the three stand types always followed this order: mixed *Pinus–Cinnamomum* stand > *Cinnamomum* stand ≥ *Pinus* stand. An exception was those in the 10–20 and 20–30 cm soil layers of the 10-year-old stand, which showed slight differences. Many researchers have demonstrated the species diversity or species identity effects on soil OC and N accumulation in forests and revealed that species diversity and/or abundance of dominant tree species exerts effects on OC and N stocks through carbon input by litter and fine roots (Berger et al., 2002; Guckland et al., 2009; Wang et al., 2013; Dawud et al., 2016). In a previous study at this site, the annual litterfall was also the highest in the mixed *Pinus–Cinnamomum* stand followed by *Pinus*; *Cinnamomum* was the lowest (Xu et al., 2013). Therefore, the higher soil organic carbon under mixed stands may be attributed to the higher annual litterfall in mixed stands. The higher soil OC in the *Cinnamomum* stand compared to *Pinus* is likely connected with the litter decomposition rate, as the needle leaves of *Pinus* accumulated on the forest floor with a lower composition rate. The higher litter input may not necessarily increase the soil carbon content (Fontaine et al., 2004), although many previous studies have also shown larger carbon stocks under coniferous species than broad-leaved species (Kasel and Bennett, 2007; Schulp et al., 2008; Wang et al., 2013). With increasing stand age, the gap in soil OC stock between the *Cinnamomum* and *Pinus* stands became more pronounced in all soil profiles investigated. Furthermore, when compared with broadleaf tree species, conifer species tended to allocate much more total organic matter production to aboveground growth, which features them as fast-growing tree species and caused a lower direct carbon input to soil (Cuevas et al., 1991). In this study, the highest OC stock presented in 24-year-old stands; in the *Cinnamomum* stands, though not significant, carbon stock still had an increasing trend from 24- to 45-year-old stands. In mixed stands, though with a relatively smaller portion of *Cinnamomum*, the trend in soil OC stock is similar to *Cinnamomum* stands but not the *Pinus* stands. This confirmed that broadleaf tree species are ahead of conifers in long-term growth.

The mixed stand exhibited the highest soil N concentrations and N stocks, revealing the priority of soil N accumulation in mixed *Pinus–Cinnamomum* stands compared to corresponding monocultures. Compared to the mean value of the counterpart monoculture at the same age, the soil N stock in mixed stands increased by 8.30, 11.17 and 31.45 %, respectively, suggesting that these admixing effects were more pronounced with stand ages along the chronosequence. Whether this enhanced the admixing effect on soil N accumulation will continue needs further investigation. In monocultures, the *Pinus* stand showed priority in earlier stages, while *Cinnamomum* showed priority in the later stage in terms of soil N stocks. This is similar to the trend in soil OC stocks but less pronounced. A large amount of soil nitrogen us stored in soil organic matter, and with the decomposition of soil organic matter, nitrogen will be released and able to be taken up by plants or leaching, as reported in our previous study. Soil nitrogen concentration positively correlated to soil organic carbon concentration (Wen et al., 2014), and thereby the soil nitrogen stocks to some degree shared similar patterns of soil OC stocks under the same circumstances. As Hume et al. (2016) discussed, the accumulation of soil nitrogen may be slower than carbon because nitrogen is progressively locked up in live biomass, which is likely the explanation for why the soil nitrogen stock changed less significantly with stand ages and between different stand types compared with soil OC stocks here.

In the uppermost soil layer, soil organic carbon stocks under the three stand types all significantly increased from 10- to 24-year-old stands and then became stable thereafter; those of the *Pinus* stands slightly declined from 24- to 45-year-old stands without significant differences. This trend is inconsistent with a previous study that reported a decline in soil organic carbon in the topsoil (0–10 cm) within a certain period of time after plantation establishment (Turner and Lambert, 2000). However, it is consistent with the results from a Chinese fir plantation where the soil OC stocks increased with increasing stand ages from about 10 to 20 years and then stabilized (Chen et al., 2013). Most similar research that focuses on the changes in organic carbon stocks along the chronosequence has suggested that soil organic carbon stocks in topsoil reach a stable level after approximately 20 years (Tremblay et al., 2006; Chen et al., 2013). This can be interpreted as the formation of a balance that depends on the carbon input and organic decomposition (Hume et al., 2016). In the initial stage of afforestation, the changes in organic carbon storage differed with increasing stand age. The conversion of plantations into natural native forests will always present a decrease in soil organic carbon in topsoil within 10 years, and then some will rebound and increase until reaching an equilibrium level (Chen et al., 2013); the remaining amount will continue decreasing to the lowest level (Turner and Lambert, 2000). In our results the soil OC and N stocks decreased significantly with increasing soil depth. However, the magnitude of over-performance in the *Pinus–Cinnamomum* mixed stands over monocultures increased with soil depth (Figs. 3 and 4). Topsoils under forests are always the most susceptible to disturbance, and they can be directly impacted by carbon input through litter and fine roots, which always decline with soil depth (Wang et al., 2016). Our data for the same sites also suggested that fine roots mostly assemble in the upper soil, and fine-root overyielding occurred in the topsoil layer (0–10 cm; see Table S1 in the Supplement). Also, in our previous study (Wen et al., 2014), the ratio of soil microbial biomass to soil organic carbon was the lowest in topsoil, which may suggest a lower mineralization rate in deeper soil and make the accumulation of OC and N stocks in deeper soil along the chronosequence more pronounced over the topsoil layer. Here we only collected the soil down to 30 cm of depth, and the effects of species diversity and species identity on the deeper soil merits further investigation to improve model parameters of soil OC and N processes in the deep soil profile.

4.2 Admixing effect along the chronosequence

In the two topsoil layers (0–10 and 10–20 cm), the admixing effect on soil OC was more pronounced along the chronosequence. This likely accounted for the increasingly intensive interactions between the two species with increasing stand age. In the pure stands, intraspecific competition also become more intensive along the chronosequence, which will block carbon sequestration because of nutrient and water limits and favor the mixed forests where the interspecific competition is relatively less intensive in general (Lei et al., 2012a). Regarding the additive effect shown in the 20–30 cm soil layer under 10-year-old mixed stands, the observations were almost equal to the expected values, which confirmed is the space occupied by belowground and aboveground tree biomass was not big enough to exert significant influence when the stands are relatively young or the period of soil OC accumulation processes is relatively short. Our results also showed, to some degree, higher admixing effects in the deeper soil layer that may contribute to the lower expected values in monocultures in the deeper soil layer, which makes the (observed − expected)/expected values more sensitive to the increment of observed values.

The admixing effects of soil N stocks in the whole profile suggested a consistent increasing trend in the whole soil profile. Fine roots in mixed stands compared with pure stands are always assumed to exploit deeper soil (Cremer et al., 2016), so they may assemble more nitrogen from deeper soil (even deeper than our samples) to upper soil. Also, Lei et al. (2012b) reported higher fine-root turnover caused by higher fine-root production of mixed stands, which potentially increases nitrogen input in mixed stands. In addition, mixed stands are more resistant to environmental disturbances, are more capable of N retention and prevent leaching in soil (Tilman et al., 1996). This merits further studies in more diverse communities in forests to confirm this pattern of higher nitrogen stock under mixed stands over monocultures and these increasing positive effects along the chronosequence.

5 Conclusions

The tree species and composition as well as the stand age significantly affect soil organic carbon and nitrogen stocks. Converting pure stands into mixed stands can significantly enhance soil OC and N stocks. This positive admixing effect becomes more pronounced along the chronosequence for OC only in topsoil, while inconsistent trends present in the deeper soil. However, for soil N stock, it becomes more pronounced along the chronosequence in the whole soil profile. Tree species identification also affects the soil OC sequestration and N stock. In topsoil, *Cinnamomum* stands always contain more soil OC and N stocks than *Pinus* stands due to the different strategies of carbon and nutrient allocation and different rates of organic decomposition, but these differences are less pronounced in deeper soil layers.

Competing interests. The authors declare that they have no conflict of interest.

Acknowledgements. This study was sponsored by the National Natural Science Foundation of China (31200346) and the Introduce Talent Fund of CSUFT. We thank the forest administration of Hunan and the forest station of Taolin for permission to use the site. Furthermore, we acknowledge the assistance of Fang Jiang, Yuqin Xu and Hao Yi in the field and laboratory.

Edited by: Sébastien Fontaine

References

Ball, B. A., Hunter, M. D., Kominoski, J. S., Swan, C. M., and Bradford, M. A.: Consequences of Non-Random Species Loss for Decomposition Dynamics: Experimental Evidence for Additive and Non-Additive Effects, J. Ecol., 96, 303–313, 2008.

Balvanera, P., Pfisterer, A. B., Buchmann, N., He, J. S., Nakashizuka, T., Raffaelli, D., and Schmid, B.: Quantifying the evidence for biodiversity effects on ecosystem functioning and services, Ecol. Lett., 9, 1146–1156, 2006.

Berger, T. W., Neubauer, C., and Glatzel, G.: Factors controlling soil carbon and nitrogen stores in pure stands of Norway spruce (*Picea abies*) and mixed species stands in Austria, Forest Ecol. Manag., 159, 3–14, 2002.

Chen, G.-S., Yang, Z.-J., Gao, R., Xie, J.-S., Guo, J.-F., Huang, Z.-Q., and Yang, Y.-S.: Carbon storage in a chronosequence of Chinese fir plantations in southern China, Forest Ecol. Manag., 300, 68–76, 2013.

Cremer, M., Kern, N. V., and Prietzel, J.: Soil organic carbon and nitrogen stocks under pure and mixed stands of European beech, Douglas fir and Norway spruce, Forest Ecol. Manag., 367, 30–40, 2016.

Cuevas, E., Brown, S., and Lugo, A. E.: Above-and belowground organic matter storage and production in a tropical pine plantation and a paired broadleaf secondary forest, Plant Soil, 135, 257–268, 1991.

Cunningham, S. C., Cavagnaro, T. R., Mac Nally, R., Paul, K. I., Baker, P. J., Beringer, J., Thomson, J. R., and Thompson, R. M.: Reforestation with native mixed-species plantings in a temperate continental climate effectively sequesters and stabilizes carbon within decades, Glob. Change Biol., 21, 1552–1566, 2015.

Dawud, S. M., Raulund-Rasmussen, K., Domisch, T., Finer, L., Jaroszewicz, B., and Vesterdal, L.: Is tree species diversity or species identity the more important driver of soil carbon stocks, C/N ratio, and pH?, Ecosystems, 19, 645–660, 2016.

Felton, A., Nilsson, U., Sonesson, J., Felton, A. M., Roberge, J.-M., Ranius, T., Ahlström, M., Bergh, J., Björkman, C., and Boberg, J.: Replacing monocultures with mixed-species stands: Ecosystem service implications of two production forest alternatives in Sweden, Ambio, 45, 124–139, 2016.

Fontaine, S., Bardoux, G., Abbadie, L., and Mariotti, A.: Carbon input to soil may decrease soil carbon content, Ecol. Lett., 7, 314–320, 2004.

Forrester, D. I., Pares, A., O'Hara, C., Khanna, P. K., and Bauhus, J.: Soil Organic Carbon is Increased in Mixed-Species Plantations of Eucalyptus and Nitrogen-Fixing Acacia, Ecosystems, 16, 123–132, 2012.

Grime, J.: Benefits of plant diversity to ecosystems: immediate, filter and founder effects, J. Ecol., 86, 902–910, 1998.

Grossiord C., Granier A., Ratcliffe S, Bouriaud O, Bruelheide H, Checko E, Forrester D. I., Dawud S. M., Finér L, Pollastrini M., Scherer-Lorenzen M., Valladares F., Bonal D., and Gessler A.: Tree diversity does not always improve resistance of forest ecosystems to drought, P. Natl. Acad. Sci. USA, 111, 14812–14815, 2014.

Guckland, A., Jacob, M., Flessa, H., Thomas, F. M., and Leuschner, C.: Acidity, nutrient stocks, and organic-matter content in soils of a temperate deciduous forest with different abundance of European beech (*Fagus sylvatica* L.), J. Plant Nutr. Soil Sc., 172, 500–511, 2009.

Hume, A., Chen, H. Y. H., Taylor, A. R., Kayahara, G. J., and Man, R.: Soil C:N:P dynamics during secondary succession following fire in the boreal forest of central Canada, Forest Ecol. Manag., 369, 1–9, 2016.

Isbell, F. I., Polley, H. W., and Wilsey, B. J.: Biodiversity, productivity and the temporal stability of productivity: patterns and processes, Ecol. Lett., 12, 443–451, 2009.

Kasel, S. and Bennett, T. L.: Land-use history, forest conversion, and soil organic carbon in pine plantations and native forests of south eastern Australia, Geoderma, 137, 401–413, 2007.

Knoke, T., Ammer, C., Stimm, B., and Mosandl, R.: Admixing broadleaved to coniferous tree species: a review on yield, ecological stability and economics, Eur. J. For. Res., 127, 89–101, 2007.

Lei, P., Scherer-Lorenzen, M., and Bauhus, J.: Belowground facilitation and competition in young tree species mixtures, Forest Ecol. Manag., 265, 191–200, 2012a.

Lei, P., Scherer-Lorenzen, M., and Bauhus, J.: The effect of tree species diversity on fine-root production in a young temperate forest, Oecologia, 169, 1105–1115, 2012b.

Marquard, E., Weigelt, A., Roscher, C., Gubsch, M., Lipowsky, A., and Schmid, B.: Positive biodiversity–productivity relationship due to increased plant density, J. Ecol., 97, 696–704, 2009.

Oxbrough, A., French, V., Irwin, S., Kelly, T. C., Smiddy, P., and O'Halloran, J.: Can mixed species stands enhance arthropod diversity in plantation forests?, Forest Ecol. Manag., 270, 11–18, 2012.

Paul, K., Polglase, P., Nyakuengama, J., and Khanna, P.: Change in soil carbon following afforestation, Forest Ecol. Manag., 168, 241–257, 2002.

Peng, S., Wang, D., and Zhao, H.: Discussion the status quality of plantation and near nature forestry management in China, Journal of Northwest Forestry University, 23, 184–188, 2008.

R Development Core Team: R: A language and environment for statistical computing. R Foundation for Statistical Computing, Vienna, Austria, available at: http://www.R-project.org (last access: 2 February 2017), 2013.

Sartori, F., Lal, R., Ebinger, M. H., and Eaton, J. A.: Changes in soil carbon and nutrient pools along a chronosequence of poplar plantations in the Columbia Plateau, Oregon, USA, Agr. Ecosyst. Environ., 122, 325–339, 2007.

Schulp, C. J. E., Nabuurs, G., Verburg, P. H., and de Waal, R. W.: Effect of tree species on carbon stocks in forest floor and mineral soil and implications for soil carbon inventories, Forest. Ecol. Manag., 256, 482–490, 2008.

Six, J., Callewaert, P., Lenders, S., De Gryze, S., Morris, S., Gregorich, E., Paul, E., and Paustian, K.: Measuring and understanding carbon storage in afforested soils by physical fractionation, Soil Sci. Soc. Am. J., 66, 1981–1987, 2002.

Tilman, D., Wedin, D., and Knops, J.: Productivity and sustainability influenced by biodiversity in grassland ecosystems, Nature, 379, 718–720, 1996.

Tremblay, S., Périé, C., and Ouimet, R.: Changes in organic carbon storage in a 50 year white spruce plantation chronosequence established on fallow land in Quebec, Can. J. Forest Res., 36, 2713-2723, 2006.

Turner, J. and Lambert, M.: Change in organic carbon in forest plantation soils in eastern Australia, Forest Ecol. Manag., 133, 231–247, 2000.

Vesterdal, L., Clarke, N., Sigurdsson, B. D., and Gundersen, P.: Do tree species influence soil carbon stocks in temperate and boreal forests?, Forest Ecol. Manag., 309, 4–18, 2013.

Wang, H., Liu, S., Wang, J., Shi, Z., Lu, L., Zeng, J., Ming, A., Tang, J., and Yu, H.: Effects of tree species mixture on soil organic carbon stocks and greenhouse gas fluxes in subtropical plantations in China, Forest Ecol. Manag., 300, 4–13, 2013.

Wang, W., Wu, X., Hu, K., Liu, J., and Tao, J.: Understorey fine root mass and morphology in the litter and upper soil layers of three Chinese subtropical forests, Plant Soil, 406, 219–230, https://doi.org/10.1007/s11104-016-2878-1, 2016.

Wen, L., Lei, P., Xiang, W., Yan, W., and Liu, S.: Soil microbial biomass carbon and nitrogen in pure and mixed stands of *Pinus massoniana* and *Cinnamomum camphora* differing in stand age, Forest Ecol. Manag., 328, 150–158, 2014.

Wiesmeier, M., Prietzel, J., Barthold, F., Spörlein, P., Geuß, U., Hangen, E., Reischl, A., Schilling, B., von Lützow, M., and Kögel-Knabner, I.: Storage and drivers of organic carbon in forest soils of southeast Germany (Bavaria) – Implications for carbon sequestration, Forest Ecol. Manag., 295, 162–172, 2013.

Xu, W. M., Yan, W. D., Li, J. B., Zhao, J., and Wang, G. J.: Amount and dynamic characteristics of litterfall in four forest types in subtropical China, Acta Ecologica Sinica, 33, 7570–7575, 2013.

4

A zero-power warming chamber for investigating plant responses to rising temperature

Keith F. Lewin, Andrew M. McMahon, Kim S. Ely, Shawn P. Serbin, and Alistair Rogers

Environmental & Climate Sciences Department, Brookhaven National Laboratory, Upton, NY 11973, USA

Correspondence to: Alistair Rogers (arogers@bnl.gov)

Abstract. Advances in understanding and model representation of plant and ecosystem responses to rising temperature have typically required temperature manipulation of research plots, particularly when considering warming scenarios that exceed current climate envelopes. In remote or logistically challenging locations, passive warming using solar radiation is often the only viable approach for temperature manipulation. However, current passive warming approaches are only able to elevate the mean daily air temperature by $\sim 1.5\,°C$. Motivated by our need to understand temperature acclimation in the Arctic, where warming has been markedly greater than the global average and where future warming is projected to be ~ 2–$3\,°C$ by the middle of the century; we have developed an alternative approach to passive warming. Our zero-power warming (ZPW) chamber requires no electrical power for fully autonomous operation. It uses a novel system of internal and external heat exchangers that allow differential actuation of pistons in coupled cylinders to control chamber venting. This enables the ZPW chamber venting to respond to the difference between the external and internal air temperatures, thereby increasing the potential for warming and eliminating the risk of overheating. During the thaw season on the coastal tundra of northern Alaska our ZPW chamber was able to elevate the mean daily air temperature 2.6 °C above ambient, double the warming achieved by an adjacent passively warmed control chamber that lacked our hydraulic system. We describe the construction, evaluation and performance of our ZPW chamber and discuss the impact of potential artefacts associated with the design and its operation on the Arctic tundra. The approach we describe is highly flexible and tunable, enabling customization for use in many different environments where significantly greater temperature manipulation than that possible with existing passive warming approaches is desired.

1 Introduction

Driven primarily by rising atmospheric carbon dioxide concentration, the global mean temperature has risen by 0.85 °C since the beginning of the industrial revolution (IPCC, 2013). Terrestrial ecosystems are currently limiting the rate at which our planet is warming by absorbing approximately one-third of our CO_2 emissions, but the fate of this terrestrial carbon sink is critically uncertain (Friedlingstein et al., 2014; Lovenduski and Bonan, 2017). A key part of reducing this uncertainty is improving understanding and model representation of the processes that underlie the response and acclimation of plants and ecosystems to rising temperature (Gregory et al., 2009; Smith and Dukes, 2013; Busch, 2015; Lombardozzi et al., 2015; Kattge and Knorr, 2007). Gaining this understanding requires the study of plant and ecosystem processes at elevated temperatures, including warming scenarios that exceed current observations (Kayler et al., 2015; Cavaleri et al., 2015).

A number of experimental approaches have been developed to study the effects of elevated temperature – including space or elevation for time approaches (Elmendorf et al., 2015), passive warming with open-top chambers (Marion et al., 1997; Natali et al., 2014), terracosms (Phillips et al., 2011), active warming with open-topped chambers (Norby et al., 1997), soil warming (Hanson et al., 2011; Natali et al., 2014; Peterjohn et al., 1993), infrared lamps (Kimball and Conley, 2009; Ruiz-Vera et al., 2015; Fay et al., 2011) and large-scale above- and below-ground warming chambers (Bronson et al., 2009; Hanson et al., 2017; Barton et al., 2010). These different approaches for manipulating temperature have clear trade-offs (Amthor et al., 2010; Aronson and McNulty, 2009; Elmendorf et al., 2015). In addition, in many regions where critical uncertainty exists, such as high and low latitudes, and in challenging locations such as high-altitude or wetlands ecosystems, logistical limitations make many of these approaches impractical. In these locations, passive warming is often the only option for temperature manipulation. Enclosures relying on solar radiation for warming must either be open-topped or have some mechanism for temperature modulation to reduce treatment variability and avoid high-temperature excursions that could result in plant mortality (Aronson and McNulty, 2009; Marion et al., 1997; Wookey et al., 1993). Unattended temperature modulation of passively warmed enclosures can be easily accomplished using currently available electronic control systems in areas with reliable electric power, but is not possible without it. Open-topped passive warming enclosures have been used previously and work well at elevating mean daily temperatures by $\sim 1.5\,°C$ (Elmendorf et al., 2012, 2015; Marion et al., 1997). However, projected global temperature increases expected for the middle to end of this century will exceed the temperature elevation achievable with open-topped passively warmed enclosures. This is especially true in the Arctic, where current warming is almost double the global average and where projections for the worst-case emission scenarios include temperature increases of up to and beyond 7.5 °C by 2100 (IPCC, 2013; Kaufman et al., 2009; Melillo et al., 2014).

In order to gain critical process knowledge of plant and ecosystem responses to the higher temperatures projected for the end of the century it is necessary to conduct warming experiments that cover the range of expected temperatures. In logistically challenging environments this is currently impossible using existing passive warming technologies (Marion et al., 1997). Here we have addressed this research need by designing and testing a novel zero-power warming (ZPW) chamber capable of unattended, power-free, temperature modulation of a vented enclosure capable of elevating temperatures beyond those achievable with existing passive warming approaches. Our motivation for the development of this new warming method was improving understanding and model representation of photosynthesis and respiration in the Arctic tundra, but the approach described here could be applied to many different ecosystems.

2 Materials and methods

The design goal for our ZPW chamber was to improve the range and regulation of the achievable temperature differential in a passively warmed chamber that can be operated unattended and without electric power. The novel aspect of this design is the use of vent actuation based on the temperature differentials of hydraulic reservoirs.

2.1 Vent actuation

Our aim was to design a chamber capable of maintaining a temperature differential between the inside and the outside of the chamber over a broad range of ambient temperatures. The basis for the ZPW vent control mechanism is the expansion of an incompressible fluid in response to increasing temperature. Since this is a linear response, we can design a system that provides a consistent volume difference at a given temperature differential. The absolute temperature range is limited by the physical properties of the fluid used. The fluid must remain a free-flowing liquid throughout the expected temperature range, with a high coefficient of expansion and low vapour pressure. The system will not function correctly if the fluid solidifies or vaporizes. For practical reasons, the fluid should be non-corrosive, readily available, low cost and non-toxic. For the prototype chamber we used a vegetable-oil-based hydraulic fluid (Hydro Safe® 130 VG 68, Hydro Safe, Inc., Hartville, OH, USA) which has a low vapour pressure and stays in liquid form in the -10 to $+30\,°C$ temperature range expected during the thaw season (June–September) at Barrow, AK. Chamber deployments in regions with different temperature ranges would require the selection of an alternative hydraulic fluid.

Figure 1 illustrates the ZPW vent control mechanism. Under conditions where the temperatures inside and outside of the chamber are identical, the vent is closed (Fig. 1a). When the temperature inside the chamber is higher than the temperature outside the chamber, the vent will open (Fig. 1b). The extent of opening is proportional to the temperature differential between the inside and the outside of the chamber. The pistons in the cylinders respond to expansion and contraction of the liquid volumes in the heat exchangers due to changing liquid temperature, which follows the air temperature surrounding the heat exchange manifolds. The liquid volumes and piston positions are initially adjusted so the vent is closed when the temperatures inside and outside the chamber are equal (Fig. 1a). When the chamber interior is warmer than the ambient environment outside the chamber (C) (Fig. 1b), the liquid within the heat exchanger located inside the chamber expands more than the liquid within the heat exchanger located outside of the chamber (E), causing the piston in cylinder B to extend more than the piston in cylinder D. This differential in piston extensions causes the vent (A) to open.

The heat exchangers were constructed from stainless steel to provide rapid heat exchange with the surrounding air and a high albedo to reduce direct heating from incident solar radiation. During early prototyping we found that copper heat exchangers tarnish with time, reducing albedo, with the result that direct solar warming of the fluid caused erroneous vent actuation. The piston diameters and stroke lengths were selected to provide sufficient piston movement to actuate the

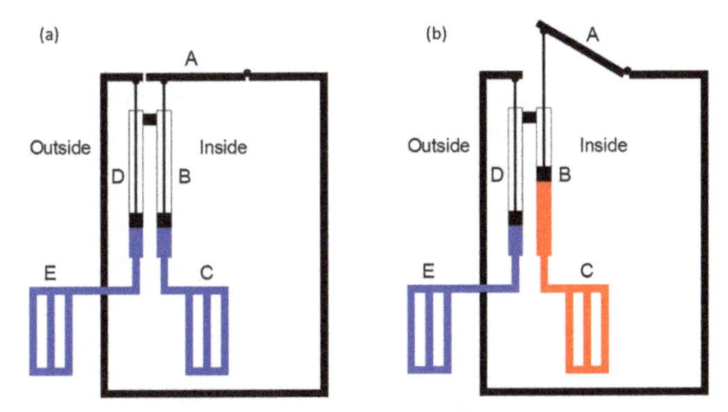

Figure 1. Schematic diagrams illustrating how the vent control system responds to air temperature differentials between the inside and outside of the chamber. Movable vent (A) is connected to a hydraulic cylinder (B) which is connected to a liquid-filled heat exchanger (C) located inside the chamber. Cylinder B is connected to another hydraulic cylinder (D). A pivoting bar (not shown) extends from the connecting link to the chamber frame to stabilize the pistons. A counterweight connected to the connecting link applies upward force on both cylinders to counteract their weight and maintain positive internal pressures on both cylinders. The piston rod on cylinder D is connected to a fixed point on the chamber and hydraulically connected to another liquid-filled heat exchanger located outside the chamber (E). The pistons in the cylinders respond to expansion and contraction of the liquid volumes in the heat exchangers due to temperature. The liquid volumes and piston positions are initially adjusted so the vent is closed when the temperatures inside and outside of the chamber are equal **(a)**. When the chamber interior is warmer than the ambient environment outside the chamber **(b)**, the liquid in the heat exchanger inside the chamber (C) will expand more than the fluid in the heat exchanger outside the chamber (E), causing the piston in cylinder B to move more than the piston in cylinder D. This differential in piston positions causes the vent (A) to open.

vent across a broad ambient temperature range. Since the liquid volumes and cylinder diameters were the same, the piston positions in both cylinders were displaced equal distances when the absolute temperatures inside and outside the chamber changed by the same amount. The piston cylinders were connected together so they moved as a unit. When both piston rods moved the same distance, the vent position did not change. Therefore, the vent position would only be affected by temperature differentials between the inside and the outside of the chamber, regardless of the absolute temperatures.

Each heat exchange manifold was equipped with valves and drain plugs to allow easy filling, draining and purging of gas bubbles during the initial setup and when changing the liquid. The manifolds were also equipped with pressure relief valves (model # 2305C-400, Kepner Products Company, Villa Park, IL, USA) to protect the hydraulic system from over-pressurization if the pistons reach the cylinder end stops or if the vent sticks in place. While not used in this prototype, a sliding connection could be installed where cylinder B connects to vent A (Fig. 1) to accommodate negative pressure events when the temperature inside the chamber is significantly lower than the outside temperature. Although this condition can occur, we have seen this only rarely in our studies. In these instances, the flexibility present in the mechanism accommodated the resultant forces. The size, and thus the achievable temperature elevation range, of the ZPW vent control mechanism can be adjusted to match the needs of a particular experimental design and location.

Tuning the ZPW chamber is straightforward and can be accomplished by one person within a few minutes. To change the magnitude of warming, i.e. the temperature differential between the inside and outside of the chamber, it is necessary to adjust the relative extensions of the pistons connected to the internal and external heat exchangers. This changes the preload on the vents, which affects how quickly the vents will begin to open as the temperature differential increases. If the preload is reduced, the vents will not immediately begin to open when the temperature differential becomes positive. If the preload is increased, the vents will begin moving as soon as the temperature differential becomes positive. This will increase the degree of vent opening for a given temperature differential, lowering the maximum differential. The preload can be increased by opening a valve to one of two oil reservoirs on the interior heat exchanger assembly while lifting the vents open to draw more oil into the cylinder as the piston rod is extended. Similarly, oil can be pushed out of the system into a reservoir by opening the valve and lowering the vents, thereby reducing the preload and increasing the temperature differential required to initiate vent opening. The sensitivity of the system to a temperature differential can also be adjusted by changing the attachment point where the piston rod is connected to the vent. Moving the attachment point closer to the fulcrum results in greater vent opening and more rapid cooling for a given change in temperature differential and piston movement. Moving the attachment point further away from the fulcrum makes the vents less responsive.

2.2 Chamber construction

Our prototype chamber (Fig. 2) was sized to allow the study of low-stature Arctic vegetation and provide easy access to the plants and monitoring equipment without having to remove the chamber from the research plot. The prototype chamber measured 2.4 m × 2.4 m × 1.4 m (L × W × maximum H). To minimize the chamber artefact of shading the vegetation and to maximize sunlight transmission, we designed this chamber to minimize structural components that would block sunlight reaching the chamber interior. The chamber walls and roof were covered with F-Clean® (AGC Chemicals, Exton, PA, USA), a 100 µm thick ethylene tetrafluoroethylene (ETFE) architectural glazing film that exhibits high light transmission throughout the solar spectrum, high structural strength, puncture resistance and low stretch, with little degradation in its optical or structural properties over time (Barton et al., 2010). The cham-

Figure 2. The ZPW chamber pictured on coastal tundra near Barrow, AK. The chamber measures 2.4 m × 2.4 m × 1.4 m (L × W × maximum H). In this picture, the two vents are partially open due to the temperature differential between the internal and external heat exchangers that are visible on the near wall of the chamber. The brass pistons and counterweight (section of white PVC pipe) can be seen opposite the door. The near corner obscures the view of most of the monitoring instrumentation but the junction box can be seen in the centre of the far wall. Outside and behind the chamber you can see a box with a solar panel that contains a data logger and battery to power the monitoring instrumentation, and an antenna for data communication.

ber framing was constructed using T-6061 alloy aluminium angle and stainless steel fasteners for mechanical strength, corrosion resistance and to minimize structural shading. The vent panels were glazed with dual-wall rigid polycarbonate sheeting (Macrolux® 10 mm twin wall, CO-EX Corporation, Wallingford, CT, USA).

2.3 Evaluation of the prototype

2.3.1 Environmental monitoring

We compared the performance of a ZPW chamber with modulated venting to a control chamber with fixed venting and an adjacent fully instrumented ambient plot. The chambers were deployed on the coastal tundra at the Barrow Environmental Observatory (BEO), near Barrow, AK (71.3° N, 156.5° W; note that on 1 December 2016 Barrow was officially renamed Utqiaġvik following the original Inupiat name). The area we used for evaluating our passive warming approach was characterized by small thaw ponds and low- and high-centred polygons with a low diversity of vascular plant species dominated by *Carex aquatilis* Wahlenb. The soils are generally classified as Gelisols and are underlain by permafrost which extends to depths of 300 m. The active-layer thickness varies between 20 and 70 cm (Bockheim et al., 1999; Shiklomanov et al., 2010). The period of evaluation ran from 15 June to 7 September 2016, which covered the thaw season (Brown et al., 1980). It is important to note that Barrow receives continuous daylight beginning prior to our period of evaluation and continuing until 1 August, when the sun first begins to set. By the end of our period of evaluation Barrow received 15 daylight hours within a 24 h period. We monitored the following parameters: air temperature and relative humidity just above canopy height using a temperature and relative humidity sensor (083E-L, Campbell Scientific, Logan, UT, USA); soil temperature at 5, 10 and 15 cm below the base of the moss layer using a stainless steel temperature probe (109SS-L, Campbell Scientific, Logan, UT, USA); volumetric soil water content below the base of the moss layer with a soil moisture sensor (GS3, Decagon Devices, Pullman, WA, USA); solar radiation using a pyranometer (LI-200R, LI-COR, Lincoln, NE, USA); and movement of the vents in the chambers using a string potentiometer (model SP2-25, Celesco Transducer Products, Inc. Chatsworth, CA, USA). Air temperature was also recorded every 15 min using stand-alone temperature loggers (UA-001-64, Onset Computer Corporation, Bourne, MA, USA) as a backup (data not shown). The data from these instruments were collected once a minute over the 85-day evaluation period and stored on data loggers (CR-1000, Campbell Scientific, Logan, UT, USA) located adjacent to each chamber and the ambient plot. Hourly instrument measurement digests were wirelessly collected (via the standard 802.11 WiFi protocol) by a control computer in a nearby instrument hut and from there, transmitted back to Brookhaven National Laboratory via FTP, where they were backed up on a server. These hourly instrument digests were parsed into summary figures and performance diagnostics using a custom script within the R environment (R Core Team, 2017) and served up via an external web site to provide near-real-time updates of experimental conditions and performance characteristics during the deployment. An Internet-enabled camera (StarDot NetCam SC; StarDot Technologies, Buena Park, CA, USA) was also used to monitor the experiment. All the instrument data are publicly available (Lewin et al., 2016; Serbin et al., 2016).

2.3.2 Short-term monitoring of carbon dioxide concentration

To evaluate the potential for draw-down and accumulation of carbon dioxide in the chambers we used infrared gas analysers (LI-6400XT, LI-COR, Lincoln, NE, USA) to monitor carbon dioxide concentration ($[CO_2]$) in the air inside the ZPW chamber and in the ambient plot. The instruments were zeroed at the field site with a common nitrogen standard (99.9998 % nitrogen, CO_2 <0.5 ppm, H_2O <0.5 ppm; ALPHAGAZ 2, Air Liquide American Specialty Gases LLC, Anchorage, AK, USA). Measurements were taken using an open leaf chamber with all environmental controls turned off. The $[CO_2]$ was logged every 60 s for 2 days.

2.3.3 Performance of the chamber skin

To quantify the performance of the F-Clean® film, we measured the transmittance of brand-new film and film that had been deployed on an earlier prototype at Brookhaven National Laboratory for approximately 1 year. The transmissivity was quantified using a full-range (i.e. 350 to 2500 nm) spectroradiometer (HR-1024i, Spectra Vista Corporation, Poughkeepsie, NY, USA) together with a fibre optic light guide attached to a measurement probe with an internal, full-spectrum calibrated light source. We measured the transmission by placing either a section of the new or deployed film between the lens of the measurement probe and a 99.99 % reflective Spectralon® standard (Labsphere, North Sutton, NH, USA) to capture the energy transmitted through the film across each of the 1024 wavelengths measured by the instrument representing the visible (VIS), near-infrared (NIR), and shortwave infrared (SWIR) spectral regions. This allowed us to examine the possible reduction of the transmission of solar energy into the chamber by the F-Clean® film across the full shortwave spectral region, and the extent to which this was impacted by an extended field deployment. We also measured the first-surface reflectivity of the film material by placing a black absorbing material behind the film within the measurement probe.

3 Results and discussion

3.1 Chamber operational overview

Figure 3 shows a sample time series of data that illustrates how the ZPW functions. As solar radiation warms the air inside the ZPW chamber, the piston attached to the internal heat exchanger begins to open the vents and prevents overheating, when solar radiation is lower or intermittent the vents remain closed or open only slightly. During this week all days had intermittent cloud cover except day of year (DOY) 203 (Fig. 3a). The resulting influence of solar radiation on ambient air temperature and the air temperature inside the ZPW can be seen in Fig. 3b. On these 5 days the air temperature in the ZPW was clearly higher than the ambient air temperature, especially during the periods of high irradiance. On DOY 201 and 205 the incoming solar radiation was not as high as on DOY 202, 203 and 204 or highly variable due to intermittent cloud cover. Figure 3c shows how much the vents in the ZPW chamber opened during these days. On DOY 201 and 205 there was very little venting, whereas on days with higher irradiance and less cloud cover there was near-continual active venting that peaked at solar noon. This differential venting enabled the ZPW chamber to maintain a similar warming profile with respect to the ambient plot on days with varying solar radiation (Fig. 3d). Despite having very similar solar radiation profiles there were different venting profiles on DOY 202, 203 and 204 which likely reflects variable wind conditions (not measured) which would affect the exchange of warm chamber air with ambient air and hence the air temperature inside the ZPW chamber.

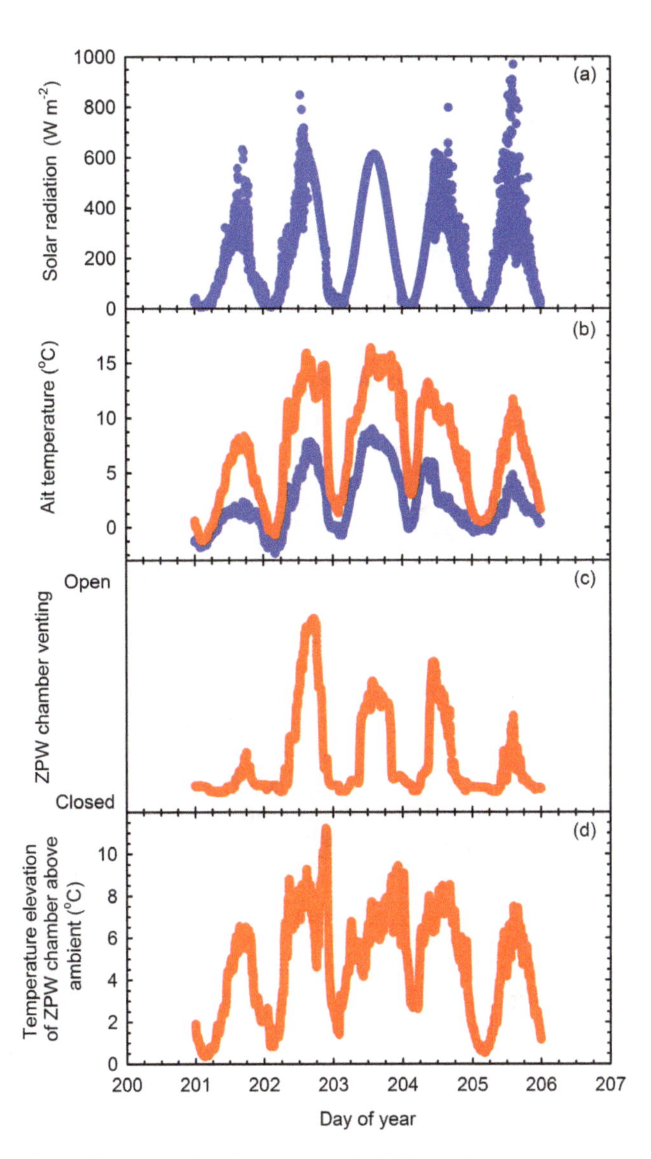

Figure 3. A sample time series of ZPW performance. Energy for warming comes from ambient solar radiation shown in **(a)**; **(b)** shows the air temperature in the ambient plot (blue) and inside ZPW chamber (red); **(c)** shows the degree to which the vents opened due to the temperature differential between the internal and external heat exchangers as measured by a string potentiometer connected to the edge of the vent; **(d)** shows the air temperature differential between the ZPW and the ambient plot. Plots show 1 min data.

Figure 4 shows an additional 4-day time series of data that included manipulation of the control chamber vent. Half way through the morning of DOY 190 we closed the vents on the control chamber (Fig. 4b) to better understand the impact of our modulated venting technology on the air temperature inside a chamber. On DOY 188 and 189 we see typical ZPW warming (Fig. 4c). On these days the ZPW chamber

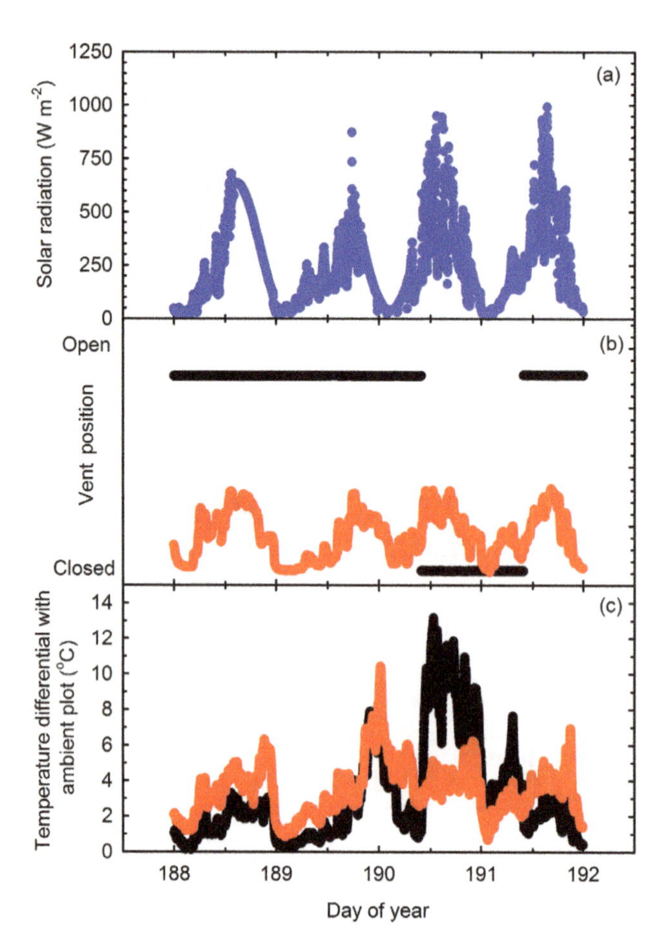

Figure 4. A 4-day sample time series of data demonstrating how the ZPW chamber can modulate the air temperature inside a chamber: **(a)** shows ambient solar radiation; **(b)** shows the degree of active venting by the ZPW chamber (red symbols) and the position of the fixed vent on the control chamber (black); **(c)** shows the air temperature differential between the ZPW chamber and the ambient plot (red) and the control chamber and the ambient plot (black).

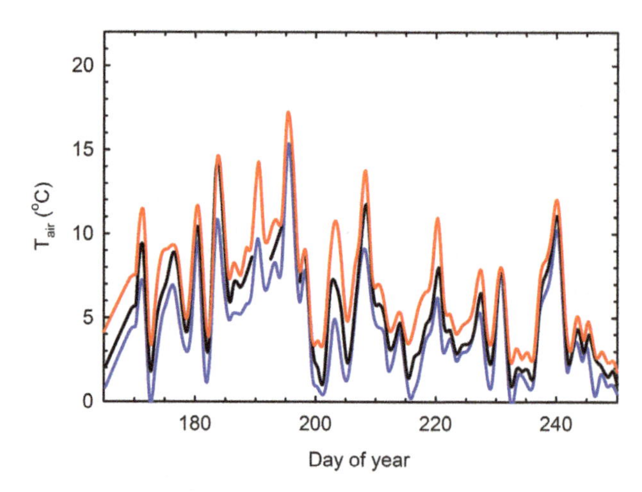

Figure 5. Mean daily air temperature measured at canopy height in the ambient plot (blue line), the control chamber, with fixed venting (black line), and the ZPW chamber, with modulated venting (red line). In order to evaluate chamber effects the control chamber was closed on DOY 190, 191, 195 and 196; data for these days are not shown.

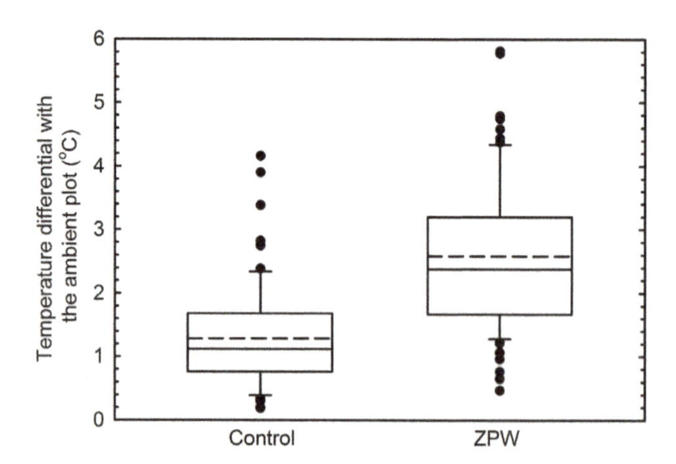

Figure 6. Tukey box plots showing the temperature differential of the mean daily air temperature between the control chamber and the ambient plot and between the ZPW chamber and the ambient plot. Warming in the ZPW chamber was double the control chamber. Box plots show the interquartile range (box), median (solid line) and mean (broken line). The whiskers show lowest and highest datum still within 1.5× interquartile range of the lower and upper quartiles. Outliers are shown as black dots.

is warmer than the control chamber and solar radiation is sufficient to induce active venting (Fig. 4a and b). Following closure of the control chamber vent we see an immediate spike in the control chamber air temperature, which rises to a peak of 13 °C above ambient. During the same day the ZPW chamber was venting and able to purge excess heat but still maintain a 4–5 °C temperature elevation above the ambient plot, thus enabling a modulated warming of air temperature. These data clearly demonstrate that ZPW chambers are capable of modulating chamber temperature passively (without the need for electric power) and can avoid overheating problems that can occur with closed or minimally vented chambers. Avoiding sustained high-temperature excursions on days associated with warmer temperatures and higher irradiance is essential for experimental manipulations where such temperature excursions have the potential to terminate the experiment by damaging the vegetation inside the chamber or affect the experiment in other ways that could negatively affect the applicability of the results to real-world conditions.

3.2 Chamber warming

3.2.1 Air temperature

We collected data on the performance of the ZPW for most of the thaw season, which is the period when we would anticipate deploying passive warming in the tundra. Figure 5 shows the daily mean air temperature in the ambient plot,

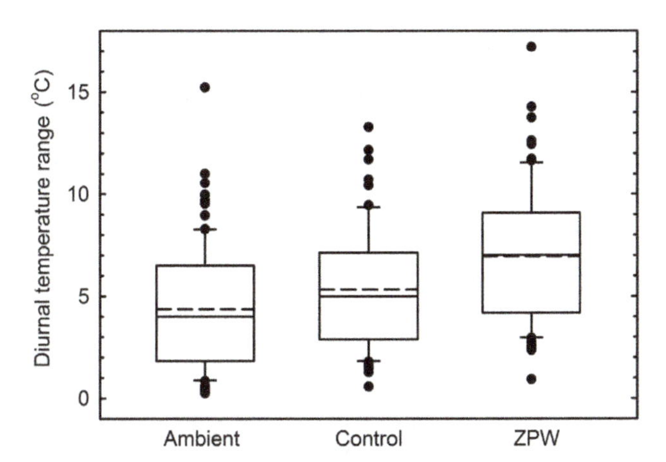

Figure 7. Tukey box plots showing the mean daily diurnal temperature range in the ambient plot and the control and ZPW chambers. The diurnal temperature range was calculated for each day by subtracting the minimum temperature from the maximum temperature recorded on a given day. Data from DOY 190, 191, 195 and 196 were omitted due to chamber manipulations. Box plots show the interquartile range (box), median (solid line) and mean (broken line). The whiskers show lowest and highest datum still within 1.5× interquartile range of the lower and upper quartiles. Outliers are shown as black dots.

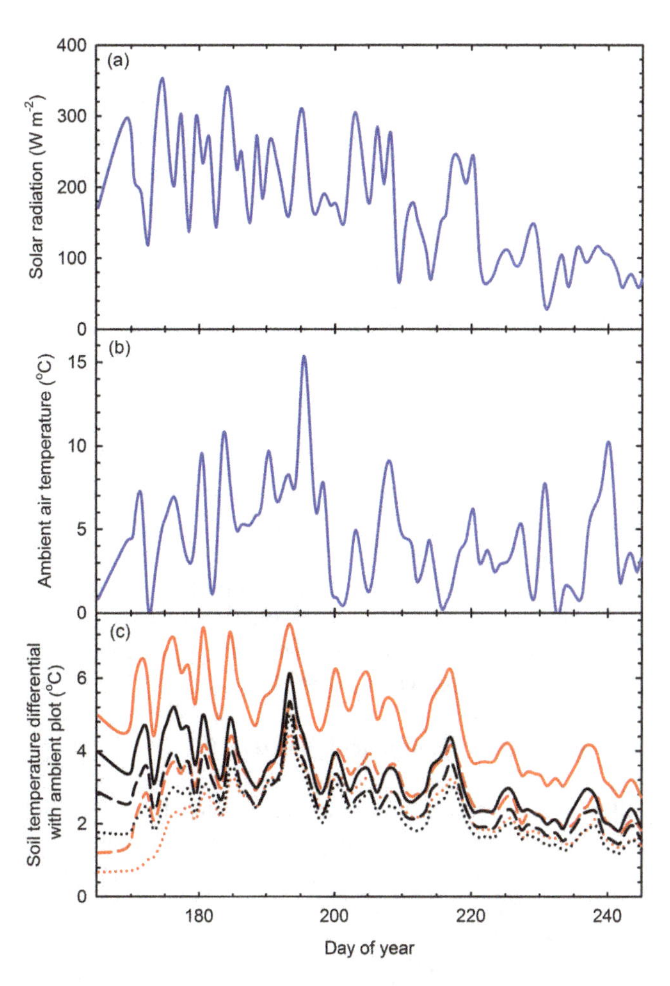

Figure 8. The influence of ambient (blue) solar radiation **(a)** and ambient air temperature **(b)** on soil temperature in the control (black, **c**) and ZPW chambers (red, **c**). Panel **(c)** shows the temperature differential between the chambers and the ambient plot. Soil temperature was measured at 5 cm (solid line), 10 cm (broken line) and 15 cm (dotted line) below the base of the moss layer.

control chamber and ZPW chamber. Air temperature fluctuated markedly over the thaw season with temperatures ranging from 0 to 15 °C. However, the air temperature inside both the control and ZPW chambers was consistently and significantly higher than the ambient air temperature ($t_{2,78}$, $P < 0.001$, Fig. 5) and the daily mean temperature differential (2.6 °C) between the ZPW and the ambient plot was double the differential between the control chamber and the ambient plot (1.3 °C, $t_{2,78}$, $P < 0.001$, Fig. 6).

It is not possible to conduct a meaningful direct comparison of the ZPW chamber performance with other Arctic passive warming approaches due to differences in location, size, seasonal variation in weather, the measurement location, use of shielded and unshielded thermocouples and the length of the trial period. However, the ZPW chamber had a warming effect that was double the adjacent partially enclosed, passively warmed control chamber with fixed vents and double the mean daily air temperature warming reported in other passive warming studies (Marion et al., 1997; Welker et al., 2004; Jonsdottir et al., 2005; Wahren et al., 2005; Bokhorst et al., 2013). A recent global assessment of these passive warming approaches and Arctic greenhouse and infrared heating experiments reported that in these experiments the mean daily summer air temperature was elevated by 1.5 °C (Elmendorf et al., 2012). Given that estimated increases in mean annual surface air temperature in the Arctic has been projected to be 2.5 °C by 2060 (ACIA, 2005), and for Alaska 1.9–3.1 °C by 2050 (Melillo et al., 2014), it is imperative that warming approaches, such as the one presented here, are developed that are capable of greater temperature elevation.

3.2.2 Diurnal temperature range

The ZPW chamber, like any enclosure that relies on solar radiation for heating, has the potential to affect the diurnal temperature range. Both the control and the ZPW chambers had a significantly higher diurnal temperature range than the ambient plot as well as significantly higher daily temperature minima and maxima ($t_{(2),78}$, $P < 0.001$, Fig. 7). On average the daily minimum temperatures were 0.58 °C (control chamber) and 1.32 °C (ZPW chamber) greater than the average daily minimum temperature in the ambient plot. The temperature maxima were 2.50 °C (control chamber) and 4.43 °C (ZPW chamber) greater than the average daily maximum temperature in the ambient plot. As expected, the increase in the diurnal temperature range was driven mostly by increases in the maximum temperature which were associated with days with high solar radiation.

3.2.3 Soil warming

Warming the air also warmed the soil. Figure 8 shows the influence of ambient solar radiation (Fig. 8a) and ambient air temperature (Fig. 8b) on the soil temperature differential at 5, 10 and 15 cm below the moss layer (Fig. 8c). The soil temperature at all three depths shows the same dynamic response that largely mirrors the pattern in air temperature and solar radiation. The difference in the soil temperature differential (with the ambient plot) between the ZPW chamber and the control chamber decreases with soil depth. Warming of the soil was greater than warming of air (Figs. 5, 6 and 8c). This likely reflects the fact that the soil acts as a heat sink for the elevated air temperature and is less influenced by air exchanges that will rapidly decrease the air temperature. This lag was expected as the soil takes longer to warm and is slower to cool than the air. However, we sited this prototype on the same footprint as a 2015 prototype (no data), and thaw and degradation of the permafrost that occurred in 2015 thickened the active layer, which may have influenced soil temperature profiles in 2016.

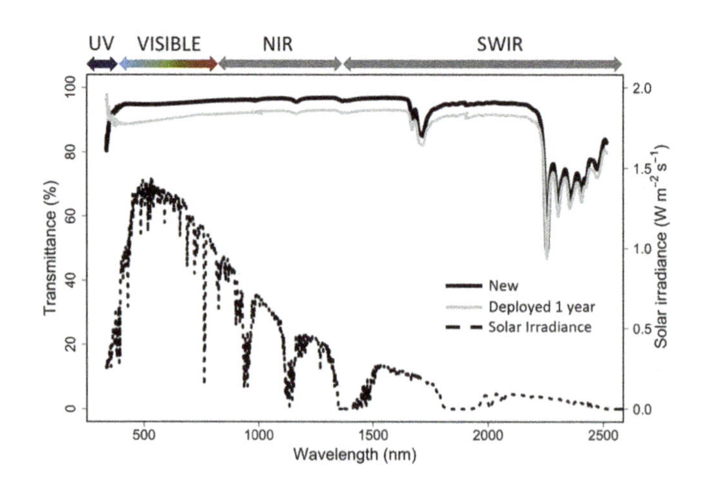

Figure 9. The full shortwave (350 to 2500 nm) transmittance of the F-Clean® film material. The two solid lines depict the measured transmittance through brand-new film (black) and film deployed for 1 full year (grey). To better understand the impact of the film on transmission of solar energy into the ZPW chamber, we also provide the American Society for Testing and Materials (ASTM) Reference Solar Spectral Irradiance for an absolute air mass of 1.5 (black dashed line, http://rredc.nrel.gov/solar/spectra/am1.5/) for the same spectral region.

3.3 Consideration of potential artefacts

All passive warming approaches have a chamber effect and the impact of chambers on plant growth has been considered in depth previously (Long et al., 2004; Marion et al., 1997). While it is possible to effectively warm plants and ecosystems without the use of enclosures, alternative approaches such as infrared heating are often not practical in remote, logistically challenging locations. Therefore, if we want to understand how plants and ecosystems will respond to rising temperature in such locations we need to use passive warming approaches, but do so with a full understanding of their limitations. In short, selection of a passive warming approach requires a balance between the degree of warming and the potential artefacts – the greater the desired warming, the larger the potential for unwanted chamber effects. We have considered and quantified some of the artefacts associated with the ZPW design. We have not considered all the issues that may be of importance to the broader community, e.g. restricted access for pollinators and herbivores, but focused our attention on key variables that impact plant physiology – i.e. light, vapour pressure deficit, [CO_2] and soil moisture content.

3.3.1 Attenuation of solar radiation

The transmittance of the new and deployed F-Clean® film showed relatively stable values in the visible and near-infrared wavelengths, with minor reductions between 1630 and 1750 nm followed by marked reductions in the far-SWIR wavelength region (Fig. 9). The average transmittance values of new film was 94 % (±0.3 %) and 95 % (±0.5 %) for the visible and NIR regions, respectively, while the transmittance values for film deployed for 1 year averaged 90 % (±0.8 %) and 92 % (±0.5 %). Wavelengths >2200 nm displayed a significant drop in transmittance (an increase in absorption by the film), with an average transmission of 74 % (±8.5 %) for the new and 71 % (±8.2 %) for the deployed film. Overall, the film exposed to the elements for 1 year displayed an average of a 4 % (±0.8 %) reduction in transmittance of shortwave radiation (i.e. 350 to 2500 nm) compared to unused film. We also found that the reflectivity of the film material was minimal, averaging 3 % for both the new and deployed F-Clean® film. Figure 9 also shows the solar spectral irradiance; comparing this spectrum with the transmission of the film clearly shows that the regions of maximum solar output correspond to the regions of highest transmission, including the important 400 to 700 nm region utilized in plant photosynthesis. The F-Clean® film performed considerably better than fibreglass and Lexan which have been typically used for passive warming chamber construction. Those materials have transmittance values of 86 % (fibreglass) and 90 % (Lexan) when new. To our knowledge, no data for the transmittance of these materials in deployed chambers have been reported (Molau and Mølgaard, 1996).

Measuring transmittance through the chamber walls, particularly of brand-new materials, does not account for other sources of attenuation such as the chamber frames. Therefore, we measured solar radiation at canopy height inside the control and ZPW chambers and compared 1 min readings to the solar radiation measured in the ambient plot. During the daytime (solar radiation >5 W m^{-2}) the transmission of solar radiation inside the control and ZPW chambers relative to the

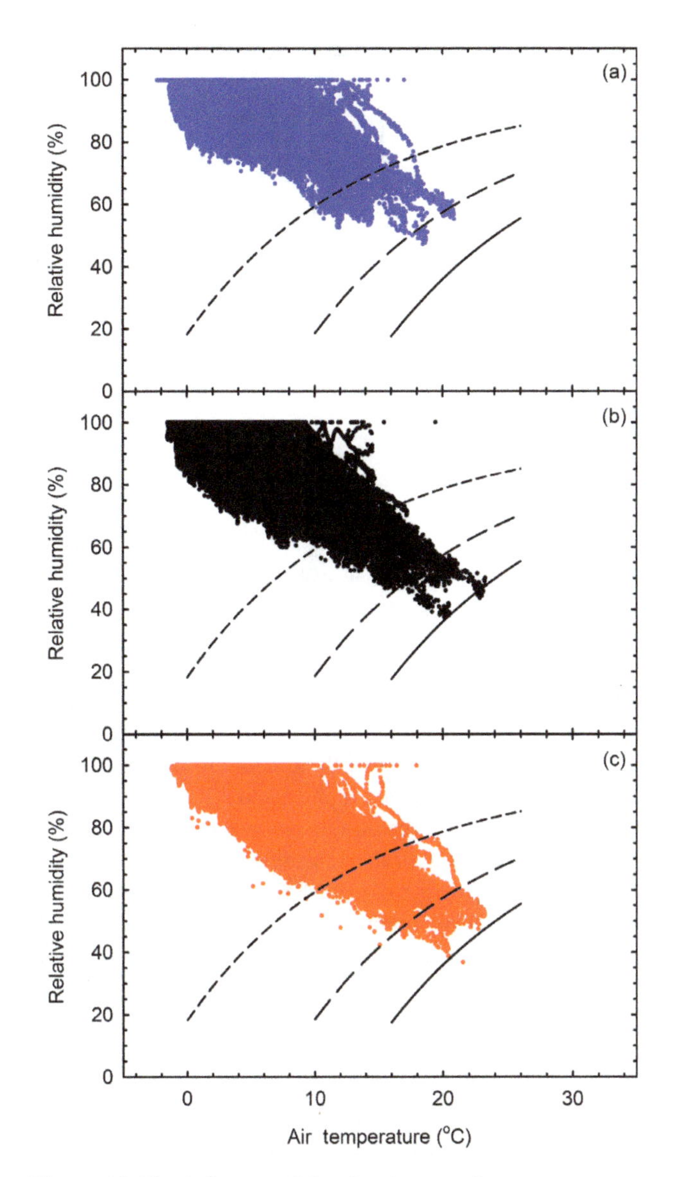

Figure 10. The influence of the chambers on the vapour pressure deficit of the air inside the chambers; 1 min readings of relative humidity plotted against air temperature measured in the ambient plot (**a**, blue), the control chamber (**b**, black) and the ZPW chamber (**c**, red) over our trial period. The vapour pressure deficit is shown for 0.5 kPa (broken line – short dashes), 1.0 kPa (broken line – long dashes) and 1.5 kPa (solid line). Data from DOY 190, 191, 195 and 196 were excluded due to chamber manipulations.

solar radiation measured in the ambient plot was 78 % ± 15 and 76 % ± 15 % SD respectively. This ratio accounts for loss of light transmission associated with the chamber frames, the ZPW apparatus, aging of the film, variation in solar angle and potential dirt, condensation and raindrops. These transmission values are the true, and rarely reported, values for attenuation of solar radiation in our field enclosures.

3.3.2 Potential for reducing vapour pressure deficit

Any warming experiment will raise the vapour pressure deficit (VPD) of the air unless water vapour is added to the warmed air. However, the volume of potable water required for such an endeavour is enormous and the artefacts associated with maintaining VPD can be considerable (Hanson et al., 2017). As a result, warming experiments rarely attempt to control VPD (Bronson et al., 2009). Elevation of VPD is a concern for plant physiology experiments if the warming treatment moves the VPD above 1.5 kPa as limitations on stomatal conductance and photosynthesis are possible. In our prototype the mean daily relative humidity (RH) was lower in both the control chamber (88 % RH) and the ZPW chamber (86 % RH) compared to the ambient plot (96 % RH). However, the average 9 % reduction in RH in the ZPW chamber was physiologically insignificant in terms of VPD. Figure 10 shows the 1 min data from the ambient plot, control chamber and ZPW chamber. During the period of study the VPD was only above 1.5 kPa on a few occasions and the majority of the time the VPD was < 0.5 kPa (Fig. 10). Therefore, the potential for negative impacts of an elevated VPD on stomatal conductance and photosynthesis in this ecosystem is minimal.

3.3.3 Potential carbon dioxide draw down and build up

When a plot is enclosed by a chamber there is potential for the vegetation and soil to influence the [CO_2] inside the enclosure, either through draw down of the [CO_2] through photosynthesis or elevation of the [CO_2] through respiration. Since the ZPW chambers are partially enclosed, we investigated the potential for a chamber effect on [CO_2]. We monitored the ambient [CO_2] and the [CO_2] inside the ZPW over 2 days (Fig. 11). We found that during periods of high solar radiation when the ZPW chamber was actively venting the [CO_2] inside the chamber was $\sim 4\,\mu mol\,mol^{-1}$ below ambient [CO_2]. When solar radiation dropped, and vents were typically closed, the [CO_2] inside the ZPW chamber rose by $\sim 8\,\mu mol\,mol^{-1}$. At the start of the day on DOY 190 the vents were closed and as solar radiation increased the vegetation was able to draw down the [CO_2] inside the chamber by $\sim 20\,\mu mol\,mol^{-1}$ before warming led to venting and this larger [CO_2] differential vanished (Fig. 11). The effect of the enclosure on the internal [CO_2] at this location was minimal and does not present a serious concern and is comparable to [CO_2] changes seen during still-air conditions in other systems. However, the potential for altering the [CO_2] should be evaluated in other locations prior to initiating a full experiment. If changes in [CO_2] did present a problem they could be mitigated by tuning the system to maintain a permanent minimum venting, or through the addition of a small solar-powered fan which could be actuated by a [CO_2] sensor.

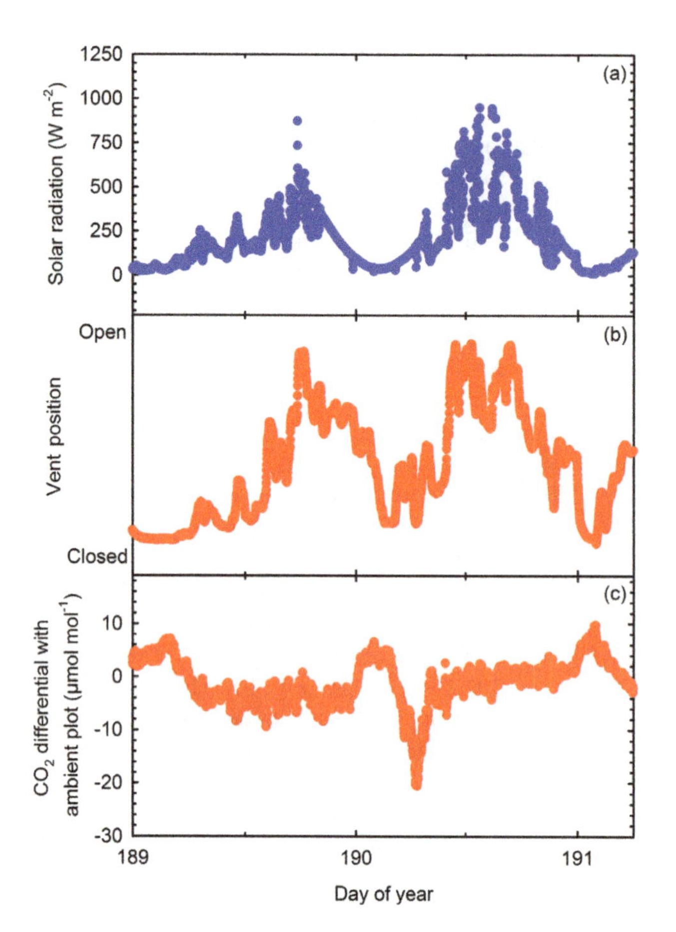

Figure 11. The influence of solar radiation **(a)** and venting of the ZPW chamber **(b)** on the carbon dioxide concentration differential between the inside of the ZPW chamber and the ambient plot **(c)**.

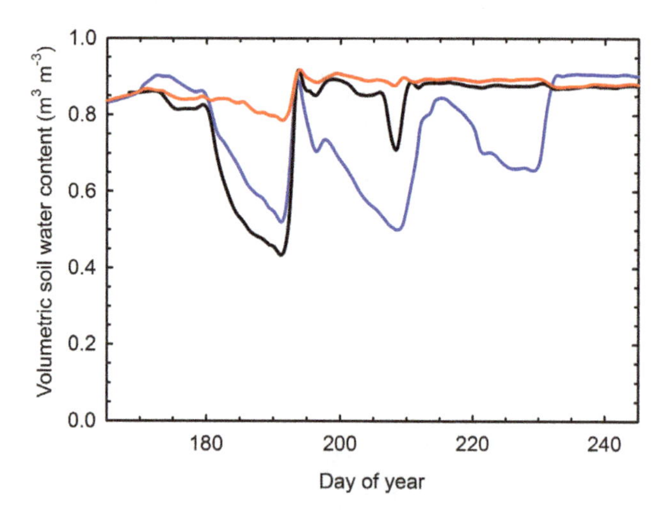

Figure 12. Volumetric soil water content measured in the centre of the ambient plot (blue), the control chamber (black) and the ZPW chamber (red).

3.3.4 Soil water content

A major concern with any chamber system is exclusion of precipitation and dewfall. Even open-topped chambers without frustums can exclude a significant amount of windblown precipitation. When coupled with a high rate of evapotranspiration there is significant potential to dry out the soil inside the enclosures. Our Arctic location presents a best-case scenario for the use of partially closed field enclosures. The presence of permafrost and an active layer thickness of 20–70 cm results in a landscape that is poorly drained. Mean annual precipitation is only 106 mm but poor drainage coupled with low temperatures and high humidity means that evapotranspiration is very low and much of the landscape is covered by standing water (Brown et al., 1980; Shiklomanov et al., 2010).

Our measurements of volumetric soil water content measured in the centre of the ambient plot and the two chambers showed marked plot-to-plot variation that was not consistent with the warming treatment. This could reflect plot-to-plot variation in topography, but also the evolution of local drainage patterns as the thaw season progressed. A heavy rainstorm on DOY 191 demonstrated how quickly lateral flow can recharge the soil water content inside the chambers (Fig. 12). However, note that these data are confounded by the use of the same footprint in 2015 and 2016, as permafrost thaw and degradation resulting from our treatment in 2015 may have altered local drainage patterns.

When considering the use of ZPW chambers in other systems exclusion of precipitation should be carefully considered. In wetlands or other locations with saturated soils, exclusion of precipitation may not be a major concern, but elsewhere external collection and immediate reapplication of precipitation into the enclosures would be an option to mitigate this artefact (Nippert et al., 2009). Another option would be solar-charged, battery-operated motors that could be used to open the vents in the event of a rainstorm, the opposite of approaches used to deploy rainfall exclusion treatments (Gray et al., 2016). For short-term deployments, for example to investigate enhanced heat wave effects, the exclusion of precipitation might not be an issue. However, we strongly recommend monitoring soil water content and providing daily reports that can be used to inform potential manual watering efforts.

4 Conclusions

Here we demonstrate the successful design, construction and testing of a novel hydraulic system capable of elevating and modulating the temperature inside a passively warmed field enclosure. Our design was able to raise the mean daily air temperature to 2.6 °C above an adjacent ambient plot, twice the temperature elevation attained in a control chamber that lacked our hydraulic system. Our fully autonomous chamber required no power and was able to avoid high-temperature excursions that occur in fully, or near fully, enclosed chambers. The design of the ZPW chambers is highly flexible and can be adjusted to meet research needs in a wide range of logistically challenging environments, including tuning of the venting system to attain higher temperature differentials. This advance opens up the possibility for markedly higher warming in remote and logistically challenging environments and the ability to conduct short-term heat stress experiments.

In a location with reliable electric power it would be possible to install an electronic control system in place of our hydraulic control system that could improve temperature control and allow the addition of other desirable features. However, greater sophistication brings additional complexity and additional points of failure. In harsh environments remote from technical support there are advantages to the reduced complexity of our design. As discussed above, there are many different approaches that can be used to understand the response and acclimation of plants and ecosystems to warming. All of these approaches have advantages and drawbacks and all approaches have artefacts. The influence of potential artefacts on the ability to address a specific hypothesis in a given ecosystem needs to be carefully considered, and should be used to determine the manipulative approach most suitable for the research environment and the scientific question.

Competing interests. The authors declare that they have no conflict of interest.

Acknowledgements. This work was supported in part by The Next Generation Ecosystem Experiments Arctic project, supported by the Office of Biological and Environmental Research in the Department of Energy, Office of Science, and through United States Department of Energy contract number DE-SC0012704 to Brookhaven National Laboratory.

Edited by: Paul Stoy

References

ACIA: Arctic Climatic Impact Assessment – Scientific Report, Cambridge University Press, Cambrige, UK, 2005.

Amthor, J. S., Hanson, P. J., Norby, R. J., and Wullschleger, S. D.: A comment on "Appropriate experimental ecosystem warming methods by ecosystem, objective, and practicality" by Aronson and McNulty, Agr. Forest Meteorol., 150, 497–498, https://doi.org/10.1016/j.agrformet.2009.11.020, 2010.

Aronson, E. L. and McNulty, S. G.: Appropriate experimental ecosystem warming methods by ecosystem, objective, and practicality, Agr. Forest Meteorol., 149, 1791–1799, https://doi.org/10.1016/j.agrformet.2009.06.007, 2009.

Barton, C. V. M., Ellsworth, D. S., Medlyn, B. E., Duursma, R. A., Tissue, D. T., Adams, M. A., Eamus, D., Conroy, J. P., McMurtrie, R. E., Parsby, J., and Linder, S.: Whole-tree chambers for elevated atmospheric CO_2 experimentation and tree scale flux measurements in south-eastern Australia: The Hawkesbury Forest Experiment, Agr. Forest Meteorol., 150, 941–951, https://doi.org/10.1016/j.agrformet.2010.03.001, 2010.

Bockheim, J. G., Everett, L. R., Hinkel, K. M., Nelson, F. E., and Brown, J.: Soil organic carbon storage and distribution in Arctic Tundra, Barrow, Alaska, Soil Sci. Soc. Am. J., 63, 934–940, 1999.

Bokhorst, S., Huiskes, A., Aerts, R., Convey, P., Cooper, E. J., Dalen, L., Erschbamer, B., Gudmundsson, J., Hofgaard, A., Hollister, R. D., Johnstone, J., Jonsdottir, I. S., Lebouvier, M., Van De Vijver, B., Wahren, C. H., and Dorrepaal, E.: Variable temperature effects of Open Top Chambers at polar and alpine sites explained by irradiance and snow depth, Glob. Change Biol., 19, 64–74, https://doi.org/10.1111/gcb.12028, 2013.

Bronson, D. R., Gower, S. T., Tanner, M., and Van Herk, I.: Effect of ecosystem warming on boreal black spruce bud burst and shoot growth, Glob. Change Biol., 15, 1534–1543, https://doi.org/10.1111/j.1365-2486.2009.01845.x, 2009.

Brown, J., Everett, K. R., Webber, P. J., MacLean, S. F., and Murray, D. F.: The Coastal Tundra at Barrow, in: An Arctic Ecosystem: the Coastal Tundra at Barrow, Alaska, edited by: Brown, J., Miller, P. C., Tiezen, L. L., and Bunnell, F. L., Dowden, Hutchinson & Ross, Inc., Stroudsburg, PA, 571, 1–19, 1980.

Busch, F. A.: Reducing the gaps in our understanding of the global terrestrial carbon cycle, New Phytol., 206, 886–888, https://doi.org/10.1111/nph.13399, 2015.

Cavaleri, M. A., Reed, S. C., Smith, W. K., and Wood, T. E.: Urgent need for warming experiments in tropical forests, Glob. Change Biol., 21, 2111–2121, https://doi.org/10.1111/gcb.12860, 2015.

Elmendorf, S. C., Henry, G. H. R., Hollister, R. D., Fosaa, A. M., Gould, W. A., Hermanutz, L., Hofgaard, A., Jonsdottir, I. I., Jorgenson, J. C., Levesque, E., Magnusson, B., Molau, U., Myers-Smith, I. H., Oberbauer, S. F., Rixen, C., Tweedie, C. E., and Walker, M.: Experiment, monitoring, and gradient methods used to infer climate change effects on plant communities yield consistent patterns, P. Natl. Acad. Sci. USA, 112, 448–452, https://doi.org/10.1073/pnas.1410088112, 2015.

Fay, P. A., Blair, J. M., Smith, M. D., Nippert, J. B., Carlisle, J. D., and Knapp, A. K.: Relative effects of precipitation variability and warming on tallgrass prairie ecosystem function, Biogeosciences, 8, 3053–3068, https://doi.org/10.5194/bg-8-3053-2011, 2011.

Friedlingstein, P., Meinshausen, M., Arora, V. K., Jones, C. D., Anav, A., Liddicoat, S. K., and Knutti, R.: Uncertainties in CMIP5 Climate Projections due to Carbon Cycle Feedbacks, J. Clim., 27, 511–526, https://doi.org/10.1175/jcli-d-12-00579.1, 2014.

Gray, S. B., Dermody, O., Klein, S. P., Locke, A. M., McGrath, J. M., Paul, R. E., Rosenthal, D. M., Ruiz-Vera, U. M., Siebers, M. H., Strellner, R., Ainsworth, E. A., Bernacchi, C. J., Long, S. P., Ort, D. R., and Leakey, A. D. B.: Intensifying drought eliminates the expected benefits of elevated carbon dioxide for soybean, Nature Plants, 2, 16132, https://doi.org/10.1038/nplants.2016.132, 2016.

Gregory, J. M., Jones, C. D., Cadule, P., and Friedlingstein, P.: Quantifying Carbon Cycle Feedbacks, J. Clim., 22, 5232–5250, https://doi.org/10.1175/2009jcli2949.1, 2009.

Hanson, P. J., Childs, K. W., Wullschleger, S. D., Riggs, J. S., Thomas, W. K., Todd, D. E., and Warren, J. M.: A method for experimental heating of intact soil profiles for application to climate change experiments, Glob. Change Biol., 17, 1083–1096, https://doi.org/10.1111/j.1365-2486.2010.02221.x, 2011.

Hanson, P. J., Riggs, J. S., Nettles, W. R., Phillips, J. R., Krassovski, M. B., Hook, L. A., Gu, L. H., Richardson, A. D., Aubrecht, D. M., Ricciuto, D. M., Warren, J. M., and Barbier, C.: Attaining whole-ecosystem warming using air and deep-soil heating methods with an elevated CO_2 atmosphere, Biogeosciences, 14, 861–883, https://doi.org/10.5194/bg-14-861-2017, 2017.

IPCC: Climate Change 2013: The Physical Science Basis. Contribution of Working Group I to the Fifth Assessment Report of the Intergovernmental Panel on Climate Change, Cambridge, United Kingdom and New York, NY, USA, 1535 pp., 2013.

Kattge, J. and Knorr, W.: Temperature acclimation in a biochemical model of photosynthesis: a reanalysis of data from 36 species, Plant Cell Environ., 30, 1176–1190, https://doi.org/10.1111/j.1365-3040.2007.01690.x, 2007.

Kaufman, D. S., Schneider, D. P., McKay, N. P., Ammann, C. M., Bradley, R. S., Briffa, K. R., Miller, G. H., Otto-Bliesner, B. L., Overpeck, J. T., Vinther, B. M., and Arctic Lakes 2k Project, M.: Recent Warming Reverses Long-Term Arctic Cooling, Science, 325, 1236–1239, https://doi.org/10.1126/science.1173983, 2009.

Kayler, Z. E., De Boeck, H. J., Fatichi, S., Grunzweig, J. M., Merbold, L., Beier, C., McDowell, N., and Dukes, J. S.: Experiments to confront the environmental extremes of climate change, Front. Ecol. Environ., 13, 219–225, https://doi.org/10.1890/140174, 2015.

Kimball, B. A. and Conley, M. M.: Infrared heater arrays for warming field plots scaled up to 5-m diameter, Agr. Forest Meteorol., 149, 721–724, https://doi.org/10.1016/j.agrformet.2008.09.015, 2009.

Lewin, K. F., McMahon, A., Ely, K. S., Serbin, S. P., and Rogers, A.: Zero Power Warming (ZPW) Chamber Prototype Measurements, Barrow, Alaska, 2016, Next Generation Ecosystem Experiments Arctic Data Collection, Carbon Dioxide Information Analysis Center, Oak Ridge National Laboratory, Oak Ridge, Tennessee, USA, https://doi.org/10.5440/1343066, 2016.

Lombardozzi, D. L., Bonan, G. B., Smith, N. G., Dukes, J. S., and Fisher, R. A.: Temperature acclimation of photosynthesis and respiration: A key uncertainty in the carbon cycle-climate feedback, Geophys. Res. Lett., 42, 8624–8631, https://doi.org/10.1002/2015gl065934, 2015.

Long, S. P., Ainsworth, E. A., Rogers, A., and Ort, D. R.: Rising atmospheric carbon dioxide: Plants face the future, Ann. Rev. Plant Biol., 55, 591–628, https://doi.org/10.1146/annurev.arplant.55.031903.141610, 2004.

Lovenduski, N. S. and Bonan, G. B.: Reducing uncertainty in projections of terrestrial carbon uptake, Environ. Res. Lett., 12, 044020, https://doi.org/10.1088/1748-9326/aa66b8, 2017.

Marion, G. M., Henry, G. H. R., Freckman, D. W., Johnstone, J., Jones, G., Jones, M. H., Levesque, E., Molau, U., Molgaard, P., Parsons, A. N., Svoboda, J., and Virginia, R. A.: Open-top designs for manipulating field temperature in high-latitude ecosystems, Glob. Change Biol., 3, 20–32, https://doi.org/10.1111/j.1365-2486.1997.gcb136.x, 1997.

Melillo, J. M., Richmond, T. C., and Yohe, G. W.: Climate Change Impacts in the United States: The Third National Climate Assessment, US Global Change Research Program, 841, https://doi.org/10.7930/J0Z31WJ2, 2014.

Molau, U. and Mølgaard, P.: ITEX Manual, 2nd Edn., Danish Polar Center, Copenhagen, Denmark, 1996.

Natali, S. M., Schuur, E. A. G., Webb, E. E., Pries, C. E. H., and Crummer, K. G.: Permafrost degradation stimulates carbon loss from experimentally warmed tundra, Ecology, 95, 602–608, https://doi.org/10.1890/13-0602.1, 2014.

Nippert, J. B., Fay, P. A., Carlisle, J. D., Knapp, A. K., and Smith, M. D. Ecophysiological responses of two dominant grasses to altered temperature and precipitation regimes. Acta Oecol., 35, 400–408, 2009.

Norby, R. J., Edwards, N. T., Riggs, J. S., Abner, C. H., Wullschleger, S. D., and Gunderson, C. A.: Temperature-controlled open-top chambers for global change research, Glob. Change Biol., 3, 259–267, https://doi.org/10.1046/j.1365-2486.1997.00072.x, 1997.

Peterjohn, W. T., Melillo, J. M., Bowles, F. P., and Steudler, P. A.: Soil warming and trace gas fluxes – experimental design and preliminary flux results, Oecologia, 93, 18–24, 1993.

Phillips, C. L., Gregg, J. W., and Wilson, J. K.: Reduced diurnal temperature range does not change warming impacts on ecosystem carbon balance of Mediterranean grassland mesocosms, Glob. Change Biol., 17, 3263–3273, https://doi.org/10.1111/j.1365-2486.2011.02483.x, 2011.

R Core Team: A language and environment for statistical computing. R Foundation for Statistical Computing, Vienna, Austria, https://www.R-project.org/ (last access: April 2016), 2017.

Ruiz-Vera, U. M., Siebers, M. H., Drag, D. W., Ort, D. R., and Bernacchi, C. J.: Canopy warming caused photosynthetic acclimation and reduced seed yield in maize grown at ambient and elevated CO_2, Glob. Change Biol., 21, 4237–4249, https://doi.org/10.1111/gcb.13013, 2015.

Serbin, S. P., McMahon, A., Lewin, K. F., Ely K. S., and Rogers, A.: NGEE Arctic Zero Power Warming PhenoCamera Images, Barrow, Alaska, 2016, Next Generation Ecosystem Experiments Arctic Data Collection, Carbon Dioxide Information Analysis Center, Oak Ridge National Laboratory, Oak Ridge, Tennessee, USA, https://doi.org/10.5440/1358195, 2016.

Shiklomanov, N. I., Streletskiy, D. A., Nelson, F. E., Hollister, R. D., Romanovsky, V. E., Tweedie, C. E., Bockheim, J. G., and Brown, J.: Decadal variations of active-layer thickness in moisture-controlled landscapes, Barrow, Alaska, J. Geophys. Res.-Biogeo., 115, G00I04, https://doi.org/10.1029/2009jg001248, 2010.

Smith, N. G. and Dukes, J. S.: Plant respiration and photosynthesis in global-scale models: incorporating acclimation to temperature and CO_2, Glob. Change Biol., 19, 45–63, https://doi.org/10.1111/j.1365-2486.2012.02797.x, 2013.

Wahren, C. H. A., Walker, M. D., and Bret-Harte, M. S.: Vegetation responses in Alaskan arctic tundra after 8 years of a summer warming and winter snow manipulation experiment, Glob. Change Biol., 11, 537–552, https://doi.org/10.1111/j.1365-2486.2005.00927.x, 2005.

Welker, J. M., Fahnestock, J. T., Henry, G. H. R., O'Dea, K. W., and Chimner, R. A.: CO_2 exchange in three Canadian High Arctic ecosystems: response to long-term experimental warming, Glob. Change Biol., 10, 1981–1995, https://doi.org/10.1111/j.1365-2486.2004.00857.x, 2004.

Wookey, P. A., Parsons, A. N., Welker, J. M., Potter, J. A., Callaghan, T. V., Lee, J. A., and Press, M. C.: Comparative responses of phenology and reproductive development to simulated environmental change in sub-Arctic and high Arctic plants, Oikos, 67, 490–502, https://doi.org/10.2307/3545361, 1993.

5

Interplay of community dynamics, temperature, and productivity on the hydrogen isotope signatures of lipid biomarkers

S. Nemiah Ladd[1,2], **Nathalie Dubois**[2,3], **and Carsten J. Schubert**[1,4]

[1]Department of Surface Waters – Research and Management, Eawag, Swiss Federal Institute of Aquatic Science and Technology, 6047 Kastanienbaum, Switzerland
[2]Department of Earth Sciences, ETH Zürich, 8092 Zürich, Switzerland
[3]Department of Surface Waters – Research and Management, Eawag, Swiss Federal Institute of Aquatic Science and Technology, 8600 Dübendorf, Switzerland
[4]Institute of Biogeochemistry and Pollutant Dynamics, ETH Zürich, 8092 Zürich, Switzerland

Correspondence to: S. Nemiah Ladd (nemiah.ladd@eawag.ch)

Abstract. The hydrogen isotopic composition (δ^2H) of lipid biomarkers has diverse applications in the fields of paleoclimatology, biogeochemistry, and microbial community dynamics. Large changes in hydrogen isotope fractionation have been observed among microbes with differing core metabolisms, while environmental factors including temperature and nutrient availability can affect isotope fractionation by photoautotrophs. Much effort has gone into studying these effects under laboratory conditions with single species cultures. Moving beyond controlled environments and quantifying the natural extent of these changes in freshwater lacustrine settings and identifying their causes is essential for robust application of δ^2H values of common short-chain fatty acids as a proxy of net community metabolism and of phytoplankton-specific biomarkers as a paleohydrologic proxy.

This work targets the effect of community dynamics, temperature, and productivity on ^{2}H/^{1}H fractionation in lipid biomarkers through a comparative time series in two central Swiss lakes: eutrophic Lake Greifen and oligotrophic Lake Lucerne. Particulate organic matter was collected from surface waters at six time points throughout the spring and summer of 2015, and δ^2H values of short-chain fatty acids, as well as chlorophyll-derived phytol and the diatom biomarker brassicasterol, were measured. We paired these measurements with in situ incubations conducted with NaH^{13}CO$_3$, which were used to calculate the production rates of individual lipids in lake surface water. As algal productivity increased from April to June, net discrimination against ^{2}H in Lake Greifen increased by as much as 148 ‰ for individual fatty acids. During the same time period in Lake Lucerne, net discrimination against ^{2}H increased by as much as 58 ‰ for individual fatty acids. A large portion of this signal is likely due to a greater proportion of heterotrophically derived fatty acids in the winter and early spring, which are displaced by more ^{2}H-depleted fatty acids as phytoplankton productivity increases. Smaller increases in ^{2}H discrimination for phytol and brassicasterol suggest that a portion of the signal is due to changes in net photoautotrophic ^{2}H fractionation, which may be caused by increasing temperatures, a shift from maintenance to high growth, or changes in the community assemblage. The fractionation factors for brassicasterol were significantly different between the two lakes, suggesting that its hydrogen isotope composition may be more sensitive to nutrient regime than is the case for fatty acids or phytol.

1 Introduction

Compound-specific hydrogen isotope measurements of lipid biomarkers are an emerging tool with diverse applications to microbial community dynamics (Osburn et al., 2011; Heinzelmann et al., 2016), organic matter cycling (Jones et al., 2008; Li et al., 2009), and paleoclimatology (Sachse et al., 2012, and sources therein). The hydrogen isotopic composition of source water exerts a first-order control on lipid hydrogen isotopes (expressed as $\delta^2H = (^2H/^1H_{Sample})/(^2H/^1H_{VSMOW}) - 1$) (Sessions et al., 1999; Sauer et al., 2001; Sachs, 2014). However, a number of variables can influence the offset between the δ^2H values between lipids and source water, which is typically expressed by the fractionation factor $\alpha_{lipid\text{-}water} = (^2H/^1H_{lipid})/(^2H/^1H_{water})$.

For short-chain ($C < 20$) fatty acids, which can be synthesized by a diverse range of organisms, including photoautotrophs, chemoautotrophs, and heterotrophs, core metabolism typically exerts a large control on $\alpha_{lipid\text{-}water}$, with variability in δ^2H values exceeding 500 ‰ for organisms grown on the same source water (Zhang et al., 2009a; Osburn et al., 2011; Heinzelmann et al., 2015a; Osburn et al., 2016). These metabolic differences have led to the suggestion that δ^2H values of short-chain fatty acids can be used as an indicator of net community metabolism (Zhang et al., 2009a; Osburn et al., 2011; Heinzelmann et al., 2016; Osburn et al., 2016). This application has previously been assessed in coastal marine settings (Heinzelmann et al., 2016) and hot springs (Osburn et al., 2011), but not in lakes.

The δ^2H values of lipids produced exclusively by photoautotrophs, such as alkenones and certain sterols, have received particular attention as a proxy for past water isotopes (Sessions et al., 1999; Sauer et al., 2001; Huang et al., 2004; Sachse et al., 2012; Sachs, 2014), which is useful for paleoclimatologists seeking to reconstruct changes in temperature, moisture source, and the balance of precipitation to evaporation, all of which influence the δ^2H values of water (Craig and Gordon, 1965; Gat, 1996; Henderson and Schuman, 2009; Steinmann et al., 2013). The hydrogen isotopic composition of lipids produced by cyanobacteria and eukaryotic algae is well correlated with those of source water in laboratory and field settings (Sauer et al., 2001; Huang et al., 2004; Englebrecht and Sachs, 2005; Zhang and Sachs, 2007; Sachse et al., 2012) and is stable under near-surface temperatures and pressures for carbon-bound hydrogen (Sessions et al., 2004; Schimmelmann et al., 2006). Hydrogen isotopes of biomarkers from eukaryotic algae have been successfully applied to infer changes in past climate using sediment cores from diverse lakes (Huang et al., 2002; Sachs et al., 2009; Smittenberg et al., 2011; Atwood and Sachs, 2014; Zhang et al., 2014; Nelson and Sachs, 2016; Richey and Sachs, 2016; Randlett et al., 2017) and marine settings (Pahnke et al., 2007; van der Meer et al., 2007, 2008; Leduc et al., 2013; Vasiliev et al., 2013, 2017; Kasper et al., 2014).

However, among photoautotrophs, there is increasing evidence that $\alpha_{lipid\text{-}water}$ is not constant and can change with variables such as salinity, species, light availability, growth rate, and temperature (summarized in Table 1) (Sachs, 2014 and sources therein; Chivall et al., 2014; M'boule et al., 2014; Nelson and Sachs, 2014; Heinzelmann et al., 2015b; Sachs and Kawka, 2015; van der Meer et al., 2015; Wolhowe et al., 2015; Maloney et al., 2016; Sachs et al., 2016, 2017). While the array of secondary isotope effects may appear daunting, these relationships can provide useful information about past environmental changes in their own right, and developing a thorough understanding of them is important for robust interpretations of δ^2H_{lipid} values from phytoplankton.

Most previous investigations into variability in $\alpha_{lipid\text{-}water}$ in algal lipid biosynthesis have been done with controlled cultures of eukaryotes in laboratory settings. While similar relationships between salinity and $\alpha_{lipid\text{-}water}$ have been observed for eukaryotic algal and cyanobacterial lipids in both laboratory (Schouten et al., 2006; Chivall et al., 2014; M'boule et al., 2014; Heinzelmann et al., 2015b; Maloney et al., 2016; Sachs et al., 2016) and field calibrations (Sachse and Sachs, 2008; Sachs and Schwab, 2011; Nelson and Sachs, 2014), the temperature and growth rate effects observed in cultures have yet to be assessed in lacustrine settings where photoautotrophic δ^2H_{lipid} values are likely to be applied to reconstruct past hydroclimate. In contrast to cultures, lake water contains a diverse and dynamic community of phytoplankton, most of which contribute lipids to the sediment that cannot be attributed to one particular species. The culturing data that exist are limited to a few species, many of which are only found in marine environments.

In order to evaluate the significance of temperature and growth rate effects on the hydrogen isotopic composition of algal lipids produced in lakes, we collected monthly samples of particulate organic matter in two lakes in central Switzerland throughout the spring and summer of 2015. Both lakes experience similar changes in surface water temperature during this time period, but one of them (Lake Greifen) is characterized by high nutrient availability and increasing algal productivity and biomass throughout the spring and early summer. The other lake (Lake Lucerne) is oligotrophic and had relatively low constant rates of algal productivity throughout the study period. We paired measurements of hydrogen isotope fractionation with in situ incubations designed to determine lipid production rates, allowing us to distinguish between the effects of productivity and temperature on hydrogen isotope fractionation.

In addition to measuring δ^2H values of brassicasterol (24-methyl cholest-5,22-dien-3β-ol) and phytol, lipids that are produced exclusively by photoautotrophs, we also analyzed short-chain fatty acids ($nC_{14:0}$, $nC_{16:0}$, $nC_{16:1}$, $nC_{18:1}$), which, although they are typically the most abundant lipids in algal and cyanobacterial cells, are also synthesized by heterotrophic and chemoautotrophic microbes. The time series of fatty acid δ^2H values from an oligotrophic and eutrophic lake presented here is the first opportunity to assess how changes in net community metabolism might be recorded by these compounds in lakes.

Table 1. Summary of expected changes in $\alpha_{\text{lipid-water}}$ in response to different environmental variables, based on laboratory cultures and field studies in marine settings.

Variable	Sign of correlation with $\alpha_{\text{lipid-water}}$	Magnitude	References
Temperature	Negative	2–4 ‰ °C^{-1}	Zhang et al. (2009b); Wolhowe et al. (2009)
Growth rate	Negative	~ 30 ‰ division^{-1} day^{-1}	Schouten et al. (2006); Zhang et al. (2009b); Sachs and Kawka (2015); Wolhowe et al. (2015)
Nutrient availability	Negative	~ 40 ‰ difference between nutrient limited and nutrient replete cultures	Zhang et al. (2009b); Wolhowe et al. (2015)
Light availability	Positive	Below ~ 250 µmol photons m^{-2} s^{-1}, ~ 0.2 ‰ µmol^{-1} photons m^{-2} s^{-1}	van der Meer et al. (2015); Wolhowe et al. (2015); Sachs et al. (2017)
Salinity	Positive	0.5–3 ‰ practical salinity unit (PSU)$^{-1}$	Schouten et al. (2006); Sachse and Sachs (2008); Sachs and Schwab (2011); Chivall et al. (2014); M'boule et al. (2014); Nelson and Sachs (2014); Heinzelmann et al. (2015b); Maloney et al. (2016); Sachs et al. (2016)
Species assemblage	Variable	Differences up to 160 ‰ observed for nC$_{16:0}$ fatty acid among species growing under identical conditions	Schouten et al. (2006); Zhang and Sachs (2007)

2 Methods

2.1 Site description

Lake Greifen (Greifensee) is a small perialpine lake, located in the eastern fringes of the Zurich metropolitan area at 47°21′ N and 8°40′ E (Fig. 1). The lake has a surface area of 24 km^2 and a maximum depth of 32 m. The lake is fed by three small brooks and has one main outlet, the Glatt canal. Lake Greifen experienced severe eutrophication in the mid-20th century (Hollander et al., 1992; Keller et al., 2008). Strict government regulations on nutrient inputs were imposed in the 1970s, and the water quality in the lake has since improved, but its deep water remains anoxic and nutrient levels in the upper water column are still elevated. Winter overturn in the lake brings additional nutrients to the surface water, resulting in large phytoplankton blooms in the spring and summer as temperature and light availability increase (McKenzie, 1982). All samples from Lake Greifen were collected from the northern part of the lake, near a permanent platform maintained by Eawag (at 47°21.99′ N, 8°39.89′ E).

Lake Lucerne (Vierwaldstättersee) is a large perialpine lake, located in central Switzerland at 47°0′ N and 8°30′ E (Fig. 1). The lake, which has a total surface area of 116 km^2, is formed of seven distinct basins, of which the deepest is 214 m. The lake is fed by four alpine rivers: the Reuss, Muota, Engelberger Aa, and Sarner Aa, and its primary outflow is Reuss river from the northwest tip of the lake. Although Lake Lucerne experienced a mild eutrophication event in the 1970s, it is oligotrophic today (Bürgi et al., 1999; Bührer and Ambühl, 2001; Thevenon et al., 2012). All samples from Lake Lucerne were collected from the center of Kreuztricher basin (near 8°21′ N, 47°0′ E), with a water depth of 96 m.

2.2 Sample collection

Particulate material in each lake was collected at approximately monthly intervals throughout the spring and summer of 2015 (mid-April through early September). Surface water (~ 0.5 m water depth) was filtered onto a pre-combusted 142 mm diameter GF/F filter (0.7 µm pore size) using a WTS-LV Large Volume Pump (McLane, Massachusetts, USA). Pumping began at 7 L min^{-1} and continued until the flow rate decreased to 4 L min^{-1} or until 25 min had passed. All filters were collected at midday on sunny or mostly sunny days. Filters were wrapped in combusted aluminum foil and stored in a cool box on ice until transport to the laboratory, where they were stored at −20 °C until analysis.

Water samples were collected from surface water before and after pumping began. Samples were collected in 4 mL screw cap vials, sealed with electrical tape, and stored at room temperature prior to analysis. Depth profiles of temperature, conductivity, pH, turbidity, and dissolved oxygen were

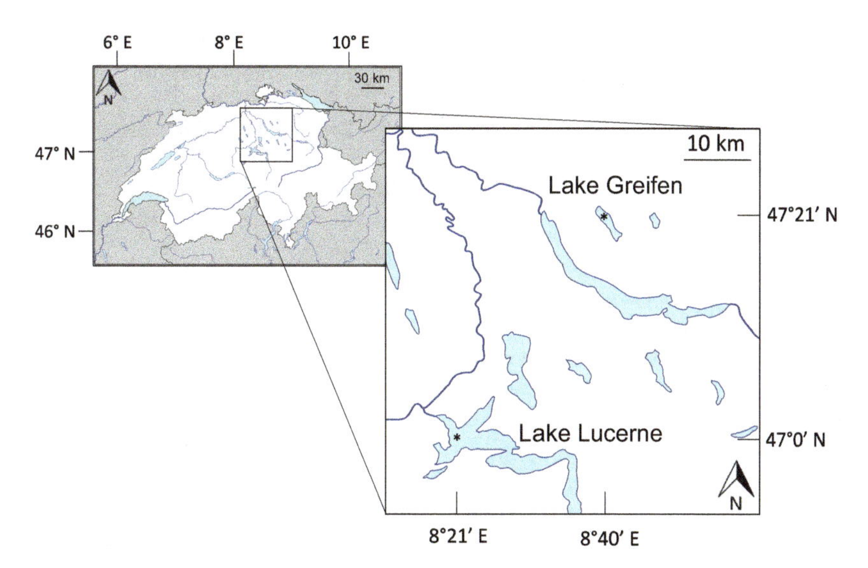

Figure 1. Map of Switzerland with locations of Lake Greifen and Lake Lucerne indicated. Base map from d-maps (http://www.d-maps.com/carte.php?num_car=2648&lang=en).

collected for the upper 20 m of the water column each sampling day at the beginning and end of filtration using a multiparameter CTD probe (75M, Sea and Sun Marine Tech, Trappenkamp, Germany).

On the morning of each sampling day, 4 × 12.5 L of surface water was collected in acid-rinsed, autoclaved, transparent carboys for in situ incubations. In two of the four carboys, 1 mL of concentrated $NaH^{13}CO_3$ solution was added. The other two carboys were not isotopically labeled. Carboys were mixed and attached to a fixed, floating line so that they stayed in the upper 50 cm of lake water throughout the day. After 6 h, they were retrieved and the contents were filtered onto a pre-combusted 142 mm diameter GF/F filter using a peristaltic pump. Water samples for DIC analyses were collected in 12 mL exetainers prior to isotopic labeling, after labeling but before incubation, and after incubation. These samples were sterile filtered through a 0.2 µm syringe filter and stored in the dark at 4 °C prior to analysis.

2.3 Water isotope measurements

Surface water isotope samples were filtered through a 25 mm syringe filter with a 0.45 µm polyethersulfone membrane to remove particulate matter. Water δ^2H and $\delta^{18}O$ values were measured by cavity ring-down spectroscopy (CRDS) on a L-2120i water isotope analyzer (Picarro, Santa Clara, CA, USA) at ETH Zurich. Each sample was injected seven times in sequence, and the first four values were discarded to avoid any memory effects from the previous sample. Three water standards with known δ^2H values of ranging from −161 to 7 ‰ and $\delta^{18}O$ values ranging from −22.5 to 0.9 ‰ were injected at the beginning and end of each sequence, as well as after every 10 samples. These standards were used to correct measured values to the VSMOW scale and to account for any instrumental drift over the course of the sequence. Average standard deviations (SDs) were 0.4 ‰ for hydrogen isotopes and 0.06 ‰ for oxygen isotopes.

2.4 DIC concentrations and $\delta^{13}C$ measurements

DIC concentrations were measured on a TOC-$L_{CSH/CHN}$ total organic carbon analyzer (Shimadzu, Kyoto, Japan). Solutions with DIC concentrations ranging from 5 to 100 mg L^{-1} were injected at the beginning of the sequence to form a calibration curve, and one standard of 50 mg L^{-1} was run after every five samples. Samples were analyzed in triplicate.

Exetainers of 3.7 mL were prepared for $\delta^{13}C$ measurements of DIC by adding 100 µL of concentrated H_3PO_4 and filling the headspace with He; 1 mL of sample water was added with a syringe through the septa of the exetainer. Samples were allowed to equilibrate overnight before analysis. Carbon isotope values were measured on an isotope ratio mass spectrometer (IRMS) (Isoprime, Stockport, UK). A standard of known isotopic composition was analyzed after every six samples. All samples were measured in duplicate.

2.5 Lipid extraction and purification

An internal standard containing nC_{19}-alkanol, nC_{19}-alkanoic acid, and 5α-cholestane was quantitatively added to freeze-dried filters, which were extracted in 30 mL of 9 : 1 dichloromethane (DCM)/methanol (MeOH) in a SOLVpro microwave reaction system (Anton Paar, Graz, Austria) at 70 °C for 5 min (Randlett et al., 2017), centrifuged, and the supernatant containing the total lipid extract (TLE) was poured off and evaporated under a gentle stream of N_2. The TLE was saponified with 3 mL of 1 N KOH in MeOH and

2 mL of solvent-extracted nanopure H_2O for 3 h at 80 °C (Randlett et al., 2017), after which the neutral fraction was extracted with hexane. Subsequently, the aqueous phase was acidified to pH = 2, and the protonated fatty acids were extracted with hexane.

Neutral fractions were further purified using silica gel column chromatography, following a scheme modified from Randlett et al. (2017). The sample was dissolved in hexane and loaded onto a 500 mg/6 mL Isolute Si gel column (Biotage, Uppsala, Sweden). *N*-alkanes were eluted in 4 mL of hexane, aldehydes and ketones in 4 mL of 1 : 1 hexane/DCM, alcohols in 4 mL of 19 : 1 DCM/MeOH, and remaining polar compounds in 4 mL of MeOH. The alcohol fraction was acetylated with 25 µL of acetic anhydride and 25 µL of pyridine for 30 min at 70 °C. The δ^2H and δ^{13}C values of the added acetyl group were determined by analyzing acetylated and unacetylated nC_{10}-alkanol.

Further purification was necessary in order to obtain base line separation of brassicasterol for δ^2H measurements. This was achieved by loading the acetylated alcohol fraction onto 500 mg of Si gel impregnated with $AgNO_3$ (10 % by weight, Sigma Aldrich) in a 6 mL glass cartridge. The first fraction, containing *n*-alkanols and phytol, was eluted with 20 mL of 4 : 1 hexane/DCM; the second fraction, containing stanols and singly unsaturated sterols (such as cholesterol) with 20 mL of 1 : 1 hexane/DCM; the third fraction, containing most doubly unsaturated sterols including brassicasterol, with 16 mL of DCM; and the remaining compounds with 4 mL of ethyl acetate.

Fatty acid fractions were methylated with 1 mL of BF_3 in MeOH (14 % by volume, Sigma Aldrich) for 2 h at 100 °C. After methylation, 2 mL of nanopure H_2O was added to the sample and the fatty acid methyl esters (FAMEs) were extracted with hexane. The δ^2H and δ^{13}C values of the added methyl group were determined by methylating phthalic acid of known isotopic composition (prepared by Arndt Schimmelmann at Indiana University).

FAMEs and brassicasterol were quantified by gas chromatography–flame ionization detection (GC-FID) (Shimazdu, Kyoto, Japan). Samples were injected by an AOC-20i autosampler (Shimadzu) through a split/splitless injector operated in splitless mode at 280 °C. The GC column was an InertCap 5MS/NP (0.25 mm × 30 m × 0.25 µm) (GL Sciences, Japan) and it was heated from 70 to 130 °C at 20 °C min^{-1}, then to 320 °C at 4 °C min^{-1}, and held at 320 °C for 20 min. FAMEs were identified by comparing their retention times to an external standard (fatty acid methyl ester mix from Sulpelco, reference no. 47885-U). Brassicasterol and phtyol were identified by comparing their retention times to those obtained by analyzing a subset of samples by gas chromatography–mass spectrometry (GC-MS) under identical conditions. In order to determine how much of the compound was in the original sample, peak areas were normalized to those of the internal standard. Peak areas were quantified relative to an external calibration curve in order to determine suitable injection volumes for isotopic analysis.

2.6 Lipid δ^2H and δ^{13}C measurements

The stable isotope values of individual FAMEs and brassicasterol were measured by gas chromatography–isotope ratio mass spectrometry (GC-IRMS). A GC-1310 gas chromatograph (Thermo Scientific, Bremen, Germany) equipped with an InertCap 5MS/NP (0.25 mm × 30 m × 0.25 µm) (GL Sciences, Japan) was interfaced to a Delta Advantage IRMS (Thermo Scientific) with a ConFlow IV interface (Thermo Scientific). Samples were injected with a TriPlusRSH autosampler to a PTV inlet operated in splitless mode at 280 °C. The oven was heated from 80 to 215 °C at 15 °C min^{-1}, then to 320 °C at 5 °C min^{-1}, and then was held at 320 °C for 10 min. Hydrogen isotope samples were pyrolyzed at 1420 °C after they eluted from the GC column. Carbon isotope samples were combusted at 1020 °C after elution.

Raw isotope values were converted to the VSMOW (hydrogen) and VPDB (carbon) scales using Thermo Isodat 3.0 software and pulses of a reference gas that was measured at the beginning and end of each analysis. Sample δ^2H and δ^{13}C values were further corrected using the slope and intercept of measured and known values of isotopic standards ($nC_{17, 19, 21, 23, 25, 28, 34}$-alkanes; Arndt Schimmelmann, Indiana University), which were run at the beginning and end of each sequence as well as after every six to eight sample injections. Offsets between measured and known values for these standards were used to correct for any drift over the course of the sequence or any isotope effects associated with peak area or retention time. The SD for these standards averaged 4 ‰ and the average offset from their known values was 2 ‰ for hydrogen isotopes. For carbon isotopes, the average SD of isotopic standards was 0.4 ‰ and the average offset from known values was 0.1 ‰ over the period of analysis.

An additional standard of nC_{29}-alkane was measured three times per sequence, corrected in the same way as the samples, and used for quality control. The SD of these measurements was 4 ‰ for hydrogen and 0.5 ‰ for carbon over the period of analysis. The H_3^+ factor was measured at the beginning of each sequence and averaged 3.6 ± 0.3 during the analysis period. Samples were corrected for hydrogen and carbon added during derivatization using isotopic mass balance, and reported errors represent propagated errors from replicate measurements and the uncertainties associated with the added hydrogen.

2.7 Calculated lipid production rates

Lipid production rates were calculated using Eq. (1) (modified from Popp et al., 2006):

$$\text{Production rate} = (\delta^{13}C_l - \delta^{13}C_n)/(\delta^{13}C_{DIC} - \delta^{13}C_n) \times (C_t/t), \quad (1)$$

Table 2. Summary of linear regression statistics; bolded relationships are significant at the $p < 0.05$ level.

	Lake Greifen					Lake Lucerne				
					α lipid water vs. temperature					
Lipid	Slope	y intercept	R^2	p	n	Slope	y intercept	R^2	p	n
All fatty acids	**-0.006 ± 0.002**	**0.95 ± 0.04**	**0.32**	**0.004**	**24**	**-0.003 ± 0.001**	**0.85 ± 0.02**	**0.24**	**0.015**	**24**
$n\mathrm{C}_{14:0}$ fatty acid	**-0.004 ± 0.001**	**0.86 ± 0.03**	**0.75**	**0.03**	**6**	-0.0012 ± 0.0009	0.80 ± 0.02	0.28	0.28	6
nC16:0 fatty acid	**-0.008 ± 0.002**	**0.98 ± 0.03**	**0.87**	**0.006**	**6**	**-0.003 ± 0.001**	**0.84 ± 0.02**	**0.66**	**0.049**	**6**
$n\mathrm{C}_{16:1}$ fatty acid	-0.005 ± 0.003	0.98 ± 0.07	0.35	0.22	6	**-0.004 ± 0.001**	**0.89 ± 0.02**	**0.74**	**0.028**	**6**
$n\mathrm{C}_{18:x}$ fatty acid	**-0.006 ± 0.0003**	**0.97 ± 0.007**	**0.99**	**< 0.0001**	**6**	**-0.004 ± 0.001**	**0.89 ± 0.02**	**0.70**	**0.037**	**6**
Phytol	-0.001 ± 0.001	0.66 ± 0.03	0.07	0.61	6	0.002 ± 0.001	0.62 ± 0.03	0.47	0.20	5
Brassicasterol	-0.002 ± 0.001	0.75 ± 0.02	0.63	0.11	5	0 ± 0.001	0.76 ± 0.02	0.00003	0.99	6
					α lipid water vs. lipid production rate					
Lipid	Slope	y intercept	R^2	p	n	Slope	y intercept	R^2	p	n
All fatty acids	-0.01 ± 0.01	0.83 ± 0.02	0.10	0.18	20	-0.04 ± 0.03	0.81 ± 0.01	0.09	0.19	20
$n\mathrm{C}_{14:0}$ fatty acid	0.00 ± 0.02	0.76 ± 0.02	0.0002	0.98	5	-0.2 ± 0.2	0.82 ± 0.05	0.25	0.39	5
$n\mathrm{C}_{16:0}$ fatty acid	0.01 ± 0.02	0.78 ± 0.05	0.02	0.83	5	-0.01 ± 0.04	0.83 ± 0.03	0.46	0.21	5
$n\mathrm{C}_{16:1}$ fatty acid	**-0.17 ± 0.04**	**0.89 ± 0.01**	**0.84**	**0.03**	**5**	0.6 ± 0.6	0.76 ± 0.06	0.22	0.42	5
$n\mathrm{C}_{18:x}$ fatty acid	-0.01 ± 0.02	0.85 ± 0.04	0.08	0.64	5	-0.09 ± 0.05	0.86 ± 0.02	0.57	0.14	5
Phytol	0 ± 1	0.63 ± 0.03	0.03	0.77	5	0 ± 7	0.66 ± 0.02	0.001	0.95	5
Brassicasterol	2 ± 2	0.70 ± 0.01	0.34	0.42	4	5 ± 43	0.76 ± 0.03	0.004	0.92	5

where $\delta^{13}C_l$ is the δ^{13}C value of the target compounds from labeled incubations, $\delta^{13}C_n$ is that from unlabeled incubations, $\delta^{13}C_{DIC}$ is the δ^{13}C value of DIC, C_t is the concentration of the lipid at the end of the incubation, and t is the duration of the incubation. Residence times – assuming a steady state, the amount of time needed to replace all molecules of a given lipid – were calculated by dividing C_t by the production rate, which reduces to Eq. (2):

$$\text{Residence time} = t \times (\delta^{13}C_{DIC} - \delta^{13}C_n)/(\delta^{13}C_l - \delta^{13}C_n). \quad (2)$$

2.8 Statistics

PRISM software (Graphpad Software Inc., La Jolla, CA, USA) was used to carry out all statistical analyses. Ordinary least-squares regression was used to determine relationships among fractionation factors, temperature, and lipid production rates. Regression lines are only shown where the slope of the regression was significantly different from 0 at the $p < 0.05$ level. The results of all linear regression analyses are presented in Table 2. Differences between the slopes of various regressions were assessed using a two-tailed test of the null hypothesis that both slopes are equal. Differences in the mean values of replicate measurements were determined using an unpaired, two-tailed t test and were considered significantly different for $p < 0.05$.

3 Results

3.1 Lipid concentrations and production rates

Lipid concentrations increased significantly in Lake Greifen from April to July and then declined slightly from July to September, except for phytol and $n\mathrm{C}_{161:1}$ fatty acid, which had increasing concentrations into the late summer (Fig. 2a). $n\mathrm{C}_{16:0}$ fatty acid had the highest concentrations, while those of $n\mathrm{C}_{16:1}$ fatty acid were usually the lowest, except in April–May, when $n\mathrm{C}_{14:0}$ was the least abundant fatty acid. Brassicasterol concentrations were 1–2 orders of magnitude smaller than those of fatty acids and phytol concentrations were intermediate (Fig. 2a). Lipid concentrations in Lake Lucerne were generally an order of magnitude lower than in Lake Greifen (Fig. 2b). Fatty acid concentrations increased significantly from April to May in Lake Lucerne and were then relatively stable throughout the rest of the time series (Fig. 2b). Again, $n\mathrm{C}_{16:0}$ fatty acid had the highest concentrations and $n\mathrm{C}_{16:1}$ was the least abundant fatty acid. Phytol concentrations were typically an order of magnitude lower than those of fatty acids in Lake Lucerne and increased slightly over the course of the time series. Brassicasterol concentrations were an order of magnitude lower still and reached a maximum in June (Fig. 2b).

In both lakes, fatty acid production rates were highest for $n\mathrm{C}_{16:0}$ (palmitic acid), followed by $n\mathrm{C}_{18:x}$ (unsaturated C_{18} fatty acids, primarily $n\mathrm{C}_{18:1n9c}$, or oleic acid), $n\mathrm{C}_{14:0}$ (myristic acid), and $n\mathrm{C}_{16:1}$ (palmitoleic acid) (Fig. 2c and d). Phytol and brassicasterol production rates were 2–3 orders of magnitude lower in both lakes than those of fatty acids (Fig. 2c and d). Lipid production rates were up to three times higher

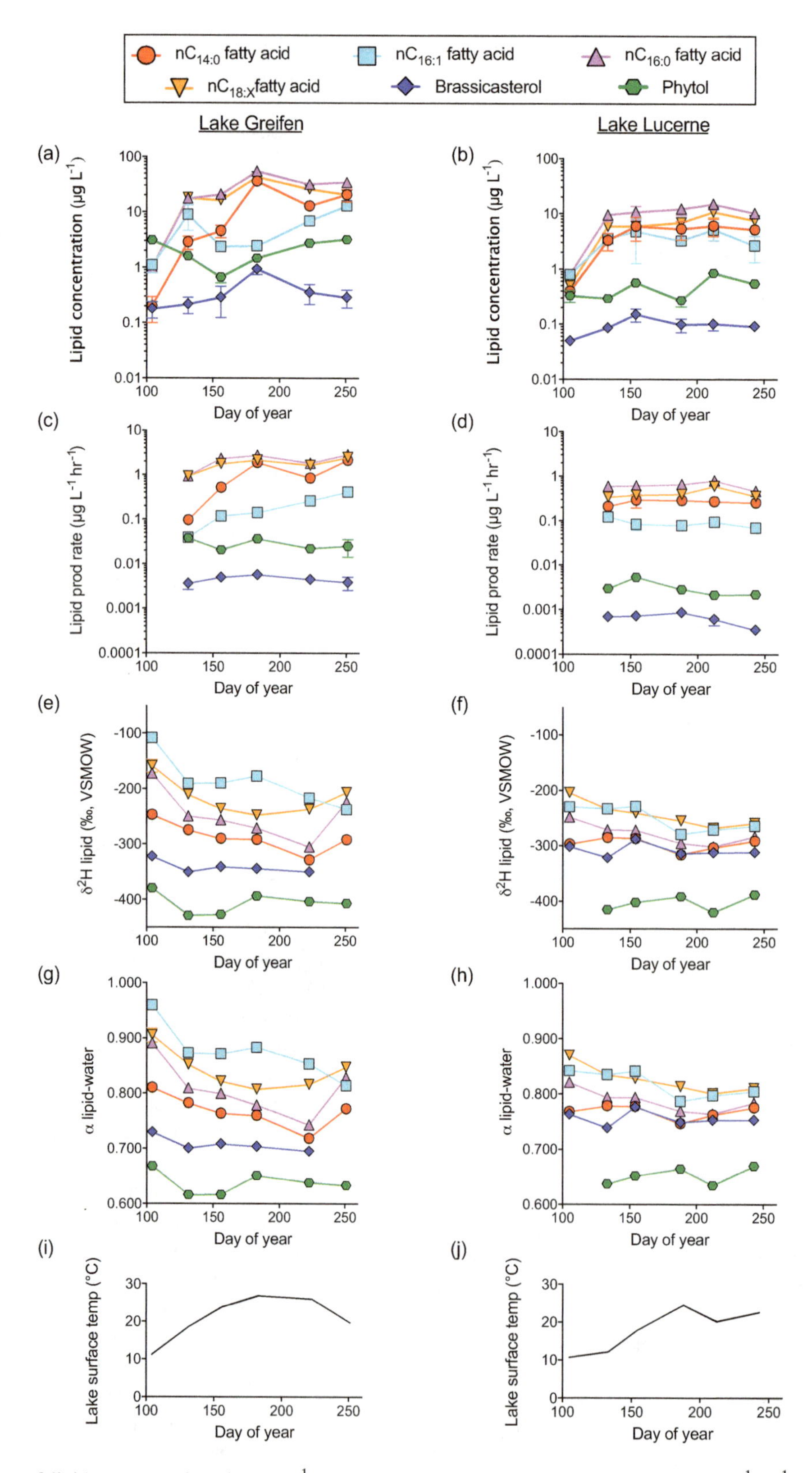

Figure 2. Time series of lipid concentrations in $\mu g L^{-1}$ (panels **a** and **b**), lipid production rates in $\mu g L^{-1} h^{-1}$ (panels **c** and **d**), lipid δ^2H values in ‰ relative to VSMOW (panels **e** and **f**, $\alpha_{lipid\text{-}water}$ values (panels **g** and **h**), and lake surface temperature (panels **i** and **j**) for Lake Greifen (left column) and Lake Lucerne (right column) during the spring and summer of 2015. Panels **(a–d)** are plotted on an exponential scale to accommodate the large range of lipid concentrations and production rates. Error bars represent 1 SD of replicate measurements and are propagated to include uncertainties from multiple sources in calculated production rates and $\alpha_{lipid\text{-}water}$ values. In cases where error bars are not visible, they are smaller than the marker size.

Table 3. Mean residence times in hours of lipids in lake surface water, calculated according to Eq. (2).

Date	$n\mathrm{C}_{14:0}$	$n\mathrm{C}_{16:0}$	$n\mathrm{C}_{16:1}$	$n\mathrm{C}_{18:x}$	Brassicasterol	Phytol
Lake Greifen						
11 May 2015	30 ± 9	19 ± 4	60 ± 13	19 ± 3	59 ± 26	42 ± 10
5 Jun 2015	9 ± 2	9 ± 1	21 ± 2	9 ± 1	58 ± 34	32 ± 8
2 Jul 2015	19 ± 1	20 ± 1	49 ± 5	20 ± 3	165 ± 51	40 ± 6
11 Aug 2015	15 ± 3	17 ± 2	41 ± 5	16 ± 2	81 ± 34	123 ± 23
8 Sept 2015	10 ± 3	12 ± 2	31 ± 8	12 ± 2	76 ± 37	127 ± 26
Lake Lucerne						
13 May 2015	16 ± 6	16 ± 5	27 ± 8	17 ± 2	124 ± 22	99 ± 19
3 Jun 2015	20 ± 12	18 ± 6	62 ± 27	16 ± 4	208 ± 56	106 ± 17
7 Jul 2015	19 ± 8	19 ± 5	34 ± 18	17 ± 5	114 ± 39	94 ± 31
31 Jul 2015	22 ± 9	19 ± 4	38 ± 8	18 ± 4	164 ± 59	399 ± 36
31 Aug 2015	20 ± 4	22 ± 3	41 ± 6	12 ± 3	258 ± 61	248 ± 42

in Lake Greifen than in Lake Lucerne (Fig. 2c and d). Lipid production rates generally increased from May to July and then remained high in Lake Greifen, while in Lake Lucerne they were relatively constant throughout the study period (Fig. 2c and d).

Residence times – or the amount of time necessary to replace all molecules of a given compound assuming steady state – of individual lipids were calculated according to Eq. (2) (Sect. 2.7) and were typically shortest for $n\mathrm{C}_{14:0}$, $n\mathrm{C}_{16:0}$, and $n\mathrm{C}_{18:x}$ fatty acids, with values as low as 9 ± 1 h in Lake Greifen in May and as low as 12 ± 3 h in Lake Lucerne in August (Table 3). Of the fatty acids, $n\mathrm{C}_{16:1}$ had the longest residence times, reaching 60 ± 13 h in Lake Greifen in May and 62 ± 27 h in Lake Lucerne in June (Table 3). Brassicasterol residence times were the longest of any lipid in the first part of the time series but were exceeded by phytol for the last two sampling dates (Table 3).

3.2 Lipid δ^2H and $\alpha_{\text{lipid-water}}$ values

In both lakes lipid δ^2H values typically decreased over the spring and summer (Fig. 2e and f). This effect was most pronounced for fatty acids in Lake Greifen. For example, $n\mathrm{C}_{16:0}$ fatty acid δ^2H values declined by 133 ‰ (from -172 to -305 ‰) from April to August in Lake Greifen, while they only declined by 53 ‰ (from -249 to -302 ‰) over the same time period in Lake Lucerne (Fig. 2e and f). During the same time period water δ^2H values increased slightly in Lake Greifen (from -73 to -65 ‰) and were relatively constant in Lake Lucerne (fluctuating between -82 and -86 ‰) (Supplement Sect. S2). Changes in the fractionation factor between fatty acids and surface water ($\alpha_{\text{lipid-water}}$) were therefore linked closely to changes in fatty acid δ^2H values. In Lake Greifen, $\alpha_{\text{lipid-water}}$ for $n\mathrm{C}_{16:0}$ fatty acid decreased from 0.891 to 0.743 from April to August (Fig. 2g), while in Lake Lucerne it decreased from 0.821 to 0.763 (Fig. 2h). Similar patterns were observed for $\alpha_{\text{lipid-water}}$ for $n\mathrm{C}_{14:0}$, $n\mathrm{C}_{16:1}$ and $n\mathrm{C}_{18:X}$ fatty acids (Fig. 2g and h). Values for $\alpha_{\text{lipid-water}}$ were less variable for phytol and brassicasterol than for fatty acids, although they also declined from April to May. Brassicasterol was always ^{2}H-depleted relative to fatty acids in Lake Greifen and was depleted in ^{2}H relative to all fatty acids except $n\mathrm{C}_{14:0}$ in Lake Lucerne (Fig. 2g and h). Phytol δ^2H values were the most ^{2}H depleted of any lipid measured in either lake (Fig. 2g and h).

Overall, fatty acid $\alpha_{\text{lipid-water}}$ values were negatively correlated with lake surface temperature (LST) in both lakes ($R^2=0.32$, $p=0.004$ in Lake Greifen; $R^2=0.24$, $p=0.01$ in Lake Lucerne) (Fig. 3; Table 2). The slope of the relationship was significantly steeper ($p=0.03$) in Lake Greifen than in Lake Lucerne ($m=-0.006\pm0.002$ in Lake Greifen and -0.003 ± 0.001 in Lake Lucerne). Significant correlations were observed between LST and $\alpha_{\text{lipid-water}}$ values for most fatty acids but not for $n\mathrm{C}_{16:1}$ in Lake Greifen and $n\mathrm{C}_{14:0}$ in Lake Lucerne (Fig. 3; Table 2). Significant relationships between LST and $\alpha_{\text{lipid-water}}$ values were not observed in either lake for brassicasterol or phytol (Fig. 3; Table 2).

Fatty acid production rates were not correlated with $\alpha_{\text{lipid-water}}$ values in either lake (Table 2). Among individual fatty acids, only $n\mathrm{C}_{16:1}$ fatty acids from Lake Greifen had a significant negative correlation between $\alpha_{\text{lipid-water}}$ values and production rate ($R^2=0.84$; $p=0.03$) (Table 2). Brassicasterol and phytol production rates were not correlated with $\alpha_{\text{lipid-water}}$ in either lake (Table 2), although brassicasterol $\alpha_{\text{lipid-water}}$ values from Lake Lucerne cluster as a significantly higher group than in Lake Greifen ($p=0.0004$).

4 Discussion

In both lakes, the most striking feature of the lipid δ^2H values in the particulate organic matter is the significant decrease that occurs for most lipids during the spring (April–June) (Fig. 2e and f). As the lake water δ^2H values increased

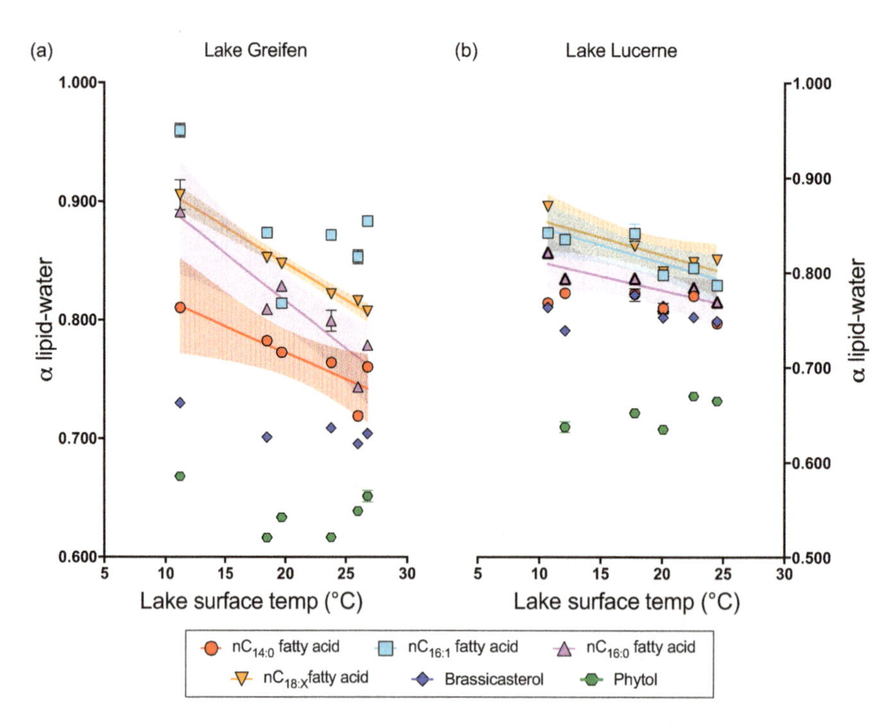

Figure 3. Relationships between $\alpha_{\text{lipid-water}}$ values and lake surface temperatures in lakes Greifen (panel **a**) and Lucerne (panel **b**) throughout the spring and summer of 2015. Error bars are propagated 1σ uncertainty from replicate measurements of surface water and lipid δ^2H values. In cases where error bars are not visible, they are smaller than the marker size. Shading represents 95 % confidence intervals of the linear regression. Statistics associated with each curve are summarized in Table 2, and plots of individual compounds are available in the Supplement.

slightly (Greifen) or remained constant (Lucerne) throughout the summer, this trend indicates a decrease in $\alpha_{\text{lipid-water}}$ (Fig. 2g and h). There are a number of factors that could contribute to this decline in $\alpha_{\text{lipid-water}}$, but they generally group into two categories: changes in lipid source or changes in environmental variables, such as temperature, light, and productivity.

4.1 Potential changes in lipid source

Hydrogen isotope fractionation for short-chain fatty acids varies significantly among organisms with different metabolisms (Zhang et al., 2009a; Osburn et al., 2011; Heinzelmann et al., 2015a). In general, fatty acids from heterotrophs grown on tricarboxylic acid (TCA) cycle precursors are most enriched in ^{2}H, followed by heterotrophs grown on sugars, then photoautotrophs, and finally chemoautotrophs (Zhang et al., 2009a; Osburn et al., 2011; Heinzelmann et al., 2015a). This variability can be greater than 500 ‰ and has led to the suggestion that the δ^2H values of ubiquitous compounds such as palmitic acid can be used as a proxy of net community metabolism. For example, at a coastal site in the North Sea, fatty-acid-chain-weighted average δ^2H values declined by more than 40 ‰ during the spring phytoplankton bloom, which was attributed to increased contributions from photoautotrophs (Heinzelmann et al., 2016).

In lakes Lucerne and Greifen, large decreases in fatty acid δ^2H values from April to May coincide with increases in fatty acid concentrations of 1–2 orders of magnitude (Fig. 2). It is therefore possible that the ^{2}H-enriched April samples represent a wintertime background of mixed heterotrophic and autotrophic derived compounds. As phytoplankton productivity ramped up with warmer temperatures, water column stratification, and longer daylight hours in the spring, newly produced fatty acids from photoautotrophs could have overwhelmed the heterotrophic signature, causing the net fatty acid δ^2H values to decrease. The increase in phytoplankton cell density (Fig. 4) from April onward in both lakes is supportive of increased contributions of fatty acids from phytoplankton as the study period progressed.

For Lake Greifen, a simple isotopic mass balance indicates that 31 % of the total $n\text{C}_{16:0}$ fatty acid would need to come from heterotrophs with $\alpha_{\text{lipid-water}}$ values of 1.200 (the maximum observed by Zhang et al., 2009a) in mid-April if the remaining $n\text{C}_{16:0}$ fatty acid was derived from phytoplankton with $\alpha_{\text{lipid-water}}$ values of 0.750 (assuming mid-summer $\alpha_{\text{lipid-water}}$ values represent the phytoplankton end member). Similar calculations suggest that 16 % of $n\text{C}_{16:0}$ fatty acid would need to come from heterotrophic bacteria in mid-April in Lake Lucerne in order to account for the 50 ‰ decrease in $n\text{C}_{16:0}$ fatty acid δ^2H values over the course of the summer. These calculations assume that all heterotrophic bacteria use

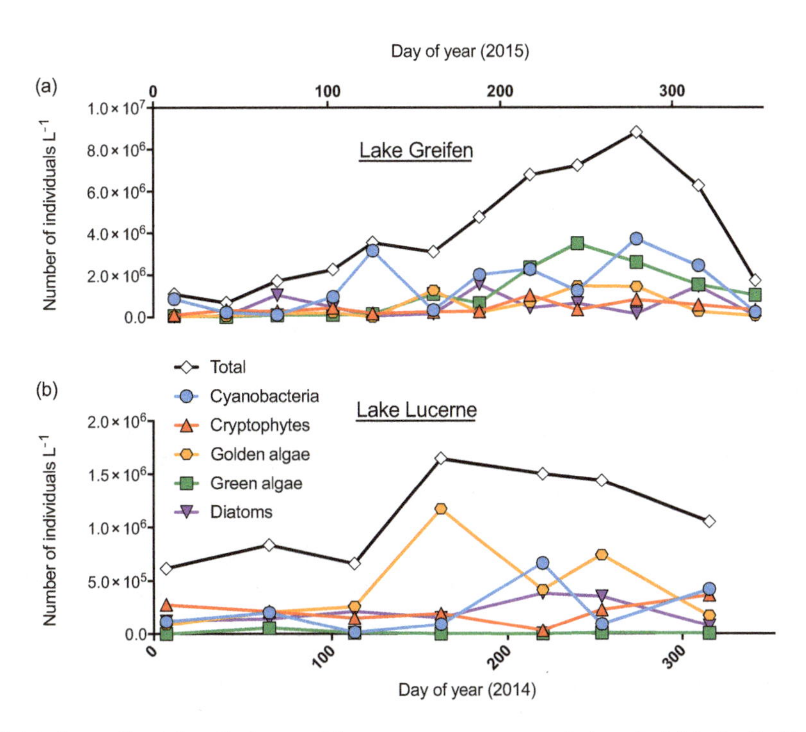

Figure 4. Cell counts (individuals per liter) for all algae and for most common taxa of algae in **(a)** Lake Greifen throughout 2015 and **(b)** Lake Lucerne in 2014. The scale of the y axis differs between the two panels. Data from long-term monitoring program are run by the department of Aquatic Ecology at Eawag.

the highest $\alpha_{\text{lipid-water}}$ value ever observed for heterotrophs and that they primarily use the TCA cycle rather than glycolysis. If, as is likely, there is diversity in $\alpha_{\text{lipid-water}}$ values for short-chain fatty acids produced by the heterotrophic bacteria, and at least some of the heterotrophs are relying primarily on glycolysis, the portion of these fatty acids from heterotrophic sources in April would need to be even higher than the values calculated above. This would necessitate a proportionally larger winter heterotrophic contribution of fatty acids than was observed in the coastal North Sea (Heinzelmann et al., 2016), and it seems likely that other variables may contribute to the springtime decline in fatty acid δ^2H values.

Contributions from heterotrophs are also an improbable explanation for ^{2}H-enriched brassicasterol and phytol in April. Brassicasterol is a sterol that is commonly used as a biomarker for diatoms, although it has also been detected in some non-diatom eukaryotic phytoplankton (Volkman et al., 1998; Volkman, 2003; Rampen et al., 2010; Taipale et al., 2016) and occasionally in plant oils (Zarrouk et al., 2009). Since brassicasterol is not produced by bacterial sources, it seems improbable that the 25 ‰ (Greifen) and 19 ‰ (Lucerne) decreases in its δ^2H values from April to May could be due to proportionately greater heterotrophic contributions during the winter, as suggested for fatty acids. The 46 ‰ April–May decrease in δ^2H values of Lake Greifen phytol, the side chain of chlorophyll molecules, is also unlikely to be caused by heterotrophic contributions in the early spring. Although phytol is produced by some photoheterotrophs, these typically have similar $\alpha_{\text{lipid-water}}$ values to photoautotrophs (Zhang et al., 2009a).

However, seasonal changes in the phytoplankton community composition alone could be a significant source of variability in $\alpha_{\text{lipid-water}}$ over the course of the study period. Hydrogen isotope fractionation for $n\text{C}_{16:0}$ fatty acid has been demonstrated to vary by over 100 ‰ among five different species of freshwater green algae grown in laboratory batch cultures (Zhang and Sachs, 2007). Such variations may be due to different enzymes involved in lipid synthesis among different species or to the colony-forming behavior of the two species with higher $\alpha_{\text{lipid-water}}$ values. Smaller species-dependent variations in $\alpha_{\text{lipid-water}}$ have been observed in cultures of haptophytes ($\sim$ 30 ‰ offset between alkenones in *G. oceanica* and in *E. huxleyi*; Schouten et al., 2006). Since there are limited data from culturing experiments, it is not possible to say how widespread such interspecies variability is. It is possible that most phytoplankton display similar magnitudes of hydrogen isotope fractionation during lipid synthesis under similar conditions. However, it is equally possible that lipid hydrogen isotope fractionation varies among species in ways that are not yet understood.

Given this uncertainty, and the significant changes in abundance of different phytoplankton taxa in Lake Greifen over the course of 2015 (Fig. 4a), contributions of lipids from different species of algae with different magnitudes of hydrogen isotope fractionation could account for some or all of the seasonal variability in $\alpha_{\text{lipid-water}}$. A comparable data set of algal species counts does not exist for Lake Lucerne from 2015, but bimonthly data have been compiled from 2014 (Fig. 4b). Some changes in relative distributions of taxa are similar between the two lakes; for example, both Lake Greifen in 2015 and Lake Lucerne in 2014 experienced a peak in golden algae (*Chrysophyceae*) in June and elevated abundance of cyanobacteria in late summer and late autumn (Fig. 4). Other trends differ starkly between the two lakes. Green algae (*Chlorophyceae*) are largely absent from Lake Lucerne, while they make up a significant portion of the algal community in Lake Greifen. Notably, the relative abundance of green algae steadily increased from May to September in Lake Greifen (Fig. 4), during which time $\alpha_{\text{lipid-water}}$ values declined at a greater rate for most compounds than they did in Lake Lucerne (Fig. 2). If green algae tend to have lower $\alpha_{\text{lipid-water}}$ values than other algal taxa, their greater abundance in Lake Greifen throughout the summer could account for the greater decline in $\alpha_{\text{lipid-water}}$ over the course of the time series than in Lake Lucerne.

4.2 Relationships between seasonal environmental gradients and $\alpha_{\text{lipid-water}}$ values

Even for lipids produced in controlled cultures of eukaryotic algae, several factors have been shown to influence hydrogen isotope fractionation, including salinity, light availability, temperature, and growth rate (summarized in Table 1) (Sachs, 2014 and sources therein; Chivall et al., 2014; M'boule et al., 2014; Heinzelmann et al., 2015b; Sachs and Kawka, 2015; van der Meer et al., 2015; Wolhowe et al., 2015; Maloney et al., 2016; Sachs et al., 2016, 2017). Of these, salinity can be excluded as a source of variability freshwater lakes such as Greifen and Lucerne. The effect of light availability on $\alpha_{\text{lipid-water}}$ has only been detected at low light levels (below 250 µmol photons $m^{-2}\,s^{-1}$) (van der Meer et al., 2015; Wolhowe et al., 2015; Sachs et al., 2017). Although photosynthetically available radiation (PAR) was not measured as part of the present study, all samples were collected from lake surface water at a midlatitude Northern Hemisphere site during boreal spring and summer, and it is unlikely that PAR was less than 250 µmol photons $m^{-2}\,s^{-1}$ at any sampling date (Pinker and Laszlo, 1992), meaning that the effect of light intensity is unlikely to be a source of the observed seasonal variability in $\alpha_{\text{lipid-water}}$ in lake surface water.

LST varied from 11 to 27 °C in Lake Greifen and from 11 to 25 °C in Lake Lucerne over the study period (Fig. 2i and j) and therefore may have contributed to the seasonal changes in lipid δ^2H values. In laboratory cultures of eukaryotic algae, $\alpha_{\text{lipid-water}}$ values for acetogenic lipids have been shown to decrease by 0.002–0.004 °C^{-1}, resulting in more depleted δ^2H values (Schouten et al., 2006; Zhang et al., 2009b; Wolhowe et al., 2009, 2015). Increased hydrogen isotope fractionation at higher temperatures has been attributed to (i) changes in the relative activity of different enzymes involved in lipid synthesis at different temperatures, (ii) changes in the relative amount of NADPH from the pentose phosphate cycle as temperature changes, and (iii) the potential for hydrogen tunneling at higher temperatures as substrate–enzyme complex vibrations increase (Sachs, 2014, and references therein). The relationship between temperature and hydrogen isotope fractionation in cultures is similar to that observed for fatty acids in Lake Lucerne, where $\alpha_{\text{lipid-water}}$ decreases by 0.003 ± 0.001 °C^{-1} (Fig. 3; Table 2). The relationship between $\alpha_{\text{lipid-water}}$ and temperature for fatty acids in Lake Greifen (-0.006 ± 0.002 °C^{-1}) (Fig. 3; Table 2) is much steeper than that observed in culturing studies.

If the influence of temperature on hydrogen isotope fractionation is consistent among laboratory cultures and lakes, warmer temperatures can account for the entire seasonal change in $\alpha_{\text{lipid-water}}$ for fatty acids in Lake Lucerne. However, increasing temperatures would only be able to explain part of the decrease in fatty acid δ^2H values in Lake Greifen over the course of the spring and summer. At most, temperature could account for half of the decrease in $\alpha_{\text{lipid-water}}$ in Lake Greifen, assuming a consistent relationship to that observed in cultures.

Brassicasterol and phytol $\alpha_{\text{lipid-water}}$ values do not have strong relationships with LST. There is no correlation between LST and $\alpha_{\text{lipid-water}}$ values for phytol in either lake nor for brassicasterol in Lake Lucerne. In Lake Greifen temperature and $\alpha_{\text{brassicasterol-water}}$ are negatively correlated, although the relationship is not quite significant (Fig. 3f; Table 2). The slope of the relationship between $\alpha_{\text{brassicasterol-water}}$ and LST in Lake Greifen is significantly shallower than that observed for fatty acids (-0.002 ± 0.001 for brassicasterol vs. $0.008 - \pm 0.002$ for $n\text{C}_{16:0}$ fatty acid). Given these relatively weak relationships, it seems unlikely that temperature influences hydrogen isotope fractionation of either brassicasterol or phytol. The negative correlations between temperature and $\alpha_{\text{lipid-water}}$ values for fatty acids may therefore be an artifact of the probable increase in photoautotrophically derived compounds as it became warmer throughout the spring (Sect. 4.1). However, the relationship between temperature and $\alpha_{\text{lipid-water}}$ values in cultures has only been observed for n-alkanoic acids and alkenones, both of which are acetogenic lipids. It is thus possible that temperature is partially responsible for the decreases in $\alpha_{\text{lipid-water}}$ values as temperature increased in lakes Lucerne and Greifen but that the responsible mechanism is specific to lipids produced acetogenically and does not affect isoprenoids, such as sterols and phytol.

Phytoplankton productivity and biomass also increased with temperature in the spring and summer in lakes Greifen and Lucerne, with a more marked effect in nutrient-rich Greifen (Fig. 2a–d, Fig. 4). This trend could also partially explain the increase in ^{2}H fractionation and decrease in lipid δ^2H values that co-occurred with rising temperatures. Increased nutrient availability and higher growth rates have been shown to result in lower $\alpha_{\text{lipid-water}}$ values during lipid synthesis for eukaryotic algae in laboratory settings, with lipids more ^{2}H depleted relative to source water as growth rate increases (Schouten et al., 2006; Zhang et al., 2009b; Sachs and Kawka, 2015; Wolhowe et al., 2015). This relationship is most likely caused by increased contributions of hydrogen from relatively enriched NADPH from the oxidative pentose phosphate cycle under low-growth, nutrient-stressed conditions, at the expense of relatively depleted hydrogen from photosystem I (Schmidt et al., 2003; Sachs and Kawka, 2015). If April samples include a higher proportion of lipids from organisms in a low-growth maintenance phase, they should therefore be relatively enriched in ^{2}H. As light availability and water column stratification became more amenable to photosynthesis later in the spring, relatively ^{2}H-depleted lipids produced with NADPH from photosystem I would be expected to become more abundant, bringing the net δ^2H and $\alpha_{\text{lipid-water}}$ values down.

Bottle incubations to determine lipid production rates were unfortunately not conducted for the first sampling in April, when the most enriched lipid δ^2H values were measured. For the remaining five sampling dates, there were not significant correlations between lipid production rate and $\alpha_{\text{lipid-water}}$, with the exception of $n\text{C}_{16:1}$ in Lake Greifen (Table 2). Lipid concentrations and production rates are not a direct proxy for growth rate, as a higher percentage of algal biomass is typically allocated to lipids under low nutrient and slow growth conditions (Roessler, 1990; Williams and Laurens, 2010). However, higher lipid production rates for the whole community (rather than on a per cell basis) will co-occur with higher growth rates. The large increase in fatty acid concentrations from April to May in both lakes, as well as smaller increases in brassicasterol concentrations, may indicate that the greatest community-wide change in growth rate occurred between those 2 months and contributes in part to the decrease in fatty acid, phytol, and brassicasterol δ^2H values from April to May.

4.3 Comparison of mean $\alpha_{\text{lipid-water}}$ in lakes with different trophic statuses

For phytol, $n\text{C}_{14:0}$, $n\text{C}_{16:0}$, and $n\text{C}_{18:x}$ fatty acids, there was no significant difference in $\alpha_{\text{lipid-water}}$ between the oligotrophic and eutrophic lake. However, significant differences in $\alpha_{\text{lipid-water}}$ do exist for brassicasterol (0.045 ± 0.008; $p = 0.0004$) and $n\text{C}_{16:1}$ fatty acid (0.058 ± 0.022; $p = 0.02$) (Fig. 2). For brassicasterol, $\alpha_{\text{lipid-water}}$ is lower in Lake Greifen (0.712 ± 0.006) than in the less productive Lake Lucerne (0.757 ± 0.005). This result would be consistent with decreased hydrogen isotope fractionation (higher α values) for sterols in more nutrient-limited systems, as predicted by culturing experiments (Zhang et al., 2009b; Sachs and Kawka, 2015). The fact that this strong difference in fractionation between the two lakes is observed only for brassicasterol may be because it is the most source specific of the biomarkers that were analyzed. Phytol and short-chain fatty acids are produced by all photoautotrophs and may be dominated by phytoplankton that are optimized to grow under the nutrient regimes of each system, while relatively more of the brassicasterol may come from taxa that are nutrient-stressed and relying more on the pentose phosphate pathway than photosystem I.

Alternatively, the difference in $\alpha_{\text{brassicasterol-water}}$ values between the two lakes could be due to variable contributions of brassicasterol from different phytoplankton sources. Even though brassicasterol is produced by fewer organisms than short-chain fatty acids and phytol, it still has multiple sources (Volkman et al., 1998; Volkman, 2003; Rampen et al., 2010; Taipale et al., 2016). Species-specific differences in hydrogen isotope fractionation have not been observed for sterols but have been reported for fatty acids and alkenones (Schouten et al., 2006; Zhang and Sachs, 2007), making this an unconstrained possibility that could be responsible for the difference in $\alpha_{\text{brassicasterol-water}}$ between the oligotrophic and eutrophic lake. Different sources could also account for the difference in $\alpha_{\text{lipid-water}}$ for $n\text{C}_{16:1}$ fatty acid, which displays higher $\alpha_{\text{lipid-water}}$ values in the more productive lake, and therefore cannot be explained by the nutrient effect observed in cultures.

5 Conclusions

We measured δ^2H values of short-chain fatty acids, phytol, and the diatom biomarker brassicasterol in surface water particulate organic matter in two lakes in central Switzerland with different trophic states at six time points throughout the spring and summer of 2015. Measurements were paired with in situ incubations with ^{13}C-enriched DIC that allowed us to calculate lipid production rates.

In April in both lakes, lipid concentrations were at their lowest and lipid δ^2H values were at their highest. In the case of short-chain fatty acids, which are produced by both photoautotrophic and heterotrophic microbes, the relatively high fractionation factors observed in the spring are consistent with a greater proportion of these compounds being derived from heterotrophs (Zhang et al., 2009a; Osburn et al., 2011; Heinzelmann et al., 2015a). As phytoplankton productivity increased throughout the springtime, net $\alpha_{\text{lipid-water}}$ values declined to the range more commonly associated with photoautotrophs. The observed decline in $\alpha_{\text{lipid-water}}$ for fatty acids in oligotrophic Lake Lucerne was similar to that observed during the spring bloom in the North Sea (Heinzelmann et al., 2016) but was nearly three times as large in eutrophic Lake Greifen.

Changing contributions from heterotrophs cannot explain all of the decline in $\alpha_{\text{lipid-water}}$ from April to May, since this was also observed to a lesser extent in phytol and brassicasterol, compounds produced exclusively by photoautotrophs. Several factors could be responsible for changes in photoautotrophic $\alpha_{\text{lipid-water}}$ throughout the spring, including temperature, growth rate, and species assemblage. Fractionation factors were inversely correlated with temperature for most fatty acids in each lake, and the slope of this relationship in Lake Lucerne (Fig. 3) was consistent with laboratory cultures, which suggest that $\alpha_{\text{lipid-water}}$ decreases with temperature by 0.002–0.004 $°C^{-1}$ for acetogenic lipids (Zhang et al., 2009b; Wolhowe et al., 2009). Slower growth rates in the early spring could also result in higher $\alpha_{\text{lipid-water}}$ values at this time, as low growth rates correlate with higher $\alpha_{\text{lipid-water}}$ values in cultures (Schouten et al., 2006; Zhang et al., 2009b; Sachs and Kawka, 2015). Finally, changes in phytoplankton species assemblage could have contributed to changes in $\alpha_{\text{lipid-water}}$ over time, as hydrogen isotope fractionation has been observed to vary among eukaryotic algal species grown in culture (Schouten et al., 2006; Zhang and Sachs, 2007; Heinzelmann et al., 2015a)

While average fractionation factors for most lipids were consistent between the two lakes, average $\alpha_{\text{lipid-water}}$ values for brassicasterol were 0.045 ± 0.008 lower in Lake Greifen relative to Lake Lucerne, suggesting that sterol hydrogen isotopes may be more sensitive to nutrient availability than those of fatty acids and phytol.

Author contributions. SNL designed the study with input from ND and CJS. SNL and ND collected the samples. SNL processed and measured the samples. SNL, ND, and CJS contributed to data interpretation. SNL prepared the manuscript with contributions from ND and CJS.

Competing interests. The authors declare that they have no conflict of interest.

Acknowledgements. This research was funded by a National Science Foundation Earth Sciences Postdoctoral Fellowship (award no. 1452254) to NL and Eawag internal funds. Alois Zwyssig and Alfred Lück assisted with sample collection. Serge Robert and Julian Stauffer assisted with sample preparation and laboratory analyses. Daniel Montluçon at ETH Zürich measured the water isotopes. Algal counts were conducted by Esther Keller as part of Eawag's Department of Aquatic Ecology's long-term monitoring program. We had productive conversations with Ashley Maloney, Daniel Nelson, Julian Sachs, Blake Matthews, and Romana Limberger that improved the study design and interpretation of results. Magdalena Osburn and Rienk Smittenberg provided helpful reviews that significantly strengthened the manuscript. We are grateful for all of their contributions.

Edited by: Marcel van der Meer

References

Atwood, A. R. and Sachs, J. P.: Separating ITCZ-and ENSO-related rainfall changes in the Galápagos over the last 3 kyr using D/H ratios of multiple lipid biomarkers, Earth Planet. Sc. Lett., 404, 408–419, https://doi.org/10.1016/j.epsl.2014.07.038, 2014.

Bührer, H. and Ambühl, H.: Lake Lucerne, Switzerland, a long term study of 1961–1992, Aquat. Sci., 63, 432–456, https://doi.org/10.1007/s00027-001-8043-8, 2001.

Bürgi, H. R., Heller, C., Gaebel, S., Mookerji, N., and Ward, J. V.: Strength of coupling between phyto-and zooplankton in Lake Lucerne (Switzerland) during phosphorus abatement subsequent to a weak eutrophication, J. Plankton Res., 21, 485–507, https://doi.org/10.1093/plankt/21.3.485, 1999.

Chivall, D., M'Boule, D., Sinke-Schoen, D., Damsté J. S. S., Schouten, S., and van der Meer, M. T. J.: The effects of growth phase and salinity on the hydrogen isotopic composition of alkenones produced by coastal haptophyte algae, Geochim. Cosmochim. Ac., 140, 381–390, https://doi.org/10.1016/j.gca.2014.05.043, 2014.

Craig, H. and Gordon, L.: Deuterium and oxygen 18 variations in the ocean and the marine atmosphere, in: Proceedings of a Conference on Stable Isotopes in Oceanographic Studies and Paleotemperatures, edited by: Tongiori, E., CNR-Laboratorio di Geologia Nucleare, Pisa, Italy, 9–130, 1965.

Englebrecht, A. C. and Sachs, J. P.: Determination of sediment provenance at drift sites using hydrogen isotopes and unsaturation ratios in alkenones, Geochim. Cosmochim. Ac., 69, 4253–4265, https://doi.org/10.1016/j.gca.2005.04.011, 2005.

Gat, J. R.: Oxygen and hydrogen isotopes in the hydrologic cycle, Annu. Rev. Earth Pl. Sc., 24, 225–262, https://doi.org/10.1146/annurev.earth.24.1.225, 1996.

Heinzelmann, S. M., Villanueva, L., Sinke-Schoen, D., Damsté, J. S. S., Schouten, S., and van der Meer, M. T. J.: Impact of metabolism and growth phase on the hydrogen isotopic composition of microbial fatty acids, Front. Microbiol., 6, 408, https://doi.org/10.3389/fmicb.2015.00408, 2015a.

Heinzelmann, S. M., Chivall, D., M'Boule, D., Sinke-Schoen, D., Villanueva, L., Damsté, J. S. S., Schouten, S., and Van der Meer, M. T. J.: Comparison of the effect of salinity on the D/H ratio of fatty acids of heterotrophic and photoautotrophic microorganisms, FEMS Microbiol. Lett., 362, , fnv065, https://doi.org/10.1093/femsle/fnv065, 2015b.

Heinzelmann, S. M., Bale, N. J., Villanueva, L., Sinke-Schoen, D., Philippart, C. J. M., Damsté, J. S. S., Smede, J., Schouten, S., and van der Meer, M. T. J.: Seasonal changes in the D/H ratio of fatty acids of pelagic microorganisms in the coastal North Sea, Biogeosciences, 13, 5527–5539, https://doi.org/10.5194/bg-13-5527-2016, 2016.

Henderson, A. K. and Shuman, B. N.: Hydrogen and oxygen isotopic compositions of lake water in the western United States, Geol. Soc. Am. Bull., 121, 1179–1189, https://doi.org/10.1130/B26441.1, 2009.

Hollander, D. J., McKenzie, J. A., and Ten Haven, H. L.: A 200 year sedimentary record of progressive eutrophication in Lake Greifen (Switzerland): implications for the origin of organic-carbon-rich sediments, Geology, 20, 825–828, https://doi.org/10.1130/0091-7613(1992)020<0825:AYSROP>2.3.CO;2, 1992.

Huang, Y., Shuman, B., Wang, Y., and Webb III, T.: Hydrogen isotope ratios of palmitic acid in lacustrine sediments record late Quaternary climate variations, Geology, 30, 1103–1106, https://doi.org/10.1130/0091-7613(2002)030<1103:HIROPA>2.0.CO;2, 2002.

Huang, Y., Shuman, B., Wang, Y., and Webb, T.: Hydrogen isotope ratios of individual lipids in lake sediments as novel tracers of climatic and environmental change: a surface sediment test, J. Paleolimnol., 31, 363–375, https://doi.org/10.1023/B:JOPL.0000021855.80535.13, 2004.

Jones, A. A., Sessions, A. L., Campbell, B. J., Li, C., and Valentine, D. L.: D/H ratios of fatty acids from marine particulate organic matter in the California Borderland Basins, Org. Geochem., 39, 485–500, https://doi.org/10.1016/j.orggeochem.2007.11.001, 2008.

Kasper, S., van der Meer, M. T. J., Metz, A., Zahn, R., Damsté J. S. S., and Schouten, S.: Salinity changes in the Agulhas leakage area recorded by stable hydrogen isotopes of C37 alkenones during Termination I and II, Clim. Past., 10, 251–260, https://doi.org/10.5194/cp-10-251-2014, 2014.

Keller, B., Wolinska, J., Manca, M., and Spaak, P.: Spatial, environmental and anthropogenic effects on the taxon composition of hybridizing Daphnia, P. R. Soc. B., 363, 2943–2952, https://doi.org/10.1098/rstb.2008.0044, 2008.

Ladd, S. N., Dubois, N., and Schubert, C. J.: Data set for paper "Interplay of community dynamics, temperature, and productivity on the hydrogen isotope signatures of lipid biomarkers", https://doi.org/10.3929/ethz-b-000176730, 2017.

Leduc, G., Sachs, J. P., Kawka, O. E., and Schneider, R. R.: Holocene changes in eastern equatorial Atlantic salinity as estimated by water isotopologues, Earth Planet. Sc. Lett., 362, 151–162, https://doi.org/10.1016/j.epsl.2012.12.003, 2013.

Li, C., Sessions, A. L., Kinnaman, F. S., and Valentine, D. L.: Hydrogen-isotopic variability in lipids from Santa Barbara Basin sediments, Geochim. Cosmochim. Ac., 73, 4803–4823, https://doi.org/10.1016/j.gca.2009.05.056, 2009.

Maloney, A. E., Shinneman, A. L., Hemeon, K., and Sachs, J. P.: Exploring lipid $^2H/^1H$ fractionation mechanisms in response to salinity with continuous cultures of the diatom *Thalassiosira pseudonana*, Org. Geochem., 101, 154–165, https://doi.org/10.1016/j.orggeochem.2016.08.015, 2016.

M'boule, D., Chivall, D., Sinke-Schoen, D., Sinninghe Damsté, J. S. S., Schouten, S., and van der Meer, M. T. J.: Salinity dependent hydrogen isotope fractionation in alkenones produced by coastal and open ocean haptophyte algae, Geochim. Cosmochim. Ac., 130, 126–135, https://doi.org/10.1016/j.gca.2014.01.029, 2014.

McKenzie, J. A.: Carbon-13 cycle in Lake Greifen: a model for restricted ocean basins, in: Nature and Origin of Cretaceous Carbon-Rich Facies, edited by: Schlanger, S. O., and Cita, M., Academic Press, London, UK, 197–208, 1982.

Nelson, D. B. and Sachs, J. P.: The influence of salinity on D/H fractionation in dinosterol and brassicasterol from globally distributed saline and hypersaline lakes, Geochim. Cosmochim. Ac., 133, 325–339, https://doi.org/10.1016/j.gca.2014.03.007, 2014.

Nelson, D. B. and Sachs, J. P.: Galápagos hydroclimate of the Common Era from paired microalgal and mangrove biomarker $^2H/^1H$ values, P. Natl. Acad. Sci. USA, 113, 3476–3481, https://doi.org/10.1073/pnas.1516271113, 2016.

Osburn, M. R., Sessions, A. L., Pepe-Ranney, C., and Spear, J. R.: Hydrogen-isotopic variability in fatty acids from Yellowstone National Park hot spring microbial communities, Geochim. Cosmochim. Ac., 75, 4830–4845, https://doi.org/10.1016/j.gca.2011.05.038, 2011.

Osburn, M. R., Dawson, K. S., Fogel, M. L., and Sessions, A. L.: Fractionation of hydrogen isotopes by sulphate- and nitrate-reducing bacteria, Front. Microbiol., 7, 1166 m https://doi.org/10.3389/fmicb.2016.01166, 2016.

Pahnke, K., Sachs, J. P., Keigwin, L., Timmermann, A., and Xie, S.: Eastern tropical Pacific hydrologic changes during the past 27 000 years from D/H ratios in alkenones, Paleoceanography, 22, PA4214, https://doi.org/10.1029/2007PA001468, 2007.

Pinker, R. T. and Laszlo, I.: Modeling surface solar irradiance for satellite applications on a global scale, J. Appl. Meteorol., 31, 194–211, https://doi.org/10.1175/1520-0450(1992)031<0194:MSSIFS>2.0.CO;2, 1992.

Popp, B. N., Prahl, F. G., Wallsgrove, R. J., and Tanimoto, J. K.: Seasonal patterns of alkenone production in the subtropical oligotrophic North Pacific, Paleoceanography, 21, PA1004, https://doi.org/10.1029/2005PA001165, 2006.

Rampen, S. W., Abbas, B. A., Schouten, S., and Damsté J. S. S.: A comprehensive study of sterols in marine diatoms (*Bacillariophyta*): implications for their use as tracers for diatom productivity, Limnol. Oceanogr., 55, 91–105, https://doi.org/10.4319/lo.2010.55.1.0091, 2010.

Randlett, M. E., Bechtel, A., van der Meer, M. T., Peterse, F., Litt, T., Pickarski, N., Kwiecien, O., Stockhecke, M., Wehrli, B., and Schubert, C. J.: Biomarkers in Lake Van sediments reveal dry conditions in eastern Anatolia during 110 000–10 000 years BP, Geochem. Geophy. Geosy., 18, 571–583, https://doi.org/10.1002/2016GC006621, 2017.

Richey, J. N. and Sachs, J. P.: Precipitation changes in the western tropical Pacific over the past millennium, Geology, 44, 671–674, https://doi.org/10.1130/G37822.1, 2016.

Roessler, P. G.: Environmental control of glycerolipid metabolism in microalgae: commercial implications and future research directions, J. Phycol., 26, 393–399, https://doi.org/10.1111/j.0022-3646.1990.00393.x, 1990.

Sachs, J. P.: Hydrogen isotope signatures in the lipids of phytoplankton, in: Treatise on Geochemistry, 2nd Edition, edited by: Holland, H. D., and Turekian, K. K., Elsevier, Oxford, UK, 79–94, 2014.

Sachs, J. P. and Kawka, O. E.: The influence of growth rate on $^2H/^1H$ fractionation in continuous cultures of the coccolithophorid Emiliania huxleyi and the diatom *Thalassiosira pseudonana*, Plos one, 10, e0141643, https://doi.org/10.1371/journal.pone.0141643, 2015.

Sachs, J. P. and Schwab, V. F.: Hydrogen isotopes in dinosterol from the Chesapeake Bay estuary, Geochim. Cosmochim. Ac., 75, 444–459, https://doi.org/10.1016/j.gca.2010.10.013, 2011.

Sachs, J. P., Sachse, D., Smittenberg, R. H., Zhang, Z., Battisti, D. S., and Golubic, S.: Southward movement of the Pacific intertropical convergence zone AD 1400–1850, Nat. Geosci., 2, 519–525, https://doi.org/10.1038/ngeo554, 2009.

Sachs, J. P., Maloney, A. E., Gregersen, J., and Paschall, C. Effect of salinity on $^2H/^1H$ fractionation in lipids from continuous cultures of the coccolithophorid Emiliania huxleyi, Geochim. Cosmochim. Ac., 189, 96–109, https://doi.org/10.1016/j.gca.2016.05/041, 2016.

Sachs, J. P., Maloney, A. E., and Gregersen, J.: Effect of light on $^2H/^1H$ fractionation in lipids from continuous cultures of the diatom *Thalassiosira pseudonana*, Geochim. Cosmochim. Ac., 209, 204–215, https://doi.org/10.1016/j.gca.2017.04.008, 2017.

Sachse, D. and Sachs, J. P.: Inverse relationship between D/H fractionation in cyanobacterial lipids and salinity in Christmas Island saline ponds, Geochim. Cosmochim. Ac., 72, 793–806, https://doi.org/10.1016/j.gca.2007.11.022, 2008.

Sachse, D., Billault, I., Bowen, G. J., Chikaraishi, Y., Dawson, T. E., Feakins, S. J., Freeman, K. H., Magill, C. R., McInerney, F. A., van der Meer, M. T. J., Polissar, P., Robins, R. J., Sachs, J. P., Schmidt, H. L., Sessions, A. L., White, J. W. C., West, J. B., and Kahmen, A.: Molecular paleohydrology: interpreting the hydrogen-isotopic composition of lipid biomarkers from photosynthesizing organisms, Annu. Rev. Earth Pl. Sc., 40, 212–249, https://doi.org/10.1146/annurev-earth-042711-105535, 2012.

Sauer, P. E., Eglinton, T. I., Hayes, J. M., Schimmelmann, A., and Sessions, A. L.: Compound-specific D/H ratios of lipid biomarkers from sediments as a proxy for environmental and climatic conditions, Geochim. Cosmochim. Ac., 65, 213–222, https://doi.org/10.1016/S0016-7037(00)00520-2, 2001.

Schimmelmann, A., Sessions, A. L., and Mastalerz, M.: Hydrogen isotopic (D/H) composition of organic matter during diagenesis and thermal maturation, Annu. Rev. Earth Pl. Sc., 34, 501–533, https://doi.org/10.1146/annurev.earth.34.031405.125011, 2006.

Schmidt, H. L., Werner, R. A., and Eisenreich, W.: Systematics of ^{2}H patterns in natural compounds and its importance for the elucidation of biosynthetic pathways, Phytochem. Rev., 2, 61–85, https://doi.org/10.1023/B:PHYT.0000004185.92648.ae, 2003.

Schouten, S., Ossebaar, J., Schreiber, K., Kienhuis, M. V. M., Langer, G., Benthien, A., and Bijma, J.: The effect of temperature, salinity and growth rate on the stable hydrogen isotopic composition of long chain alkenones produced by *Emiliania huxleyi* and *Gephyrocapsa oceanica*, Biogeosciences, 3, 113–119, https://doi.org/10.5194/bg-3-113-2006, 2006.

Sessions, A. L., Burgoyne, T. W., Schimmelmann, A., and Hayes J. M.: Fractionation of hydrogen isotopes in lipid biosynthesis, Org. Geochem., 30, 1193–1200, https://doi.org/10.1016/S0146-6380(99)00094-7, 1999.

Sessions, A. L., Sylva, S. P., Summons, R. E., and Hayes, J. M.: Isotopic exchange of carbon-bound hydrogen over geologic timescales, Geochim. Cosmochim. Ac., 68, 1545–1559, https://doi.org/10.1016/j.gca.2003.06.004, 2004.

Smittenberg, R. H., Saenger, C., Dawson, M. N., and Sachs, J. P.: Compound-specific D/H ratios of the marine lakes of Palau as proxies for West Pacific Warm Pool hydrologic variability, Quaternary Sci. Rev., 30, 921–933, https://doi.org/10.1016/j.quascirev.2011.01.012, 2011.

Steinman, B. A., Abbott, M. B., Nelson, D. B., Stansell, N. D., Finney, B. P., Bain, D. J., and Rosenmeier, M. F.: Isotopic and hydrologic responses of small, closed lakes to climate variability: comparison of measured and modeled lake level and sediment core oxygen isotope records, Geochim. Cosmochim. Ac., 105, 455–471, https://doi.org/10.1016/j.gca.2012.11.027, 2013.

Taipale, S. J., Hiltunen, M., Vuorio, K., and Petomaa, E.: Suitability of phytosterols alongside fatty acids as chemotaxonomic biomarkers for phytoplankton, Front. Plant Sci., 7, 212, https://doi.org/10.3389/fpls.2016.00212, 2016.

Thevenon, F., Adatte, T., Poté, J., and Spangenberg, J. E.: Recent human-induced trophic change in the large and deep perialpine Lake Lucerne (Switzerland) compared to historical geochemical variations, Palaeogeogr. Palaeocl., 363, 37–47, https://doi.org/10.1016/j.palaeo.2012.08.010, 2012.

van der Meer, M., Baas, M., Rijpstra, W., Marino, G., Rohling, E., Damsté, J. S. S., and Schouten, S.: Hydrogen isotopic compositions of long-chain alkenones record freshwater flooding of the Eastern Mediterranean at the onset of sapropel deposition, Earth Planet. Sc. Lett., 262, 594–600, https://doi.org/10.1016/j.epsl.2007.08.014, 2007.

van der Meer, M., Sangiorgi, F., Baas, M., Brinkhuis, H., Damsté J. S. S., and Schouten, S.: Molecular isotopic and dinoflagellate evidence for Late Holocene freshening of the Black Sea, Earth Planet. Sc. Lett., 267, 426–434, https://doi.org/10.1016/j.epsl.2007.12.001, 2008.

van der Meer, M. T., Benthien, A., French, K. L., Epping, E., Zondervan, I., Reichart, G. J., Bijma, J., Damsté J. S. S., and Schouten, S.: Large effect of irradiance on hydrogen isotope fractionation of alkenones in *Emiliania huxleyi*, Geochim. Cosmochim. Ac., 160, 16–24, https://doi.org/10.1016/j.gca.2015.03.024, 2015.

Vasiliev, I., Reichart, G. J., and Krijgsman, W.: Impact of the Messinian Salinity Crisis on Black Sea hydrology – insights from hydrogen isotopes analysis on biomarkers, Earth Planet. Sc. Lett., 362, 272–282, https://doi.org/10.1016/j.epsl.2012.11.038, 2013.

Vasiliev, I., Mezger, E. M., Lugli, S., Reichart, G. J., Manzi, V., and Roveri, M.: How dry was the Mediterranean during the Messinian salinity crisis?, Palaeogeogr. Palaeocl., 471, 120–133, https://doi.org/10.1016/j.palaeo.2017.01.032, 2017.

Volkman, J. K.: Sterols in microorganisms, Appl. Microbiol. Biot., 60, 495–506, https://doi.org/10.1007/s00253-002-1172-8, 2003.

Volkman, J. K., Barrett, S. M., Blackburn, S. I., Mansour, M. P., Sikes, E. L., and Gelin, F.: Microalgal biomarkers: a review of recent research developments, Org. Geochem., 29, 1163–1179, https://doi.org/10.1016/S0146-6380(98)00062-X, 1998.

Williams, P. J. L. B. and Laurens, L. M.: Microalgae as biodiesel and biomass feedstocks: review and analysis of the biochemistry, energetics and economics, Energy and Environmental Science, 3, 554–590, https://doi.org/10.1039/B924978H, 2010.

Wolhowe, M. D., Prahl, F. G., Probert, I., and Maldonado, M.: Growth phase dependent hydrogen isotopic fractionation in alkenone-producing haptophytes, Biogeosciences, 8, 1681–1694, https://doi.org/10.5194/bg-6-1681-2009, 2009.

Wolhowe, M. D., Prahl, F. G., Langer, G., Oviedo, A. M., and Ziveri, P.: Alkenone δD as an ecological indicator: a culture and field study of physiologically-controlled chemical and hydrogen-isotopic variation in C37 alkenones, Geochim. Cosmochim. Ac., 162, 166–182, https://doi.org/10.1016/j.gca.2015.04.034, 2015.

Zarrouk, W., Carrasco-Pancorbo, A., Zarrouk, M., Segura-Carretero, A., and Fernandez-Gutierrez, A.: Multi-component analysis (sterols, tocopherols and triterpenic dialcohols) of the unsaponifiable fraction of vegetable oils by liquid chromatography–atmospheric pressure chemical ionization–ion trap mass spectrometry, Talanta, 80, 924–934, https://doi.org/10.1016/j.talanta.2009.08.022, 2009.

Zhang, X., Gillespie, A., and Sessions, A.: Large D/H variations in bacterial lipids reflect central metabolic pathways, P. Natl. Acad. Sci. USA, 106, 12580–12586, https://doi.org/10.1073/pnas.0903030106, 2009a.

Zhang, Z. and Sachs, J. P.: Hydrogen isotope fractionation in freshwater algae: 1. Variations among lipids and species, Org. Geochem., 38, 582–608, https://doi.org/10.1016/j.orggeochem.2006.12.004, 2007.

Zhang, Z., Sachs, J. P., and Marchetti, A.: Hydrogen isotope fractionation in freshwater and marine algae: II. Temperature and nitrogen limited growth rate effects, Org. Geochem., 40, 428–439, https://doi.org/10.1016/j.orggeochem.2008.11.002, 2009b.

Zhang, Z., Leduc, G., and Sachs, J. P.: El Niño evolution during the Holocene revealed by a biomarker rain gauge in the Galápagos Island, Earth Planet. Sc. Lett., 404, 420–434, https://doi.org/10.1016/j.epsl.2014.07.013, 2014.

6

Morphological plasticity of root growth under mild water stress increases water use efficiency without reducing yield in maize

Qian Cai[1,2], Yulong Zhang[1], Zhanxiang Sun[2], Jiaming Zheng[2], Wei Bai[2], Yue Zhang[3], Yang Liu[2], Liangshan Feng[2], Chen Feng[2], Zhe Zhang[2], Ning Yang[2], Jochem B. Evers[4], and Lizhen Zhang[3]

[1]College of Land and Environment, Shenyang Agricultural University, Shenyang, 110161, Liaoning, China
[2]Tillage and Cultivation Research Institute, Liaoning Academy of Agricultural Sciences, Shenyang, 110161, Liaoning, China
[3]College of Resources and Environmental Sciences, China Agricultural University, Beijing, 100193, China
[4]Wageningen University, Centre for Crop Systems Analysis (CSA), Droevendaalsesteeg 1, 6708 PB Wageningen, the Netherlands

Correspondence to: Yulong Zhang (ylzsau@163.com) and Zhanxiang Sun (sunzhanxiang@sohu.com)

Abstract. A large yield gap exists in rain-fed maize (*Zea mays* L.) production in semi-arid regions, mainly caused by frequent droughts halfway through the crop-growing period due to uneven distribution of rainfall. It is questionable whether irrigation systems are economically required in such a region since the total amount of rainfall does generally meet crop requirements. This study aimed to quantitatively determine the effects of water stress from jointing to grain filling on root and shoot growth and the consequences for maize grain yield, above- and below-ground dry matter, water uptake (WU) and water use efficiency (WUE). Pot experiments were conducted in 2014 and 2015 with a mobile rain shelter to achieve conditions of no, mild or severe water stress. Maize yield was not affected by mild water stress over 2 years, while severe stress reduced yield by 56 %. Both water stress levels decreased root biomass slightly but shoot biomass substantially. Mild water stress decreased root length but increased root diameter, resulting in no effect on root surface area. Due to the morphological plasticity in root growth and the increase in root / shoot ratio, WU under water stress was decreased, and overall WUE for both above-ground dry matter and grain yield increased. Our results demonstrate that an irrigation system might be not economically and ecologically necessary because the frequently occurring mild water stress did not reduce crop yield much. The study helps us to understand crop responses to water stress during a critical water-sensitive period (middle of the crop-growing season) and to mitigate drought risk in dry-land agriculture.

1 Introduction

Maize (*Zea mays* L.) is the most important crop globally, and also a major food crop in north-eastern China with an average yield around 5.3 $\mathrm{t\,ha^{-1}}$ (Dong et al., 2017). However, the yield gap to the potential of 10.9 $\mathrm{t\,ha^{-1}}$ is still large (Liu et al., 2012), mainly due to frequent summer droughts. Due to the increasing probability of extreme climate events (IPCC, 2007), water stress for agricultural production is likely to increase in this region (Song et al., 2014; Yu et al., 2014) which is detrimental for crop photosynthesis and yield (Richards, 2000).

Although the averaged total rainfall during the crop-growing season can meet the requirements of rain-fed maize in the semi-arid north-east of China, the yearly and seasonal variations often cause droughts (mostly mild water stress) during summer, resulting in yield loss. Since quantitative information on the effects of water stress on maize performance is lacking, it can be questioned whether irrigation systems using underground water are economically and ecologically required in this situation.

Yield reduction by water stress has been attributed to decreased crop growth (Payero et al., 2006), canopy height (Traore et al., 2000), leaf area index (NeSmith and Ritchie, 1992) and root growth (Gavloski et al., 1992). Crop shoot development and biomass accumulation are greatly reduced by soil water deficit at seeding stage (Kang et al., 2000). Short-duration water deficit during the rapid vegetative growth period causes around 30 % loss in final dry matter (Cakir,

2004). The reduction in maize yield by water stress can be observed in all yield components such as ear density, number of kernels per ear and kernel weight (Ge et al., 2012), especially for stress during or before the maize silk and pollination period (Claassen and Shaw, 1970). Biomass and harvest index (the ratio of grain yield over total above-ground dry matter) are decreased under water stress during anthesis (Traore et al., 2000).

Water use efficiency (WUE, expressed in kg yield obtained per m^3 of water) is notably reduced by severe water stress. However, a moderate water stress at V16 (with 16 fully expanded leaves) and R1 (silking) stages in maize increased WUE (Ge et al., 2012). Intentional irrigation deficits before the maize tasselling stage are often used for improving WUE in regions with serious water scarcity, e.g. the North China Plain (Qiu et al., 2008; Zhang et al., 2017). Under water stress, plant photosynthesis and transpiration decrease due to a decrease in stomatal conductance (Killi et al., 2017) induced by increasing concentration of abscisic acid (ABA) (Beis and Patakas, 2015). However, limited knowledge exists on how much the growth and biomass partitioning between shoot and root in maize is affected by water stress during the middle and late growing stages, and whether changes in root growth and morphology caused by water stress could affect maize yielding and water use efficiency.

Since field experiments that aim to quantify the effects of water stress are difficult to carry out in rain-fed agriculture, a mobile rain shelter is often used in studies to control water stress in the field (NeSmith and Ritchie, 1992). The objective of this study was to quantify maize shoot and root growth, grain yield and WUE under different water stress levels during the middle of the crop-growing season with a well-controlled mobile rain shelter to understand the crop response to water stress.

2 Materials and methods

2.1 Experimental design

The experiments were conducted at Shenyang (41°48′ N, 123°23′ E), Liaoning province, north-eastern China in 2014 and 2015. The experimental site is 45 m above sea level. On average from 1965 to 2015, annual potential evaporation is 1445 mm, with a total precipitation 720 mm, and mean air temperature 8 °C. The frost-free period is 150–170 days. Average relative humidity is 63 %. Annual mean wind speed is 3.1 m s^{-1}. The climate is a typical continental monsoon climate with four distinct seasons, characterized by a hot summer and cold winter. The annual mean air temperature was 9.5 °C in 2014 and 9.1 °C in 2015. The mean air temperature during the crop-growing season (May to September) was 20.2 °C in 2014 and 19.4 °C in 2015 (Fig. 1).

Maize plants were grown in pots in three treatments: (1) no water stress, (2) mild water stress and (3) severe water stress

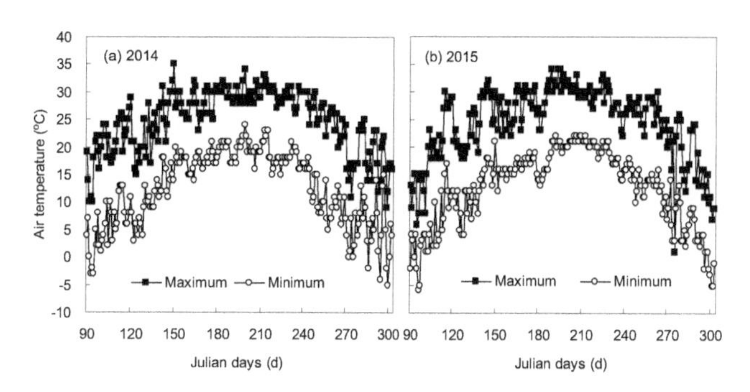

Figure 1. Daily maximum and minimum air temperatures in 2014 and 2015 in Shengyang, Liaoning, China.

(Table 1). The levels of water stress were based on historical rainfall frequency analysis. The water supply was controlled by a mobile rain shelter with a steel frame and transparent PVC cover. The mobile rain shelter is built on a mechanical movement track equipped with an electricity motor to move the shelter with a remote control. The shelter was moved away from the experimental plots on no rain days and covered before the rain came; therefore the effect of shelter on incoming radiation could be ignored. The mobile rain shelter is 9 m in width, 30 m in length and 4.5 m in height. The top and both sides of the shelter have transparent PVC boards to prevent outside rainfall from entering. There is a water gutter outside the movement track to drain the rainwater. Therefore the rainwater intrusion can be avoided. Water treatments began from maize jointing (V6, with 6 fully expended leaves) to filling stages (R3, milk) (Abendroth et al., 2011). Water treatments were conducted by supplying irrigation once every 5 days before starting water treatments with the same amount for all pots, and once every 3 days during the period of water treatment. The amount of water supplied to each treatment is listed in Table 1. The experiments entailed a completely randomized block design with three replicates. Each treatment consisted of 12 pots (one plant per pot) and was divided into 3 replicates (4 pots each). At each sampling (4 samplings in total at an interval of approximately 30 days), one pot was used.

Each pot was 40 cm in diameter and 50 cm in height, filled with 40 kg naturally dried soil with a bulk density of 1.31 g cm^{-3}. The large size of pots in the experiments effectively avoided the space effect for growing good maize. The soil was sandy loam with a pH of 6.15, total N of 1.46 g kg^{-1}, total of P 0.46 g kg^{-1} and total K of 12.96 g kg^{-1}. 46.5 g compound fertilizer (N 15 %, P_2O_5 15 % and K_2O 15 %) and 15.5 g diammonium phosphate (N 18 % and P_2O_5 46 %) were applied to each pot before sowing. No other fertilizer was applied during the maize-growing season. Maize cultivar used in both years was Liaodan 565, a local commonly used drought-resistant cultivar. One plant was grown in each pot. Maize was sown on 13 May and harvested on 30 September in both 2014 and 2015.

Table 1. Water treatments during crop-growing seasons from 2014 to 2015.

Year	Water treatment	Initial volumetric soil moisture content (%)	Actual water supply at three growing periods (mm)			
			Early (16–29 DAS*)	Middle (30–102 DAS)	Late (103–121 DAS)	Total
2014	No stress	24.4	11.9	478	56	545
	Mild stress	24.8	11.9	299	56	366
	Severe stress	24.9	11.9	122	56	190
2015	No stress	25.3	11.9	510	32	553
	Mild stress	25.3	11.9	334	32	378
	Severe stress	24.4	11.9	159	32	203

* DAS refers to days after maize sowing.

2.2 Dry matter and grain yield measurements

To determine maize dry matter, four plants were harvested on 49 (V6, jointing), 77 (VT, tasselling), 113 (R3, milk) and 141 (R5, dent) days after sowing (DAS) in 2014, and one sampling was done on 132 DAS in 2015. The samples were separated into roots and shoots and oven-dried at 80 °C for 48 h until they reached a constant weight. The shoot / root ratio was calculated using measured organ-specific dry matter.

Grain yield was measured by harvesting all cobs in a pot at maize-harvesting time. The grain was sun-dried to a water content of 15 %. Yield components, i.e. ear (cob) numbers per plant, kernel numbers per ear and thousand kernel weight were measured for each plot.

2.3 Root measurements

Root growth and morphological traits (root length, diameter and surface area) were measured four times during the crop-growing season on 49, 77, 113, 141 DAS in 2014. All of the roots were collected for each pot at the time of dry matter measurements. Root samples were carefully washed with tap water to remove soil. The cleaned roots were placed on the glass plate of a root system scanner. Scanned root images were analysed by a plant root image analyser WinRHIZO PRO 2009 (Regent Instruments Inc., Canada) to quantify total root length (m), diameter (mm) and surface area (m^2) per plant (pot).

2.4 Measuring soil moisture content, water uptake and water use efficiency

Soil moisture contents were measured by a soil auger at sowing and harvesting times for each plot (three replicates per treatment). Soil cores were taken from the middle pot for each 10 cm soil layer. After measuring fresh soil weight, soil samples were oven-dried at 105 °C for approximately 48 h until a constant weight was reached. The gravimetric soil moisture contents (%, $g\,g^{-1}$) measured by soil auger were calculated into volumetric soil moisture content (%, $m^3\,m^{-3}$) by multiplying them with soil bulk density.

Water uptake (WU) of maize was calculated using a simplified soil water balance equation (Kang et al., 2002). Because the experiments were sheltered, rainfall, drainage and capillary rise of water did not occur in this situation and therefore were not taken into account in the calculation:

$$\mathrm{WU} = I + \Delta S, \quad (1)$$

where WU (mm) is crop water uptake (mm) during the whole of the crop-growing season, I is the amount of water supplied to each pot (mm). ΔS is the change of total soil water between sowing and harvesting dates.

Water use efficiency (WUE) was calculated by measuring final yield or above-ground dry matter and total WU during the crop-growing season (Zhang et al., 2007).

$$\mathrm{WUE} = Y/\mathrm{WU}, \quad (2)$$

where WUE ($g\,m^{-2}\,mm^{-1}$ or $kg\,m^{-3}$) is water use efficiency expressed in gain yield WUE_Y or dry matter $\mathrm{WUE}_{\mathrm{DM}}$. Y ($g\,m^{-2}$) is grain yield or dry matter.

2.5 Statistical analysis

Analysis of variance on yield, WU, WUE and dry matter for shoot and root were performed using a general linear model of SPSS 20 (SPSS Inc., Chicago, USA). The differences between means were evaluated through least significant difference multiple comparison tests at a significant level of 0.05.

3 Results

3.1 Variation and frequency distribution of rainfall

The average rainfall during the maize-growing season (May to September) at an experimental site from 1965 to 2015 was 531 mm with a standard deviation of 134 mm (Fig. 2a).

Table 2. Yield and yield components affected by different water stress from 2014 to 2015.

		Ear number	Kernel number	Thousand kernel weight	Yield plant	Harvest index
Year	Water treatment	ears plant^{-1}	kernels ear^{-1}	g	g plant^{-1}	g g^{-1}
2014	No stress	2.0 ± 0.0a	354 ± 32a	440 ± 6.8a	301 ± 33a	0.36 ± 0.01a
	Mild stress	2.0 ± 0.0a	350 ± 16a	416 ± 1.2b	276 ± 14a	0.37 ± 0.01a
	Severe stress	2.0 ± 0.0a	245 ± 35b	412 ± 3.7b	166 ± 25b	0.27 ± 0.02b
2015	No stress	2.0 ± 0.0a	341 ± 67a	426 ± 12a	240 ± 60a	0.29 ± 0.04a
	Mild stress	2.0 ± 0.0a	244 ± 53a	427 ± 22a	168 ± 42ab	0.25 ± 0.03a
	Severe stress	1.3 ± 0.3b	172 ± 46a	412 ± 16a	81 ± 22b	0.17 ± 0.04a
mean	No stress	2.0 ± 0.2a	347 ± 38a	432 ± 7.5a	266 ± 36a	0.32 ± 0.03a
	Mild stress	2.0 ± 0.0a	289 ± 36ab	422 ± 12a	214 ± 32a	0.30 ± 0.03ab
	Severe stress	1.6 ± 0.0b	203 ± 31b	412 ± 8.5a	118 ± 23b	0.21 ± 0.03b
P	Treatment	0.021	0.003	0.556	0.005	0.013
	Year	0.184	0.514	0.889	0.237	0.039
	Treatment × year	0.111	0.664	0.555	0.835	0.758

The same lower-case letters indicate no significant difference between water treatments within the same year at $a = 0.05$.

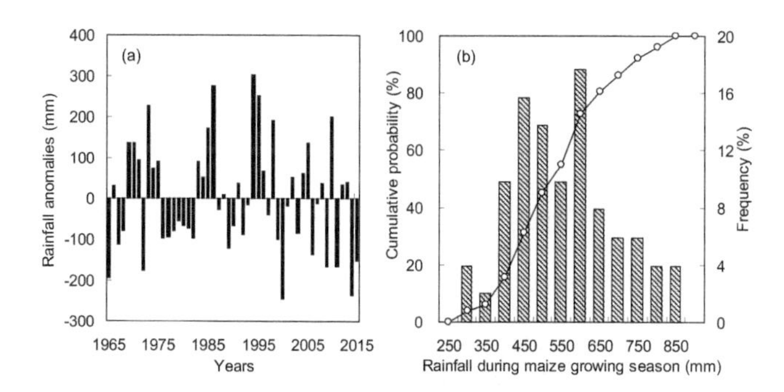

Figure 2. Anomalies and cumulative frequency of rainfall during the maize-growing season (May to September) from 1965 to 2015 at Shengyang, Liaoning.

Rainfall in the experimental years was much less than in a normal year, 296 mm in 2014 and 379 mm in 2015. The frequency of years with rainfall above 500 mm was 68.6 % over the past 51 years. For years with mild drought stress (350–450 mm), this was 27.5 % and with severe drought stress (200–300 mm) it was 3.9 % (Fig. 2b), indicating that maize growing in this region mainly suffered from mild water stress.

3.2 Yield and yield components

The maize yield under mild water stress over 2 years was not significantly different, while in severe stress the yield was 55.6 % lower than in the no water stress control (Table 2). The decrease of maize yield in severe water treatment was due to the decreases in ear and kernel numbers as well as the harvest index (HI). However, water stress did not affect kernel weight, while other yield components were decreased. The yearly effect was only significant for HI, which was likely caused by the variation in air temperature: the cooler weather in 2015 during the maize-growing season decreased the HI compared with a warmer year in 2014. There were no interactions between year and treatment.

3.3 Above- and below-ground dry matter

Mild water stress did not reduce root dry matter (Fig. 3a, b), but greatly reduced shoot dry matter, especially at grain-filling stage (113 DAS) (Fig. 3c, d). The severe water stress decreased both root and shoot dry matter compared with no stress control, but the magnitude of the decrease in shoot was much larger than in root. At maize tasselling stage (77 DAS), as taproots reached their maximum size, root dry matter under severe water stress was much lower than mild and no water stress treatments. However, it was less different later in the season, which indicated a strong complementarily growth of root system under water stress. Due to the different responses of shoot and root to water stress, the root/shoot ratios under water stress increased (Fig. 3e, f), especially during crop rapid growing period (77 to 113 DAS).

3.4 Root length, diameter and total surface area affected by water stress

Root length per plant was much lower under severe water stress than in the control, especially at the tasselling stage (77 DAS). The decrease of root length under mild water stress during the middle of the maize-growing season was much smaller than under severe stress (Fig. 4a). Root diameters under both mild and severe water stress treatments were

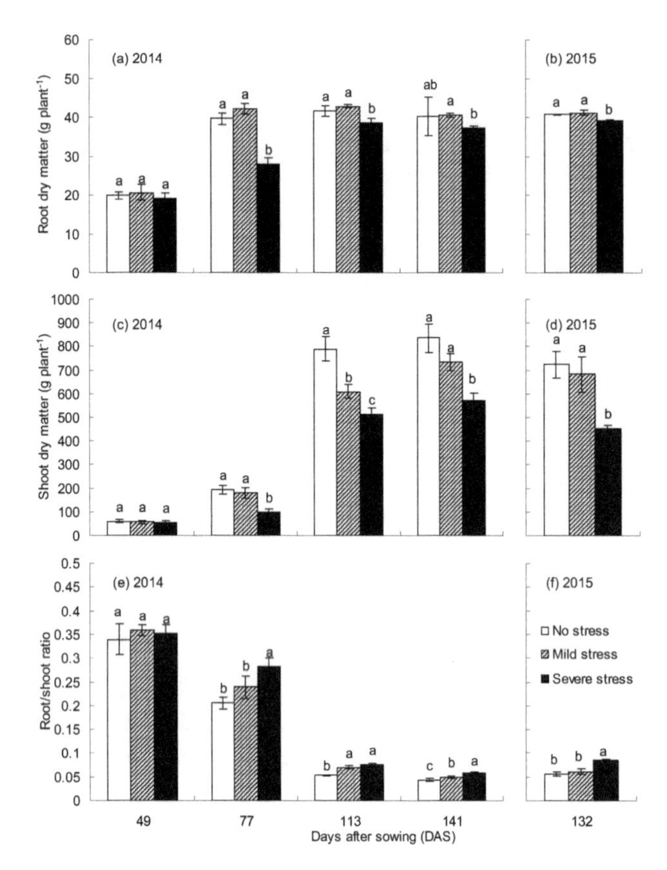

Figure 3. Root and shoot dry matter of maize under water stress at different growing stages in 2014–2015.

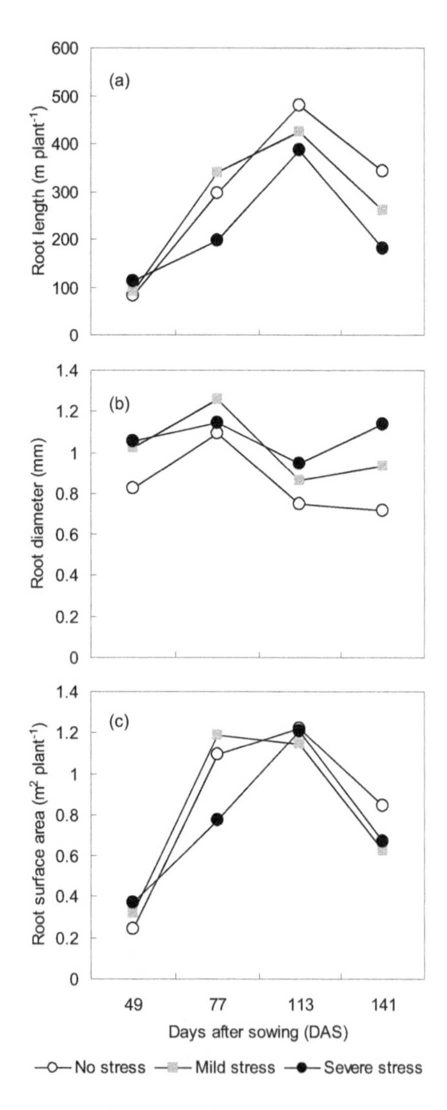

Figure 4. Total root length, average diameter and total surface area per plant affected by water stress in 2014.

much higher than under the no water stress control (Fig. 4b), especially during the late growing season. The total root surface area was less changed (Fig. 4c), especially during the reproductive growth period (113 DAS).

3.5 Water uptake and use efficiency

Total water uptake (WU) reduced by 28.9 % under mild water stress and by 54.6 % under severe stress compared with no stress control (588 mm) (Fig. 5). Water use efficiency for maize above-ground dry matter (WUE_{DM}) under both water stress treatments across all years increased by 31.2 % compared with no stress control (Fig. 5b). The WUE_{DM} in severe water stress was the highest (14.4 kg m^{-3}), which was 42.2 % higher than the control, while that in mild stress increased by 20.2 %. However, WUE for grain yield under severe water stress (3.51 kg m^{-3}) was not significantly different from that in the control (3.38 kg m^{-3}), while WUE_Y in mild water stress over 2 years increased by 17.3 % (Fig. 5c). The difference between WUEs in dry matter and grain yield was due to the extent of decreasing HI under the levels of water stress (Table 2).

4 Discussion

Mild water stress from maize jointing (V6) to filling stages (R3) did not significantly reduce maize grain yield. This is different from a previous report which claimed that maize yield is much more affected by water stress during the flowering stage than at other stages (Doorenbos et al., 1979). Our result differed from a previous study, which showed that mild water stress seriously reduced crop production (Kang et al., 2000). This is likely due to our choice of a drought-resistant variety (Zhengdan 565) and the difference in ecological zones. Genotype-dependent relationships between yield and crop growth rate would be stronger under water stress than under the no stress condition (Lake and Sadras, 2016).

Mild water stress during the middle of the crop-growing period can maintain maize yield but substantially reduces the water consumption at the same time in our study. Thus, the

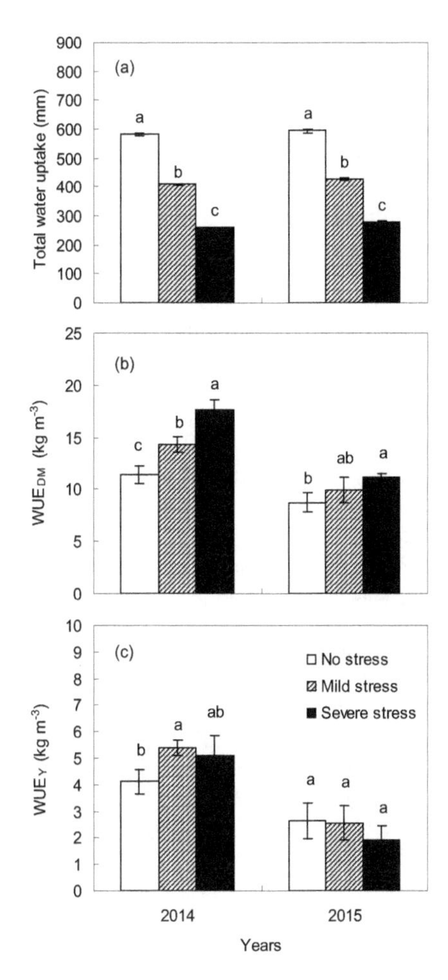

Figure 5. Total water uptake (WU) during the crop-growing season and water use efficiency for above-ground dry matter (WUE_{DM}) and grain yield (WUE_{Y}) under water stress in 2014–2015.

water use efficiency was increased (Liu et al., 2016). Mild water stress reduced total water uptake, resulting in a 20.2 % higher WUE in dry matter and 17.3 % in yield. The increase in WUE under mild water stress benefitted from the morphological responses of shoot and root growth to water stress with an increase in root / shoot ratio. The water stress reduced root length; however, this reduction was compensated by an increase in root diameter. The maintenance of crop growth under water deficit was limited by the severity of the stress. Under severe water stress, maize growth fails to be compensated by plant plasticity.

Severe water stress greatly reduced both shoot and root biomass. A large decrease in shoot growth, i.e. less biomass and leaf area, reduces the light interception and transpiration (Monteith, 1981). Under mild water stress during vegetative and tasselling stages, the shoot growth was not significantly reduced in this study but was in a previous report, e.g. in plant height and leaf area (Cakir, 2004). Mild soil water deficit may also reduce water loss of plants through physiological regulation (Davies and Zhang, 1991). Moderate soil drying at the vegetative stage encourages root growth and distribution in deep soil (Jupp and Newman, 1987; Zhang and Davies, 1989), which is consistent with our findings. A large root system with deep distribution is beneficial for water-limited agriculture (McIntyre et al., 1995). These mechanisms explained why maize yield under mild water stress did not decrease in our study.

We found an increase in root diameter under water stress. This result indicated that there were fewer lateral roots under water stress than under no water stress. This may limit water absorption since the lateral roots is younger and more active in uptake function (Lynch, 1995). Average root diameters in all treatments decreased from 77 to 113 DAS, which was caused by highly emerged lateral roots after the taproot reached its maximum (VT stage). The higher root diameter under water stress than in the no water stress control at 141 DAS was probably due to a fast senescence of late-developed lateral roots.

Our results on root morphological plasticity under mild water deficit provided more evidence for the explanation of enhancing WUE and maintaining yielding in relation to the crop–water response. However, the mechanism that determines the crop response to water stress may also involve other processes, e.g. intercellular CO_2, stomatal conductance, photosynthetic rate, oxidative stress, sugar signaling, membrane stability and root chemical signals (Xue et al., 2006; Dodd, 2009). The relationship between carbon assimilation and water stress has been widely explored to understand the physiological mechanism for improving WUE (Ennahli and Earl, 2005; Xue et al., 2006; Zhang et al., 2013). The abscisic acid (ABA)-based drought stress chemical signals regulate crop vegetative and reproductive development and contribute to crop drought adaptation (Killi et al., 2017). Increased concentration of ABA in the root induced by soil drying may maintain root growth and increase root hydraulic conductivity, thus alleviating the water deficit in the shoot (Liu et al., 2005). The increase of ABA can also induce stomatal closure and reduce crop transpiration (Haworth et al., 2016), net photosynthesis and crop growth (Killi et al., 2017).

The maize yield in 2015 was much lower than in 2014 independent of water stress. That might be caused by a higher maximum air temperature in 2015 (32.0 °C) than in 2014 (29.1 °C) during the flowering period. High air temperature reduces maize pollination (Muller and Rieu, 2016) and directly affects yield formation and HI.

5 Conclusions

This study clearly demonstrates that the maize yield under mild water stress during summer does not decrease but the water use efficiency increases due to changes in root and shoot growth. A higher root / shoot ratio under mild water stress allows plants to efficiently use limited soil water. In the studied region (Liaoning province), maize mainly grows in rain-fed conditions (2.4 million ha), covering 73 % of the total area for grain crops. To reduce the possible effect of drought on maize production, a well system that pipes ground-water to irrigate crops has recently been planned. The wells need to be 60 to 70 m deep and have an average cost of 12 000 Yuan each. Each well can only irrigate 9 to 10 ha of maize. According to our results, only severe water stress significantly reduces maize yield by 55.6 % across two experimental years (Table 2), which occurs only 3.9 % during 1965 to 2015. Mild water stress occurs much frequently (27.5 % of years); however, it does not significantly affect maize yield. Our study suggested that the well system in this region might not be economically and ecologically necessary. Other agronomy practices such as intercropping maize with crops requiring less water (e.g. peanut), cultivar selection, adjusting sowing windows (Liu et al., 2013; Lu et al., 2017) and ridge-furrow with covering plastic film (Dong et al., 2017) are likely more applicable in optimizing crop yield and regional sustainability. Our study provides more evidence to understand crop responses to water stress, especially in relation to root morphological plasticity in a drought environment. The results can be further applied by combining them with a crop model (Mao et al., 2015) to mitigate climate risk in dry-land agriculture.

Author contributions. ZS, YZ, JZ and QC conceived and designed the experiments. QC, WB, YZ, YL, LF, CF, ZZ and NY performed the experiments. LZ, QC and JBE analysed the data and wrote the paper.

Competing interests. The authors declare that they have no conflict of interest.

Special issue statement. This article is part of the special issue "Ecosystem processes and functioning across current and future dryness gradients in arid and semi-arid lands". It is not associated with a conference.

Acknowledgements. This research was supported by the National key research and development programme of China (2016YFD0300204), the International Cooperation and Exchange (31461143025) and the Youth Fund (31501269) of the National Science Foundation of China, Liaoning BaiQianWan Talent Program (201746), Outstanding Young Scholars of National High-level Talent Special Support Program of China.

Edited by: Zisheng Xing

References

Abendroth, L. J., Elmore, R. W., Boyer, M. J., and Marlay, S. K.: Corn Growth and Development. PMR 1009. Iowa State University Extension, Ames, Iowa, USA, available at: https://store.extension.iastate.edu/Product/Corn-Growth-and-Development (last access: 8 June 2017), 2011.

Beis, A. and Patakas, A.: Differential physiological and biochemical responses to drought in grapevines subjected to partial root drying and deficit irrigation, Eur. J. Agron., 62, 90–97, 2015.

Cakir, R.: Effect of water stress at different development stages on vegetative and reproductive growth of corn, Field Crop Res., 89, 1–16, 2004.

Claassen, M. M. and Shaw, R. H.: Water deficit effects on corn. II. Grain components, Agron. J., 62, 652–655, 1970.

Davies, W. J. and Zhang, J.: Root signals and the regulation of growth and development of plants in drying soil, Ann. Rev. Plant Phys., 42, 55–76, 1991.

Dodd, I. C.: Rhizosphere manipulations to maximize "crop per drop" during deficit irrigation, J. Exp. Bot., 60, 2454–2459, 2009.

Dong, W., Zhang, L., Duan, Y., Sun, L., Zhao, P., van der Werf, W., Evers, J. B., Wang, Q., Wang, R., and Sun, Z.: Ridge and furrow systems with film cover increase maize yields and mitigate climate risks of cold and drought stress in continental climates, Field Crops Res., 207, 71–78, 2017.

Doorenbos, J., Kassam, A. H., Bentvelsen, C., and Uittenbogaard, G.: Yield response to water, FAO Irrigation and Drainage Paper No. 33, FAO, Rome, Italy, 193 pp., 1979.

Ennahli, S. and Earl, H. J.: Physiological limitations to photosynthetic carbon assimilation in cotton under water stress, Crop Sci., 45, 2374–2382, 2005.

Gavloski, J. E., Whitfield, G. H., and Ellis, C. R.: Effect of restricted watering on sap flow and growth in corn (*Zea mays* L.), Can. J. Plant Sci., 72, 361–368, 1992.

Ge, T., Sui, F., Bai, L., Tong, C., and Sun, N.: Effects of water stress on growth, biomass partitioning, and water-use efficiency in summer maize (*Zea mays* L.) throughout the growth cycle, Acta Physiol. Plant., 34, 1043–1053, 2012.

Haworth, M., Cosentino, S.L., Marino, G., Brunetti, C., Scordia, D., Testa, G., Riggi, E., Avola, G., Loreto, F., and Centritto, M.: Physiological responses of *Arundo donax* ecotypes to drought: a common garden study, Glob. Change Biol. Bioenergy, 9, 132–143, https://doi.org/10.1111/gcbb.12348, 2016.

IPCC: Climate change, 2007, The physical science basis, in: Contribution of working group I to the fourth assessment report of the intergovernmental panel on climate change, edited by: Solomon, S., Qin, D., Manning, M., Chen, Z., Marquis, M., Averyt, K. B., Tignor, M., Miller, H. L., Cambridge University Press, Cambridge, 996 pp., 2007.

Jupp, A. P. and Newman, E. I.: Morphological and anatomical effects of severe drought on the roots of *Lolium perenne* L., New Phytol., 105, 393–402, 1987.

Kang, S. Z., Shi, W. J., and Zhang, J. H.: An improved water-use efficiency for maize grown under regulated deficit irrigation, Field Crops Res., 67, 207–214, 2000.

Kang, S. Z., Zhang, L., Liang, Y. L., Hu, X. T., Cai, H. J., and Gu, B. J.: Effects of limited irrigation on yield and water use efficiency of winter wheat in the Loess Plateau of China, Agr. Water Manage., 55, 203–216, 2002.

Killi, D., Bussotti, F., Raschi A., and Haworth, M.: Adaptation to high temperature mitigates the impact of water deficit during combined heat and drought stress in C3 sunflower and C4 maize varieties with contrasting drought tolerance, Physiol. Plant., 159, 130–147, 2017.

Lake, L. and Sadras, V. O.: Screening chickpea for adaptation to water stress: Associations between yield and crop growth rate, Eur. J. Agron., 81, 86–91, 2016.

Liu, E. K., Mei, X. R., Yan, C. R., Gong, D. Z., and Zhang, Y. Q.: Effects of water stress on photosynthetic characteristics, dry matter translocation and WUE in two winter wheat genotypes, Agr. Water Manage., 167, 75–85, 2016.

Liu, F., Jensen, C. R., and Andersen, M. N.: A review of drought adaptation in crop plants: changes in vegetative and reproductive physiology induced by ABA-based chemical signals, Australian J. Agr. Resour., 56, 1245–1252, 2005.

Liu, Z. J., Yang, X. G., Hubbard, K. G., and Lin, X. M.: Maize potential yields and yield gaps in the changing climate of northeast China, Glob. Change Biol., 18, 3441–3454, 2012.

Liu, Z. J., Hubbard, K. G., Lin, X. M., and Yang, X. G.: Negative effects of climate warming on maize yield are reversed by the changing of sowing date and cultivar selection in Northeast China, Glob. Change Biol., 19, 3481–3492, 2013.

Lu, H., Xue, J., and Guo, D.: Efficacy of planting date adjustment as a cultivation strategy to cope with drought stress and increase rainfed maize yield and water-use efficiency, Agr. Water Manage., 179, 227–235, 2017.

Lynch, J. P.: Root architecture and plant productivity, Plant Physiol., 109, 7–13, 1995.

Mao, L., Zhang, L., Evers, J. B., van der Werf, W., Wang, J., Sun, H., Su, Z., and Spiertz, H.: Resource use, sustainability and ecological intensification of intercropping systems, J. Integr. Agr., 14, 1442–1550, 2015.

McIntyre, B. D., Riha, S. J., and Flower, D. J.: Water uptake by pearl millet in a semiarid environment, Field Crops Res., 43, 67–76, 1995.

Monteith, J. L.: Coupling of plants to the atmosphere, in: Grace, J., Ford, E. D., and Jarvis, P. G., Plants and their Atmospheric Environment, Blackwell, Oxford, 1–29, 1981.

Muller, F. and Rieu, I.: Acclimation to high temperature during pollen development, Plant Reprod., 29, 107–118, 2016.

NeSmith, D. S. and Ritchie, J. T.: Short- and long-term responses of corn to pre-anthesis soil water deficit, Agron. J., 84, 107–113, 1992.

Payero, J. O., Melvin, S. R., Irmak, S., and Tarkalson, D.: Yield response of corn to deficit irrigation in a semiarid climate, Agr. Water Manage., 84, 101–112, 2006.

Qiu, G. Y., Wang, L. M., He, X. H., Zhang, X. Y., Chen, S. Y., Chen, J., and Yang, Y. H.: Water use efficiency and evapotranspiration of winter wheat and its response to irrigation regime in the north China plain, Agr. Forest Meteorol., 148, 1848–1859, 2008.

Richards, A.: Selectable traits to increase crop photosynthesis and yield of grain crops, J. Exp. Bot., 51, 447–458, 2000.

Song, X. Y., Li, L. J., Fu, G. B., Li, J. Y., Zhang, A. J., Liu, W. B., and Zhang, K.: Spatial-temporal variations of spring drought based on spring-composite index values for the Songnen Plain, Northeast China, Theor. Appl. Climatol., 116, 371–384, 2014.

Traore, S. B., Carlson, R. E., Pilcher, C. D., and Rice, M. E.: Bt and Non-Bt maize growth and development as affected by temperature and drought stress, Agron. J., 92, 1027–1035, 2000.

Xue, Q. W., Zhu, Z. X., Musick, J. T., Stewart, B. A., and Dusek, D. A.: Physiological mechanisms contributing to the increased water-use efficiency in winter wheat under deficit irrigation, J. Plant Physiol., 163, 154–164, 2006.

Yu, X. Y., He, X. Y., Zheng, H. F., Guo, R. C., Ren, Z. B., Zhang, D., and Lin, J. X.: Spatial and temporal analysis of drought risk during the crop-growing season over northeast China, Nat. Hazards, 71, 275–289, 2014.

Zhang, J. and Davies, W. J.: Abscisic acid produced in dehydrating roots may enable the plant to measure the water status of the soil, Plant Cell Environ., 12, 73–81, 1989.

Zhang, J., Sun, J. S., Duan, A., Wang, J. L., Shen, X. J., and Liu, X. F.: Effects of different planting patterns on water use and yield performance of winter wheat in the Huang-Huai-Hai plain of China, Agr. Water Manage., 92, 41–47, 2007.

Zhang, X., Qin, W., Chen, S., Shao, L., and Sun, H.: Responses of yield and WUE of winter wheat to water stress during the past three decades – A case study in the North China Plain, Agr. Water Mange., 179, 47–54, 2017.

Zhang, X., Wang, Y., Sun, H., Chen, S., and Shao, L.: Optimizing the yield of winter wheat by regulating water consumption during vegetative and reproductive stages under limited water supply, Irrig. Sci., 31, 1103–1112, 2013.

7

Tracing the origin of the oxygen-consuming organic matter in the hypoxic zone in a large eutrophic estuary: The lower reach of the Pearl River Estuary, China

Jianzhong Su[1], Minhan Dai[1], Biyan He[1,2], Lifang Wang[1], Jianping Gan[3], Xianghui Guo[1], Huade Zhao[1], and Fengling Yu[1]

[1]State Key Laboratory of Marine Environmental Science, Xiamen University, Xiamen, China
[2]College of Food and Biological Engineering, Jimei University, Xiamen, China
[3]Department of Mathematics and Division of Environment, Hong Kong University of Science and Technology, Kowloon, Hong Kong SAR, China

Correspondence to: Minhan Dai (mdai@xmu.edu.cn)

Abstract. We assess the relative contributions of different sources of organic matter, marine vs. terrestrial, to oxygen consumption in an emerging hypoxic zone in the lower Pearl River Estuary (PRE), a large eutrophic estuary located in Southern China. Our cruise, conducted in July 2014, consisted of two legs before and after the passing of Typhoon Rammasun, which completely de-stratified the water column. The stratification recovered rapidly, within 1 day after the typhoon. We observed algal blooms in the upper layer of the water column and hypoxia underneath in bottom water during both legs. Repeat sampling at the initial hypoxic station showed severe oxygen depletion down to 30 µmol kg^{-1} before the typhoon and a clear drawdown of dissolved oxygen after the typhoon. Based on a three endmember mixing model and the mass balance of dissolved inorganic carbon and its isotopic composition, the $\delta^{13}C$ of organic carbon remineralized in the hypoxic zone was -23.2 ± 1.1 ‰. We estimated that 65 ± 16 % of the oxygen-consuming organic matter was derived from marine sources, and the rest (35 ± 16 %) was derived from the continent. In contrast to a recently studied hypoxic zone in the East China Sea off the Changjiang Estuary where marine organic matter dominated oxygen consumption, here terrestrial organic matter significantly contributed to the formation and maintenance of hypoxia. How varying amounts of these organic matter sources drive oxygen consumption has important implications for better understanding hypoxia and its mitigation in bottom waters.

1 Introduction

The occurrence of hypoxia has been exacerbated worldwide (Nixon, 1995; Diaz and Rosenberg, 2008; Rabalais et al., 2010; Zhang et al., 2013). In recent decades, more than 400 coastal hypoxic systems have been reported with an exponential growth rate of 5.5 ± 0.23 % yr^{-1}, demonstrating their persistence and complexity with respect to both science and management (Diaz and Rosenberg, 2008; Vaquer-Sunyer and Duarte, 2008). Hypoxia may not only reduce biodiversity and endanger aquatic and benthic habitats but also alter the redox chemistry in both the water column and the underlying sediments, triggering the release of secondary pollutants (Breitburg, 2002; Rutger et al., 2002). Moreover, the management and recovery of these systems are complicated due to the hysteresis of hypoxic conditions, and the varying timescales of biological loss (within hours to weeks) and recovery from hypoxia (from months to years) (Steckbauer et al., 2011).

Coastal hypoxia usually occurs in stratified water columns where the downward mixing of oxygen from the surface is impeded (Kemp et al., 2009). Below the pycnocline, aerobic respiration is usually the predominant sink of oxygen. Organic matter, which consumes dissolved oxygen (DO) as it becomes oxidized, is thus the ultimate cause of hypoxia under favourable physical settings (Rabouille et al., 2008; Rabalais et al., 2014; Qian et al., 2016). The organic carbon (OC) that fuels respiration-driven reduction of oxygen in these systems could originate from either eutrophication-induced primary production (marine OC; OC_{mar}) or naturally and/or anthropogenically driven delivery from terrestrial environments (terrestrial OC; OC_{terr}) (Paerl, 2006; Rabalais et al., 2010).

The question of how much OC in hypoxic zones is supplied from on-site primary production versus the quantity derived from terrestrial sources has been an issue of debate (Wang et al., 2016). A proportion of the phytoplankton-centric hypoxia literature suggests that OC_{mar} dominates oxygen consumption in hypoxic zones, owing to its higher microbial availability than OC_{terr} (Zimmerman and Canuel, 2000; Boesch et al., 2009; Carstensen et al., 2014). Wang et al. (2016) quantified for the first time the relative contributions of particulate OC_{mar} (POC_{mar}) and particulate OC_{terr} (POC_{terr}) in consuming DO in the bottom waters of the East China Sea (ECS) off the Changjiang Estuary (CJE), and found that POC_{mar} dominated DO consumption. However, other studies suggest that POC_{terr} may also play an important role (Swarzenski et al., 2008; Bianchi, 2011a, b). It is thus very important to quantify the relative contributions of organic matter (OC_{mar} vs. OC_{terr}) driving the onset and maintenance of hypoxia in coastal systems, since reducing organic matter vs. nutrient inputs requires a different set of management strategies.

The Pearl River Estuary (PRE; 21.2–23.1° N, 113.0–114.5° E) is surrounded by several large cities including Hong Kong, Shenzhen and Guangzhou and has received very high loads of nutrients from the drainage basin in the last three decades. As such, eutrophication has increasingly become an issue of concern (Huang et al., 2003; Ye et al., 2012). Dissolved inorganic nitrogen (DIN) concentrations in the PRE increased approximately 4-fold from 1986 ($19.3\,\mu mol\,L^{-1}$) to 2002 ($76.1\,\mu mol\,L^{-1}$) (He and Yuan, 2007). This DIN increase has been attributed to increased inputs of domestic sewage, industrial wastewater, agricultural runoff and aquaculture in the watershed (Huang et al., 2003).

Recent observations based on monthly surveys between April 2010 and March 2011 and long-term monitoring data from 1990 to 2014 have suggested that the lower PRE has emerged as a seasonal hypoxic zone (Qian et al., 2017). This is supported by our current study, as two relatively large hypoxic zones ($> 300\,km^2$) were observed in the lower PRE with DO $< 2\,mg\,L^{-1}$. However, the origin of the organic matter driving hypoxia in the lower PRE has not previously been examined. Here, we quantified the relative proportions of OC_{mar} and OC_{terr} contributing to DO drawdown in bottom waters of the lower PRE, an economically important coastal region. This study has important biological, societal and managerial implications for the region, particularly relating to water quality in the vicinity of Hong Kong in the lower PRE. For example, the government of Hong Kong is examining the efficacy of its costly Harbour Area Treatment Scheme project and whether additional treatment should be implemented (http://www.gov.hk/en/residents/environment/water/harbourarea.htm).

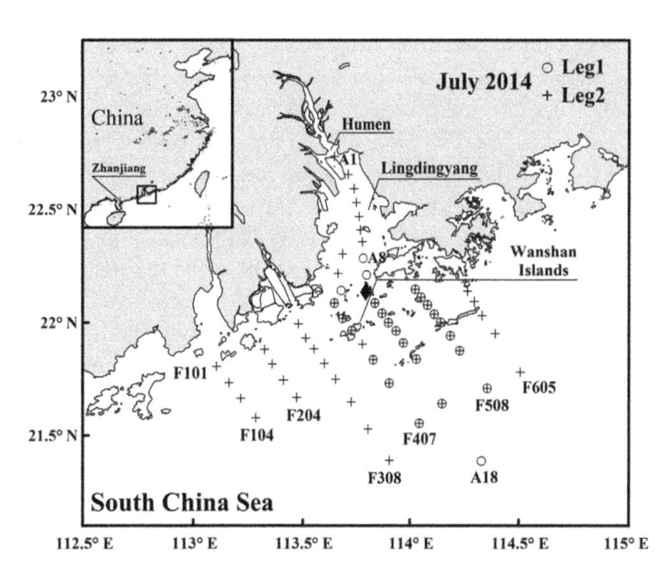

Figure 1. Map of the Pearl River Estuary and adjacent coastal waters. The open circles denote Leg 1 stations visited on 13–16 July 2014, and the crosses represent Leg 2 stations visited on 19–27 July 2014. Note that the filled diamond is the location of Station A10.

2 Materials and methods

2.1 Sampling and analysis

Interrupted by Typhoon Rammasun during the period 17–18 July 2014, our cruise was divided into two legs (Fig. 1). During Leg 1 between 13 and 16 July, we sampled transects F4, F5 and stations A08–A18. During Leg 2 between 19 and 27 July, we sampled stations A01–A10, transects F3 and F4, stations A11–A17 and transects F5, F6, F1 and F2, in sequence. In order to monitor the development of hypoxia before and after the passage of the typhoon, we revisited Station A10 three more times (13, 20 and 27 July).

According to the gauge in the upper Pearl River, water discharge peaked in June and July. Typhoon Rammasun increased discharge during the period 15–18 July, with daily average values of 19 480, 26 115, 22 981 and 17 540 $m^3\,s^{-1}$, respectively. Nevertheless, the freshwater discharge was 18 908 $m^3\,s^{-1}$ in Leg 1 and 15 698 $m^3\,s^{-1}$ in Leg 2, comparable to the long-term (2000–2011) monthly average.

Temperature and salinity were determined with a SBE 25 conductivity–temperature–depth/pressure unit (Sea-Bird Co.). Water samples were collected using 4 L Go-Flo bottles (General Oceanics). DIC and DO were measured at all stations with depth profiles. Samples for $\delta^{13}C_{DIC}$ were collected primarily along Transect A as well as at depth in low-oxygen layers.

The DO concentrations in discrete water samples were measured on board within 8 h using the classic Winkler titration method (Dai et al., 2006). In addition, we conducted on-deck incubation experiments using unfiltered water taken from the hypoxic zone on 27 July 2014 following He et al. (2014). Bottom water from ~ 2 m above the sediment surface was collected and incubated for 24 h in 65 mL BOD bottles in the dark at ambient temperature controlled by the flowing surface water. Note that the maximum difference in temperature between the bottom and surface water was 3 °C during the incubation. Total oxygen consumption rate was determined by comparing the DO concentration at the initial and end point of the experiment.

DIC was measured with an infrared detector after acidifying 0.5–0.7 mL of water sample with a precision of 0.1 % for estuarine and sea waters (Cai et al., 2004). Dissolved calcium concentrations (Ca^{2+}) were determined using an EGTA titration with a Metrohm 809 TITRANDO potentiometer, which has a precision better than ±5 µmol kg^{-1} (Cao et al., 2011).

For $\delta^{13}C_{DIC}$ analysis, an ~ 20 mL DIC sample was converted into gaseous CO_2 and progressively purified through a vacuum line. The pure CO_2 sample was analysed with an isotope ratio mass spectrometer (IRMS, Finnigan MAT 252, Bremen, Germany). The analytical precision was better than 0.1 ‰.

Water samples for TSM (total suspended matter), POC and $\delta^{13}C_{POC}$ analysis were concentrated onto preweighed and pre-combusted 0.7 µm Whatman GF/F filters after filtering 0.2–1.0 L of water under a mild vacuum (~ 25 kPa). Filters were washed with distilled water and stored at −20 °C. Prior to analysis, all filters were freeze-dried. TSM was determined using the net weight increment on the filter and the filtration volume. Filters were decarbonated with 1.0 mol L^{-1} HCl and dried at 40 °C for 48 h (Kao et al., 2012) and analysed for POC and $\delta^{13}C_{POC}$ on an elemental analyser coupled with an IRMS (EA-IRMS). The analytical precision for $\delta^{13}C_{POC}$ was better than 0.1 ‰. Chl *a* was measured with a Turner fluorometer after extracting filters with 90 % acetone (He et al., 2010b). Calibrations were performed using a Sigma Chl *a* standard.

2.2 Three endmember mixing model

We adopted a three endmember mixing model to construct the conservative mixing scheme among different water masses (Cao et al., 2011; Han et al., 2012):

$$F_{RI} + F_{SW} + F_{SUB} = 1, \quad (1)$$

$$\theta_{RI} \times F_{RI} + \theta_{SW} \times F_{SW} + \theta_{SUB} \times F_{SUB} = \theta, \quad (2)$$

$$S_{RI} \times F_{RI} + S_{SW} \times F_{SW} + S_{SUB} \times F_{SUB} = S, \quad (3)$$

where θ and S represent potential temperature and salinity; the subscripts RI, SW and SUB denote the three different water masses (Pearl River plume water, offshore surface seawater and upwelled subsurface water); and F_{RI}, F_{SW} and F_{SUB} represent the fractions that each endmember contributes to the in situ samples. These fractions were applied to predict conservative concentrations of DIC (DIC_{con}) and its isotopic composition ($\delta^{13}C_{DICcon}$) resulting solely from conservative mixing:

$$DIC_{RI} \times F_{RI} + DIC_{SW} \times F_{SW} + DIC_{SUB} \times F_{SUB} = DIC_{con}, \quad (4)$$

$$\frac{\delta^{13}C_{DICRI} \times DIC_{RI} \times F_{RI} + \delta^{13}C_{DICSW} \times DIC_{SW} \times F_{SW} + \delta^{13}C_{DICSUB} \times DIC_{SUB} \times F_{SUB}}{DIC_{con}} = \delta^{13}C_{DICcon}. \quad (5)$$

The difference (Δ) between measured and conservative DIC values represents the magnitude of the biological alteration of DIC (Wang et al., 2016).

3 Results

3.1 Horizontal distribution

Although the average freshwater discharge rate during our sampling period (16 369 $m^3 s^{-1}$) was slightly higher than the multi-year (2000–2011) monthly average (15 671 $m^3 s^{-1}$), Typhoon Rammasun modified the system to some extent as shown from the evolution of chemical species at Station A10 before and after the typhoon (see Sect. 3.4). The interruption of Leg 1 due to the typhoon (17–18 July) led to a smaller survey area, covering only outside Lingdingyang Bay (traditionally regarded as the PRE), while Leg 2 covered Lingdingyang Bay from the Humen Outlet to the adjacent coastal sea.

As depicted in Fig. 2, the sea surface temperature (SST) during Leg 1 (28.9–32.2 °C) was slightly higher than during Leg 2 (28.9–31.0 °C). Sea surface salinity (SSS) measurements showed that plume water was restricted more landward during Leg 2 than Leg 1. However, a steeper gradient to higher SST offshore during Leg 1 was likely induced by the upwelling of bottom water, featuring by relatively high SSS (18.6), high DIC (1789 µmol kg^{-1}) and low DO saturation (DO %, 86 %). During Leg 1, the region with the most productivity was found east of the Wanshan Islands, characterized by high concentrations of Chl *a* (8.0 µg kg^{-1}), low concentrations of DIC (1607 µmol kg^{-1}), and DO supersaturation, with the highest DO % greater than 160 % at Station F503. During Leg 2, there were three patches of high productivity, south of Huangmaohai, at the PRE entrance, and off Hong Kong. The central region of high productivity had the highest DO %, greater than 140 % at Station A14, and was characterized by relatively high concentrations of Chl *a* (7.8 µg kg^{-1}) and low concentrations of DIC (1737 µmol kg^{-1}).

As shown in Fig. 3, bottom water hypoxia during Leg 1 was located more centrally in the study area relative to the surface phytoplankton bloom. The centre of the hypoxic zone

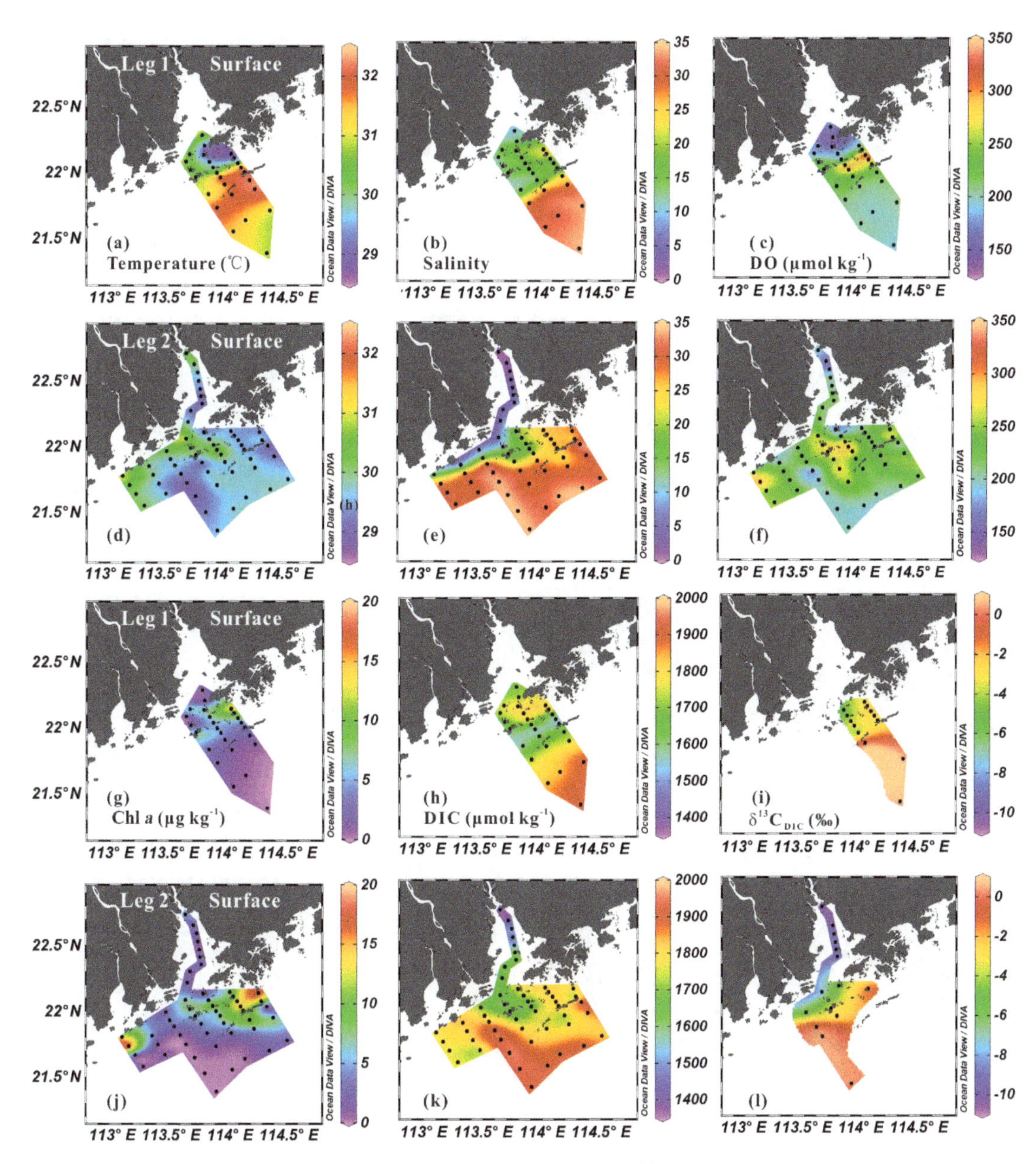

Figure 2. Surface water distribution of temperature, salinity, DO, Chl *a*, DIC and $\delta^{13}C_{DIC}$ during Leg 1 **(a–c, g–i)** and Leg 2 **(d–f, j–l)**.

was found at Station A10, characterized by the lowest observed DO concentrations (as low as 30 μmol kg^{-1}) and a relatively high concentration of DIC (2075 μmol kg^{-1}). During Leg 2, hypoxic conditions were no longer found at Station A10, and instead the largest hypoxic zone was discovered to the southwest of the Wanshan Islands, where the lowest DO values were observed (as low as 7 μmol kg^{-1} at F304), and once again coincided with relatively high concentrations of DIC (2146 μmol kg^{-1}). We were unable to precisely constrain the areas of the regions impacted by bottom water hypoxia due to the limited spatial coverage, but our results suggest it covered an area of > 280 km^2 during Leg 1 and > 290 km^2 during Leg 2 according to the definition of hypoxia as DO < 2 mg L^{-1} or 63 μM, or an area of > 900 km^2 during Leg 1 and > 800 km^2 during Leg 2 assuming the threshold of the oxygen-deficit zone was < 3 mg L^{-1} or 95 μM (Rabalais et al., 2010; Zhao et al., 2017).

3.2 Vertical distribution

During Leg 1, plume water reached 50 km offshore from the entrance of the PRE, forming a 5–10 m thick surface layer (Fig. 4b). Both the thermocline and halocline contributed to the stability of the water column structure, which favoured the formation of bottom water hypoxia. The thickness of the bottom water hypoxic layer was ~ 5 m. The region of highest

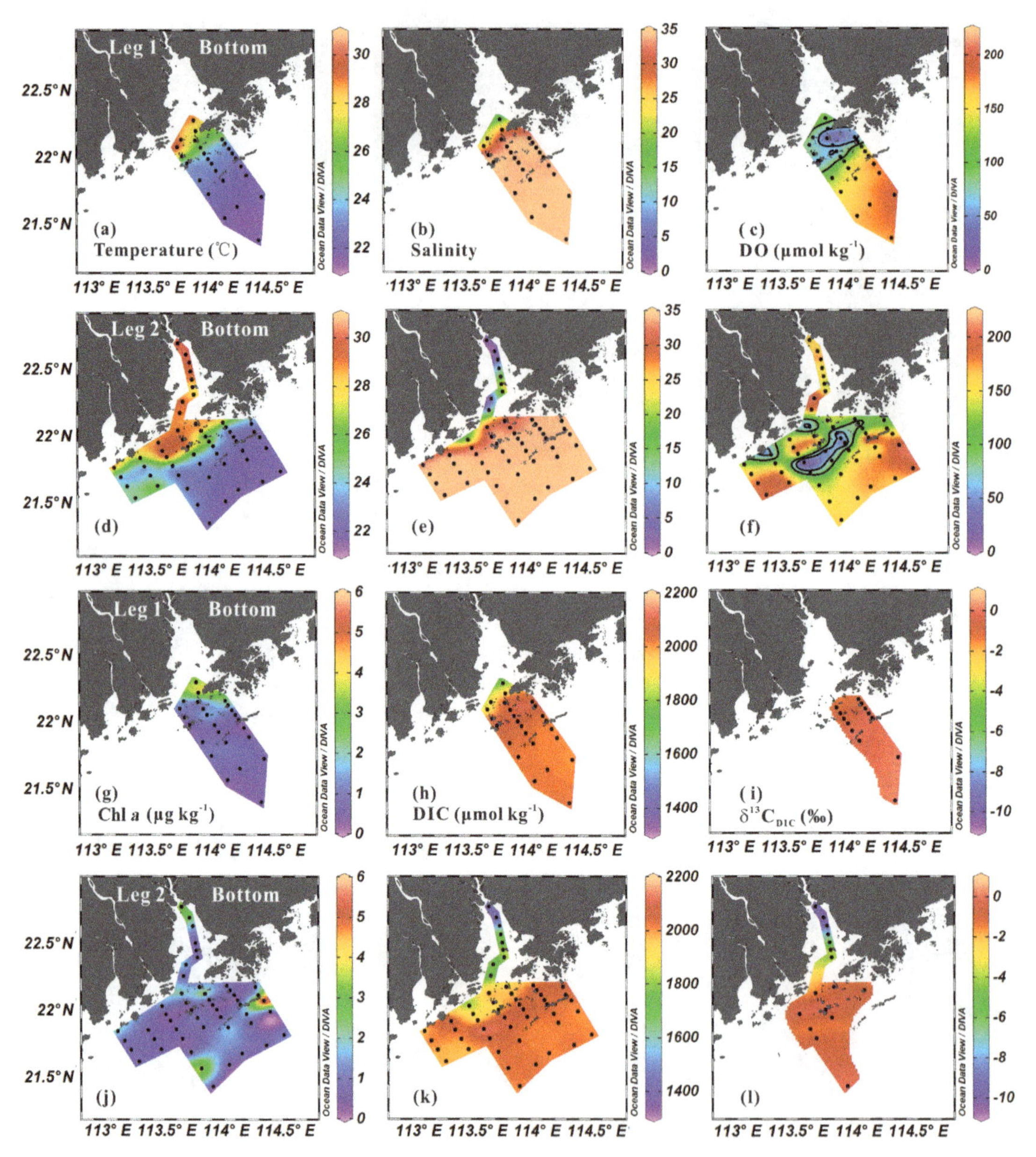

Figure 3. Bottom water distribution of temperature, salinity, DO, Chl *a*, DIC and $\delta^{13}C_{DIC}$ during Leg 1 **(a–c, g–i)** and Leg 2 **(d–f, j–l)**. Note that the black lines in **(c)** and **(f)** indicate DO contours of 63 and 95 μM.

productivity, however, was not observed in the same location as the hypoxic zone, but further offshore.

During Leg 2, although the passing of the typhoon would be expected to absorb large amounts of potential heat and cause extensive mixing of the water column, the enhanced freshwater discharge could rapidly re-stratify the water column and facilitate the re-formation of hypoxia. This time, the primary region of hypoxia was observed directly below the bloom, with a thickness of 3 m (Fig. 4i). Additionally, near the Humen Outlet we observed low DIC (1466 $\mu mol\,kg^{-1}$) and moderately low DO (89 $\mu mol\,kg^{-1}$), which reflected the input of the low DO water mass from upstream as reported previously (Dai et al., 2006, 2008a; He et al., 2014).

3.3 Isotopic composition of DIC and POC

The $\delta^{13}C$ values of DIC became progressively heavier from stations dominated by freshwater (~ -11.4 ‰) to offshore seawater (~ -0.6 ‰), with a relatively wide range of values beyond a salinity of 13 (Fig. 5). Owing to a malfunction of the instrument, $\delta^{13}C_{POC}$ data from our cruise were not available. Instead, we reported a valid $\delta^{13}C_{POC}$ dataset from a 2015 summer cruise in approximately the same region. $\delta^{13}C_{POC}$ values showed a similar trend with $\delta^{13}C_{DIC}$,

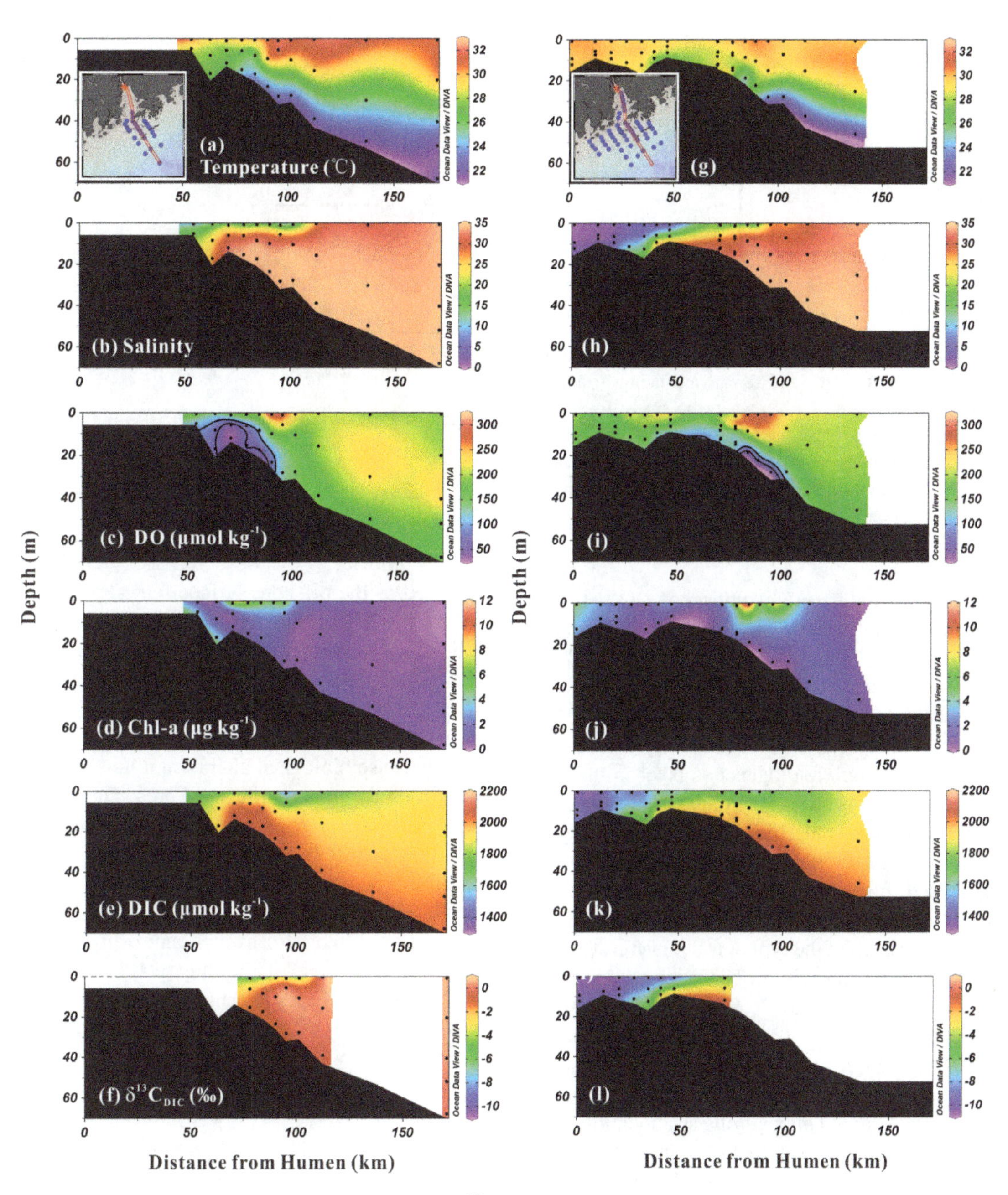

Figure 4. Profiles of temperature, salinity, DO, Chl a, DIC and $\delta^{13}C_{DIC}$ along Transect A during Leg 1 **(a–f)** and Leg 2 **(g–l)**. Note that the black lines in **(c)** and **(i)** indicate DO contours of 63 and 95 μM.

i.e. ^{13}C-enriched seaward, from ~ -28 to ~ -20 ‰. In the bloom, where the DO % was above 125 %, the mean $\delta^{13}C$ value for POC was -19.4 ± 0.8 ‰ ($n = 8$), which was within the typical range of marine phytoplankton (Peterson and Fry, 1987). As shown in Fig. 5, there was a large $\delta^{13}C_{POC}$ decrease near a salinity of 15. Geographically, it was located at the mixing-dominated zone in inner Lingdingyang Bay, where intense resuspension of ^{13}C-depleted sediments may occur (Guo et al., 2009).

3.4 Reinstatement of the hypoxic station after Typhoon Rammasun

Typhoon Rammasun made landfall at Zhanjiang, located 400 km to the southwest of the PRE, at 20:00 LT (local time) on 18 July, and was dissipated by 05:00 LT on 20 July. The typhoon completely de-stratified the water column during its passing. However, the associated heavy precipitation and runoff appeared to re-establish stratification rather quickly, within 1 day, as suggested by the salinity gradient (18–30) from 0 to 10 m depth during Leg 2 at 15:20 LT on 20 July

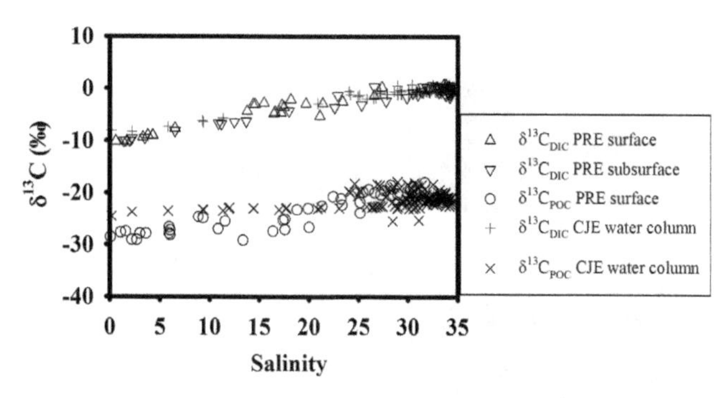

Figure 5. Distribution of $\delta^{13}C_{DIC}$ and $\delta^{13}C_{POC}$ with respect to salinity in the PRE. The upward- and downward-facing triangles denote surface and subsurface $\delta^{13}C_{DIC}$ data, respectively, from July 2014, while the open circles represent $\delta^{13}C_{POC}$ values in surface water from July 2015. Additionally, the plus signs and crosses show the $\delta^{13}C_{DIC}$ and $\delta^{13}C_{POC}$ data, respectively, from the Changjiang Estuary (CJE) in Wang et al. (2016).

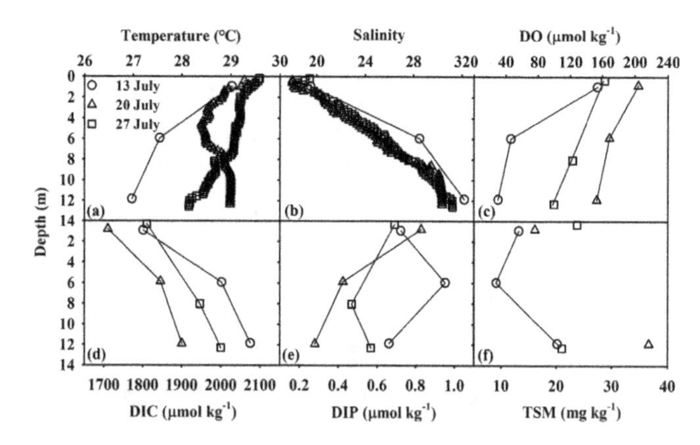

Figure 6. Profiles of **(a)** temperature, **(b)** salinity, **(c)** DO, **(d)** DIC, **(e)** DIP, **(f)** TSM and their evolution during repeated sampling at Station A10.

(Fig. 6b). In order to capture the evolution of DO between the disruption and reinstatement of stratification, we resumed our cruise and revisited Station A10 (Fig. 6). On 13 July, the bottom water at Station A10 was the hypoxic core, with the lowest observed DO (30 μmol kg^{-1}) and highest DIC (2075 μmol kg^{-1}) concentrations. On 20 July, the results showed that the temperature homogeneous layer in the bottom water (9–13 m) might reflect the remnants of typhoon-induced mixing (Fig. 6a), while the reduction in salinity at < 9 m depicted the rapid re-establishment of stratification as a result of enhanced freshwater discharge (Fig. 6b). Bottom water DO increased to 153 μmol kg^{-1} and DIC decreased to 1901 μmol kg^{-1} as a result of the typhoon-induced water column mixing and aeration. In addition, TSM increased sharply from 20.2 before the typhoon to 36.6 mg kg^{-1}, suggesting that large volumes of sediment had been resuspended during its passing. On 27 July, 1 week after the typhoon, strong thermohaline stratification was re-established in the whole water column. Along with the intensifying stratification, bottom water DO decreased to 99 μmol kg^{-1}, indicating continuous DO depletion and the potential for hypoxia formation. Meanwhile, bottom water DIC concentrations increased to 2000 μmol kg^{-1} and dissolved inorganic phosphate (DIP) rose from 0.28 to 0.57 μmol kg^{-1}. Moreover, bottom water TSM returned to pre-typhoon (13 July) levels.

4 Discussion

4.1 Selection of endmembers and model validation

The potential temperature–salinity plot displayed a three endmember mixing scheme over the PRE and adjacent coastal waters (Fig. 7a), consisting of Pearl River plume water, offshore surface seawater and upwelled subsurface water. During the summer, a DIC concentration of ~ 1917 μmol kg^{-1} was observed at $S = 33.7$, which can be regarded as the offshore surface seawater endmember (Guo and Wong, 2015). Here, by choosing $S = 34.6$ as the offshore subsurface water salinity endmember, we obtained a DIC value of ~ 2023 μmol kg^{-1}, similar to the value at ~ 100 m depth adopted by Guo and Wong (2015). For the plume endmember, it was difficult to directly select from the field data, because biological alteration might lead to altered values within the plume-influenced area. Therefore, we first assumed that the plume water observed on the shelf consisted of a mixture of freshwater and offshore surface seawater. Then, we compiled 3 years of surface data from the summer (August 2012, July 2014 and July 2015) to extrapolate the relatively stable freshwater endmember and examine the biological effect on DIC–salinity relationships. By constraining DIC endmembers (freshwater and offshore surface seawater), we observed that DIC remained overall conservative when salinity was < 10.8 but showed removal when salinity was > 10.8 (Han et al., 2012). Thus, we derived plume endmember values (1670 ± 50 μmol kg^{-1}) from the DIC–salinity conservative mixing curve at $S = 10.8$. Furthermore, $S = 10.8$ was observed at the innermost station (A08) during Leg 1, which agreed well with the spatial and temporal scale of the actual water mass mixing in our survey. To confirm our results, we also used a freshwater endmember ($S = 0$), but the output of the model showed little difference from that based on the plume endmember at $S = 10.8$.

The $\delta^{13}C_{DIC}$ value was 0.6 ± 0.2 ‰ in the offshore surface seawater at $S = \sim 33.7$, where nutrient ($NO_3^- + NO_2^-$ and DIP) concentrations were close to their detection limits and DO was nearly saturated, indicating little biological activity. As DIC remained overall conservative when salinity was < 10.8, the $\delta^{13}C_{DIC}$ value of -11.4 ± 0.2 ‰ at $S < 0.4$ is representative of the freshwater source. Assuming the plume water is a mixture of freshwater and offshore surface seawater, the initial plume endmember of $\delta^{13}C_{DIC}$ at $S = 10.8$ can be calculated via an isotopic mass balance (-7.0 ± 0.8 ‰).

A summary of the endmember values used in this study is listed in Table 1.

We calculated the fractions of the three water masses based on potential temperature and salinity equations, so as to predict conservative DIC (DIC_{con}) and its isotopic composition ($\delta^{13}C_{DICcon}$) solely from conservative mixing. We chose the concentration of Ca^{2+} as a conservative tracer to validate our model prediction, assuming that $CaCO_3$ precipitation or dissolution is not significant. This assumption is supported by a strong linear relationship between surface water Ca^{2+} and salinity, and aragonite oversaturation ($\Omega_{arag} = 2.6 \pm 0.7$) in the subsurface water. Our model-derived values were in good accordance with the field-observed values (Fig. 7b), which strongly supported our model prediction.

As shown in Fig. 7c, most of the observed DIC concentrations in the subsurface water were higher than the conservative values, as a result of DIC production via OC oxidation. This coincided with lighter $\delta^{13}C_{DIC}$ values than conservative, owing to the accumulation of isotopically lighter carbon entering the DIC pool from remineralized organic matter (Fig. 7d). Based on the differences between the observed and conservative values of DIC and $\delta^{13}C_{DIC}$, the carbon isotopic composition of the oxygen-consuming organic matter could be traced precisely (see details in Sect. 4.2).

In the subsurface water, the bulk of ΔDIC values varied from 0 to 132 μmol kg^{-1}, coupled with a range of apparent oxygen utilization (AOU) values from 0 to 179 μmol kg^{-1}. ΔDIC values positively correlated with AOU (Fig. 7e), corresponding to the fact that the additional DIC was supplied by organic matter remineralization via aerobic respiration. The slope of ΔDIC vs. AOU in the subsurface water was 0.71 ± 0.03, which agrees well with classic Redfield stoichiometry (i.e. $106/138 = 0.77$), providing further evidence for aerobic respiration as the source of added DIC. As a first-order comparison, the water column total oxygen consumption rate of 9.8 μmol L^{-1} d^{-1} could well support the oxygen decline rate observed at Station A10 in the hypoxic zone between 20 and 27 July (Fig. 6), which was 7.7 μmol L^{-1} d^{-1}. This comparison along with the stoichiometry between ΔDIC and AOU strongly suggests that water column aerobic respiration may predominate in the formation of the hypoxia in the present case.

4.2 Isotopic composition of the oxygen-consuming OC

The DIC isotopic mass balance (Wang et al., 2016) is expressed as

$$\delta^{13}C_{DICobs} \times DIC_{obs} = \delta^{13}C_{DICcon} \times DIC_{con} + \delta^{13}C_{DICbio} \times DIC_{bio}, \quad (6)$$

where the subscripts obs, con and bio refer to the field-observed, conservative and biologically altered values.

Degradation of OC typically produces DIC with minor isotopic fractionation from the OC substrate (Hullar et al., 1996; Breteler et al., 2002). Thus, the isotopic composition of DIC_{bio} (i.e. $\delta^{13}C_{DICbio}$) should be identical to the $\delta^{13}C$ of the OC ($\delta^{13}C_{OC_x}$), which consumed oxygen and produced DIC_{bio}. The $\delta^{13}C_{OC_x}$ was derived from the mass balance equations of both DIC and its stable isotope:

$$\delta^{13}C_{OC_x} = \frac{\delta^{13}C_{obs} \times DIC_{obs} - \delta^{13}C_{con} \times DIC_{con}}{DIC_{obs} - DIC_{con}}, \quad (7)$$

which can be rearranged into

$$\Delta(\delta^{13}C_{DIC} \times DIC) = \delta^{13}C_{OC_x} \times \Delta DIC. \quad (8)$$

As shown in Fig. 8, the slope of the linear regression represents $\delta^{13}C_{OC_x}$ or $\delta^{13}C_{DICbio}$, which here is equal to -23.2 ± 1.1 ‰. This value reflects the original $\delta^{13}C$ signature of the remineralized organic matter contributing to the observed addition of DIC.

Although studies have shown selective diagenesis of isotopically heavy or light pools of organic matter (Marthur et al., 1992; Lehmann et al., 2002), these effects are small compared to the isotopic differences among different sources of organic matter (Meyers, 1997). It is thus reasonable to assume that the isotopic ratios are conservative and that physical mixing of the endmember sources determine the isotopic composition of organic matter in natural systems (Gearing et al., 1984; Cifuentes et al., 1988; Thornton and McManus, 1994). The relative contributions of marine and terrestrial sources to oxygen-consuming organic matter in our study area could be estimated based on the following equation (Shultz and Calder, 1976; Hu et al., 2006):

$$f\,(\%) = \frac{\delta^{13}C_{mar} - \delta^{13}C_{OC_x}}{\delta^{13}C_{mar} - \delta^{13}C_{terr}} \times 100. \quad (9)$$

Here, for the terrestrial endmember ($\delta^{13}C_{terr}$), we adopted the average $\delta^{13}C$ value of POC sampled near the Humen Outlet ($S < 4$), which represents the predominant source of riverine material entering the estuary (He et al., 2010b). The mean $\delta^{13}C_{POC}$ value, -28.3 ± 0.7 ‰ ($n = 7$), is very similar to the freshwater $\delta^{13}C_{POC}$ value of -28.7 ‰ reported by Yu et al. (2010), which reflected a terrigenous mixture of C_3 plant fragments and forest soils. For the marine endmember ($\delta^{13}C_{mar}$), we calculated the mean surface water $\delta^{13}C_{POC}$ value (-19.4 ± 0.8 ‰, $n = 8$) from stations with $S > 26$ where significant phytoplankton blooms were observed, as indicated by DO supersaturation (DO % > 125 %) and relatively high pH values (> 8.3) and POC content (5.3 ± 2.4 %). This value is similar to, although slightly heavier than, the marine endmember used by Chen et al. (2008), who measured a $\delta^{13}C$ value of -20.9 ‰ in tow-net phytoplankton samples from outer Lingdingyang Bay, in the same region as this study. Additionally, He et al. (2010a) reported a $\delta^{13}C$ value of -20.8 ± 0.4 ‰ in phytoplankton collected from the northern South China Sea. These values are consistent enough for us to compile and use an average $\delta^{13}C_{mar}$

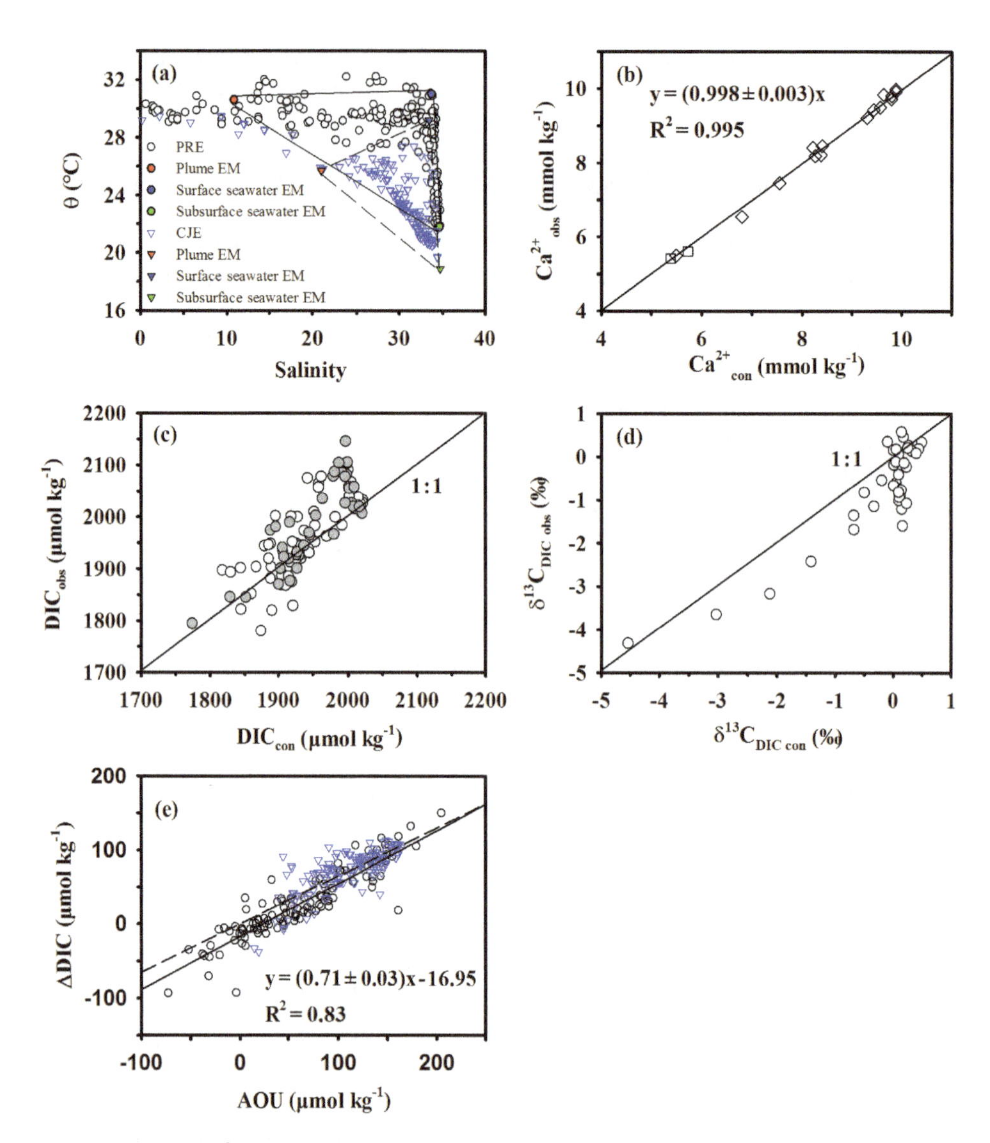

Figure 7. (a) Potential temperature (θ) (°C) vs. salinity in the PRE and adjacent coastal waters (open circles) based on data collected during the July 2014 cruise. The three endmembers are shown as different coloured symbols. The blue triangles represent data collected during the August 2011 cruise in the Changjiang Estuary (CJE) (Wang et al., 2016). **(b)** Correlation between the field-observed Ca^{2+} (Ca^{2+}_{obs}) and conservative Ca^{2+} (Ca^{2+}_{con}). The straight line denotes a linear regression line of both surface (square) and subsurface (diamond) data. **(c, d)** Relationship between observed and conservative DIC and $\delta^{13}C_{DIC}$ values. The straight line represents a 1 : 1 reference line. Note that the grey dots in **(c)** identify data also in **(d)**. **(e)** Correlation of ΔDIC vs. AOU for all subsurface water data. ΔDIC is the difference between the field-observed and conservative DIC concentrations. Also shown are the data from Wang et al. (2016). The straight and dashed lines indicate linear regressions of data from the PRE and CJE, respectively.

value of -20.5 ± 0.9 ‰. This value agrees well with the reported stable carbon isotopic signature of marine organic matter in other coastal regions. For example, mean isotopic values of phytoplankton were reported as -20.3 ± 0.6 ‰ in Narragansett Bay (Gearing et al., 1984), -20.3 ± 0.9 ‰ in Auke Bay and Fritz Cove (Goering et al., 1990), and -20.1 ± 0.8 ‰ in the Gulf of Lion (Harmelin-Vivien et al., 2008).

Our model results suggest that marine organic matter contributed 65 ± 16 % of the observed oxygen consumption, while terrestrial organic matter accounted for the remaining 35 ± 16 %. It is thus clear that marine organic matter from eutrophication-induced primary production dominated oxygen consumption in the hypoxic zone; however, terrestrial organic matter also contributed significantly to the formation and maintenance of hypoxia in the lower PRE and adjacent coastal waters.

Table 1. Summary of endmember values and their uncertainties adopted in the three endmember mixing model.

Water mass	θ (°C)	Salinity	DIC (μmol kg^{-1})	$\delta^{13}C_{DIC}$ (‰)	Ca^{2+} (μmol kg^{-1})
Plume	30.6 ± 1.0	10.8	1670 ± 50[a]	-7.0 ± 0.8[b]	3670 ± 16[c]
Surface	31.0 ± 1.0	33.7 ± 0.2	1917 ± 3	0.6 ± 0.2	9776 ± 132[c]
Subsurface	21.8 ± 1.0	34.6 ± 0.1	2023 ± 6	0.1 ± 0.1	10053

[a] In order to derive a proper plume endmember value, we took advantage of 3 years of surface dataset from summer cruises (see Sect. 4.1). For DIC, the data are from cruises during August 2012, July 2014 and July 2015. [b] See details in Sect. 4.1. [c] The Ca^{2+} values of the plume and surface seawater endmember are derived from a conservative mixing calculation (Ca^{2+} vs. S) based on 3 years of surface data during the summer (August 2012, July 2014 and July 2015).

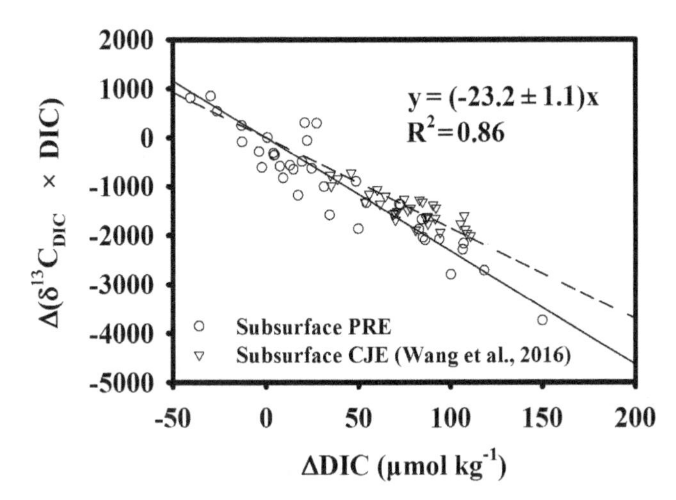

Figure 8. Δ ($\delta^{13}C_{DIC} \times$ DIC) vs. ΔDIC in the PRE. Samples were collected from subsurface water (> 5 m). Δ is the difference between the field-observed and conservative values. Also shown are data from the Changjiang Estuary (CJE) reported by Wang et al. (2016). The straight and dashed lines indicate linear regression lines of data from the PRE and CJE, respectively. The regression equation is shown for the PRE.

4.3 Comparison with hypoxia in the East China Sea off the Changjiang Estuary

As one of the largest rivers in the world, the Changjiang has been suffering from eutrophication for the past few decades (Zhang et al., 1999; Wang et al., 2014). In summer, sharp density gradients with frequent algal blooms and subsequent organic matter decomposition cause seasonal hypoxia in the bottom water of the ECS off the CJE. Wang et al. (2016) revealed that the remineralization of marine organic matter (OC_{mar}) overwhelmingly (nearly 100 %) contributed to DO consumption in the ECS off the CJE. However, our present study showed that less OC_{mar} contributed to the oxygen depletion (65 ± 16 %) in the hypoxic zone of the lower PRE.

As shown in Fig. 5, there is little difference between $\delta^{13}C_{DIC}$ and $\delta^{13}C_{POC}$ values of the marine endmember. However, the $\delta^{13}C_{DIC}$ and $\delta^{13}C_{POC}$ values of the freshwater endmember showed some dissimilarity, with lighter values in the PRE (-11.4 ± 0.2 ‰, -28.3 ± 0.7 ‰) than those in the CJE (-8.8 ‰, -24.4 ± 0.2 ‰) at $S < 0.2$. In Fig. 7e, the amplitudes of ΔDIC and AOU values suggest a similar intensity of organic matter (OM) biodegradation, and the slope of ΔDIC vs. AOU (0.71 ± 0.03 vs. 0.65 ± 0.04) indicates a predominance of aerobic respiration in the two systems. As seen from Table 2, there is no significant difference between the $\delta^{13}C$ values of surface sediments within the hypoxic zones of the PRE and CJE. However, data in Fig. 7a show generally higher water temperatures in the PRE than in the CJE. For instance, the temperatures of surface and subsurface seawater endmembers in the PRE were 2–3°C higher than in the CJE. From a spatial point of view, the distance from the river mouth to the hypoxic zone in the CJE is 2–3 times longer than in the PRE, possibly resulting in a longer travel time of OC_{terr}. Therefore, we contend that the difference in the predicted distributions of marine and terrestrial sources of organic matter contributing to oxygen consumption in and off the PRE and CJE is likely related to differences in the bioavailability of OC_{terr} and OC_{mar}, the microbial community structures and the physical settings between these two hypoxic systems.

Although C_3 plants dominate and C_4 plants are minor in both the Pearl River and Changjiang drainage basins (Hu et al., 2006; Zhu et al., 2011a), the OC_{terr} delivered from these two watersheds experiences varying degrees of degradation within the estuaries before being transported into the coastal hypoxic zones. In the CJE, approximately 50 % of OC_{terr} becomes remineralized during transport through the estuary, likely due to efficient OM unloading from mineral surfaces (Zhu et al., 2011a) and longer residence times within the estuary, facilitating microbial transformation and degradation. In contrast, the PRE appears to be a somewhat intermediate site with the export of OC_{terr} being closely associated with sedimentary regimes and not characterized by extensive degradative loss (Strong et al., 2012). Thus, the bioavailability of OC_{terr} that reached the hypoxic zone is likely higher in the PRE than in the CJE. Moreover, the increased precipitation and runoff during the typhoon may have mobilized additional fresh anthropogenic OM from surrounding megacities (e.g. Guangzhou, Shenzhen and Zhuhai) deposited in

Table 2. Comparison of $\delta^{13}C$ values in surface sediments within the hypoxic zone[a] between the PRE and CJE.

$\delta^{13}C$ (‰)	Mean ± SD	Stations involved	References
		Pearl River Estuary	
−23.4 to −22.1	−22.9 ± 0.5	A4, A5, C1-C4, D1	Hu et al. (2006)
−23.2 to −22.3	−22.7 ± 0.5	28, 29, 30	Zong et al. (2006)
−23.6 to −21.5	−22.5 ± 1.1	E8-1, E7A, S7-1, S7-2	He et al. (2010a)
–[b]	−23.1 ± 0.6	Clustering groups G6 and G7	Yu et al. (2010)
Average	−22.8 ± 0.6		
		Changjiang Estuary	
−22.9 to −20.9	−21.8 ± 0.6	–[c]	Tan et al. (1991)
−22.4 to −19.9	−21.2 ± 1.0	32, 37, 38, 42, 48, 49, 54, 56, 64	Kao et al. (2003)
−22.7 to −20.8	−22.0 ± 0.8	H1-12, H2-10, H2-11, S1-2, S2-4	Xing et al. (2011)
−23.5 to −20.4	−22.6 ± 1.0	3, 12, 13, 20–25	Yao et al. (2014)
Average	−21.9 ± 1.0		

[a] In the PRE, the data are from similar sites to our present study, which is in the northeast (Leg 1) and southwest (Leg 2) of the Wanshan Islands. While in the CJE, the hypoxic zone is located around 30.0–32.0° N, 122.7–123.2° E, which is frequently reported in previous studies (Li et al., 2002; Zhu et al., 2011b; Wang et al., 2016). [b] The authors provide an average value of clustering groups instead of individual data from each site. [c] In Fig. 7 of Tan et al. (1991), the sampling sites are shown without numbers.

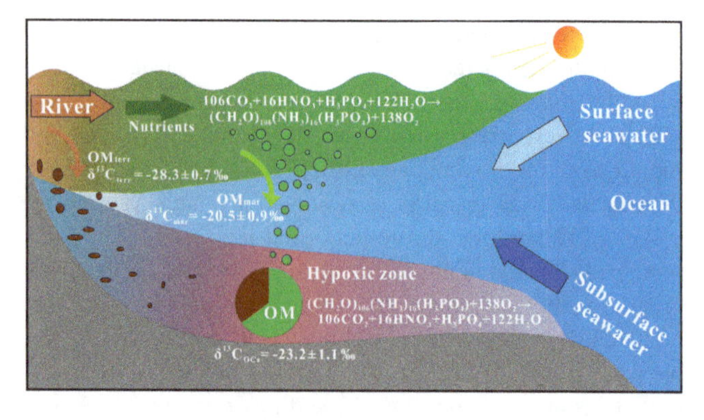

Figure 9. A conceptual diagram illustrating the partitioning of oxygen-consuming organic matter (OC_{mar} vs. OC_{terr}) within the hypoxic zone in the lower PRE and the adjacent coastal area. See Sect. 5 for explanations.

the river channel, which could lead to more labile OC_{terr} in the PRE. Additionally, the difference in bacterial community structure between the two systems may have played a role. Recent studies have demonstrated that the bacterial community in the PRE is characterized by higher relative abundances of Actinobacteria and lower relative abundances of *Cytophaga*–Flavobacteria–*Bacteroides* (CFB) than in the CJE (Liu et al., 2012; Zhang et al., 2016). Whether such differences would promote the degradation of OC_{terr} in the PRE relative to the CJE remains unknown. Finally, the temperature of the bottom water in the PRE hypoxic zone (23.7–27.0 °C) was higher than in the CJE hypoxic zone (20.5–23 °C), which may have accelerated the rates of bacterial growth and OM decomposition (Brown et al., 2004).

5 Conclusions

Based on a three endmember mixing model and the mass balance of DIC and its isotopic composition, we demonstrated that the organic matter decomposed via aerobic respiration in the stratified subsurface waters of the lower PRE and adjacent coastal waters was predominantly OC_{mar} (49–81 %, mean 65 %), with a significant portion of OC_{terr} also decomposed (19–51 %, mean 35 %). The relative distribution of organic matter sources contributing to oxygen drawdown differs in the hypoxic zone off the CJE, where it is caused almost entirely by OC_{mar}. These differences have important implications for better understanding the controls on hypoxia and its mitigation. Nevertheless, with respect to increasing coastal nutrient levels, a significant implication of the present study is that reducing and managing nutrients is critical to control eutrophication and, subsequently, to mitigate hypoxia (Conley et al., 2009; Paerl, 2009; Mercedes et al., 2015; Stefan et al., 2016). Given that OC_{terr} also contributes to the consumption of oxygen in the lower PRE hypoxic zone, it is crucial to characterize the source of this oxygen-consuming terrestrial organic matter, whether from natural soil leaching and/or anthropogenic wastewater discharge, so as to make proper policies for hypoxia remediation.

The processes involved in the partitioning of organic matter sources, their isotopic signals and their subsequent biogeochemical transformations in the PRE hypoxic zone are illustrated in the conceptual diagram in Fig. 9. The river delivers a significant amount of nutrients and terrestrial organic matter to the estuary, stimulating phytoplankton blooms in the surface water at the lower reaches of the estuary where turbidity is relatively low and conditions are favourable for phytoplankton growth (Gaston et al., 2006; Dai et al., 2008b; Guo et al., 2009). The subsequent sinking of this biomass along with terrestrial organic matter below the pycnocline consumes oxygen and adds respired DIC to subsurface waters, resulting in coastal hypoxia. Therefore, we conclude that within the PRE and adjacent coastal areas, the most important biological process with respect to forming and maintaining hypoxic conditions is aerobic respiration.

Competing interests. The authors declare that they have no conflict of interest.

Acknowledgements. This research was funded by the National Natural Science Foundation of China through grants 41130857, 41576085 and 41361164001, and by the Hong Kong Research Grants Council under the Theme-based Research Scheme (TRS) through project no. T21-602/16R. We thank Tengxiang Xie, Li Ma, Shengyao Sun, Chenhe Zheng and Liangrong Zou for their assistance in sample collections; Yan Li and Yawen Wei for providing the calcium concentration data; and Liguo Guo, Tao Huang and Dawei Li for assisting on the measurements of DIC, nutrients and $\delta^{13}C_{POC}$. The captain and the crew of R/V *Kediao 8* are acknowledged for their cooperation during the cruise. Finally, we express our gratitude to the editor and two anonymous referees for their insightful and constructive comments and input.

Edited by: Koji Suzuki

References

Bianchi, T. S.: The role of terrestrially derived organic carbon in the coastal ocean: A changing paradigm and the priming effect, P. Natl. Acad. Sci. USA, 108, 19473–19481, https://doi.org/10.1073/pnas.1017982108, 2011a.

Bianchi, T. S., Wysocki, L. A., Schreiner, K. M., Filley, T. R., Corbett, D. R., and Kolker, A. S.: Sources of terrestrial organic carbon in the Mississippi plume region: evidence for the importance of coastal marsh inputs, Aquat. Geochem., 17, 431–456, https://doi.org/10.1007/s10498-010-9110-3, 2011b.

Boesch, D. F., Boynton, W. R., Crowder, L. B., Diaz, R. J., Howarth, R. W., Mee, L. D., Nixon, S. W., Rabalais, N. N., Rosenberg, R., Sanders, J. G., Scavia, D., and Turner, R. E.: Nutrient Enrichment Drives Gulf of Mexico Hypoxia, Eos, Trans. Amer. Geophys. Union, 90, 117–118, https://doi.org/10.1029/2009EO140001, 2009.

Breitburg, D.: Effects of hypoxia, and the balance between hypoxia and enrichment, on coastal fishes and fisheries, Estuaries, 25, 767–781, https://doi.org/10.1007/BF02804904, 2002.

Breteler, W. C. K., Grice, K., Schouten, S., Kloosterhuis, H. T., and Damsté, J. S. S.: Stable carbon isotope fractionation in the marine copepod Temora longicornis: unexpectedly low $\delta^{13}C$ value of faecal pellets, Mar. Ecol. Prog. Ser., 240, 195–204, https://doi.org/10.3354/meps240195, 2002.

Brown, J. H., Gillooly, J. F., Allen, A. P., Savage, V. M., and West, G. B.: Toward a metabolic theory of ecology, Ecology, 85, 1771–1789, https://doi.org/10.1890/03-9000, 2004.

Cai, W.-J., Dai, M., Wang, Y., Zhai, W., Huang, T., Chen, S., Zhang, F., Chen, Z., and Wang, Z.: The biogeochemistry of inorganic carbon and nutrients in the Pearl River estuary and the adjacent Northern South China Sea, Cont. Shelf Res., 24, 1301–1319, https://doi.org/10.1016/j.csr.2004.04.005, 2004.

Cao, Z., Dai, M., Zheng, N., Wang, D., Li, Q., Zhai, W., Meng, F., and Gan, J.: Dynamics of the carbonate system in a large continental shelf system under the influence of both a river plume and coastal upwelling, J. Geophys. Res.-Biogeo., 116, G02010, https://doi.org/10.1029/2010JG001596, 2011.

Carstensen, J., Andersen, J. H., Gustafsson, B. G., and Conley, D. J.: Deoxygenation of the Baltic Sea during the last century, P. Natl. Acad. Sci. USA, 111, 5628–5633, https://doi.org/10.1073/pnas.1323156111, 2014.

Chen, F., Zhang, L., Yang, Y., and Zhang, D.: Chemical and isotopic alteration of organic matter during early diagenesis: Evidence from the coastal area off-shore the Pearl River estuary, south China, J. Mar. Syst., 74, 372–380, https://doi.org/10.1016/j.jmarsys.2008.02.004, 2008.

Cifuentes, L., Sharp, J., and Fogel, M. L.: Stable carbon and nitrogen isotope biogeochemistry in the Delaware estuary, Limnol. Oceanogr., 33, 1102–1115, https://doi.org/10.4319/lo.1988.33.5.1102, 1988.

Conley, D. J., Paerl, H. W., Howarth, R. W., Boesch, D. F., Seitzinger, S. P., Karl, E., Karl, E., Lancelot, C., Gene, E., and Gene, E.: Controlling eutrophication: nitrogen and phosphorus, Science, 123, 1014–1015, https://doi.org/10.1126/science.1167755, 2009.

Dai, M., Guo, X., Zhai, W., Yuan, L., Wang, B., Wang, L., Cai, P., Tang, T., and Cai, W.-J.: Oxygen depletion in the upper reach of the Pearl River estuary during a winter drought, Mar. Chem., 102, 159–169, https://doi.org/10.1016/j.marchem.2005.09.020, 2006.

Dai, M., Wang, L., Guo, X., Zhai, W., Li, Q., He, B., and Kao, S.-J.: Nitrification and inorganic nitrogen distribution in a large perturbed river/estuarine system: the Pearl River Estuary, China, Biogeosciences, 5, 1227–1244, https://doi.org/10.5194/bg-5-1227-2008, 2008a.

Dai, M., Zhai, W., Cai, W.-J., Callahan, J., Huang, B., Shang, S., Huang, T., Li, X., Lu, Z., Chen, W., and Chen, Z.: Effects of an estuarine plume-associated bloom on the carbonate system in the lower reaches of the Pearl River estuary and the coastal zone of the northern South China Sea, Cont. Shelf Res., 28, 1416–1423, https://doi.org/10.1016/j.csr.2007.04.018, 2008b.

Diaz, R. J. and Rosenberg, R.: Spreading dead zones and consequences for marine ecosystems, Science, 321, 926–929, https://doi.org/10.1126/science.1156401, 2008.

Gaston, T. F., Schlacher, T. A., and Connolly, R. M.: Flood discharges of a small river into open coastal waters: Plume traits and material fate, Estuar. Coast. Shelf Sci., 69, 4–9, https://doi.org/10.1016/j.ecss.2006.03.015, 2006.

Gearing, J. N., Gearing, P. J., Rudnick, D. T., Requejo, A. G., and Hutchins, M. J.: Isotopic variability of organic carbon in a phytoplankton-based, temperate estuary, Geochim. Cosmochim. Ac., 48, 1089–1098, https://doi.org/10.1016/0016-7037(84)90199-6, 1984.

Goering, J., Alexander, V., and Haubenstock, N.: Seasonal variability of stable carbon and nitrogen isotope ratios of organisms in a North Pacific Bay, Estuar. Coast. Shelf Sci., 30, 239–260, https://doi.org/10.1016/0272-7714(90)90050-2, 1990.

Guo, X. and Wong, G. T.: Carbonate chemistry in the northern South China Sea shelf-sea in June 2010, Deep-Sea Res. Pt. II, 117, 119–130, https://doi.org/10.1016/j.dsr2.2015.02.024, 2015.

Guo, X., Dai, M., Zhai, W., Cai, W.-J., and Chen, B.: CO^2 flux and seasonal variability in a large subtropical estuarine system, the Pearl River Estuary, China, J. Geophys. Res.-Biogeo., 114, G03013, https://doi.org/10.1029/2008JG000905, 2009.

Han, A., Dai, M., Kao, S.-J., Gan, J., Li, Q., Wang, L., Zhai, W., and Wang, L.: Nutrient dynamics and biological consumption in a large continental shelf system under the influence of both a river plume and coastal upwelling, Limnol. Oceanogr., 57, 486–502, https://doi.org/10.4319/lo.2012.57.2.0486, 2012.

Harmelin-Vivien, M., Loizeau, V., Mellon, C., Beker, B., Arlhac, D., Bodiguel, X., Ferraton, F., Hermand, R., Philippon, X., and Salen-Picard, C.: Comparison of C and N stable isotope ratios between surface particulate organic matter and microphytoplankton in the Gulf of Lions (NW Mediterranean), Cont. Shelf Res., 28, 1911–1919, 2008.

He, B., Dai, M., Huang, W., Liu, Q., Chen, H., and Xu, L.: Sources and accumulation of organic carbon in the Pearl River Estuary surface sediment as indicated by elemental, stable carbon isotopic, and carbohydrate compositions, Biogeosciences, 7, 3343-3362, https://doi.org/10.5194/bg-7-3343-2010, 2010a.

He, B., Dai, M., Zhai, W., Wang, L., Wang, K., Chen, J., Lin, J., Han, A., and Xu, Y.: Distribution, degradation and dynamics of dissolved organic carbon and its major compound classes in the Pearl River estuary, China, Mar. Chem., 119, 52–64, https://doi.org/10.1016/j.marchem.2009.12.006, 2010b.

He, B., Dai, M., Zhai, W., Guo, X., and Wang, L.: Hypoxia in the upper reaches of the Pearl River Estuary and its maintenance mechanisms: A synthesis based on multiple year observations during 2000–2008, Mar. Chem., 167, 13–24, https://doi.org/10.1016/j.marchem.2014.07.003, 2014.

He, G.-F. and Yuan, G.-M.: Assessment of the water quality by fuzzy mathematics for last 20 years in Zhujiang Estuary, Mar. Environ. Sci., 26, 53–57, 2007.

Hu, J., Peng, P. a., Jia, G., Mai, B., and Zhang, G.: Distribution and sources of organic carbon, nitrogen and their isotopes in sediments of the subtropical Pearl River estuary and adjacent shelf, Southern China, Mar. Chem., 98, 274–285, https://doi.org/10.1016/j.marchem.2005.03.008, 2006.

Huang, X., Huang, L., and Yue, W.: The characteristics of nutrients and eutrophication in the Pearl River estuary, South China, Mar. Pollut. Bull., 47, 30–36, https://doi.org/10.1016/S0025-326X(02)00474-5, 2003.

Hullar, M., Fry, B., Peterson, B., and Wright, R.: Microbial utilization of estuarine dissolved organic carbon: a stable isotope tracer approach tested by mass balance, Appl. Environ. Microbiol., 62, 2489–2493, 1996.

Kao, S. J., Lin, F. J., and Liu, K. K.: Organic carbon and nitrogen contents and their isotopic compositions in surficial sediments from the East China Sea shelf and the southern Okinawa Trough, Deep Sea Res. Pt. II, 50, 1203–1217, https://doi.org/10.1016/S0967-0645(03)00018-3, 2003.

Kao, S.-J., Terence Yang, J.-Y., Liu, K.-K., Dai, M., Chou, W.-C., Lin, H.-L., and Ren, H.: Isotope constraints on particulate nitrogen source and dynamics in the upper water column of the oligotrophic South China Sea, Global Biogeochem. Cy., 26, GB2033, https://doi.org/10.1029/2011GB004091, 2012.

Kemp, W. M., Testa, J. M., Conley, D. J., Gilbert, D., and Hagy, J. D.: Temporal responses of coastal hypoxia to nutrient loading and physical controls, Biogeosciences, 6, 2985–3008, https://doi.org/10.5194/bg-6-2985-2009, 2009.

Lehmann, M. F., Bernasconi, S. M., Barbieri, A., and McKenzie, J. A.: Preservation of organic matter and alteration of its carbon and nitrogen isotope composition during simulated and in situ early sedimentary diagenesis, Geochim. Cosmochim. Ac., 66, 3573–3584, https://doi.org/10.1016/S0016-7037(02)00968-7, 2002.

Li, D., Zhang, J., Huang, D., Wu, Y., and Liang, J.: Oxygen depletion off the Changjiang (Yangtze River) estuary, Sci. China Ser. D, 45, 1137–1146, https://doi.org/10.1360/02yd9110, 2002.

Liu, M., Xiao, T., Wu, Y., Zhou, F., Huang, H., Bao, S., and Zhang, W.: Temporal distribution of bacterial community structure in the Changjiang Estuary hypoxia area and the adjacent East China Sea, Environ. Res. Lett., 7, 025001, https://doi.org/10.1088/1748-9326/7/2/025001, 2012.

Marthur, J. M., Tyson, R. V., Thomson, J., and Mattey, D.: Early diagenesis of marine organic matter: Alteration of the carbon isotopic composition, Mar. Geol., 105, 51–61, https://doi.org/10.1016/0025-3227(92)90181-G, 1992.

Mercedes, M. C. B., Luiz Antonio, M., Tibisay, P., Rafael, R., Jean Pierre, H. B. O., Felipe Siqueira, P., Silvia Rafaela Machado, L., and Sorena, M.: Nitrogen management challenges in major watersheds of South America, Environ. Res. Lett., 10, 065007, https://doi.org/10.1088/1748-9326/10/6/065007, 2015.

Meyers, P. A.: Organic geochemical proxies of paleoceanographic, paleolimnologic, and paleoclimatic processes, Org. Geochem., 27, 213–250, https://doi.org/10.1016/S0146-6380(97)00049-1, 1997.

Nixon, S. W.: Coastal marine eutrophication: A definition, social causes, and future concerns, Ophelia, 41, 199-2-19, https://doi.org/10.1080/00785236.1995.10422044, 1995.

Paerl, H. W.: Assessing and managing nutrient-enhanced eutrophication in estuarine and coastal waters: Interactive effects of human and climatic perturbations, Ecol. Eng., 26, 40–54, https://doi.org/10.1016/j.ecoleng.2005.09.006, 2006.

Paerl, H. W.: Controlling Eutrophication along the Freshwater–Marine Continuum: Dual Nutrient (N and P) Reductions are Essential, Estuar. Coast., 32, 593–601, https://doi.org/10.1007/s12237-009-9158-8, 2009.

Peterson, B. J. and Fry, B.: Stable Isotopes in Ecosystem Studies, Annu. Rev. Ecol. Syst., 18, 293–320, 1987.

Qian, W., Dai, M., Xu, M., Kao, S.-j., Du, C., Liu, J., Wang, H., Guo, L., and Wang, L.: Non-local drivers of the summer hypoxia in the East China Sea off the Changjiang Estuary, Estuar. Coast. Shelf Sci., https://doi.org/10.1016/j.ecss.2016.08.032, in press, 2016.

Qian, W., Gan, J., Liu, J., He, B., Lu, Z., Guo, X., Wang, D., Guo, L., Huang, T., and Dai, M.: Current status of emerging hypoxia in a large eutrophic estuary: the lower reach of Pearl River estuary, China, Estuar. Coast. Shelf Sci., submitted, 2017.

Rabalais, N., Cai, W.-J., Carstensen, J., Conley, D., Fry, B., Hu, X., Quiñones-Rivera, Z., Rosenberg, R., Slomp, C., Turner, E., Voss, M., Wissel, B., and Zhang, J.: Eutrophication-Driven Deoxygenation in the Coastal Ocean, Oceanography, 27, 172–183, https://doi.org/10.5670/oceanog.2014.21, 2014.

Rabalais, N. N., Díaz, R. J., Levin, L. A., Turner, R. E., Gilbert, D., and Zhang, J.: Dynamics and distribution of natural and human-caused hypoxia, Biogeosciences, 7, 585–619, https://doi.org/10.5194/bg-7-585-2010, 2010.

Rabouille, C., Conley, D. J., Dai, M. H., Cai, W. J., Chen, C. T. A., Lansard, B., Green, R., Yin, K., Harrison, P. J., Dagg, M., and McKee, B.: Comparison of hypoxia among four river-dominated ocean margins: The Changjiang (Yangtze), Mississippi, Pearl, and Rhône rivers, Cont. Shelf Res., 28, 1527–1537, https://doi.org/10.1016/j.csr.2008.01.020, 2008.

Rutger, R., Stefan, A., Birthe, H., Hans, C. N., and Karl, N.: Recovery of marine benthic habitats and fauna in a Swedish fjord following improved oxygen conditions, Mar. Ecol. Prog. Ser., 234, 43–53, https://doi.org/10.3354/meps234043, 2002.

Shultz, D. J. and Calder, J. A.: Organic carbon $^{13}C/^{12}C$ variations in estuarine sediments, Geochim. Cosmochim. Ac., 40, 381–385, https://doi.org/10.1016/0016-7037(76)90002-8, 1976.

Steckbauer, A., Duarte, C. M., Carstensen, J., Vaquer-Sunyer, R., and Conley, D. J.: Ecosystem impacts of hypoxia: thresholds of hypoxia and pathways to recovery, Environ. Res. Lett., 6, 025003, https://doi.org/10.1088/1748-9326/6/2/025003, 2011.

Stefan, R., Mateete, B., Clare, M. H., Nancy, K., Wilfried, W., Xiaoyuan, Y., Albert, B., and Mark, A. S.: Synthesis and review: Tackling the nitrogen management challenge: from global to local scales, Environ. Res. Lett., 11, 120205, https://doi.org/10.1088/1748-9326/11/12/120205, 2016.

Strong, D. J., Flecker, R., Valdes, P. J., Wilkinson, I. P., Rees, J. G., Zong, Y. Q., Lloyd, J. M., Garrett, E., and Pancost, R. D.: Organic matter distribution in the modern sediments of the Pearl River Estuary, Org. Geochem., 49, 68–82, https://doi.org/10.1016/j.orggeochem.2012.04.011, 2012.

Swarzenski, P., Campbell, P., Osterman, L., and Poore, R.: A 1000-year sediment record of recurring hypoxia off the Mississippi River: The potential role of terrestrially-derived organic matter inputs, Mar. Chem., 109, 130–142, https://doi.org/10.1016/j.marchem.2008.01.003, 2008.

Tan, F. C., Cai, D. L., and Edmond, J. M.: Carbon isotope geochemistry of the Changjiang estuary, Estuar. Coast. Shelf Sci., 32, 395–403, https://doi.org/10.1016/0272-7714(91)90051-C, 1991.

Thornton, S. F. and McManus, J.: Application of Organic Carbon and Nitrogen Stable Isotope and C/N Ratios as Source Indicators of Organic Matter Provenance in Estuarine Systems: Evidence from the Tay Estuary, Scotland, Estuar. Coast. Shelf Sci., 38, 219–233, https://doi.org/10.1006/ecss.1994.1015, 1994.

Vaquer-Sunyer, R. and Duarte, C. M.: Thresholds of hypoxia for marine biodiversity, P. Natl. Acad. Sci. USA, 105, 15452–15457, https://doi.org/10.1073/pnas.0803833105, 2008.

Wang, H., Dai, M., Liu, J., Kao, S.-J., Zhang, C., Cai, W.-J., Wang, G., Qian, W., Zhao, M., and Sun, Z.: Eutrophication-Driven Hypoxia in the East China Sea off the Changjiang Estuary, Environ. Sci. Technol., 50, 2255–2263, https://doi.org/10.1021/acs.est.5b06211, 2016.

Wang, Q., Koshikawa, H., Liu, C., and Otsubo, K.: 30-year changes in the nitrogen inputs to the Yangtze River Basin, Environ. Res. Lett., 9, 115005, https://doi.org/10.1088/1748-9326/9/11/115005, 2014.

Xing, L., Zhang, H., Yuan, Z., Sun, Y., and Zhao, M.: Terrestrial and marine biomarker estimates of organic matter sources and distributions in surface sediments from the East China Sea shelf, Cont. Shelf Res., 31, 1106–1115, https://doi.org/10.1016/j.csr.2011.04.003, 2011.

Yao, P., Zhao, B., Bianchi, T. S., Guo, Z., Zhao, M., Li, D., Pan, H., Wang, J., Zhang, T., and Yu, Z.: Remineralization of sedimentary organic carbon in mud deposits of the Changjiang Estuary and adjacent shelf: Implications for carbon preservation and authigenic mineral formation, Cont. Shelf Res., 91, 1-11, https://doi.org/10.1016/j.csr.2014.08.010, 2014.

Ye, F., Huang, X., Zhang, X., Zhang, D., Zeng, Y., and Tian, L.: Recent oxygen depletion in the Pearl River Estuary, South China: geochemical and microfaunal evidence, J. Oceanogr., 68, 387–400, https://doi.org/10.1007/s10872-012-0104-1, 2012.

Yu, F., Zong, Y., Lloyd, J. M., Huang, G., Leng, M. J., Kendrick, C., Lamb, A. L., and Yim, W. W.-S.: Bulk organic $\delta^{13}C$ and C/N as indicators for sediment sources in the Pearl River delta and estuary, southern China, Estuar. Coast. Shelf Sci., 87, 618–630, https://doi.org/10.1016/j.ecss.2010.02.018, 2010.

Zhang, J., Zhang, Z. F., Liu, S. M., Wu, Y., Xiong, H., and Chen, H. T.: Human impacts on the large world rivers: Would the Changjiang (Yangtze River) be an illustration?, Global Biogeochem. Cy., 13, 1099–1105, https://doi.org/10.1029/1999GB900044, 1999.

Zhang, J., Cowie, G., and Naqvi, S. W. A.: Hypoxia in the changing marine environment, Environ. Res. Lett., 8, 015025, https://doi.org/10.1088/1748-9326/8/1/015025, 2013.

Zhang, Y., Xiao, W., and Jiao, N.: Linking biochemical properties of particles to particle-attached and free-living bacterial community structure along the particle density gradient from freshwater to open ocean, J. Geophys. Res.-Biogeo., 121, 2261–2274, https://doi.org/10.1002/2016JG003390, 2016.

Zhao, H.-D., Kao, S.-J., Zhai, W.-D., Zang, K.-P., Zheng, N., Xu, X.-M., Huo, C., and Wang, J.-Y.: Effects of stratification, organic matter remineralization and bathymetry on summertime oxygen distribution in the Bohai Sea, China, Cont. Shelf Res., 134, 15–25, https://doi.org/10.1016/j.csr.2016.12.004, 2017.

Zhu, C., Wagner, T., Pan, J.-M., and Pancost, R. D.: Multiple sources and extensive degradation of terrestrial sedimentary organic matter across an energetic, wide continental shelf, Geochem. Geophy. Geosy., 12, Q08011, https://doi.org/10.1029/2011GC003506, 2011a.

Zhu, Z.-Y., Zhang, J., Wu, Y., Zhang, Y.-Y., Lin, J., and Liu, S.-M.: Hypoxia off the Changjiang (Yangtze River) Estuary: oxygen depletion and organic matter decomposition, Mar. Chem., 125, 108-116, https://doi.org/10.1016/j.marchem.2011.03.005, 2011b.

Zimmerman, A. R. and Canuel, E. A.: A geochemical record of eutrophication and anoxia in Chesapeake Bay sediments: anthropogenic influence on organic matter composition, Mar. Chem., 69, 117-137, https://doi.org/10.1016/S0304-4203(99)00100-0, 2000.

Zong, Y., Lloyd, J., Leng, M., Yim, W.-S., and Huang, G.: Reconstruction of Holocene monsoon history from the Pearl River Estuary, southern China, using diatoms and carbon isotope ratios, Holocene, 16, 251–263, https://doi.org/10.1191/0959683606hl911rp, 2006.

8

Initial shifts in nitrogen impact on ecosystem carbon fluxes in an alpine meadow: Patterns and causes

Bing Song[1,2], Jian Sun[1], Qingping Zhou[3], Ning Zong[1], Linghao Li[2], and Shuli Niu[1,4]

[1]Key Laboratory of Ecosystem Network Observation and Modeling, Institute of Geographic Sciences and Natural Resources Research, Chinese Academy of Sciences, Datun Road, Beijing 100101, China
[2]State Key Laboratory of Vegetation and Environmental Change, Institute of Botany, Chinese Academy of Sciences, Xiangshan, Beijing 100093, China
[3]Institute of Qinghai–Tibetan Plateau, Southwest Minzu University, Chengdu 610041, China
[4]Department of Resources and Environment, University of Chinese Academy of Sciences, Beijing 100049, China

Correspondence to: Shuli Niu (sniu@igsnrr.ac.cn)

Abstract. Increases in nitrogen (N) deposition can greatly stimulate ecosystem net carbon (C) sequestration through positive N-induced effects on plant productivity. However, how net ecosystem CO_2 exchange (NEE) and its components respond to different N addition rates remains unclear. Using an N addition gradient experiment (six levels: 0, 2, 4, 8, 16, 32 $\mathrm{g\,N\,m^{-2}\,yr^{-1}}$) in an alpine meadow on the Qinghai–Tibetan Plateau, we explored the responses of different ecosystem C fluxes to an N addition gradient and revealed mechanisms underlying the dynamic responses. Results showed that NEE, ecosystem respiration (ER), and gross ecosystem production (GEP) all increased linearly with N addition rates in the first year of treatment but shifted to N saturation responses in the second year with the highest NEE ($-7.77 \pm 0.48\,\mu\mathrm{mol\,m^{-2}\,s^{-1}}$) occurring under an N addition rate of 8 $\mathrm{g\,N\,m^{-2}\,yr^{-1}}$. The saturation responses of NEE and GEP were caused by N-induced accumulation of standing litter, which limited light availability for plant growth under high N addition. The saturation response of ER was mainly due to an N-induced saturation response of aboveground plant respiration and decreasing soil microbial respiration along the N addition gradient, while decreases in soil microbial respiration under high N addition were caused by N-induced reductions in soil pH. We also found that various components of ER, including aboveground plant respiration, soil respiration, root respiration, and microbial respiration, responded differentially to the N addition gradient. These results reveal temporal dynamics of N impacts and the rapid shift in ecosystem C fluxes from N limitation to N saturation. Our findings bring evidence of short-term initial shifts in responses of ecosystem C fluxes to increases in N deposition, which should be considered when predicting long-term changes in ecosystem net C sequestration.

1 Introduction

Anthropogenic reactive nitrogen (N) inputs to the terrestrial biosphere has increased more than 3-fold over the past century and is predicted to increase further (Lamarque et al., 2005; Galloway et al., 2008). Because of the strong coupling of ecosystem carbon (C) and N cycles, excess N deposition could have significant impacts on the ecosystem C cycle (LeBauer and Treseder, 2008; Liu and Greaver, 2010; Lu et al., 2011). Ecosystem net C sequestration is predicted to increase or have no significant change under rising N deposition (Nadelhoffer et al., 1999; Magnani et al., 2007; Reay et al., 2008; Niu et al., 2010; Lu et al., 2011; Fernandez-Martinez et al., 2014). However, we have a limited understanding of the dynamic N responses of C sequestration in terrestrial ecosystems, which is crucial for model projection of the future terrestrial C cycle under rising N deposition (Reay et al., 2008).

Although N addition generally enhances plant growth and ecosystem net primary productivity (NPP) based on global syntheses of N addition experiments (LeBauer and Treseder,

2008; Xia and Wan, 2008; Lu et al., 2011), the responses of ecosystem C fluxes vary with N loading rates (Liu and Greaver, 2010; Lu et al., 2011). According to the N saturation hypothesis, NPP is assumed to slowly increase with N addition rates first, then reach its maximum value at the N saturation point, and finally decline with further increase in N input (Aber et al., 1989; Lovett and Goodale, 2011). During this process, NPP shifts from an N-limited, an N intermediate, to an N saturation stage as N deposition increases. Similarly, net ecosystem CO_2 exchange (NEE) and its components of gross ecosystem production (GEP) and ecosystem respiration (ER) may also respond nonlinearly to increasing N loading rates (Fleischer et al., 2013; Gomez-Casanovas et al., 2016; Tian et al., 2016). In the N-limited stage, low rates of N addition could stimulate ecosystem productivity (Aber et al., 1989), GEP (Fleischer et al., 2013; Gomez-Casanovas et al., 2016), and ER (Hasselquist et al., 2012; Zhu et al., 2016), while in the N saturation stage, high doses of N addition could have negative effects on GEP and ER (Treseder, 2008; Janssens et al., 2010; Maaroufi et al., 2015). The unbalanced responses of GEP and ER may lead to changes in NEE.

Moreover, ER can be divided into aboveground plant respiration, belowground plant respiration (root respiration), and soil microbial respiration. These components of ER could be affected by plant aboveground biomass, root biomass, soil organic matter, and microbial biomass C, which may respond variously to N addition (Phillips and Fahey, 2007; Hasselquist et al., 2012). For example, root respiration would be enhanced or not significantly changed under N addition, while soil microbial respiration may be suppressed by N addition (Zhou et al., 2014). The different responses of various components of ER to N addition will also consequently change the response of NEE. Nevertheless, there is limited knowledge on how various components of NEE respond differentially to N addition gradient. In addition, the N responses of ecosystem C fluxes may shift over time because of changes in plant community structure and other limiting factors (Niu et al., 2010). It is not clear when and how ecosystem C fluxes get N saturated under increasing N input. The mechanisms underlying the saturation response of C fluxes are even far from clear, which prevent us from accurately predicting the C cycle in response to rising N deposition.

In this study, we explored the responses of various ecosystem C cycle processes to an N addition gradient in an alpine meadow on the Qinghai–Tibetan Plateau. The Qinghai–Tibetan Plateau has an area of 2.5 million km^2 with alpine meadow covering 35 % of this area, and it is sensitive to environmental change and human activities (Chen et al., 2013). The objectives of this study were to explore how different components of ecosystem C fluxes respond to increasing N loading gradient. Specifically, we addressed the following questions. (i) How do NEE and its components respond to the N addition gradient in the alpine meadow? (ii) Can various C cycle processes get N saturated? If so, at which N addition level are they saturated and how do the responses shift with time? (iii) What are the mechanisms underlying N saturation responses of different C cycle processes?

2 Materials and methods

2.1 Study site

The study site is located in an alpine meadow in Hongyuan County, Sichuan Province, China, which is on the eastern Qinghai–Tibetan Plateau (32°48′ N, 102°33′ E). The altitude is ~3500 m. Long-term (1961–2013) mean annual precipitation is 747 mm with approximately 80 % occurring in May to September. Long-term mean annual temperature is 1.5 °C with monthly mean temperature ranging from −9.7 °C in January to 11.1 °C in July. The dominant species in this alpine meadow are *Deschampsia caespitosa* (Linn.) Beauv., *Kobresia setchwanensis* Hand.-Mazz., *Carex schneideri* Nelmes, and *Anemone rivularis* Buch.-Ham. The vegetation cover of this grassland is over 90 %. The soil in the study site is classified as Mat Cry-gelic Cambisol according to the Chinese classification, with surface soil bulk density being 0.89 $g\,cm^{-3}$. The soil organic C content and total N content are 37 $gC\,kg^{-1}$ and 3.5 $gN\,kg^{-1}$, respectively. The background N deposition ranges from 0.87 to 1.38 $gN\,m^{-2}\,yr^{-1}$ on the eastern Qinghai–Tibetan Plateau, and the natural N deposition rate in China ranges from 0.11 to 6.35 $gN\,m^{-2}\,yr^{-1}$ (Lü and Tian, 2007).

2.2 Experimental design

We conducted an N addition experiment with six levels of N addition rate (0, 2, 4, 8, 16, 32 $gN\,m^{-2}\,yr^{-1}$) in early 2014. The six N treatments were represented by N0 (control), N2, N4, N8, N16, and N32. The treatments were randomly assigned with five replications, so there were totally 30 plots. Each plot was 8 × 8 m, and the distance between any two adjacent plots was 3 m. The N addition treatments started from May 2014. In 2014 and 2015, N was applied by hand as NH_4NO_3 (>99 %) every month from May to September (i.e., during the growing season) before rainfall. The N amount was same in each month. In order to distribute dry NH_4NO_3 evenly in the plots, we mixed dry NH_4NO_3 with enough amounts of soil to apply.

2.3 Ecosystem C cycle properties and soil pH measurement

Ecosystem C fluxes were measured using a transparent static chamber (0.5 × 0.5 × 0.5 m) attached to an infrared gas analyzer (LI-6400XT; LI-COR Environmental, Lincoln, Nebraska, USA) in the field. During each measurement, the chamber was positioned over a square steel frame, which was permanently inserted into soil and offered a flat base for the chamber. Inside the chamber, two electric fans were mounted in order to mix the chamber atmosphere. The measurements were conducted twice per month on clear, sunny days from May to September in 2014 and 2015. Nine consecutive recordings of CO_2 concentration were taken at each base at 10 s intervals. CO_2 flux rates were determined from the time courses of the concentrations to calculate NEE. After the measurement of NEE, the chamber was covered by an opaque cloth and the CO_2 measurement was repeated. As the second measurement eliminated light, the CO_2 flux value obtained represented ER. GEP was calculated as the difference between NEE and ER. Negative or positive NEE and GEP values represent net C uptake or release, respectively. The detailed methods have also been described in Niu et al. (2008, 2013).

Soil respiration (SR) was assessed following the measurement of NEE and ER. It was also measured with LI-6400XT attaching a soil CO_2 flux chamber (991 cm^3 in total volume; LI-6400-09; LI-COR Environmental, Lincoln, Nebraska, USA). A PVC collar (10.5 cm in diameter and 5 cm in height) was permanently installed 2–3 cm into the soil. The soil respiration chamber attached to LI-6400XT was placed on each PVC collar for 1–2 min to measure SR. Living plants inside the collars were removed regularly by hand to eliminate aboveground plant respiration. Soil heterotrophic respiration (i.e., soil microbial respiration, R_{mic}) was measured using the same method as soil respiration. Differently, the PVC collar was 40 cm in height and installed 36–38 cm into the soil. As > 90 % of plant roots were distributed in the topsoil (0–20 cm), 40 cm long PVC collars could cut off old plant roots and prevented new roots from growing inside the collars. Plants in the collars were completely removed by hand to exclude C supply. The experiment was conducted in early 2014 and the measurements of CO_2 fluxes above these 40 cm long PVC collars began in late July 2014, leaving enough time for the remaining plant roots inside the collars to die. Thus CO_2 fluxes in those deep collars represented R_{mic}. The method was the same as in Wan et al. (2005) and Zhou et al. (2007). Root respiration (R_{root}) was calculated by the value of SR minus R_{mic}. Aboveground plant respiration (R_{above}) was calculated by ER minus SR, and ecosystem plant respiration (R_{plant}) was calculated as the difference between ER and R_{mic}. All the measurements of ecosystem C fluxes were simultaneous.

Soil samples were collected from the topsoil (0–10 cm) of the 30 plots on 15 August 2014 and 14 August 2015. Two soil cores (8 cm in diameter and 10 cm in depth) were taken at least 1 m from the edge in each plot, and then completely mixed to get a composite sample. The soil samples were sieved by a 2 mm mesh and then were air-dried for chemical analysis. Soil pH was determined with a glass electrode in a 1 : 2.5 soil / water solution (w/v).

2.4 Statistical analysis

Repeated-measures ANOVA was used to examine N addition effects on each ecosystem C flux over the growing season in 2014 and 2015. When we evaluate N addition effects on the different components of ER and their proportions, we averaged their values across the year and then used one-way ANOVA to test the differences among treatments. To test the response pattern of ecosystem C cycle properties to the N addition gradient, we fitted the response parameter to linear or quadratic functions which had higher R^2. We also compared the Akaike information criterion (AIC) between the functions. Simple linear regression analyses were used to evaluate relationships of ER with its components and NEE across the 2 years. R_{mic} and pH were calculated by data in different N addition treatments minus data in the control treatment. Stepwise multiple linear regressions were used to test the drivers which best predict R_{mic}. All data were tested for normal distribution before statistical analysis. The a posteriori comparisons were performed by Duncan test, and the effects were considered to be significantly different if $P < 0.05$. All statistical analyses were conducted with SAS V.8.1 software (SAS Institute Inc., Cary, North Carolina, USA).

3 Results

3.1 NEE and its components in response to the N addition gradient

NEE varied throughout the growing seasons in both 2014 and 2015. The maximum rates of net CO_2 uptake (indicated by large negative values of NEE) occurred in July in both years (Fig. 1a, d). N addition had a significant impact on NEE in 2014 ($P = 0.020$) and a marginally significant effect in 2015 ($P = 0.059$) (Table 1). Mean NEE across months had different responses to the N addition gradient between the 2 years (Fig. 1a, d). It increased linearly with N addition rates in 2014 (Table 2; Fig. 2a) but shifted to a saturating response with N addition rates in 2015 (Table 2; Fig. 2d). The largest NEE was $-7.77 \pm 0.48\,\mu mol\,m^{-2}\,s^{-1}$ under an 8 $gN\,m^{-2}\,yr^{-1}$ addition rate (N8) in 2015.

The N addition gradient had significant effects on ER ($P = 0.033$ and 0.006, respectively) and GEP ($P = 0.002$ and 0.038, respectively) in both 2014 and 2015 (Table 1). Similar to NEE, both ER and GEP showed linear responses to N addition rates in 2014 but shifted to saturation responses in 2015 (Table 2; Fig. 2). On average, ER was enhanced by 0.9–16.1 % in 2014 and 7.9–23.7 % in 2015 under different N addition treatments. GEP was increased by 2.4–19.2 % in 2014 and 6.7–20.5 % in 2015 under different N addition levels, with maximal values being $-24.40 \pm 0.48\,\mu mol\,m^{-2}\,s^{-1}$ under 32 $gN\,m^{-2}\,yr^{-1}$ in 2014 and $-15.38 \pm 0.72\,\mu mol\,m^{-2}\,s^{-1}$ under 16 $gN\,m^{-2}\,yr^{-1}$ in 2015 (Fig. 2).

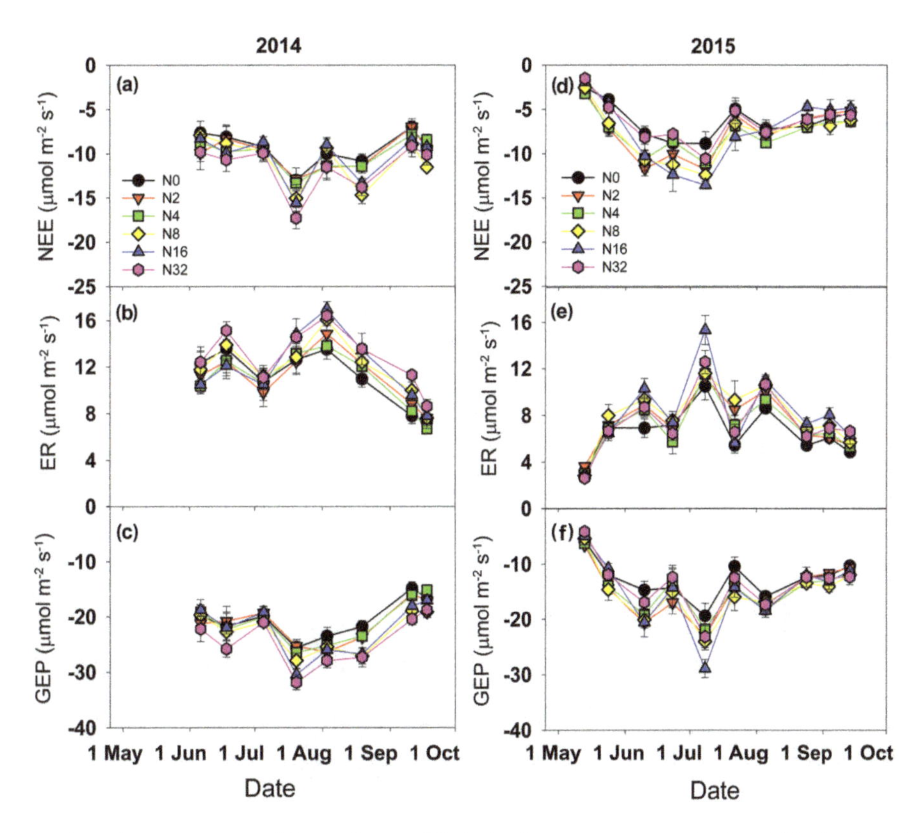

Figure 1. Seasonal dynamics of net ecosystem CO_2 exchange (NEE) **(a, d)**, ecosystem respiration (ER) **(b, e)**, and gross ecosystem production (GEP) **(c, f)** in 2014 and 2015. N0, N2, N4, N8, N16, and N32 represent N addition rates of 0, 2, 4, 8, 16, and 32 gN m^{-2} yr^{-1}, respectively.

Table 1. Results (*F* and *P* values) of one-way ANOVA on the effects of nitrogen addition on ecosystem C fluxes in 2014 and 2015. NEE: net ecosystem CO_2 exchange; ER: ecosystem respiration; GEP: gross ecosystem production; SR: soil respiration; R_{mic}: soil microbial respiration; R_{plant}: plant respiration; R_{above}: aboveground plant respiration; R_{root}: plant root respiration.

		NEE		ER		GEP		SR		R_{mic}	
	df	*F*	*P*	*F*	*P*	*F*	*P*	*F*	*P*	*F*	*P*
2014	5	3.35	0.020	2.95	0.033	5.37	0.002	1.56	0.209	1.49	0.246
2015	5	2.50	0.059	4.35	0.006	2.83	0.038	3.94	0.010	1.40	0.259

		R_{plant}		R_{above}		R_{root}		R_{above}/ER		R_{root}/ER		R_{mic}/ER	
	df	*F*	*P*	*F*	*P*	*F*	*P*	*F*	*F*	*F*	*P*	*F*	*P*
2014	5	1.06	0.409	3.84	0.011	2.64	0.049	3.08	0.027	3.56	0.015	0.28	0.919
2015	5	3.25	0.022	5.38	0.002	0.78	0.573	5.54	0.002	0.97	0.456	2.46	0.062

3.2 Components of ecosystem respiration in response to the N addition gradient

We divided ER into R_{above}, SR, R_{root}, and R_{mic}, and found that different ER components showed diverse responses to the N addition gradient. Mean SR across months was not significantly changed by the N addition gradient in 2014 (Table 1; Fig. 3). However, in 2015, it ranged from 4.98 ± 0.33 to 6.23 ± 0.23 μmol m^{-2} s^{-1} under different N addition levels, with a significant reduction under high N addition levels of 16 and 32 gN m^{-2} yr^{-1} ($P = 0.010$; Fig. 3). Additionally, the relationship between SR and N addition rates was not significant in 2014 (Fig. 3a), while SR leveled off under high N addition rates in 2015 (Fig. 3c). Interestingly, R_{mic} increased linearly with N addition rates in 2014 (Table 2; Fig. 3b), while it decreased with N addition rates in 2015 (Table 2; Fig. 3d). Comparing among various components of ER, only R_{mic} showed distinctively inverse responses to N

Table 2. Comparisons of the Akaike information criterion (AIC) among functions describing the relationships between NEE, ER, GEP, SR, and R_{mic} (Y) and N addition rates (X). Quadratic functions work better than linear ones for ecosystem C fluxes in 2015. NEE: net ecosystem CO_2 exchange; ER: ecosystem respiration; GEP: gross ecosystem production; SR: soil respiration; R_{mic}: soil microbial respiration.

	Functions in 2014		Functions in 2015	
	Linear[1]	Quadratic[2]	Linear[1]	Quadratic[2]
NEE	78.39	80.26	88.69	84.82
ER	71.68	73.48	90.12	82.30
GEP	77.96	79.86	87.69	77.68
SR	87.88	87.34	79.43	78.18
R_{mic}	78.33	80.27	85.15	84.48

[1] Linear model: $Y = b_1 + b_2 \times X$. [2] Quadratic model: $Y = b_1 + b_2 \times X + b_3 \times X^2$.

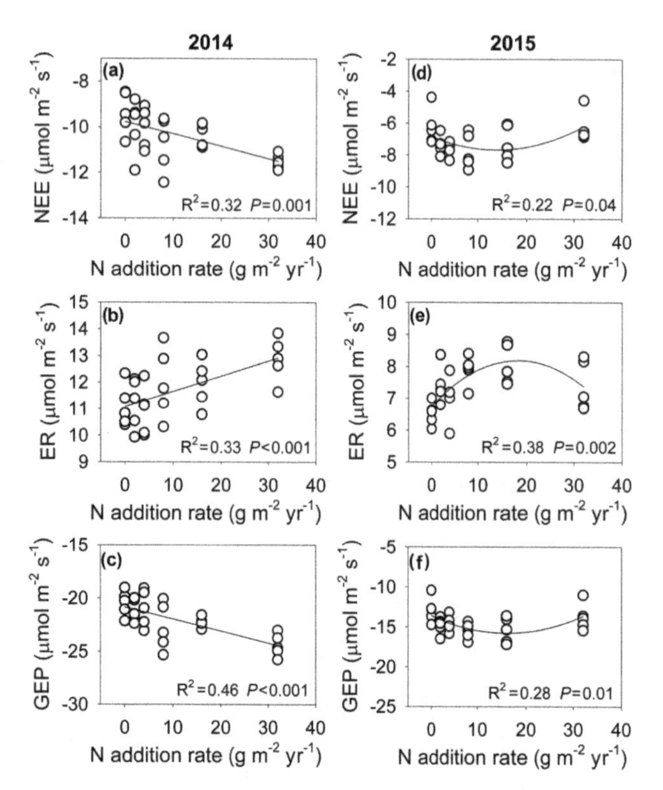

Figure 2. Relationships between N addition rate and net ecosystem CO_2 exchange (NEE) **(a, d)**, ecosystem respiration (ER) **(b, e)**, and gross ecosystem production (GEP) **(c, f)** in 2014 and 2015.

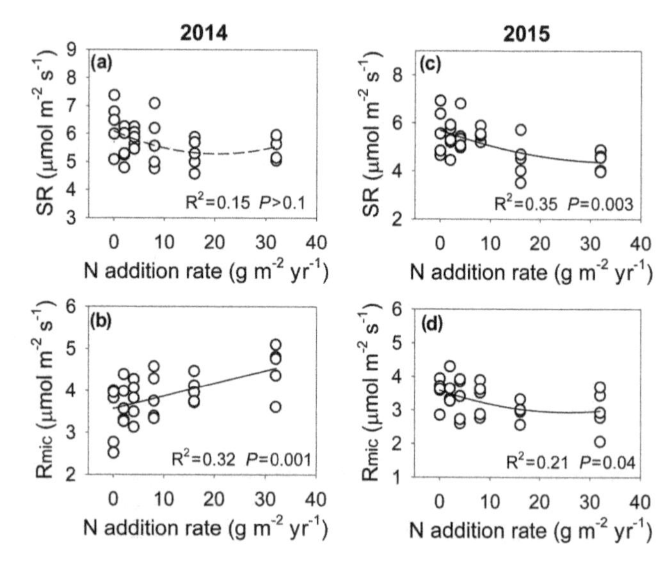

Figure 3. Relationships between N addition rate and soil respiration (SR) **(a, c)** and soil microbial respiration (R_{mic}) **(b, d)** in 2014 and 2015.

addition rates between the years. R_{above} increased with increasing N addition rates in 2014 (Fig. 4b) but reached its maximum value at N16 in 2015 (Fig. 4e). By contrast, R_{root} decreased with increasing N addition rates in 2014 (Fig. 4c), while it had no statistically significant response to the N addition gradient in 2015 (Fig. 4f).

In addition, the proportions of different efflux components to ER differed in response to the N addition gradient between years (Fig. 5). The proportions of R_{above} to ER kept increasing with N addition rates in 2014 but became saturated at N16 in 2015 (Fig. 5a, d). The proportions of R_{root} to ER ranged from 31.90 ± 6.69 % in N0 plots to 11.18 ± 1.28 % in N32 plots in 2014 (Fig. 5b), but they were not significantly different among N addition levels in 2015 (Table 1; Fig. 5e). In 2014, the contributions of R_{mic} to ER did not significantly change under N addition treatments (Table 1; Fig. 5c), whereas they declined along the N addition gradient in 2015 (Fig. 5f).

3.3 Causes of the N saturation responses of ecosystem C fluxes

In order to examine the causes of the N saturation responses of NEE and ER in 2015, we examined the relationship between ER and its various components and also NEE. The results showed that ER had a significantly positive correlation with R_{above} (Fig. 6a) but not with R_{root} (Fig. 6b) and had a significantly negative correlation with R_{mic} (Fig. 6c).

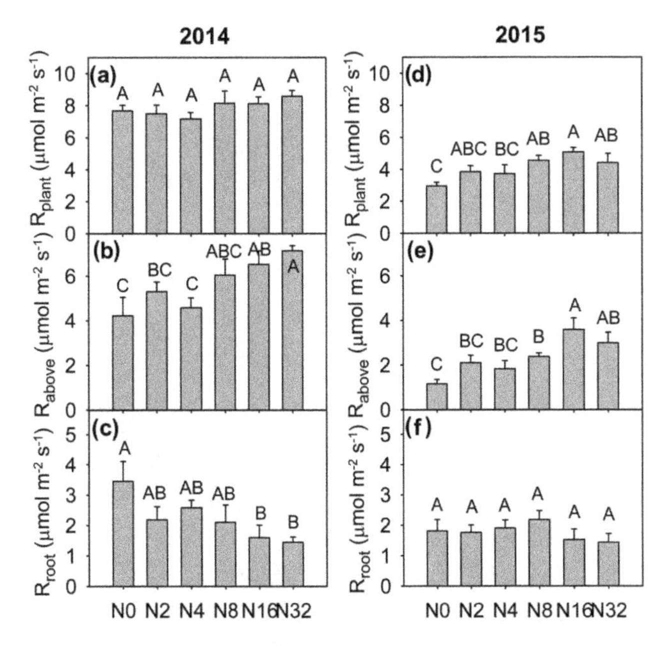

Figure 4. Plant respiration and its components in response to the N addition gradient in 2014 and 2015 (mean ± SE, $n = 5$). R_{plant}: plant respiration **(a, d)**; R_{above}: aboveground plant respiration **(b, e)**; R_{root}: plant root respiration **(c, f)**. N0, N2, N4, N8, N16, and N32 represent N addition rates of 0, 2, 4, 8, 16, and 32 gN m^{-2},yr^{-1}, respectively.

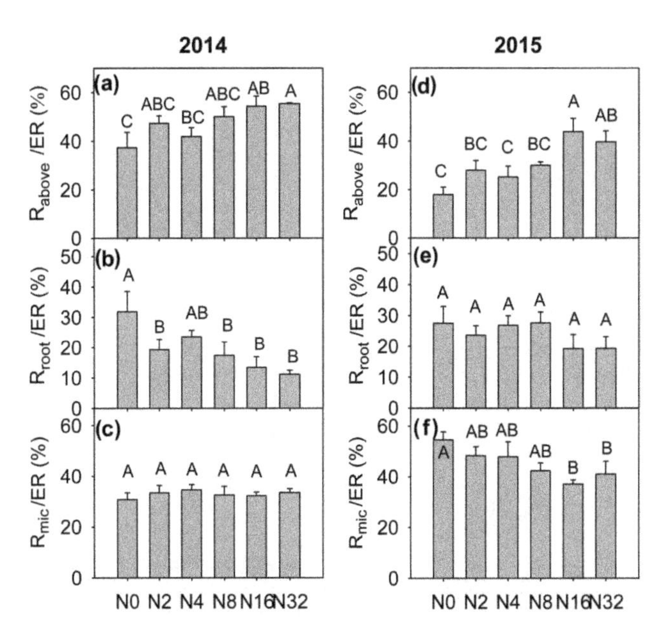

Figure 5. The contributions of different source components to ecosystem respiration (ER) in response to the N addition gradient in 2014 and 2015 (mean ± SE, $n = 5$). R_{above}: aboveground plant respiration; R_{root}: plant root respiration; R_{mic}: soil microbial respiration. N0, N2, N4, N8, N16, and N32 represent N addition rate of 0, 2, 4, 8, 16, and 32 gN m^{-2} yr^{-1}, respectively.

Moreover, NEE closely correlated with ER (Fig. 6d). Using stepwise multiple linear regressions, we further explored the causes of decreasing R_{mic} with N addition in 2015 and examined the relationships of the N-induced reduction in R_{mic} and N-induced reductions in soil pH (pH) and soil N availability (NH_4^+-N, NO_3^--N). The result showed that pH was the best factor to predict R_{mic}. N addition significantly reduced soil pH in 2015 (Fig. 7a), and R_{mic} was positively dependent on pH in 2015 (Fig. 7b) but not in 2014.

4 Discussion

4.1 Nitrogen saturation responses of ecosystem C fluxes and the causes

Our results showed that initial ecosystem C fluxes (NEE, ER, and GEP) in 2014 suggested ecosystem N limitation, whereas in 2015 these C fluxes clearly suggested N saturation under high N addition rates. These findings not only extend the N saturation hypothesis for the response of NPP to N addition (Aber et al., 1998, 1989; Lovett and Goodale, 2011) but also provide comprehensive evidence of potential relationships between various ecosystem C fluxes and ecosystem N dynamics. Previous N addition studies used only one level of N addition and found that NEE showed a positive (Niu et al., 2010; Huff et al., 2015) or no significant response (Harpole et al., 2007; Bubier et al., 2007) to N addition. Using one level of N addition only might not be enough to capture or quantify complex ecosystem responses to N addition. By using an N addition gradient experiment, this study comprehensively showed the saturation responses of NEE and its components to different N loading rates.

The N saturation response of NEE in 2015 was mainly attributed to the saturation responses of ER and GEP (Fig. 2), while the N saturation response of ER was likely caused by the saturation response of aboveground plant respiration and decreasing soil microbial respiration along the N addition gradient. The decrease in aboveground plant respiration under N32 treatment was primarily due to N addition stimulating plant growth and thus standing litter accumulation after plant senescence (Supplement Figs. S1–S2). In 2014, plant aboveground biomass (AGB) was stimulated under the high N addition treatment, especially AGB of grasses (Fig. S2). In this grassland, grasses usually have higher height than other plants. The accumulation of grasses' standing litter under the N32 treatment limited light conditions for other plants and negatively influenced plant growth in the early growing season in 2015. Therefore, GEP and NEE did not keep increasing at the highest N addition rate, leading to N saturation response. The N-induced light limitation for plant growth was also observed in other ecosystems, like temperate grassland (Niu et al., 2010; Kim and Henry, 2013). Moreover, our results showed that most components of ER had a similar response patterns between the 2 years except for soil microbial respiration, which increased in 2014 but decreased in 2015

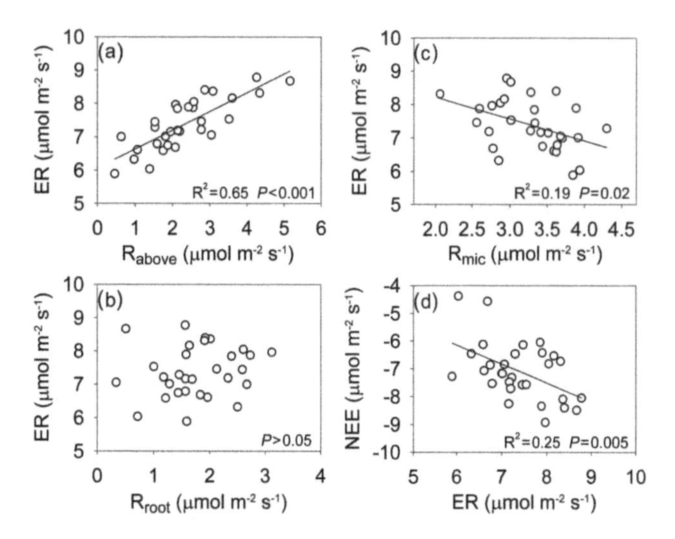

Figure 6. Relationships between aboveground plant respiration (R_{above}), root respiration (R_{root}), soil microbial respiration (R_{mic}) and ecosystem respiration (ER) **(a, b, c)**, ER, and net ecosystem CO_2 exchange(NEE) **(d)** across all plots in 2015.

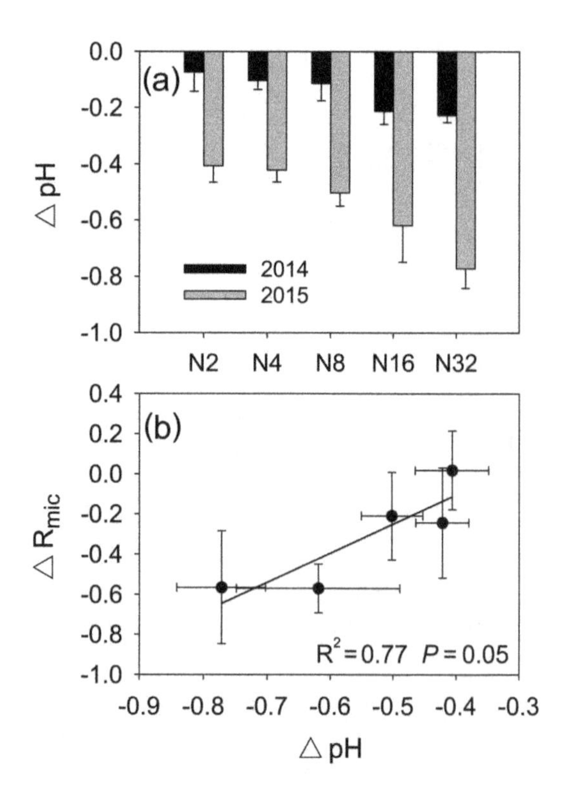

Figure 7. N-induced changes in soil pH (pH) **(a)** (mean ± SE, $n = 5$) and the dependence of N-induced changes in soil microbial respiration (R_{mic}) on N-induced changes in soil pH (pH) in 2015 **(b)**. N2, N4, N8, N16, and N32 represent N addition rate of 2, 4, 8, 16, and 32 $gN\,m^{-2}\,yr^{-1}$, respectively.

along with N addition rates. Thus, we propose that soil microbial respiration might play a key role in mediating the N saturation effects for ER and thus NEE, which is not reported in previous studies. The decline in microbial respiration under high N addition conditions was primarily due to the N-induced reduction in soil pH (Fig. 7). Although many factors can influence soil microbial respiration, such as soil N availability and microbial community structure (Janssens et al., 2010), previous studies with a similar N addition gradient suggested that soil pH was the most important driver for responses of microbes under high N addition rates (Liu et al., 2014; Song et al., 2014; Chen et al., 2016). N addition can lead to soil acidification and have negative impacts on soil microbial growth and activities (Liu et al., 2014; Tian et al., 2016). In this study, the decreased soil pH may cause toxicity effects on microbial activity (Treseder, 2008; Zhou et al., 2012) and thus reduces microbial respiration after 2 years of N addition.

4.2 The time and N threshold for the saturation responses

Our findings demonstrate that N responses of ecosystem C fluxes shifted from a linear response to a saturation response over the 2 years of treatments (Fig. 2). A recent study revealed that ecosystem C fluxes exhibited saturating responses to N addition during two consecutive measurement years in a temperate grassland (Tian et al., 2016). However, their measurement was conducted after 10 years of N addition treatments (similar N addition rates to our study), so it did not capture the early response signals of ecosystem C exchange. Results of another N addition gradient experiment carried out in three marsh ecosystems showed that aboveground plant biomass increased linearly with N addition rates after 7 months of treatment but showed saturating responses after 14 months of N addition (Vivanco et al., 2015). Taken together with our results, this suggests that N saturation of ecosystem C fluxes might happen within a couple years of N input. The different responses between years in this study are not likely due to climate differences because temperature and precipitation were not significantly different between 2014 and 2015. We acknowledge that our findings are only based on the short-term study, while a long-time experiment may capture more robust patterns in N saturation and the underlying mechanisms, but the findings of the initial shift in N responses are helpful to better understand the dynamics of the ecosystem in response to external N input.

The N saturation threshold for ecosystem C fluxes of this alpine meadow is approximately 8 $gN\,m^{-2}\,yr^{-1}$. This level is much higher than that in an alpine steppe on the Qinghai–Tibetan Plateau (Liu et al., 2013). In Liu et al. (2013), biomass N concentration, soil N_2O flux, N-uptake efficiency and N-use efficiency showed saturating responses at an N addition rate of 4 $gN\,m^{-2}\,yr^{-1}$. The discrepancy is probably caused by different precipitation at the two sites. The precipitation is 747 mm at our study site, and it is 415 mm at their study site. The lower precipitation may constrain the ecosystem's response to N addition in Liu et al. (2013). Likewise, the N saturation load in our alpine meadow is higher than that in an alpine dry meadow in Colorado (Bowman et al., 2006) and is comparable with that in a temperate steppe of Eurasian grasslands, which had a saturation N addition rate of approximately 10.5 $gN\,m^{-2}\,yr^{-1}$ (Bai et al., 2010). The differences could also relate to the initial nutrient availability at different sites. Ecosystems with high N availability may reach N saturation at a low rate of N addition if there are no other limiting factors. The higher saturation levels indicate that this alpine meadow is more limited by N compared to other resources. Furthermore, the N critical load for causing changes in ecosystem C cycle processes is around 2 $gN\,m^{-2}\,yr^{-1}$ in this alpine meadow. In the first year, ecosystem C exchanges were not significantly different between N0 and N2 treatments, but C fluxes were greater in N2 plots than in N0 plots in the second year (Fig. 1). This threshold for triggering changes in ecosystem C fluxes is comparable to that in another alpine meadow on the mid-southern part of the Tibetan Plateau (Zong et al., 2016). Considering that atmospheric wet N deposition ranges from 0.87 to 1.38 $gN\,m^{-2}\,yr^{-1}$ on the eastern Qinghai–Tibetan Plateau (Lü and Tian, 2007), our estimate of N critical load suggests that the ecosystem C cycle may be largely affected under future N deposition in the alpine meadow of the Qinghai–Tibetan Plateau.

4.3 Diverse responses of C flux components to the N addition gradient

The components of ER showed diverse responses to the N addition gradient (Figs. 4, 5). For example, in 2014, aboveground plant respiration and its proportion to ER increased, but belowground plant respiration and its proportion to ER decreased with N addition amounts (Figs. 4b, c, 5a, b). To our knowledge, no previous study examined the different components of ER in response to the N addition gradient. Some studies conducted in alpine grassland demonstrated that N addition had no significant effects on ER (Jiang et al., 2013; Gong et al., 2014), since aboveground biomass did not respond to N addition in their studies. In this study, compared to the control treatment (without N addition), greater plant growth and aboveground biomass under N addition enhanced aboveground plant respiration and thus stimulated ER. The lack of N effect on SR in 2014 may be attributed to the counteractive responses of soil microbial respiration and root respiration to N addition. In the first year, N addition ameliorated the nutrient limitation for microbes; thus, soil microbial activity and biomass increased in the short term (Treseder, 2008) and subsequently stimulated microbial respiration (Peng et al., 2011). On the other hand, N addition could reduce belowground biomass allocation (Haynes and Gower, 1995), leading to a decrease in root respiration. The increase in soil microbial respiration partly offsets the decrease in root respiration. As a result, SR had no significant difference among N treatments in the first year. However, in the second year, soil microbial respiration declined under high N addition levels, in combination with the low root respiration, resulting in decreases in SR under N16 and N32 treatments. This decrease in SR was also observed in other ecosystems under long-term or high levels of N addition (Yan et al., 2010; Zhou and Zhang, 2014; Maaroufi et al., 2015). We are fully aware that there are some limitations for the partitioning technique, in which we used deep versus shallow collars to partition root from microbial respiration. This approach cuts roots and excludes the effects of changes in plant C allocation on microbial respiration. Soil moisture content may also change in the deep collars, which likely affects microbial respiration. However, this method has the advantage of exploring mechanisms of microbial responses in the absence of plant effects; it is a common and useful technique to partition the components of ER and is widely used in previous studies (Wan et al., 2005; Zhou et al., 2007).

5 Conclusions

Based on a field N addition gradient experiment, this study tested N saturation theory against multiple C cycle processes. We found that the ecosystem C fluxes of NEE, GEP, and ER shifted from linear responses to saturation responses over 2 years of N addition. The saturation responses of NEE and ER were mainly caused by the N-induced saturation response of aboveground plant respiration and decreasing soil microbial respiration along the N addition gradient. N-induced reduction in soil pH was the main mechanism underlying declines in microbial respiration under high N addition. We also revealed that various components of ER, including aboveground plant respiration, soil respiration, root respiration, and microbial respiration, responded differentially to the N addition gradient. The findings suggest that the C cycle processes have differential responses to N addition between aboveground and belowground plant parts and between plants and microbes. Our findings provide experimental evidence for the dynamic N responses of the ecosystem C cycle, which is helpful for parameterizing biogeochemical models and guiding ecosystem management in light of future increasing N deposition.

Competing interests. The authors declare that they have no conflict of interest.

Acknowledgements. The authors thank Xiaojing Qin, Yanfang Li, Fangyue Zhang, Quan Quan, Zheng Fu, Qingxiao Yang, and Xiaoqiong Huang for their help in field measurement. We thank the staff of Institute of Qinghai–Tibetan Plateau at Southwest University for Nationalities. This study was financially supported by the National Natural Science Foundation of China (31625006, 31470528), the Ministry of Science and Technology of China (2016YFC0501803, 2013CB956300), the "Thousand Youth Talents Plan", and the West Light Foundation of the Chinese Academy of Sciences.

Edited by: Michael Bahn

References

Aber, J. D., Nadelhoffer, K. J., Steudler, P., and Melillo, J. M.: Nitrogen saturation in northern forest ecosystems, Bioscience, 39, 378–286, 1989.

Aber, J. D., McDowell, W., Nadelhoffer, K., Magill, A., Berntson, G., Kamakea, M., McNulty, S., Currie, W., Rustad, L., and Fernandez, I.: Nitrogen saturation in temperate forest ecosystems – Hypotheses revisited, Bioscience, 48, 921–934, https://doi.org/10.2307/1313296, 1998.

Bai, Y., Wu, J., Clark, C. M., Naeem, S., Pan, Q., Huang, J., Zhang, L., and Han, X.: Tradeoffs and thresholds in the effects of nitrogen addition on biodiversity and ecosystem functioning: evidence from Inner Mongolia Grasslands, Glob. Change Biol., 16, 358–372, https://doi.org/10.1111/j.1365-2486.2009.01950.x, 2010.

Bowman, W. D., Gartner, J. R., Holland, K., and Wiedermann, M.: Nitrogen critical loads for alpine vegetation and terrestrial ecosystem response: Are we there yet?, Ecol. Appl., 16, 1183–1193, 2006.

Bubier, J. L., Moore, T. R., and Bledzki, L. A.: Effects of nutrient addition on vegetation and carbon cycling in an ombrotrophic bog, Glob. Change Biol., 13, 1168–1186, https://doi.org/10.1111/j.1365-2486.2007.01346.x, 2007.

Chen, D., Li, J., Lan, Z., Hu, S., and Bai, Y.: Soil acidification exerts a greater control on soil respiration than soil nitrogen availability in grasslands subjected to long-term nitrogen enrichment, Funct. Ecol., 30, 658–669, 2016.

Chen, H., Zhu, Q., Peng, C., Wu, N., Wang, Y., Fang, X., Gao, Y., Zhu, D., Yang, G., and Tian, J.: The impacts of climate change and human activities on biogeochemical cycles on the Qinghai-Tibetan Plateau, Glob. Change Biol., 19, 2940–2955, 2013.

Fernandez-Martinez, M., Vicca, S., Janssens, I. A., Sardans, J., Luyssaert, S., Campioli, M., Chapin, F. S., Ciais, P., Malhi, Y., Obersteiner, M., Papale, D., Piao, S. L., Reichstein, M., Roda, F., and Penuelas, J.: Nutrient availability as the key regulator of global forest carbon balance, Nature Climate Change, 4, 471–476, https://doi.org/10.1038/Nclimate2177, 2014.

Fleischer, K., Rebel, K. T., van der Molen, M. K., Erisman, J. W., Wassen, M. J., van Loon, E. E., Montagnani, L., Gough, C. M., Herbst, M., Janssens, I. A., Gianelle, D., and Dolman, A. J.: The contribution of nitrogen deposition to the photosynthetic capacity of forests, Global Biogeochem. Cy., 27, 187–199, https://doi.org/10.1002/gbc.20026, 2013.

Galloway, J. N., Townsend, A. R., Erisman, J. W., Bekunda, M., Cai, Z. C., Freney, J. R., Martinelli, L. A., Seitzinger, S. P., and Sutton, M. A.: Transformation of the nitrogen cycle: Recent trends, questions, and potential solutions, Science, 320, 889–892, https://doi.org/10.1126/science.1136674, 2008.

Gomez-Casanovas, N., Hudiburg, T. W., Bernacchi, C. J., Parton, W. J., and Delucia, E. H.: Nitrogen deposition and greenhouse gas emissions from grasslands: Uncertainties and future directions, Glob. Change Biol., 22, 1348–1360, https://doi.org/10.1111/gcb.13187, 2016.

Gong, Y. M., Mohammat, A., Liu, X. J., Li, K. H., Christie, P., Fang, F., Song, W., Chang, Y. H., Han, W. X., Lü, X. T., Liu, Y. Y., and Hu, Y. K.: Response of carbon dioxide emissions to sheep grazing and N application in an alpine grassland – Part 1: Effect of sheep grazing, Biogeosciences, 11, 1743–1750, https://doi.org/10.5194/bg-11-1743-2014, 2014.

Harpole, W. S., Potts, D. L., and Suding, K. N.: Ecosystem responses to water and nitrogen amendment in a California grassland, Glob. Change Biol., 13, 2341–2348, https://doi.org/10.1111/j.1365-2486.2007.01447.x, 2007.

Hasselquist, N. J., Metcalfe, D. B., and Högberg, P.: Contrasting effects of low and high nitrogen additions on soil CO_2 flux components and ectomycorrhizal fungal sporocarp production in a boreal forest, Glob. Change Biol., 18, 3596–3605, 2012.

Haynes, B. E. and Gower, S. T.: Belowground carbon allocation in unfertilized and fertilized red pine plantations in northern Wisconsin, Tree Physiol., 15, 317–325, 1995.

Huff, L. M., Potts, D. L., and Hamerlynck, E. P.: Ecosystem CO_2 exchange in response to nitrogen and phosphorus addition in a restored, temperate Grassland, The Am. Midl. Nat., 173, 73–87, 2015.

Janssens, I., Dieleman, W., Luyssaert, S., Subke, J.-A., Reichstein, M., Ceulemans, R., Ciais, P., Dolman, A. J., Grace, J., and Matteucci, G.: Reduction of forest soil respiration in response to nitrogen deposition, Nat. Geosci., 3, 315–322, 2010.

Jiang, J., Zong, N., Song, M., Shi, P., Ma, W., Fu, G., Shen, Z., Zhang, X., and Ouyang, H.: Responses of ecosystem respiration and its components to fertilization in an alpine meadow on the Tibetan Plateau, Eur. J. Soil Biol., 56, 101–106, 2013.

Kim, M. K. and Henry, H. A.: Net ecosystem CO_2 exchange and plant biomass responses to warming and N addition in a grass-dominated system during two years of net CO_2 efflux, Plant Soil, 371, 409–421, 2013.

Lamarque, J. F., Kiehl, J. T., Brasseur, G. P., Butler, T., Cameron-Smith, P., Collins, W. D., Collins, W. J., Granier, C., Hauglustaine, D., Hess, P. G., Holland, E. A., Horowitz, L., Lawrence, M. G., McKenna, D., Merilees, P., Prather, M. J., Rasch, P. J., Rotman, D., Shindell, D., and Thornton, P.: Assessing future nitrogen deposition and carbon cycle feedback using a multimodel approach: Analysis of nitrogen deposition, J. Geophys. Res.-Atmos., 110, D19303, https://doi.org/10.1029/2005jd005825, 2005.

LeBauer, D. S. and Treseder, K. K.: Nitrogen limitation of net primary productivity in terrestrial ecosystems is globally distributed, Ecology, 89, 371–379, https://doi.org/10.1890/06-2057.1, 2008.

Liu, L. L. and Greaver, T. L.: A global perspective on belowground carbon dynamics under nitrogen enrichment, Ecol. Lett., 13, 819–828, https://doi.org/10.1111/j.1461-0248.2010.01482.x, 2010.

Liu, W., Jiang, L., Hu, S., Li, L., Liu, L., and Wan, S.: Decoupling of soil microbes and plants with increasing anthropogenic nitrogen inputs in a temperate steppe, Soil Biol. Biochem., 72, 116–122, 2014.

Liu, Y. W., Xu, X. L., Wei, D., Wang, Y. H., and Wang, Y. S.: Plant and soil responses of an alpine steppe on the Tibetan Plateau to multi-level nitrogen addition, Plant Soil, 373, 515–529, 2013.

Lovett, G. M. and Goodale, C. L.: A new conceptual model of nitrogen saturation based on experimental nitrogen addition to an oak forest, Ecosystems, 14, 615–631, https://doi.org/10.1007/s10021-011-9432-z, 2011.

Lu, M., Zhou, X. H., Luo, Y. Q., Yang, Y. H., Fang, C. M., Chen, J. K., and Li, B.: Minor stimulation of soil carbon storage by nitrogen addition: A meta-analysis, Agr. Ecosyst. Environ., 140, 234–244, https://doi.org/10.1016/j.agee.2010.12.010, 2011.

Lü, C. Q. and Tian, H. Q.: Spatial and temporal patterns of nitrogen deposition in China: synthesis of observational data, J. Geophys. Res.-Atmos., 112, D22S05, https://doi.org/10.1029/2004GB002315, 2007.

Maaroufi, N. I., Nordin, A., Hasselquist, N. J., Bach, L. H., Palmqvist, K., and Gundale, M. J.: Anthropogenic nitrogen deposition enhances carbon sequestration in boreal soils, Glob. Change Biol., 21, 3169–3180, 2015.

Magnani, F., Mencuccini, M., Borghetti, M., Berbigier, P., Berninger, F., Delzon, S., Grelle, A., Hari, P., Jarvis, P. G., and Kolari, P.: The human footprint in the carbon cycle of temperate and boreal forests, Nature, 447, 849–851, 2007.

Nadelhoffer, K. J., Emmett, B. A., Gundersen, P., Kjonaas, O. J., Koopmans, C. J., Schleppi, P., Tietema, A., and Wright, R. F.: Nitrogen deposition makes a minor contribution to carbon sequestration in temperate forests, Nature, 398, 145–148, https://doi.org/10.1038/18205, 1999.

Niu, S., Wu, M., Han, Y., Xia, J., Li, L., and Wan, S.: Water-mediated responses of ecosystem carbon fluxes to climatic change in a temperate steppe, New Phytol., 177, 209–219, 2008.

Niu, S., Sherry, R. A., Zhou, X., and Luo, Y.: Ecosystem carbon fluxes in response to warming and clipping in a tallgrass prairie, Ecosystems, 16, 948–961, 2013.

Niu, S. L., Wu, M. Y., Han, Y., Xia, J. Y., Zhang, Z., Yang, H. J., and Wan, S. Q.: Nitrogen effects on net ecosystem carbon exchange in a temperate steppe, Glob. Change Biol., 16, 144–155, https://doi.org/10.1111/j.1365-2486.2009.01894.x, 2010.

Peng, Q., Dong, Y., Qi, Y., Xiao, S., He, Y., and Ma, T.: Effects of nitrogen fertilization on soil respiration in temperate grassland in Inner Mongolia, China, Environ. Earth Sci., 62, 1163–1171, 2011.

Phillips, R. P. and Fahey, T. J.: Fertilization effects on fineroot biomass, rhizosphere microbes and respiratory fluxes in hardwood forest soils, New Phytol., 176, 655–664, 2007.

Reay, D. S., Dentener, F., Smith, P., Grace, J., and Feely, R. A.: Global nitrogen deposition and carbon sinks, Nat. Geosci., 1, 430–437, 2008.

Song, B., Niu, S. L., Li, L. H., Zhang, L. X., and Yu, G. R.: Soil carbon fractions in grasslands respond differently to various levels of nitrogen enrichments, Plant Soil, 384, 401–412, 2014.

Tian, D. S., Niu, S. L., Pan, Q. M., Ren, T. T., Chen, S. Q., Bai, Y. F., and Han, X. G.: Nonlinear responses of ecosystem carbon fluxes and water-use efficiency to nitrogen addition in Inner Mongolia grassland, Funct. Ecol., 30, 490–499, https://doi.org/10.1111/1365-2435.12513, 2016.

Treseder, K. K.: Nitrogen additions and microbial biomass: A meta-analysis of ecosystem studies, Ecol. Lett., 11, 1111–1120, https://doi.org/10.1111/j.1461-0248.2008.01230.x, 2008.

Vivanco, L., Irvine, I. C., and Martiny, J. B. H.: Nonlinear responses in salt marsh functioning to increased nitrogen addition, Ecology, 96, 936–947, https://doi.org/10.1890/13-1983.1, 2015.

Wan, S., Hui, D., Wallace, L., and Luo, Y.: Direct and indirect effects of experimental warming on ecosystem carbon processes in a tallgrass prairie, Global Biogeochem. Cy., 19, GB2014, https://doi.org/10.1029/2006JD007990, 2005.

Xia, J. Y. and Wan, S. Q.: Global response patterns of terrestrial plant species to nitrogen addition, New Phytol., 179, 428–439, https://doi.org/10.1111/j.1469-8137.2008.02488.x, 2008.

Yan, L., Chen, S., Huang, J., and Lin, G.: Differential responses of auto- and heterotrophic soil respiration to water and nitrogen addition in a semiarid temperate steppe, Glob. Change Biol., 16, 2345–2357, 2010.

Zhou, L. Y., Zhou, X. H., Zhang, B. C., Lu, M., Luo, Y. Q., Liu, L. L., and Li, B.: Different responses of soil respiration and its components to nitrogen addition among biomes: A meta-analysis, Glob. Change Biol., 20, 2332–2343, https://doi.org/10.1111/Gcb.12490, 2014.

Zhou, X. and Zhang, Y.: Seasonal pattern of soil respiration and gradual changing effects of nitrogen addition in a soil of the Gurbantunggut Desert, northwestern China, Atmos. Environ., 85, 187–194, 2014.

Zhou, X., Wan, S., and Luo, Y.: Source components and interannual variability of soil CO_2 efflux under experimental warming and clipping in a grassland ecosystem, Glob. Change Biol., 13, 761–775, 2007.

Zhou, X., Zhang, Y., and Downing, A.: Non-linear response of microbial activity across a gradient of nitrogen addition to a soil from the Gurbantunggut Desert, northwestern China, Soil Biol. Biochem., 47, 67–77, 2012.

Zhu, C., Ma, Y. P., Wu, H. H., Sun, T., La Pierre, K. J., Sun, Z. W., and Yu, Q.: Divergent Effects of nitrogen addition on soil respiration in a semiarid grassland, Sci. Rep., 6, 33541, https://doi.org/10.1038/srep33541, 2016.

Zong, N., Shi, P., Song, M., Zhang, X., Jiang, J., and Chai, X.: Nitrogen critical loads for an alpine meadow ecosystem on the Tibetan Plateau, Environ. Manage., 57, 531–542, https://doi.org/10.1007/s00267-015-0626-6, 2016.

9

Effects of shrub and tree cover increase on the near-surface atmosphere in Northern Fennoscandia

Johanne H. Rydsaa[1], Frode Stordal[1], Anders Bryn[2], and Lena M. Tallaksen[1]

[1]Department of Geosciences, University of Oslo, Oslo, Norway
[2]Natural History Museum, University of Oslo, Oslo, Norway

Correspondence to: Johanne H. Rydsaa (j.h.rydsaa@geo.uio.no)

Abstract. Increased shrub and tree cover in high latitudes is a widely observed response to climate change that can lead to positive feedbacks to the regional climate. In this study we evaluate the sensitivity of the near-surface atmosphere to a potential increase in shrub and tree cover in the northern Fennoscandia region. We have applied the Weather Research and Forecasting (WRF) model with the Noah-UA land surface module in evaluating biophysical effects of increased shrub cover on the near-surface atmosphere at a fine resolution (5.4 km × 5.4 km). Perturbation experiments are performed in which we prescribe a gradual increase in taller vegetation in the alpine shrub and tree cover according to empirically established bioclimatic zones within the study region. We focus on the spring and summer atmospheric response. To evaluate the sensitivity of the atmospheric response to inter-annual variability in climate, simulations were conducted for two contrasting years, one warm and one cold. We find that shrub and tree cover increase leads to a general increase in near-surface temperatures, with the highest influence seen during the snowmelt season and a more moderate effect during summer. We find that the warming effect is stronger in taller vegetation types, with more complex canopies leading to decreases in the surface albedo. Counteracting effects include increased evapotranspiration, which can lead to increased cloud cover, precipitation, and snow cover. We find that the strength of the atmospheric feedback is sensitive to snow cover variations and to a lesser extent to summer temperatures. Our results show that the positive feedback to high-latitude warming induced by increased shrub and tree cover is a robust feature across inter-annual differences in meteorological conditions and will likely play an important role in land–atmosphere feedback processes in the future.

1 Introduction

Arctic warming is occurring at about twice the rate as the global mean warming (IPCC, 2013; Pithan and Mauritsen, 2014). This is partly owing to land–atmosphere feedback mechanisms in high-latitude ecosystems (Beringer et al., 2001; Chapin et al., 2005; Serreze and Barry, 2011; Pearson et al., 2013), such as Arctic greening (Myneni et al., 1997; Piao et al., 2011; Snyder, 2013). Arctic greening refers to the observed increase in high-latitude biomass resulting mainly from increased temperature (Walker et al., 2006; Forbes et al., 2010; Elmendorf et al., 2012). The observed increase in biomass includes extensive increase in shrub and tree cover in areas previously covered by tundra (Tape et al., 2006; Sturm et al., 2001b; Forbes et al., 2010) and northward-migrating treelines (Soja et al., 2007; Tommervik et al., 2009; Hofgaard et al., 2013; Chapin et al., 2005).

Increased tree and shrub cover alters the biophysical properties of the surface, inducing land–atmosphere feedbacks (e.g. Bonan, 2008). With increasing canopy height and complexity (including associated variables such as leaf and steam area, shade area etc.), the overall surface albedo decreases as more of the incoming radiation is absorbed. Sturm et al. (2005a) observed the impact of shrub cover on wintertime albedo in snow-covered regions and its implications for the winter surface energy balance. They concluded that increased shrub cover caused a positive feedback to warming through lowered surface albedo. The absorbed radiation heated the canopy itself and increased the sensible heat (SH) flux to the atmosphere. They also found that an increase in shrub canopies protruding from the snow cover shaded the snow beneath the canopy from radiation. This further led to decreased melt and sublimation, as higher shrub and tree cover increased the winter snow cover beneath the shrubs and the soil temperature in winter. Other studies have shown that more shrubs act to speed both the onset and advance of the melting season through its effect on surface albedo (McFadden et al., 2001; Sturm et al., 2001a).

Enhanced leaf area index (LAI) associated with an increase in shrub and tree cover can lead to higher evapotranspiration (ET). This subsequently leads to more latent heat (LH) being transferred into the atmosphere and acts to increase air temperature (Chapin et al., 2005). The increase in LH may also lead to more cloudiness and precipitation (Bonfils et al., 2012; Liess et al., 2011). Increased cloud cover may in turn limit the effect of a lower surface albedo through lowering the short-wave (SW) radiation reaching the surface.

The height of the shrubs and trees influences the strength of the land–atmosphere feedbacks, as studied specifically by Bonfils et al. (2012). They found a higher increase in the regional temperature for taller shrubs as compared to lower ones. They explained the temperature increase by the additional lowering of albedo and increase in LH corresponding to taller and more complex canopies. In summer, increased shrub cover may also act to shade the soil beneath the shrubs, thereby lowering the temperature of the soil and thus decreasing summer permafrost thaw as observed by Blok et al. (2010). This effect was also modelled in a study by Lawrence and Swenson (2011). Their findings suggest, however, that increased temperatures due to albedo decrease more than offset the cooling of the soil by the shading effect, resulting in a net increase in soil temperatures. The studies of Bonfils et al. (2012) and Lawrence and Swenson (2011) both prescribe a 20 % increase in shrub by expanding existing shrub cover into areas of tundra or bare ground. Based on circumpolar dendroecological data and several future emission scenarios, Pearson et al. (2013) concluded that the warming effect of increased shrub cover found in these two studies was realistic; however, a shrub expansion of 20 % may be substantially underestimated. They predicted, by applying various climate scenarios, that about half of the regions defined as tundra could be covered by shrubs by 2050.

The actual extent of shrub expansion into tundra regions and the predicted increase in shrub height in coming decades are highly uncertain and determined by numerous and complex mechanisms and environmental forcers.

Several of the controlling factors regulating shrub growth and expansion have been investigated using dynamic vegetation models. Miller and Smith (2012) simulated an increase in shrub cover caused by mainly warmer temperatures and longer growing seasons. They found that the shrub cover increase was in part enhanced by shrub–atmosphere feedbacks, particularly related to a reduction in albedo with an increase in canopies protruding from the snow cover. In agreement with observations, several other modelling studies have also found increased biomass production and LAI related to shrub invasion and replacement of low shrubs by taller shrubs and trees in response to increased temperatures in tundra regions (e.g. Zhang et al., 2013; Miller and Smith, 2012; Wolf et al., 2008).

Several recent studies have aimed at isolating a few of the dominating environmental drivers of shrub expansion. Myers-Smith et al. (2015) investigated climate–shrub growth relationships and found that mean summer temperatures and soil moisture content are particularly important forcers. By examining circumpolar dendroecological data from Arctic and alpine sites, they demonstrated that the sensitivity of shrub growth to increased summer temperatures was higher at European than American sites. Furthermore, they found a higher sensitivity to climate forcing for taller shrubs at the upper or northern edges of their present domain and at sites with higher soil moisture. Based on dendroecological observations, Hallinger et al. (2010) concluded that the mean summer temperature and winter snow cover are the main climatic drivers correlated with shrub growth in subalpine areas in northern Scandinavia. Based on tundra vegetation surveys covering 30 years in 158 plant communities spread across 46 high-latitude locations, Elmendorf et al. (2012) demonstrated a biome-wide link between high-latitude vegetation increase and local summer warming.

The changes in biophysical properties associated with increased shrub cover in tundra areas are more moderate compared, for example, to an expansion of forest ecosystems, and a rather modest effect on the overlying atmosphere is expected (Beringer et al., 2005; Chapin et al., 2005; Rydsaa et al., 2015). Still, aforementioned observational and modelling studies have demonstrated notable feedbacks to the regional climate. However, large uncertainties still exist concerning the estimated extent of shrub and tree advance in response to warming and to the corresponding feedback to climate resulting in response to these ecosystem changes (Myers-Smith et al., 2015; Pearson et al., 2013).

In this study we investigate the regional atmospheric response related to biophysical changes resulting from enhanced vegetation cover in high latitudes. Our investigations are carried out on a domain covering northern Fennoscandia and northwestern Russia. This is a sensitive region for shrub expansion in response to climate forcing (Myers-Smith et al., 2015). Extensive increase in the shrub-covered area and shifts in the treeline towards higher latitudes and altitudes have been observed in this region over the past decades (Tommervik et al., 2004, 2009; Hallinger et al., 2010; Rannow, 2013). This study addresses the atmospheric response to an increase in the area covered by shrubs and low deciduous trees in northern Fennoscandia and the sensitivity to their height. The primary research questions are as follows:

a. How will the feedback be influenced by increased shrub and tree cover and height?

b. Which season will be more affected and experience the strongest feedback?

c. How sensitive is the feedback to varying climatic conditions, such as snow cover and temperatures?

d. How sensitive are the atmospheric feedbacks to the amount of shrub and tree increase?

Details of the methodology, experimental design, model used, and development of bioclimatic envelopes for redistributing shrubs and trees across the study domain are presented in Sect. 2. The results for atmospheric response for spring and summer are presented in Sect. 3, including differences in response under various climatic conditions and for varying degrees of shrub and tree cover. Finally, discussion and conclusions follow in Sects. 4 and 5.

2 Methodology and study design

2.1 Study design

Model simulations were conducted on a limited region with a state-of-the-art high spatial resolution (5.4 km × 5.4 km). This enabled us to investigate finer-scale features of vegetation changes and corresponding finer-scale atmospheric responses. To investigate the effects of increased shrub and tree cover (referring to both areal expansion and taller vegetation types; research question a), we conducted six simulations: reference simulations for two different seasons (research questions b) in two climatically contrasting years (research question c). For each year, two additional simulations with manually perturbed vegetation cover representing a gradual increase in shrub and tree cover (using two different vegetation redistributions) were conducted (research questions d). By comparing the reference and perturbed simulations, we can isolate the effect of shrub and tree cover changes on the overlying atmosphere and evaluate the feedback sensitivity to the degree of shrub and tree increase since the simulations are otherwise identical.

The spring season has been identified as the season with the strongest feedback to temperatures from increased shrub cover in previous studies due to surface albedo changes (Bonfils et al., 2012; Lawrence and Swenson, 2011). Furthermore, a large potential for growth feedbacks lies with the warming response of the atmosphere during summer. For these reasons we have chosen to focus on the atmospheric response during the spring and summer seasons.

As the atmospheric response may vary under different climatic conditions (e.g. warm vs. cold, snow rich vs. snow poor, present vs. future), we chose to run experiments for two contrasting years. The two years span the natural variability across a 10-year period with respect to temperature and snow cover in the study region. The two years were selected based on a 10-year (2001–2010) simulation by Rydsaa et al. (2015), who performed a dynamical downscaling of ERA-Interim using the Weather Research and Forecasting (WRF) model. This dataset provides the ability to search through relevant variables to identify suitable years and keep consistency in model set-up and boundary conditions with this study. By averaging the response across two climatically contrasting years, we achieve a robust result representing the meteorological variability across this period, without simulating many years. Secondly, by investigating the contrasting response between the two years, this set-up provides us with valuable information of how the contrasting climatic conditions influence the atmospheric feedbacks (research question c). The year 2003 was chosen as it represented a low-snow-cover spring season and a warm summer season in this region (hereafter referred to as the warm spring and summer season). The year 2008 represented a snow-rich spring season and a cold summer season in this region (hereafter referred to as the cold spring and summer season).

2.2 Land cover and redistribution

Two different vegetation redistributions were applied to account for some of the uncertainties inherent in the shrubs' response to summer temperatures. They are based on the concept of bioclimatic zones. By applying two different redistributions, one more moderate and one more drastic, we account for some of the uncertainties related to the atmosphere's influence on the shrub cover growth. The more drastic vegetation change may represent a scenario in which the response of the shrub cover to warmer conditions is faster, or it may alternatively represent some future distribution of shrubs. Furthermore, combining findings of the atmospheric response in two different vegetation distributions and the response in the two contrasting years (warm and cold) allows us to identify potential responses in future climate conditions.

The land cover data in the reference simulations (RefVeg) are based on the newly available 20 class MODIS 15 s resolution dataset (Broxton et al., 2014). In this dataset most of the Arctic and alpine part of our study area is covered by the dominant vegetation category "open shrubland", consisting of low shrubs of less than 0.5 m height. This land use category was split into three shrub categories to distinguish the atmospheric sensitivity to shrubs and low deciduous trees of various heights. The study domain was divided into bioclimatic zones based on mean JJA temperatures and redistributed shrubs and low trees following the approach of Bakkestuen et al. (2008). The shrub and tree vegetation was redistributed across the study domain by applying bioclimatic envelopes, which were derived from empirically determined vegetation–climate relationships for the region. In order to prevent shrubs from being distributed in areas unsuitable for growth despite favourable climatic conditions, the

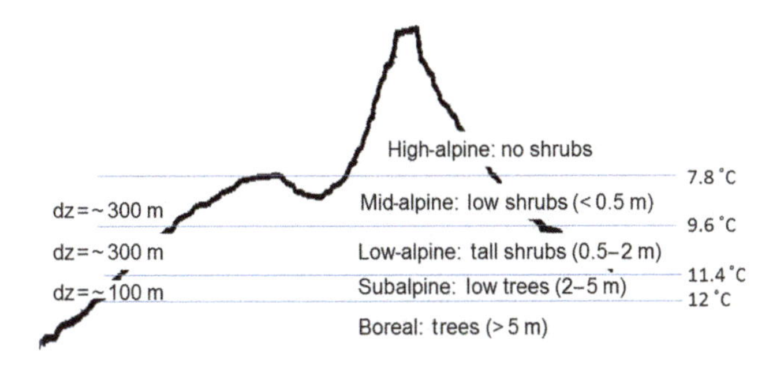

Figure 1. Illustration of alpine zones and corresponding dominating shrub vegetation. The altitudinal extent of each alpine zone is indicated by the values of elevation differences (dz) and corresponding mean JJA temperatures dividing the zones based on mean summer lapse rates in the area.

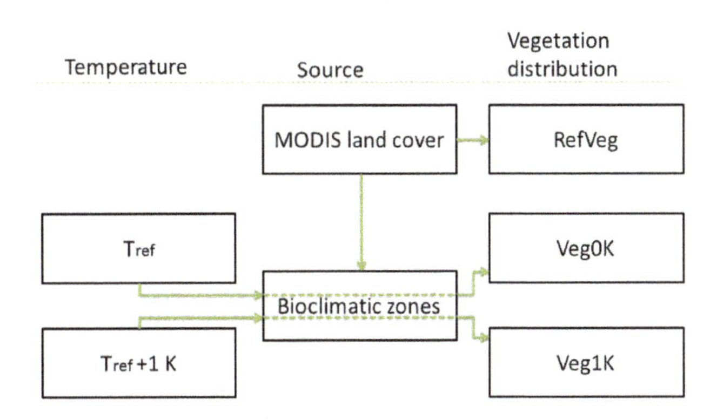

Figure 2. Illustration of the procedure applied to derive the vegetation distributions used in each simulation. T_{ref} is the mean summer (JJA) temperature distribution in the area as averaged across 2001–2010 (from Rydsaa et al., 2015). T_{ref}+ 1 K is the same temperature distribution, with a 1 K increase. Each of the three distributions has been simulated for two climatically contrasting years (cold and warm), yielding in total six simulations.

area extent of vegetation categories other than open shrubland was kept unaltered. In this way, the heterogeneity in the vegetation distribution across the domain was kept similar to the original dataset.

The bioclimatic zones for each shrub category were derived using some general features of vegetation distribution that have been determined for this area. Gottfried et al. (2012) defined various alpine zones as altitude-dependent belts of vegetation above the forest line, and each alpine zone represents a bioclimatic envelope in this study. Although the altitudinal extent of each alpine zone is determined by the local mean temperature lapse rate, in addition to various geographical and climatic features, the altitudinal extent of each zone remains rather constant across the domain, as illustrated in Fig. 1. The altitudinal extent of each alpine zone used in this study is based on Moen et al. (1999) but is also confirmed by a new dataset from the region (Bjørklund et al., 2015).

Following the vegetation categorization of Moen et al. (1999) and Bakkestuen et al. (2008), we defined tall shrubs and boreal deciduous trees with a height from 2 to 5 m (Aune et al., 2011) to belong to the subalpine zone, shrubs with a height from 0.5 to 2 m to belong to the low-alpine zone, and low shrubs with a height of up to 0.5 m to belong to the mid-alpine zone (Fig. 1). The high-alpine zone contains no shrubs and is characterized by barren ground, boulder fields, or scattered vegetation (Moen et al., 1999). High mountain tops were regarded as high-alpine (largely in agreement with the defined climatic limits), and vegetation cover in these areas was adjusted accordingly (e.g. see Karlsen et al., 2005).

The climatic forest line was used to separate the boreal forest from the subalpine region, which is characterized by scattered mountain birch (Aas and Faarlund, 2000). The last mountain birches in this region stretching towards higher elevations are approximately 2 m tall and define the so-called boreal–tundra or treeline ecotone (Hofgaard, 1997; Bryn et al., 2013; de Wit et al., 2014). This ecotone was determined here to be above the line at which the fraction of boreal trees exceeds 25 % in each grid cell. This line furthermore defines the baseline temperature above which the alpine vegetation zones at higher elevations are derived and was found to correspond well with the mean summer 12 °C isotherm (in our domain). This is slightly higher than what is found in southern parts of mountainous Scandinavia (Aas and Faarlund, 2000; Bryn, 2008). The upper limit of the subalpine zone was then determined based on an average altitudinal extent of 100 m (Aas and Faarlund, 2000). The low-alpine and mid-alpine zones were both estimated to be on average 300 m in altitudinal extent, and vegetation cover at higher elevations was defined as high-alpine zone (Moen et al., 1999), as illustrated in Fig. 1.

Based on temperature simulations by Rydsaa et al. (2015), the mean tropospheric JJA lapse rate for the area was found to be 6.0 K km^{-1}. This value was used together with the average zone heights to find the potential summer temperature ranges for each vegetation zone (Fig. 1, right). The interpolated mean JJA 2 m temperature was then used to distribute each shrub category across the domain, in accordance with their potential temperature range (i.e. their climatic envelope). This vegetation distribution is referred to as Veg0K. Revised bioclimatic zones with a 1 K increase in JJA 2 m temperatures and the same zone heights were derived in the same way and vegetation categories were redistributed, resulting in an upward and northward shift in the distribution of shrub categories across the domain. This distribution is referred to as Veg1K and represents a more drastic change in shrub distribution compared to the reference simulation. A schematic overview of the simulations and how they were derived from existing datasets is shown in Fig. 2.

The reference vegetation distribution (RefVeg) and the two perturbed distributions (Veg0K and Veg1K) are shown in Fig. 3. To represent each alpine shrub type in the model,

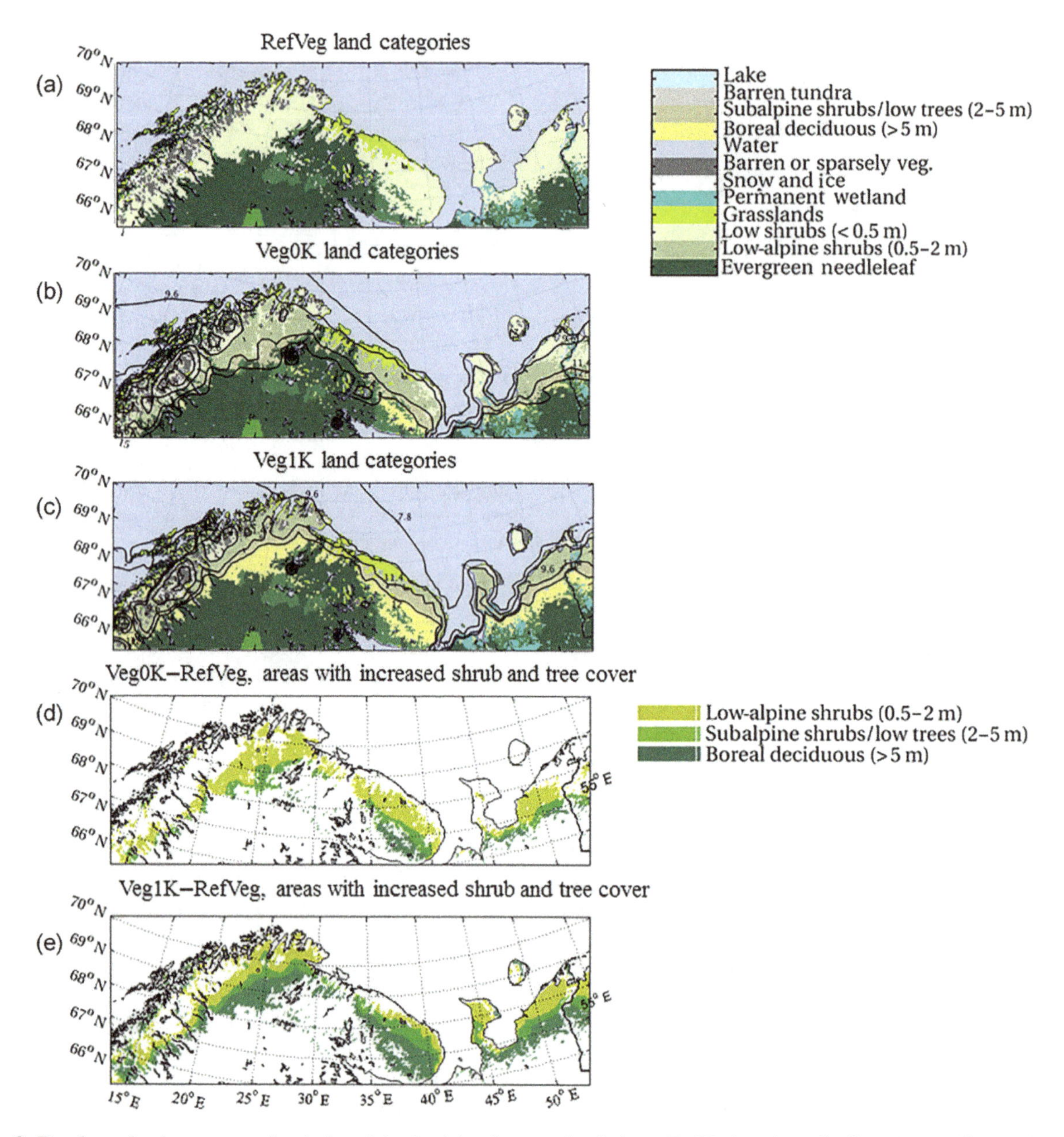

Figure 3. Dominant land use categories (colours) in the **(a)** reference simulations (RefVeg) and as distributed according to the derived bioclimatic zones (indicated by contour lines) in each of the perturbed simulations. Panel **(b)** shows Veg0K distribution, and panel **(c)** shows Veg1K vegetation distribution. Only the temperature contour lines calculated to distinguish between the various alpine zones are shown. In the bottom panels only areas with increased shrub and tree cover are coloured to show the difference in vegetation cover between the perturbed and the reference simulations. Panel **(d)** shows Veg0K–RefVeg vegetation changes; panel **(e)** shows Veg1K–RefVeg vegetation changes.

we chose suitable vegetation categories (and corresponding parameter values) from the ones already defined within the satellite dataset provided and thus tested within the framework of the model system. The categories were chosen based on vegetation types already present in the domain. Special emphasis was given to decreasing LAI and canopy height for vegetation distributed towards higher altitudes and latitudes and further based on a recent mapping of vegetation types in the region (Bjørklund et al., 2015). A list of the shrub categories and their corresponding parameter values is presented in Supplement Table S1. With two exceptions (see Table S1, bold), parameter values were left unaltered to keep consistency between and within each vegetation category.

The only alteration between the reference simulations (RefVeg) and perturbed simulations (Veg0K, Veg1K) is the land cover. Any differences in atmospheric and soil variable values result from the land cover changes, as simulations are otherwise identical with respect to set-up and meteorological forcing. The difference between Veg0K and RefVeg shows the effects of an increase in shrub and tree cover in which shrub heights are in equilibrium with the climatic potential (as defined by the bioclimatic zones and 10-year mean JJA temperatures). The difference between Veg1K and RefVeg, in comparison, shows the sensitivity to a potential vegetation shift derived from a 1 K increase in mean JJA temperatures.

2.3 Model

WRF v3.7.1 (Skamarock et al., 2008) is a non-hydrostatic weather prediction system with a wide variety of applications ranging from local-scale domains of a few hundred metres in resolution to global simulations. With a range of physical parameterization schemes, the set-up may be adjusted to simulate case-specific short-term weather events or long decadal climate simulations. The current set-up is based on available literature (refer to the NCAR choices for physical parametrizations for high-latitude domains) and a consideration of the polar WRF set-up and validation studies (Hines and Bromwich, 2008; Hines et al., 2011). Key physical schemes applied include the Mellor–Yamada–Janjić planetary boundary scheme (Janjic, 1994), the Morrison two-moment microphysics scheme (Morrison et al., 2009), and rapid radiative transfer model with GCM (RRTMG) SW and long-wave (LW) radiation (Iacono et al., 2008).

As initial and boundary conditions we used the ERA-Interim 6 h reanalysis. The model was run for two domains, in which the outer domain with a resolution of 27 km × 27 km (90 × 49 grid cells) serves purely as a bridge between the coarse-resolution boundary conditions and the finer inner domain (330 × 130 grid cells) with a resolution of 5.4 km × 5.4 km used for analysis. The model was run with 42 vertical layers and 3 h outputs. Each simulation spans the snow accumulation season (starting in November); however, only the spring (MAM) and summer (JJA) seasons are included in the analyses.

The model was run with the Noah-UA land surface model (LSM), which is the widely used Noah LSM (Tewari et al., 2004), with added parameterization for snow–vegetation interactions by Wang et al. (2010), including vegetation shading effect on snow sublimation and snowmelt, under-canopy resistance, improvements to the ground heat flux computation when snow is deep, and revision of the momentum roughness length computation when snow is present. The soil is divided into four layers of varying thickness, in total 2 m. The LSM controls the soil and surface energy and water budgets and computes the water and energy fluxes to the atmosphere, depending on air temperature and moisture, wind speed, and surface properties. The dominant vegetation category in a given grid cell determines a range of biophysical parameters that control its interaction with the atmosphere. These parameters include the height and density of the canopy; the number of soil layers available to the plants' roots; minimum canopy resistance; snow-depth water equivalent required for total snow cover; and ranges for values of LAI, albedo, emissivity, and surface roughness length. A list of parameter values used to represent the relevant vegetation categories in our simulation is presented in Table S1. The value within each range is scaled according to the vegetation greenness factor, which is based on a prescribed monthly dataset provided with the WRF model.

This model set-up is able to capture changes in surface properties following a redistribution of vegetation classes and the corresponding atmospheric response. It will not simulate the vegetation's response to environmental forcing, such as changes in surface temperature or soil moisture. Only prescribed changes to the vegetation as described in the next section differ in reference versus perturbed simulations. Differences in the atmosphere result from the biophysical changes accompanying the applied vegetation changes only.

3 Results

Section 3.1–3.3 present the seasonal effects on the overlying atmosphere of increased shrub and tree cover. Results are presented as mean anomalies between the reference and perturbed simulations (Veg0K–RefVeg), as averaged over the warm and the cold year. Special emphasis is on how the increased shrub and tree cover alters the feedback to atmospheric near-surface temperatures. Changes in other variables are presented largely to explain variations in temperature. We start presenting the results as averages over the two spring (MAM) seasons and the two summer (JJA) seasons (Sect. 3.1). This gives an estimate of the mean response of the atmosphere across a wide range in meteorological conditions and thus represents a robust estimate of shrub-induced effects across inter-annual variations. To show the sensitivity in the atmospheric response to differing meteorological conditions, results comparing the response in the warm versus the cold spring and summer seasons are presented next in Sect. 3.2 and 3.3. Section 3.2 focuses on the effect of variation in spring snow cover between the two years, and Sect. 3.3 focuses on the effect of variation in summer near-surface temperatures. Finally, in order to account for the sensitivity of the shrub and tree cover to JJA temperatures, the atmospheric response to the more extensive vegetation redistribution (Veg1K–RefVeg) is presented in Sect. 3.4.

3.1 Atmospheric effects of shrub and tree cover increase

Responses in surface fluxes and near-surface atmospheric variables as averaged over all areas with vegetation changes and across the warm and cold years (Veg0K–RefVeg), and for each year (in parentheses), are presented in Table 1. Effects of shrub and tree cover increase as averaged over the two spring seasons (Veg0K–RefVeg) are presented in Fig. 4.

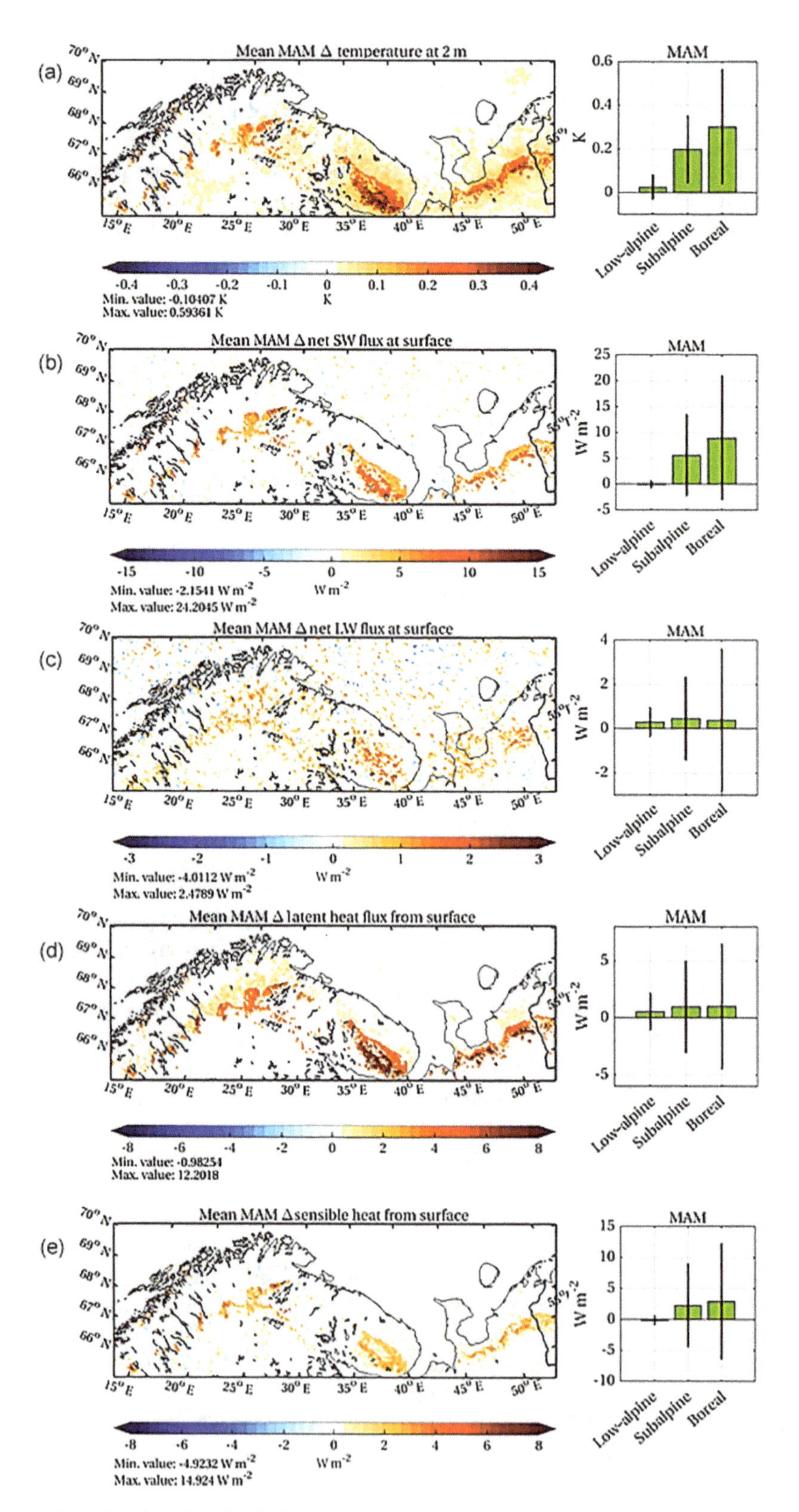

Figure 4. Effects of increased shrub cover (Veg0K–RefVeg) on the MAM season **(a)** 2 m temperature, surface fluxes of **(b)** net SW and **(c)** LW radiation (both direction downward). Fluxes of **(d)** LH and **(e)** SH (direction upward from surface). The minimum and maximum in mean seasonal values are shown below each map to present the full spatial variation in the average seasonal response. Colours only show significant results at the 95 % confidence level based on a Mann–Whitney test of equal medians. Bar plots indicate the mean response as averaged over the separate areas with vegetation changes (black lines indicate 1 σ range about the mean). Note that scales differ among variables.

Table 1. Mean response in surface fluxes and near-surface atmospheric variables as averaged over all areas with vegetation changes.

	RefVeg mean value		ΔVeg0K–RefVeg		ΔVeg1K–RefVeg	
	MAM (Warm, cold)	JJA (Warm, cold)	MAM (Warm, cold)	JJA (Warm, cold)	MAM (Warm, cold)	JJA (Warm, cold)
Near-surface temperature (K)	−5.77 (−4.28 , −7.25)	10.02 (11.0, 9.06)	0.10 (0.13,0.07)	0.05 (0.06, 0.03)	0.23 (0.28, 0.18)	0.16 (0.16, 0.15)
Upward sensible heat flux ($W m^{-2}$)	0.3 (0.1, 0.5)	52.3 (59.2, 45.5)	0.8 (1.1, 0.6)	1.8 (2.2, 1.5)	1.9 (2.4, 1.3)	4.2 (4.5, 3.8)
Upward latent heat flux ($W m^{-2}$)	6.1 (7.7, 4.5)	33.7 (34.7, 32.7)	2.3 (2.3, 2,3)	2.5 (2.8, 2.2)	3.7 (3.7, 3.7)	3.8 (4.2, 3.5)
Net short-wave down ($W m^{-2}$)	54.2 (60.2, 48.3)	153.2 (165.4, 141.0)	2.45 (3.18, 1.73)	3.6 (4.26, 2.99)	4.93 (5.98, 3.88)	7.22 (7.86, 6.58)
Net long-wave down ($W m^{-2}$)	−38.0 (−40.3, −35.7)	−55.45 (−60.8, −50.1)	0.35 (0.09, 0.60)	0.64 (0.59, 0.69)	0.60 (0.16, 1.04)	0.47 (0.53, 0.42)
Precipitation[1] (mm day^{-1})	5865 (6496, 5234)	8446 (8090, 8801)	1.07 % (1.1 %,1.01 %)	2.2 % (2.4 %, 2.06 %)	2.5 % (2.7 %, 1.6 %)	4.3 % (5.0, 3.7) %
Snowfall[1] (mm day^{-1})	4477 (4289, 4666)	274 (328, 220)	1.4 % (1.5 %, 1.3 %)	2.3 %[2] (3.04 %, 1.4 %)[2]	2.8 % (3.0 %, 2.4 %)	3.0 %[2] (3.5 %, 1.2 %)[2]
Low cloud coverage (<3 km) (fraction)[3]	0.31 (0.29,0.29)	0.16 (0.14, 0.19)	1.92 % (2.06 %, 1.85 %)	0.81 % (1.0 %, 0.7 %)	3.2 % 3.3 %, 3.4 %)	0.71 % (1.0 %, 0.5 %)
Vegetation buried by snow (fraction)	0.87 (0.78, 0.95)	0.01 (0.00, 0.02)	−0.42 (−0.43, −0.42)	–	−0.52 (−0.49, −0.55)	–

[1] Accumulated values over areas with vegetation changes. [2] Not statistically significant. [3] Average fraction over model layers below 3 km.

Figure 4a shows the spatial distribution in 2 m temperature anomalies (left) and mean values for each bioclimatic zone in the bar plot (right).

In spring, an overall increase in near-surface temperatures is seen for all areas in which shrub and tree cover increases (Fig. 4a). The higher anomaly values are seen in areas with an increase in taller shrubs and trees (as indicated in the bar plots). The average increase in 2 m temperature over the spring season is 0.1 K (Table 1); however, there are large spatial differences (Fig. 4a, bar plot). Values close to 0.6 K are seen in some areas with taller vegetation. There is also large temporal variability within the season, and the increase as averaged over all areas with vegetation changes peaks during the melting season in mid-May with 0.8 K (not shown).

The highest increase in net SW radiation is seen during the spring season (Fig. 4b), mainly due to decreased surface albedo caused by increased shrub and tree cover and its effect on earlier snowmelt (Sect. 3.2). There is a slight decrease in downwelling SW (not shown) caused by enhanced cloud cover (Table 1), but the reduction in downwelling SW is more than compensated for by the albedo decrease in areas with subalpine vegetation (taller vegetation). The net value is close to zero in areas with low-alpine shrub increase (lower vegetation) due to smaller albedo changes (4b, and bar plot). The LW radiation slightly increases (Fig. 4c) in response to enhanced cloud cover and atmospheric humidity (Table 1). The increase in LW radiation is more evenly distributed across the region than changes in SW radiation, as it is not as directly linked to the vegetation changes.

The heating associated with the increase in SW radiation is partly balanced by an increase in ET, shown as the LH flux (Fig. 4d). The increased LAI caused by more shrub and tree cover (Table S1, and Supplement Fig. S4) results in increased ET and correspondingly higher LH. The effect is larger in areas with a larger LAI increase, i.e. in areas with taller vegetation. The increase is largest towards the end of the spring season (not shown), much owing to larger above-snow canopy fraction due to the canopy height increase associated with more shrubs and trees and reduced snow cover (Figs. S2, S3). An increase in SH flux (Fig. 4e) from the surface and from canopies protruding from the snow cover is seen in areas with taller vegetation where net SW is positive. This adds to the effect of increasing LH in balancing the surplus of SW energy at the surface.

In the summer season (Fig. 5) the 2 m temperature increases in areas with taller vegetation and decreases in areas with low-alpine shrub increase (lower vegetation; Fig. 5a). The latter areas are characterized by a lowering of net SW radiation in this season, which results in a decreased SH flux and less warming of the lower atmosphere. The negative net SW radiation (Fig. 5b) is related to a slight albedo increase in early summer (early to mid-June, not shown) caused by enhanced snow cover in these areas (Figs. S3 and S4). The enhanced snow cover is a result of increased precipitation (including snow fall; Table 1). In addition, the summer sea-

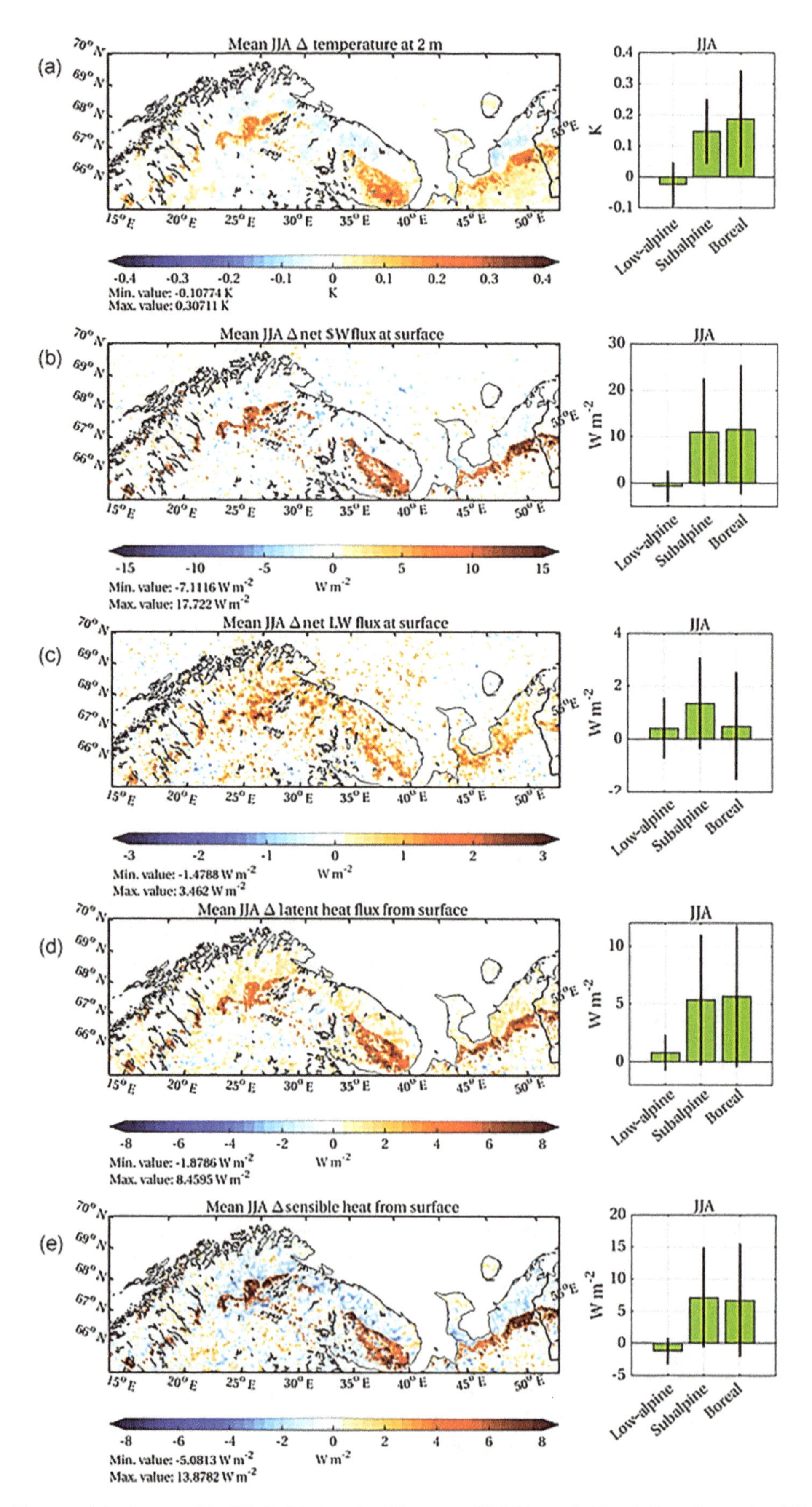

Figure 5. Effects of increased shrub cover (Veg0K–RefVeg) on the JJA season. Variables as in Fig. 4. Note that scales differ among variables.

son SW downwelling is decreased due to an increase in cloud cover (Table 1), as confirmed by the increased LW radiation to the surface (Fig. 5c). Conversely, in areas with taller shrubs and trees, the stronger albedo decrease dominates, leading to a decrease in snow cover throughout the spring and summer (albedo changes are shown in Fig. S4).

The increased SH mainly acts to heat the planetary boundary layer (PBL), while the LH is mainly released above the PBL height. The LH therefore does not affect the 2 m temperature to the same degree as the SH, as the heat is released as the water condenses, which may well be higher up in the atmosphere. The vertical structure of the lower-atmosphere heating along a cross section is shown in Fig. S6, along with changes in PBL height and turbulent fluxes of SH and LH. The atmospheric humidity increase associated with increased shrub cover results in more clouds and total precipitation in both seasons (Table 1).

The spatial distribution of mean changes in the low cloud fraction (here defined as below 3 km) and precipitation anomalies in the two seasons is shown in Fig. 6. The top panels show the relative change in low cloud cover resulting from increased shrub and tree cover. Here the change in cloud cover is shown as the difference in fractional cloud cover averaged over the lower 3 km of the atmosphere (further details about this variable in the Supplement). The increased cloud cover acts to decrease the SW radiation reaching the surface in both seasons (shown only as net SW radiation, Figs. 4 and 5) and increase the amount of LW radiation towards the surface (shown only as net LW radiation, Figs. 4 and 5). The effect is largest in areas in which the humidity increases the most through enhanced LH, i.e. in areas with an increase in taller vegetation.

The most prominent increase in low cloud cover occurs in spring (Fig. 6, upper left panel), largely covering areas with vegetation changes. The summer season's response is patchier, although a tendency towards increased cloud cover in areas with vegetation change (refer to Fig. 3) is recognizable. The second row shows the relative increase in precipitation (in percent), as accumulated over the season. For both variables only areas with significant changes are shown. The relative change in precipitation is based on daily accumulated values. As with the cloud cover, the spatial distribution of (significant) precipitation changes is somewhat patchy, particularly for the summer season. However, the significance is higher in areas with vegetation changes, as compared to the total area (cells with significant differences in areas with vegetation changes is 8.3 %, versus 5.7 % in the total domain). The increase in accumulated precipitation is most prominent in summer, amounting to 186 mm in areas with vegetation changes, corresponding to a 2.2 % increase (p value based on the Mann–Whitney significance test is 1.2×10^{-5}). For spring, the increase in precipitation is 1.07 %, and for precipitation in the form of snow and ice it is 1.4 %.

3.2 Sensitivity to snow cover

The two contrasting spring seasons are characterized by large differences in snow cover, albedo, and near-surface temperatures. In the reference simulations, the warm spring season ($\text{RefVeg}_{\text{warm}}$) has 16 % less snow cover than the cold one ($\text{RefVeg}_{\text{cold}}$), resulting in a decreased albedo of 12 % and an average 2 m temperature that is 3.1 K warmer (numbers are averages over the land area of the total study domain). Total precipitation is similar for the two years, although the rain-to-snow ratio is larger in the warm spring due to higher temperatures. The snowmelt also starts earlier in the warm spring season ($\text{RefVeg}_{\text{warm}}$) (more than 2 weeks) and a faster rate of snowmelt is seen as compared to the cold spring season ($\text{RefVeg}_{\text{cold}}$), with the largest difference in snow cover in May (Fig. 7). It is worth noting that the most pronounced effects of increased shrub cover on the atmosphere are during the melting season, i.e. May–June.

The warm spring season experiences up to 0.38 K higher increases in 2 m temperature in response to shrub and tree cover increase as compared to the cold one (Fig. 8). As seen in panel (b) of Fig. 8, the anomaly distribution is shifted towards overall higher values in the warm season. The shrubs act to enhance warming more in the warm than in the cold spring season. This represents a positive feedback to warm conditions and early snowmelt.

The increased shrub and tree cover leads to a reduction in snow depth in spring as averaged over all areas with vegetation changes, as seen in Fig. 9a (the spatial distribution of snow cover is shown in Fig. S3). An exception is seen in late spring (and early cold summer, not shown). This is related to the late spring and early summer increase in snow cover found in areas with low-alpine shrub increase. These areas experienced an increase in snow fall in the cold summer season and subsequently a shortening of the snow-free season (a grid cell is considered snow free if the fraction of ground covered by snow is less than 0.1; Fig. 9b). In the cold season the shortening is only about half a day averaged over the areas with vegetation changes. The warming effect of shrub cover in the warm season, however, acts to prolong the snow-free season by just over 1 day; however, it speeds the onset of melting by several days.

Also, increased shrub and tree cover acts to enhance soil temperature (Fig. 9c), with maximum impact on the upper layers of the soil (not shown). The increased precipitation during both spring and summer also influences the soil moisture. Soil moisture (Fig. 9d) increases in areas with increased shrub and tree cover throughout the warm spring. A notable increase in soil moisture, and a corresponding decrease in surface run-off, is seen in mid-May at the time of maximum snowmelt (Fig. 9e), for both the cold and warm melting seasons. However, before the main snowmelt starts, run-off is slightly higher during the warm spring season because of the increased snowmelt earlier in spring for areas with increased shrub and tree cover.

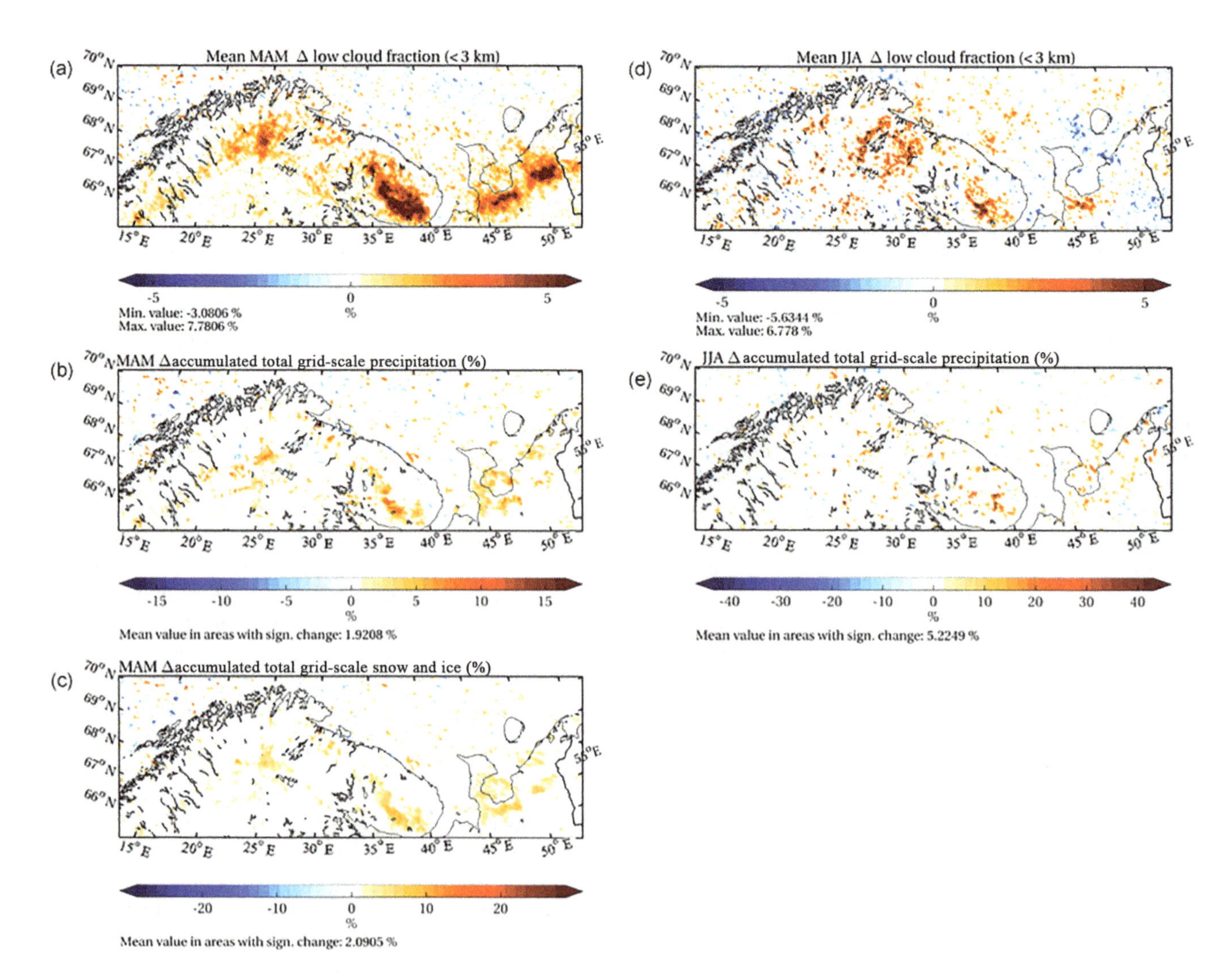

Figure 6. Effects of increased shrub cover (Veg0K–RefVeg) on low-level (< 3 km) cloud cover fraction **(a, d)**, relative change in accumulated seasonal precipitation **(b, e)**, and spring season snow and ice precipitation **(c)**. Only showing significant changes at the 95 % confidence level, as in Fig. 4. For precipitation, significance tests are conducted on daily values of accumulated precipitation rather than 3-hourly values. Mean over spring seasons in the left column and summer seasons in the right column. Note that scales differ among panels.

3.3 Sensitivity to summer temperatures

The warm and cold summer seasons encompass a large range in inter-annual temperature variability. For the reference vegetation (RefVeg), the mean JJA 2 m temperature (averaged over land areas in the domain) for the warm summer season ($RefVeg_{warm}$) was 11.7 °C, while the cold summer ($RefVeg_{cold}$) represents a lower-than-usual mean temperature of 9.7 °C. In some areas the difference reached 3.3 °C. The corresponding increase in atmospheric absolute humidity at 2 m is 6.9 %. The warm summer also represents drier conditions with less precipitation (Table 1).

The difference in atmospheric temperature response to increased shrub and tree cover between the two summers is shown in Fig. 10. The response of the atmosphere to increased shrub cover (Veg0K–RefVeg) shows more similarity across the warm and cold summer seasons as compared to the warm and cold spring seasons. For the summer seasons, the mean difference in 2 m temperature response is smaller and rather evenly distributed around zero (Fig. 10, panel b). Positive values over areas with low-alpine shrub expansion indicate less cooling in the warm as compared with the cold summer season, during which these areas were partially covered by snow. The tall vegetation changes contribute to similar warming in the summer seasons. The temperature response in the warm season is slightly shifted towards warmer anomalies (Fig. 10, panel b), indicating a slightly larger vegetation feedback to warmer summer temperatures in the warm summer season when compared with the cold.

The difference in atmospheric temperature response is larger between the warm and cold spring seasons than between the warm and cold summer seasons. Thus, it seems

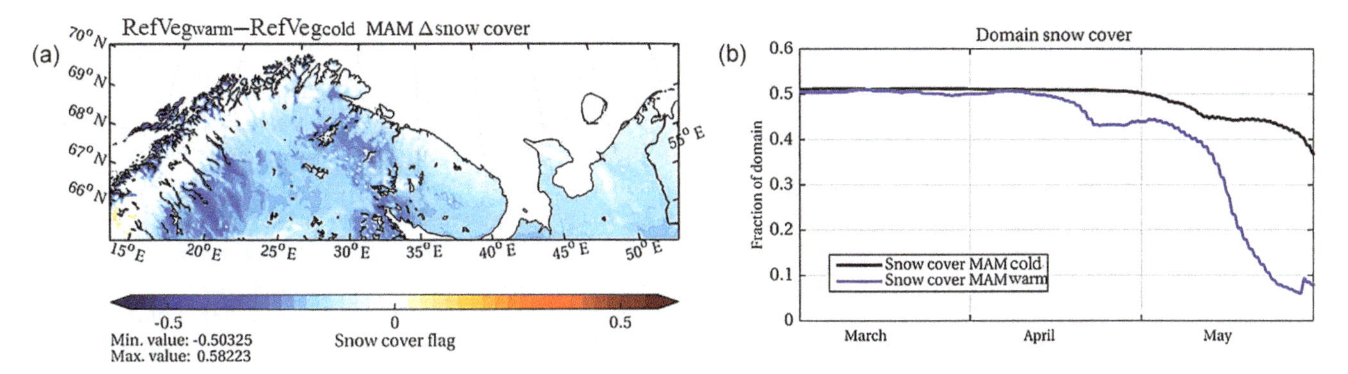

Figure 7. Difference in mean seasonal snow cover between the warm and cold spring seasons ($RefVeg_{warm}$–$RefVeg_{cold}$). Mean seasonal spatial differences are shown in panel **(a)**, and the temporal development over the seasons is shown in panel **(b)**.

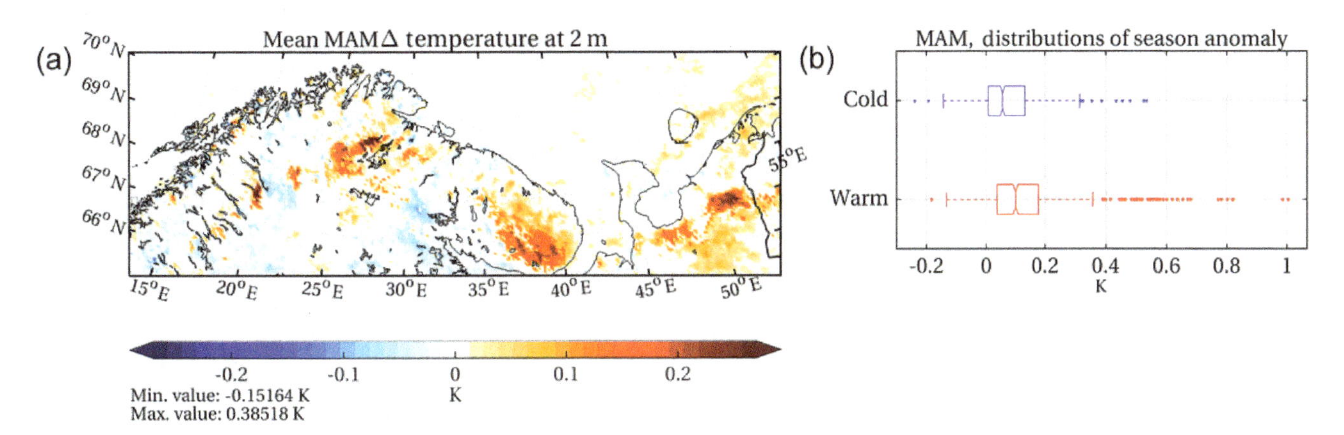

Figure 8. Difference in temperature response due to increased shrub cover (Veg0K–RefVeg) between the warm and cold years (only showing significant results at the 95 % confidence level). The distributions of shrub-induced anomaly values are shown in box plots; the red box shows warm season anomalies and the blue box shows cold season anomalies in areas with vegetation changes.

that the shrub cover feedback is more sensitive to meteorological conditions in spring than summer. This is likely due to the feedback being closely linked to albedo changes, which are heavily dependent on snow cover. Therefore, the feedback is more sensitive to temperature in the melting season.

3.4 Sensitivity to the degree of vegetation changes

The shift in shrub and tree distribution according to the theoretical 1 K increase in summer temperature (Veg1K vegetation distribution) results largely in a northward shift in the boreal treeline ecotone, replacing low-alpine shrubs with small trees across most of the shrub-covered areas, as compared to the Veg0K distribution. It also acts to increase the low-alpine shrub cover in higher latitudes and altitudes (Fig. 3). The increased cover of trees at the expense of shrubs, with a corresponding strong decrease in albedo and increase in LAI, enhances the net SW radiation absorbed by the surface. This is balanced by strong increases in SH and LH (Table 1, and Fig. S5). In addition, the vegetation changes result in increasing precipitation and cloud cover (Table 1).

The mean seasonal response in 2 m temperature caused by this vegetation shift (Veg1K–RefVeg) is shown in Fig. 11. The warming at 2 m on average more than doubles as compared to that of the more moderate shrub and tree cover distribution (Veg0K–RefVeg), in both seasons (Table 1). This is due to the more extensive changes in biophysical properties related to the shift towards taller vegetation. The warming is most prominent in the spring season, particularly in late spring when the increased vegetation cover notably affects the snowmelt and corresponding albedo and surface heat fluxes. The average spring warming is therefore strongest in areas with the tallest vegetation. However, although highly localized, the highest peak values, up to 0.71 K, are found in summer (Fig. 11). Increased LH also leads to enhanced atmospheric moisture and more summer precipitation (Table 1) and a corresponding greenhouse effect of up to 5 $W\,m^{-2}$ (not shown). The response of the Veg1K vegetation change also differs between the warm and cold summer and spring seasons. In contrast to the response of Veg0K, the strongest warming is found in the cold summer in most areas.

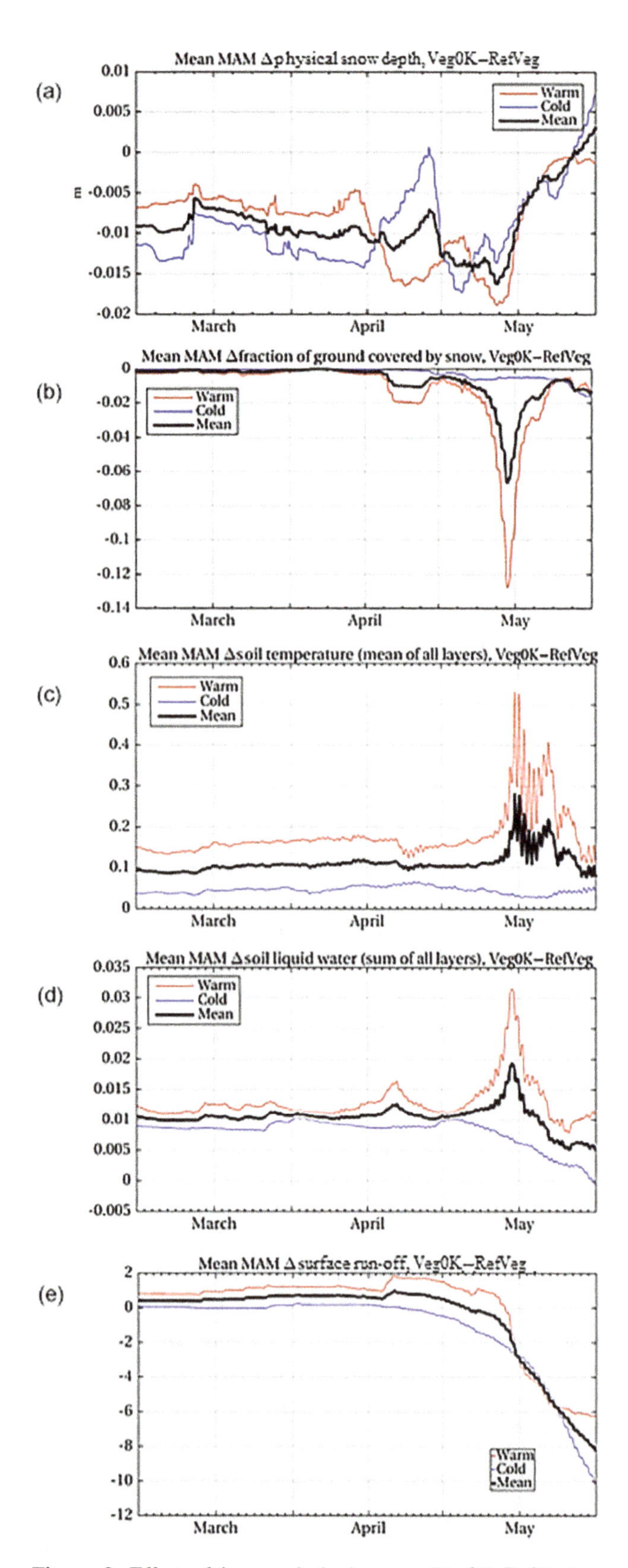

Figure 9. Effect of increased shrub cover (Veg0K–RefVeg) on spring snow depth and cover, soil temperatures, and moisture content and surface run-off, as averaged over all areas with vegetation changes. Red and blue lines indicate warm and cold season response, respectively. Black lines indicate inter-seasonal means.

4 Discussion

The spring albedo effect is often regarded as the most important effect of increased vegetation cover in high latitudes (Arora and Montenegro, 2011; Bonan, 2008), and our results confirm this as the main cause of warming during the spring season. Our findings show that the net SW radiation is highly sensitive to the vegetation properties such as the height of the vegetation. We find that competing effects of increased ET (resulting in more cloud cover, precipitation and snowfall, less downward SW radiation), versus the effect of albedo decrease (more absorbed SW radiation), determine the net SW radiation and corresponding near-surface temperatures.

In the most moderate vegetation redistribution case (Veg0K–RefVeg) the seasonal average spring temperature increase reached 0.59 K in the areas with the tallest vegetation. The warming as averaged over the entire area with vegetation changes reached 1.0 K during the melting season in the warmest of the two years studied, due to the strong impact of shrubs and trees under snow-free conditions. These peak values represent the warming potential of the vegetation changes applied in this experiment. The albedo decrease related to more complex canopies and enhanced snowmelt dominates over competing effects and causes warming in spring in areas with increased tall vegetation, but this dominance is smaller and sometimes reversed in areas with increased low shrub cover. In the large areas with increased low-alpine shrub cover, the average summer warming was only 0.1 K, reflecting an increased early summer snow cover and albedo in these areas caused by increased snowfall. This, combined with the weak counteracting effect of small albedo decreases associated with the low-alpine shrubs, resulting in a decrease in the net SW radiation and 2 m temperatures. In areas with taller vegetation, the summer maximum increase in near-surface temperature reached 0.39 K. This contrasting pattern in summer warming confirms the strong dependence of the atmospheric response on vegetation height as was also found by Bonfils et al. (2012). They applied a 20 % increase in shrub cover in bare ground areas north of 60° N in order to simulate the influence of shrubs on climate. They found a regional annual mean temperature increase of 0.66 K for shrubs up to 0.5 m high, which was most prominent during the spring melting season. To investigate the sensitivity of height and stature of shrubs, they performed a second experiment, increasing the shrub heights to 2 m. This caused the regional annual warming to increase to 1.84 K by 2100. Furthermore, they found increases in both SH and LH, the latter mainly resulting from an increase in ET. Similar to our results, they also found an increase in summer precipitation, particularly in the case of tall shrubs.

Lawrence and Swenson (2011) also applied a 20 % increase in shrub cover north of 60° N. In their case this led to a moderate increase in mean annual temperatures of 0.49–0.59 K, with a peak of 1–2 K during the melting season in May. They also found an increase of 3–5 K in soil tempera-

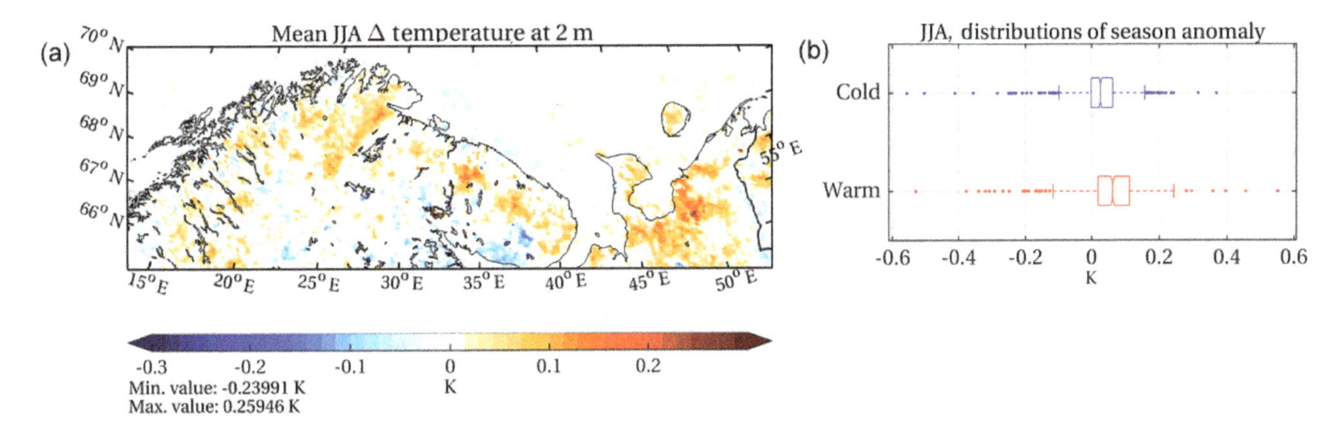

Figure 10. Difference ($RefVeg_{warm}$–$RefVeg_{cold}$) in temperature response due to increased shrub cover (Veg0K–RefVeg; only showing significant results at the 95 % confidence level, as in Fig. 4). The anomaly distribution across the domain is shown (panel **(b)**). The red box shows warm season anomalies and the blue box shows cold season anomalies in areas with vegetation changes.

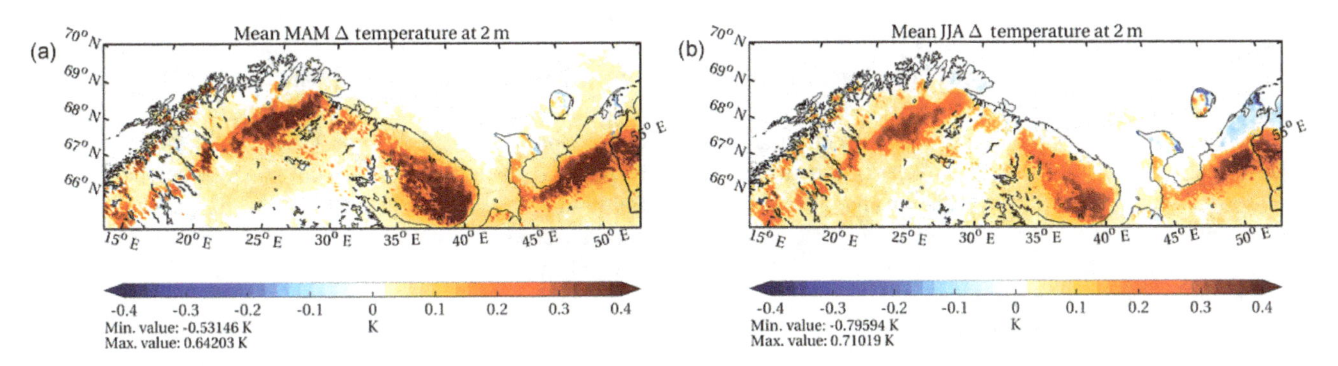

Figure 11. Effects of increased shrub cover (Veg1K–RefVeg) on the 2 m temperature resulting from a shrub and tree cover increase corresponding to a 1 K warming of JJA temperatures (only showing significant results at the 95 % confidence level, as in Fig. 4). Mean spring season response is shown in panel **(a)** and mean summer season response is shown in panel **(b)**.

tures in winter and spring following added shrub cover and redistributed snow cover. Although not directly comparable, we note that their results were substantially larger than the soil temperature response in our results, with maximum values reaching up to 1.5 K in the top soil layer during the warm melting season. This difference is probably related to intermodel differences in soil and vegetation properties and particularly to differences in simulation domain and extension of shrub and snow cover increase. Their analyses did not include effects on cloud cover and precipitation. Also, Swann et al. (2010) applied a 20 % increase in shrub cover north of 60° N and found an annual warming of 0.2 K and a decrease in low-level clouds despite increased vapour content due to increased ET. Similar to our study, they also found an increase in summer precipitation, but not in spring.

The atmospheric response to shrub cover increase in our simulations was larger in the warm than in the cold year, both in the spring and summer seasons. However, the difference in response between warm and cold summers was more moderate as compared to the warm and cold springs. Based on these results, we might expect that in a warmer climate, shrub expansion would increase spring surface temperatures more than summer temperatures. The areas with the strongest feedback to the summer season warming were related to an increase in taller vegetation (subalpine and boreal).

The sensitivity of shrub expansion to summer temperatures is not well known, and for this reason, we applied a second set of simulations with vegetation distribution based on a 1 K increase in JJA temperatures (Veg1K). When interpreted with care, the atmospheric response to this vegetation change as compared to the more moderate one may serve as a simplified proxy as a future vegetation redistribution scenario. However, precautions should be made, as the time delay related to such a vegetation shift could be substantial (Corlett and Westcott, 2013) and because the actual vegetation redistribution according to such a shift in summer temperatures could be limited by other environmental and ecological factors, as mentioned in the introduction and discussed by Svenning and Sandel (2013) and Myers-Smith et al. (2011). Also, the warmer climate might influence the response itself, with responses even falling outside the range of climatic conditions represented by the two contrasting years in this study.

Keeping all this is in mind, a careful interpretation of the results as representing some future state can still be beneficial. The Veg1K redistribution was largely dominated by extended areas of subalpine and boreal deciduous vegetation cover, consisting of tall shrubs and low trees. The northward migration of taller trees and the subalpine ecotone more than doubled the warming in both seasons, but to a larger degree in summer (on average 0.16 K in Veg1K–RefVeg, as compared to 0.05 in Veg0K–RefVeg; Table 1). Peak seasonal anomalies in this experiment were also higher in the summer season as compared to the spring season.

Combining our findings, we find that the main summer temperature feedbacks are mainly related to increases in taller vegetation. The surface albedo decrease is largest in summer in areas with boreal and subalpine deciduous trees, despite the snow masking effect of snow-protruding canopies in spring. This is mostly owing to the deciduous nature of the northward expanding shrubs and trees in this study, which is based on what is observed in the study region (Hofgaard et al., 2013; Aune et al., 2011). This would be different if we allowed for expansion of evergreen needle-leaved trees (Rydsaa et al., 2015; Arora and Montenegro, 2011; Betts and Ball, 1997), which would more strongly affect the albedo across all seasons. Allowing for such a vegetation change could certainly be interesting in this type of investigation. However, in this study, the main focus has been on the relatively fast shrub and (deciduous) tree cover increase.

As the mean summer temperature is assumed here to be the main environmental driver of shrub expansion, our results lead us to conclude that a warming effect on summer temperature strong enough to lead to a positive feedback to shrub and tree growth would depend on establishment of taller shrubs and subalpine trees in tundra areas, rather than an increase in lower shrub types. This also supports the findings by de Wit et al. (2014).

As the differences in atmospheric response between the warm and cold summers in these experiments are rather small, a positive feedback to summer warming is a robust feature across inter-annual variations. Given the strong impact of the northward migrating subalpine ecotone on the summer temperature shown here, we find the possibility for a future ecological "tipping point" in this area possible, and this would be an interesting topic to investigate further. The term refers to the level of vegetation response at which the atmospheric warming resulting from increased shrub and tree cover feedbacks enhances the further growth to such a degree that the response becomes nonlinear in relation to the initial warming (Brook et al., 2013). However, other factors will also influence the future shrub growth. As highlighted by Myers-Smith et al. (2011), climatic forcers (e.g. air temperature, incoming solar radiation, precipitation) and soil properties (e.g. soil moisture, soil temperature and active layer depth), coupled with biochemical factors such as the availability of soil nutrients and atmospheric CO_2 concentrations, all influence the rate of shrub growth. In addition, disturbances, such as fires, heavy snowpack, and biotic interactions including herbivory, make accurate estimates of future shrub distribution challenging (Milbau et al., 2013). Tape et al. (2012) highlighted the importance of soil properties in estimating likely areas of shrub expansion and shrub–climate sensitivity, and they argued that this factor increases the geographic heterogeneity of shrub expansion. In addition, increased shrub cover has also been suggested to trigger feedback loops that further induce shrub growth by shrub–snow interactions, for example (Sturm et al., 2005a, b, 2001a). Positive feedbacks include lowering of spring albedo causing earlier snowmelt, longer growing seasons, and increased soil temperatures, which are all favourable for growth. Also, thicker wintertime snowpack in shrub areas acts to insulate the ground during winter and increase the soil temperatures (Sturm et al., 2001a).

The temperature increases in our results, both for the peak melting seasons and in seasonal means, are below the seasonal estimates of some similar studies. This was expected given the comparatively more moderate vegetation shifts (both on an areal scale and partly in vegetation properties) in our simulations. Also, large variations in the atmospheric response with regard to cloud cover and precipitation were found among other modelling studies, despite qualitatively similar responses of enhanced ET and LH related to increased shrub cover. The vegetation perturbations applied to represent shrub and tree cover increase in this study are moderate in both areal extent and in vegetation property changes, as compared to other studies with a similar purpose (e.g. Bonfils et al., 2012; Lawrence and Swenson, 2011). We have altered shrub properties only in areas already covered by tundra and low shrubs and only within empirically based suitable climatic zones (Figs. 1 and 3). Shrub properties were selected from predefined vegetation categories within the modelling system employed to represent high-latitude vegetation. Only minimal alterations were made to the existing categories in order to keep consistency within and between the vegetation categories applied in the modelling domain. This approach does inherit some uncertainty regarding the suitability of single-parameter values. However, we judged that further alterations might lead to unintended biases within the modelling system. A complete review of the parameter values applied for each vegetation category within the modelling system is beyond the scope of this study.

Since we have chosen to focus on biophysical aspects of the effects of increased shrub and tree cover, there have been no atmospheric or soil chemistry changes included, nor effects of aerosols. These factors may substantially alter atmospheric composition and possibly impact the response to vegetation changes. However, other studies have concluded that the main impact of changes in the high-latitude ecosystems results from biophysical effects (Pearson et al., 2013; Bonan, 2008).

Our investigations are based on simulations using a relatively fine spatial resolution. This has enabled a more realistic representation of finer-scale features of the shrub–atmosphere feedbacks as compared to previous modelling studies. However, this comes with the price of having to reduce the size of the domain. Due to its limited size, and the proximity to warm waters along the coast of Norway, our domain is largely influenced by the incoming marine air from the west. This advection of weather into the domain acts to diffuse the effects of shrubs and trees on the atmosphere. As such, our results for impacts on upper-atmospheric features, such as cloud cover and precipitation, are heavily influenced by the meteorological boundary conditions and not only near-surface variables. This effect could influence our results for atmospheric response to be more modest when compared to results of similar studies on circumpolar domains (e.g. Bonfils et al., 2012; Liess et al., 2011).

Vegetation dynamics were not included in this study to account for the vegetation's response to the changing environmental conditions. This represents a limitation in our simulations, particularly with regard to differing responses among the cold and warm seasons. However, it is hard to predict whether this represents an overestimation or underestimation of our results. In this model version, the daily interpolated greenness factor (based on monthly values), acts to scale between maximum and minimum parameter values representing each vegetation category (such as the LAI and vegetation albedo etc.). This gives rise to the seasonal variation in vegetation in these simulations. The greenness fraction describes the vegetation density distribution within each grid cell. Since we have made no assumptions about changes in the density of vegetation, only about the type of dominant vegetation, this variable was left unaltered in our perturbations. Although it can be argued that an assumption of enhanced vegetation density (i.e. greenness) is reasonable, we considered it beyond the scope of this study to estimate scales and predictions regarding such changes. In addition, recent reports on arctic browning suggest high uncertainty related to enhanced vegetation density (Phoenix and Bjerke, 2016). Also, limiting the perturbations to affect only the vegetation types and heights, not the density, is beneficial for the interpretation of the results. We do however acknowledge that this choice might influence the results for the atmospheric response. Particularly the partitioning between LH and SH flux could be affected by the choice of perturbations applied.

5 Summary and conclusions

We have applied the Weather Research and Forecasting model coupled with the Noah-UA land surface model to evaluate biophysical effects of shrub expansion and increase in shrub height on the near-surface atmosphere at a state-of-the-art fine resolution. We first applied an increase in shrub and deciduous tree cover with heights varying in line with the present climate potential according to empirical temperature–vegetation limits for the region (bioclimatic envelopes). To evaluate the sensitivity of the atmospheric response to climatic variations, simulations were conducted for two contrasting years, one with warmer and one with colder spring and summer conditions. The response across the different years represents an atmospheric response across a broad range in temperature and snow cover conditions. To evaluate the sensitivity to a potential further expansion in shrub and tree cover, we conducted additional simulations for each year, applying a second vegetation cover shifted according to bioclimatic envelopes corresponding to a 1 K increase in mean summer temperature.

Our results show that shrub and tree cover increase leads to a general increase in near-surface temperatures, enhanced surface fluxes of heat and moisture, and an increase in precipitation and cloud cover across both warm and cold years and seasons. A notable exception are areas with subalpine shrubs, where increased atmospheric moisture resulting from shrub expansion leads to increased snowfall and surface albedo early in the colder summer season. This highlights that the net SW radiation absorbed by the surface strongly depends on the strength of the albedo decrease due to enhanced canopies versus albedo changes related to increased ET causing enhanced cloud cover and precipitation (including snow fall). The atmospheric responses in all variables strongly depend on the shrub and tree heights. However, increased LAI leads to a persistent increase in LH in all areas with shrub expansion in all seasons investigated.

We find that the effects of increased shrub and tree cover are more sensitive towards snow cover variations than summer temperatures. Increased shrub cover has the largest effect in spring, leading to an earlier onset of the melting season, particularly in the warmer spring season. This represents a positive feedback to warm spring temperatures. Taller vegetation influences summer temperatures more than spring temperatures in most areas. The response is not affected by variations in summer temperatures to any large degree and is a robust signal across inter-annual variations.

Summer temperatures have been estimated to be one of the strongest drivers of vegetation expansion in high latitudes. Here, we find that the strongest feedbacks to the summer temperatures are related to the expansion of taller vegetation rather than shorter shrubs. Due to large areas with small elevation gradients within this domain as well as the rest of the circumpolar tundra-covered areas, the temperature zones as derived here are highly sensitive to increases in summer temperatures. Small increases in mean temperatures will as such make vast areas climatically available for shrubs and tree growth. Our results show that the positive feedback to summer temperatures induced by increased tall shrub and tree cover is a consistent feature across inter-annual variability in summer temperatures. In combination with the vast area that is made available for taller shrubs and trees by relatively small increases in temperature, this represents a potential for a so-called vegetation-feedback tipping point. This is a possibility that we find to be an interesting subject for further research.

Competing interests. The authors declare that they have no conflict of interest.

Acknowledgements. This work is part of LATICE, which is a strategic research area funded by the Faculty of Mathematics and Natural Sciences at the University of Oslo. Discussions and collaboration with members of LATICE have greatly improved this paper. In particular we thank James Stagge for his valuable advice regarding the statistical analysis. We would also like to express our gratitude towards the two anonymous referees and the associate editor for their constructive comments and suggestions that led to improvements of our paper.

Edited by: Sönke Zaehle

References

Aas, B. and Faarlund, T.: Forest limits and the subalpine birch belt in North Europe with a focus on Norway, AmS-Varia, 37, 103–147, 2000.

Arora, V. K. and Montenegro, A.: Small temperature benefits provided by realistic afforestation efforts, Nat. Geosci., 4, 514–518, https://doi.org/10.1038/Ngeo1182, 2011.

Aune, S., Hofgaard, A., and Soderstrom, L.: Contrasting climate- and land-use-driven tree encroachment patterns of subarctic tundra in northern Norway and the Kola Peninsula, Can. J. Forest Res., 41, 437–449, https://doi.org/10.1139/X10-086, 2011.

Bakkestuen, V., Erikstad, L., and Halvorsen, R.: Step-less models for regional environmental variation in Norway, J. Biogeogr., 35, 1906–1922, https://doi.org/10.1111/j.1365-2699.2008.01941.x, 2008.

Beringer, J., Tapper, N. J., McHugh, I., Chapin, F. S., Lynch, A. H., Serreze, M. C., and Slater, A.: Impact of Arctic treeline on synoptic climate, Geophys. Res. Lett., 28, 4247–4250, https://doi.org/10.1029/2001gl012914, 2001.

Beringer, J., Chapin Iii, F. S., Thompson, C. C., and McGuire, A. D.: Surface energy exchanges along a tundra-forest transition and feedbacks to climate, Agr. Forest Meteorol., 131, 143–161, https://doi.org/10.1016/j.agrformet.2005.05.006, 2005.

Betts, A. K. and Ball, J. H.: Albedo over the boreal forest, J. Geophys. Res.-Atmos., 102, 28901–28909, https://doi.org/10.1029/96JD03876, 1997.

Bjørklund, P. K., Rekdal, Y., and Strand, G.-H.: Arealregnskap for utmark, Arealstatistikk for Finnmark 01/2015, 2015.

Blok, D., Heijmans, M. M. P. D., Schaepman-Strub, G., Kononov, A. V., Maximov, T. C., and Berendse, F.: Shrub expansion may reduce summer permafrost thaw in Siberian tundra, Glob. Change Biol., 16, 1296–1305, https://doi.org/10.1111/j.1365-2486.2009.02110.x, 2010.

Bonan, G. B.: Forests and climate change: Forcings, feedbacks, and the climate benefits of forests, Science, 320, 1444–1449, https://doi.org/10.1126/science.1155121, 2008.

Bonfils, C. J. W., Phillips, T. J., Lawrence, D. M., Cameron-Smith, P., Riley, W. J., and Subin, Z. M.: On the influence of shrub height and expansion on northern high latitude climate, Environ. Res. Lett., 7, 015503, https://doi.org/10.1088/1748-9326/7/1/015503, 2012.

Brook, B. W., Ellis, E. C., Perring, M. P., Mackay, A. W., and Blomqvist, L.: Does the terrestrial biosphere have planetary tipping points?, Trends Ecol. Evol., 28, 396–401, https://doi.org/10.1016/j.tree.2013.01.016, 2013.

Broxton, P. D., Zeng, X. B., Sulla-Menashe, D., and Troch, P. A.: A Global Land Cover Climatology Using MODIS Data, J. Appl. Meteorol. Clim., 53, 1593–1605, https://doi.org/10.1175/Jamc-D-13-0270.1, 2014.

Bryn, A.: Recent forest limit changes in south-east Norway: Effects of climate change or regrowth after abandoned utilisation?, Norsk Geogr. Tidsskr., 62, 251–270, https://doi.org/10.1080/00291950802517551, 2008.

Bryn, A., Dourojeanni, P., Hemsing, L. O., and O'Donnell, S.: A high-resolution GIS null model of potential forest expansion following land use changes in Norway, Scand. J. Forest Res., 28, 81–98, https://doi.org/10.1080/02827581.2012.689005, 2013.

Chapin, F. S., Sturm, M., Serreze, M. C., McFadden, J. P., Key, J. R., Lloyd, A. H., McGuire, A. D., Rupp, T. S., Lynch, A. H., Schimel, J. P., Beringer, J., Chapman, W. L., Epstein, H. E., Euskirchen, E. S., Hinzman, L. D., Jia, G., Ping, C. L., Tape, K. D., Thompson, C. D. C., Walker, D. A., and Welker, J. M.: Role of land-surface changes in Arctic summer warming, Science, 310, 657–660, https://doi.org/10.1126/science.1117368, 2005.

Corlett, R. T. and Westcott, D. A.: Will plant movements keep up with climate change?, Trends Ecol. Evol., 28, 482–488, https://doi.org/10.1016/j.tree.2013.04.003, 2013.

de Wit, H. A., Bryn, A., Hofgaard, A., Karstensen, J., Kvalevag, M. M., and Peters, G. P.: Climate warming feedback from mountain birch forest expansion: reduced albedo dominates carbon uptake, Glob. Change Biol., 20, 2344–2355, https://doi.org/10.1111/gcb.12483, 2014.

Elmendorf, S. C., Henry, G. H. R., Hollister, R. D., Bjork, R. G., Boulanger-Lapointe, N., Cooper, E. J., Cornelissen, J. H. C., Day, T. A., Dorrepaal, E., Elumeeva, T. G., Gill, M., Gould, W. A., Harte, J., Hik, D. S., Hofgaard, A., Johnson, D. R., Johnstone, J. F., Jonsdottir, I. S., Jorgenson, J. C., Klanderud, K., Klein, J. A., Koh, S., Kudo, G., Lara, M., Levesque, E., Magnusson, B., May, J. L., Mercado-Diaz, J. A., Michelsen, A., Molau, U., Myers-Smith, I. H., Oberbauer, S. F., Onipchenko, V. G., Rixen, C., Schmidt, N. M., Shaver, G. R., Spasojevic, M. J., Porhallsdottir, P. E., Tolvanen, A., Troxler, T., Tweedie, C. E., Villareal, S., Wahren, C. H., Walker, X., Webber, P. J., Welker, J. M., and Wipf, S.: Plot-scale evidence of tundra vegetation change and links to recent summer warming, Nature Climate Change, 2, 453–457, https://doi.org/10.1038/Nclimate1465, 2012.

Forbes, B. C., Fauria, M. M., and Zetterberg, P.: Russian Arctic warming and "greening" are closely tracked by tundra shrub willows, Glob. Change Biol., 16, 1542–1554, https://doi.org/10.1111/j.1365-2486.2009.02047.x, 2010.

Gottfried, M., Pauli, H., Futschik, A. et al.: Continent-wide response of mountain vegetation to climate change, Nature Climate Change, 2, 111–115, https://doi.org/10.1038/nclimate1329, 2012.

Hallinger, M., Manthey, M., and Wilmking, M.: Establishing a missing link: warm summers and winter snow cover promote shrub expansion into alpine tundra in Scandinavia, New Phytol., 186, 890–899, https://doi.org/10.1111/j.1469-8137.2010.03223.x, 2010.

Hines, K. M. and Bromwich, D. H.: Development and testing of Polar Weather Research and Forecasting (WRF) Model. Part I: Greenland ice sheet meteorology, Mon. Weather Rev., 136, 1971–1989, https://doi.org/10.1175/2007mwr2112.1, 2008.

Hines, K. M., Bromwich, D. H., Bai, L. S., Barlage, M., and Slater, A. G.: Development and Testing of Polar WRF. Part III: Arctic Land, J. Climate, 24, 26–48, https://doi.org/10.1175/2010jcli3460.1, 2011.

Hofgaard, A.: Inter-relationships between treeline position, species diversity, land use and climate change in the central Scandes Mountains of Norway, Global Ecol. Biogeogr., 6, 419–429, https://doi.org/10.2307/2997351, 1997.

Hofgaard, A., Tømmervik, H., Rees, G., and Hanssen, F.: Latitudinal forest advance in northernmost Norway since the early 20th century, J. Biogeogr., 40, 938–949, https://doi.org/10.1111/jbi.12053, 2013.

Iacono, M. J., Delamere, J. S., Mlawer, E. J., Shephard, M. W., Clough, S. A., and Collins, W. D.: Radiative forcing by long-lived greenhouse gases: Calculations with the AER radiative transfer models, J. Geophys. Res.-Atmos., 113, D13103, https://doi.org/10.1029/2008JD009944, 2008.

IPCC: Climate Change 2013: The Physical Science Basis, Contribution of Working Group I to the Fifth Assessment Report of the Intergovernmental Panel on Climate Change, Cambridge University Press, Cambridge, United Kingdom and New York, NY, USA, 1535 pp., 2013.

Janjic, Z. I.: The Step-Mountain Eta Coordinate Model – Further Developments of the Convection, Viscous Sublayer, and Turbulence Closure Schemes, Mon. Weather Rev., 122, 927–945, https://doi.org/10.1175/1520-0493(1994)122<0927:Tsmecm>2.0.Co; 2, 1994.

Karlsen, S. R., Elvebakk, A., and Johansen, B.: A vegetation-based method to map climatic variation in the arctic-boreal transition area of Finnmark, north-easternmost Norway, J. Biogeogr., 32, 1161–1186, https://doi.org/10.1111/j.1365-2699.2004.01199.x, 2005.

Lawrence, D. M. and Swenson, S. C.: Permafrost response to increasing Arctic shrub abundance depends on the relative influence of shrubs on local soil cooling versus large-scale climate warming, Environ. Res. Lett., 6, 045504, https://doi.org/10.1088/1748-9326/6/4/045504, 2011.

Liess, S., Snyder, P. K., and Harding, K. J.: The effects of boreal expansion on the summer Arctic frontal zone, Clim. Dynam., https://doi.org/10.1007/s00382-011-1064-7, 2011.

McFadden, J. P., Liston, G. E., Sturm, M., Pielke, R. A., and Chapin, F. S.: Interactions of shrubs and snow in arctic tundra: measurements and models, Iahs-Aish P., 317–325, 2001.

Milbau, A., Shevtsova, A., Osler, N., Mooshammer, M., and Graae, B. J.: Plant community type and small-scale disturbances, but not altitude, influence the invasibility in subarctic ecosystems, New Phytol., 197, 1002–1011, https://doi.org/10.1111/nph.12054, 2013.

Miller, P. A. and Smith, B.: Modelling Tundra Vegetation Response to Recent Arctic Warming, Ambio, 41, 281–291, https://doi.org/10.1007/s13280-012-0306-1, 2012.

Moen, A., Lillethun, A., and Odland, A.: Vegetation: National Atlas of Norway, Norwegian Mapping Authority, Hønefoss, Norway, 1999.

Morrison, H., Thompson, G., and Tatarskii, V.: Impact of Cloud Microphysics on the Development of Trailing Stratiform Precipitation in a Simulated Squall Line: Comparison of One- and Two-Moment Schemes, Mon. Weather Rev., 137, 991–1007, https://doi.org/10.1175/2008MWR2556.1, 2009.

Myers-Smith, I. H., Forbes, B. C., Wilmking, M., Hallinger, M., Lantz, T., Blok, D., Tape, K. D., Macias-Fauria, M., Sass-Klaassen, U., Levesque, E., Boudreau, S., Ropars, P., Hermanutz, L., Trant, A., Collier, L. S., Weijers, S., Rozema, J., Rayback, S. A., Schmidt, N. M., Schaepman-Strub, G., Wipf, S., Rixen, C., Menard, C. B., Venn, S., Goetz, S., Andreu-Hayles, L., Elmendorf, S., Ravolainen, V., Welker, J., Grogan, P., Epstein, H. E., and Hik, D. S.: Shrub expansion in tundra ecosystems: dynamics, impacts and research priorities, Environ. Res. Lett., 6, 045509, https://doi.org/10.1088/1748-9326/6/4/045509, 2011.

Myers-Smith, I. H., Elmendorf, S. C., Beck, P. S. A., Wilmking, M., Hallinger, M., Blok, D., Tape, K. D., Rayback, S. A., Macias-Fauria, M., Forbes, B. C., Speed, J. D. M., Boulanger-Lapointe, N., Rixen, C., Levesque, E., Schmidt, N. M., Baittinger, C., Trant, A. J., Hermanutz, L., Collier, L. S., Dawes, M. A., Lantz, T. C., Weijers, S., Jorgensen, R. H., Buchwal, A., Buras, A., Naito, A. T., Ravolainen, V., Schaepman-Strub, G., Wheeler, J. A., Wipf, S., Guay, K. C., Hik, D. S., and Vellend, M.: Climate sensitivity of shrub growth across the tundra biome, Nature Climate Change, 5, https://doi.org/10.1038/nclimate2697, 2015.

Myneni, R. B., Keeling, C. D., Tucker, C. J., Asrar, G., and Nemani, R. R.: Increased plant growth in the northern high latitudes from 1981 to 1991, Nature, 386, 698–702, 1997.

Pearson, R. G., Phillips, S. J., Loranty, M. M., Beck, P. S. A., Damoulas, T., Knight, S. J., and Goetz, S. J.: Shifts in Arctic vegetation and associated feedbacks under climate change, Nature Climate Change, 3, 673–677, https://doi.org/10.1038/Nclimate1858, 2013.

Phoenix, G. K. and Bjerke, J. W.: Arctic browning: extreme events and trends reversing arctic greening, Glob. Change Biol., 22, 2960–2962, https://doi.org/10.1111/gcb.13261, 2016.

Piao, S., Wang, X., Ciais, P., Zhu, B., Wang, T. A. O., and Liu, J. I. E.: Changes in satellite-derived vegetation growth trend in temperate and boreal Eurasia from 1982 to 2006, Glob. Change Biol., 17, 3228–3239, https://doi.org/10.1111/j.1365-2486.2011.02419.x, 2011.

Pithan, F. and Mauritsen, T.: Arctic amplification dominated by temperature feedbacks in contemporary climate models, Nat. Geosci., 7, 181–184, https://doi.org/10.1038/Ngeo2071, 2014.

Rannow, S.: Do shifting forest limits in south-west Norway keep up with climate change?, Scand. J. Forest Res., 28, 574–580, https://doi.org/10.1080/02827581.2013.793776, 2013.

Rydsaa, J. H., Stordal, F., and Tallaksen, L. M.: Sensitivity of the regional European boreal climate to changes in surface properties resulting from structural vegetation perturbations, Biogeosciences, 12, 3071–3087, https://doi.org/10.5194/bg-12-3071-2015, 2015.

Rydsaa, J. H.: Effects of shrub cover increase on the near surface atmosphere in northern Fennoscandia, Norstore, https://doi.org/10.11582/2017.00013, 2017.

Serreze, M. C. and Barry, R. G.: Processes and impacts of Arctic amplification: A research synthesis, Global Planet. Change, 77, 85–96, https://doi.org/10.1016/j.gloplacha.2011.03.004, 2011.

Skamarock, W. C., Klemp, J. B., Dudhia, J., Gill, D. O., Barker, D. M., Duda, M. G., Huang, X.-Y., Wang, W., and Powers, J. G.: A Description of the Advanced Research WRF Version 3, National Center for Atmospheric Research, Boulder, Colorado, USA, 2008.

Snyder, P. K.: ARCTIC GREENING Concerns over Arctic warming grow, Nature Climate Change, 3, 539–540, 2013.

Soja, A. J., Tchebakova, N. M., French, N. H. F., Flannigan, M. D., Shugart, H. H., Stocks, B. J., Sukhinin, A. I., Parfenova, E. I., Chapin, F. S., and Stackhouse, P. W.: Climate-induced boreal forest change: Predictions versus current observations, Global Planet. Change, 56, 274–296, https://doi.org/10.1016/j.gloplacha.2006.07.028, 2007.

Sturm, M., McFadden, J. P., Liston, G. E., Chapin, F. S., Racine, C. H., and Holmgren, J.: Snow-shrub interactions in Arctic tundra: A hypothesis with climatic implications, J. Climate, 14, 336–344, https://doi.org/10.1175/1520-0442(2001)014<0336:Ssiiat>2.0.Co; 2, 2001a.

Sturm, M., Racine, C., and Tape, K.: Climate change – Increasing shrub abundance in the Arctic, Nature, 411, 546–547, https://doi.org/10.1038/35079180, 2001b.

Sturm, M., Douglas, T., Racine, C., and Liston, G. E.: Changing snow and shrub conditions affect albedo with global implications, J. Geophys. Res.-Biogeo., 110, G01004, https://doi.org/10.1029/2005jg000013, 2005a.

Sturm, M., Schimel, J., Michaelson, G., Welker, J. M., Oberbauer, S. F., Liston, G. E., Fahnestock, J., and Romanovsky, V. E.: Winter Biological Processes Could Help Convert Arctic Tundra to Shrubland, Bioscience, 55, 17–26, https://doi.org/10.1641/0006-3568(2005)055[0017:wbpchc]2.0.co;2, 2005b.

Svenning, J. C. and Sandel, B.: Disequilibrium Vegetation Dynamics under Future Climate Change, Am. J. Bot., 100, 1266–1286, https://doi.org/10.3732/ajb.1200469, 2013.

Swann, A. L., Fung, I. Y., Levis, S., Bonan, G. B., and Doney, S. C.: Changes in Arctic vegetation amplify high-latitude warming through the greenhouse effect, P. Natl. Acad. Sci. USA, 107, 1295–1300, https://doi.org/10.1073/pnas.0913846107, 2010.

Tape, K., Sturm, M., and Racine, C.: The evidence for shrub expansion in Northern Alaska and the Pan-Arctic, Glob. Change Biol., 12, 686–702, https://doi.org/10.1111/j.1365-2486.2006.01128.x, 2006.

Tape, K. D., Hallinger, M., Welker, J. M., and Ruess, R. W.: Landscape Heterogeneity of Shrub Expansion in Arctic Alaska, Ecosystems, 15, 711–724, https://doi.org/10.1007/s10021-012-9540-4, 2012.

Tewari, M., Chen, F., Wang, W., Dudhia, J., LeMone, M. A., Mitchell, K., Ek, M., Gayno, G., Wegiel, J., and Cuenca, R. H.: Implementation and verification of the unified NOAH land surface model in the WRF model, 20th conference on weather analysis and forecasting/16th conference on numerical weather prediction, 2004.

Tommervik, H., Johansen, B., Tombre, I., Thannheiser, D., Hogda, K. A., Gaare, E., and Wielgolaski, F. E.: Vegetation changes in the Nordic mountain birch forest: The influence of grazing and climate change, Arct. Antarct. Alp. Res., 36, 323–332, https://doi.org/10.1657/1523-0430(2004)036[0323:Vcitnm]2.0.Co;2, 2004.

Tommervik, H., Johansen, B., Riseth, J. A., Karlsen, S. R., Solberg, B., and Hogda, K. A.: Above ground biomass changes in the mountain birch forests and mountain heaths of Finnmarksvidda, northern Norway, in the period 1957-2006, Forest Ecol. Manag., 257, 244–257, https://doi.org/10.1016/j.foreco.2008.08.038, 2009.

Walker, M. D., Wahren, C. H., Hollister, R. D., Henry, G. H. R., Ahlquist, L. E., Alatalo, J. M., Bret-Harte, M. S., Calef, M. P., Callaghan, T. V., Carroll, A. B., Epstein, H. E., Jonsdottir, I. S., Klein, J. A., Magnusson, B., Molau, U., Oberbauer, S. F., Rewa, S. P., Robinson, C. H., Shaver, G. R., Suding, K. N., Thompson, C. C., Tolvanen, A., Totland, O., Turner, P. L., Tweedie, C. E., Webber, P. J., and Wookey, P. A.: Plant community responses to experimental warming across the tundra biome, P. Natl. Acad. Sci. USA, 103, 1342–1346, https://doi.org/10.1073/pnas.0503198103, 2006.

Wang, Z., Zeng, X., and Decker, M.: Improving snow processes in the Noah land model, J. Geophys. Res.-Atmos., 115, D20108, https://doi.org/10.1029/2009JD013761, 2010.

Wolf, A., Callaghan, T. V., and Larson, K.: Future changes in vegetation and ecosystem function of the Barents Region, Climatic Change, 87, 51–73, https://doi.org/10.1007/s10584-007-9342-4, 2008.

Zhang, W. X., Miller, P. A., Smith, B., Wania, R., Koenigk, T., and Doscher, R.: Tundra shrubification and tree-line advance amplify arctic climate warming: results from an individual-based dynamic vegetation model, Environ. Res. Lett., 8, 034023, https://doi.org/10.1088/1748-9326/8/3/034023, 2013.

10

Temperature and UV light affect the activity of marine cell-free enzymes

Blair Thomson[1]**, Christopher David Hepburn**[1]**, Miles Lamare**[1]**, and Federico Baltar**[1,2]

[1]Department of Marine Science, University of Otago, Dunedin, New Zealand
[2]NIWA/University of Otago Research Centre for Oceanography, Dunedin, New Zealand

Correspondence to: Federico Baltar (federico.baltar@otago.ac.nz)

Abstract. Microbial extracellular enzymatic activity (EEA) is the rate-limiting step in the degradation of organic matter in the oceans. These extracellular enzymes exist in two forms: cell-bound, which are attached to the microbial cell wall, and cell-free, which are completely free of the cell. Contrary to previous understanding, cell-free extracellular enzymes make up a substantial proportion of the total marine EEA. Little is known about these abundant cell-free enzymes, including what factors control their activity once they are away from their sites (cells). Experiments were run to assess how cell-free enzymes (excluding microbes) respond to ultraviolet radiation (UVR) and temperature manipulations, previously suggested as potential control factors for these enzymes. The experiments were done with New Zealand coastal waters and the enzymes studied were alkaline phosphatase (APase), β-glucosidase, (BGase), and leucine aminopeptidase (LAPase). Environmentally relevant UVR (i.e. in situ UVR levels measured at our site) reduced cell-free enzyme activities by up to 87 % when compared to controls, likely a consequence of photodegradation. This effect of UVR on cell-free enzymes differed depending on the UVR fraction. Ambient levels of UV radiation were shown to reduce the activity of cell-free enzymes for the first time. Elevated temperatures (15 °C) increased the activity of cell-free enzymes by up to 53 % when compared to controls (10 °C), likely by enhancing the catalytic activity of the enzymes. Our results suggest the importance of both UVR and temperature as control mechanisms for cell-free enzymes. Given the projected warming ocean environment and the variable UVR light regime, it is possible that there could be major changes in the cell-free EEA and in the enzymes contribution to organic matter remineralization in the future.

1 Introduction

Heterotrophic microbes are ubiquitous in the marine environment, recycling most of the organic matter available in the oceans. The discovery of the microbial loop made clear that heterotrophic microbes are one of the most important nutrient vectors in marine food webs (Azam et al., 1983; Azam and Cho, 1987). According to the size–reactivity model, microbes selectively prefer high-molecular-weight dissolved organic matter (HMWDOM) due to its superior nutritional value (Amon and Benner, 1996; Benner and Amon, 2015). The main obstacle for the use of HMWDOM by microbes is that these compounds are generally too large to be transported across microbial cell membranes. Enzymatic hydrolysis outside of the cell is required to break HMWDOM down to smaller size fractions (< 600 Da) before uptake can occur (Weiss et al., 1991). Thus, microbial extracellular enzymatic activity (EEA) is the process that initiates the microbial loop (Arnosti, 2011; Hoppe et al., 2002) and is recognized as the rate-limiting step in the degradation of organic matter in the oceans (Hoppe, 1991). This key role has led to extracellular enzymes being referred to as "gatekeepers of the carbon cycle" (Arnosti, 2011).

There are two forms of EEA: cell-bound, which are attached to the outside of the microbial cell wall or reside in the periplasmic space, and cell-free, which are completely free of the cell and are suspended in the water column. Cell-free enzymes can come from a variety of sources in the marine environment including the sloppy grazing behaviour of protists (Bochdansky et al., 1995; Hoppe, 1991), microbial starvation (Chròst, 1991), the lysis of cells by viruses (Kamer and Rassoulzadegan, 1995), and the direct release by microbes in response to the detection of appropriate substrates (Alderkamp et al., 2007). Up until recently, research on extracellular enzymes has been mostly on cell-bound enzymes, as they were considered to be the only abundant form (Hoppe, 1983; Hoppe et al., 2002). This led to a view that cell-bound extracellular enzymes were the only form of ecological significance (Chróst and Rai, 1993; Rego et al., 1985). However, studies have now shown that the second form, cell-free extracellular enzymes, can make up a substantial proportion of the total extracellular enzyme pool (Baltar et al., 2010, 2013, 2016; Allison et al., 2012; Duhamel et al., 2010; Kamer and Rassoulzadegan, 1995; Li et al., 1998). This has been a major conceptual shift for research in marine enzymatic activity, generating new research questions about what controls cell-free enzymes in the marine environment and how they function (Arnosti, 2011; Arnosti et al., 2014; Baltar et al., 2010, 2016).

One of the many consequences of this discovery is that cell-free enzymes can be decoupled temporally and/or spatially from the microbial community that produces them (Arnosti, 2011; Baltar et al., 2010, 2016), since cell-free enzymes have long residence times after they are released, lasting up to several weeks (Baltar et al., 2013; Steen and Arnosti, 2011). The activity of cell-free enzymes away from their sites (cells) can condition macromolecular dissolved organic compounds (DOCs) and organic surfaces for subsequent microbial growth. This action at a distance complicates discerning links between producing microbes and their enzymes expression, as cell-free enzymes have the potential to contribute to the availability of nutrients at a great distance from the releasing cell (Arnosti, 2011; Baltar et al., 2010, 2016). It has been suggested that the history of the water mass may be more informative in understanding current cell-free enzyme activities than the history of in situ microbial communities present at the time of sampling (Baltar et al., 2010, 2013, 2016; Kamer and Rassoulzadegan, 1995; Arnosti, 2011).

There are only a limited number of published investigations into the dynamics of cell-free enzymes (Baltar et al., 2010, 2013, 2016; Kim et al., 2007; Steen and Arnosti, 2011; Li et al., 1998; Kamer and Rassoulzadegan, 1995; Duhamel et al., 2010). These papers provide good evidence of the importance of cell-free enzymes in the marine environment, but the controls for cell-free enzymes (once separated from the microbial cell) are poorly understood (Arnosti, 2011). Steen and Arnosti (2011) tested the effect of ultraviolet radiation (UVR) on cell-free enzymes directly, finding a reduction in cell-free enzyme activity only at artificially high UVR doses (i.e. UV-B intensities 5–10 times higher than in situ), with natural illumination showing no significant effects of photodegradation. One recent study by Baltar et al. (2016) in the Baltic Sea revealed strong correlations between seasonal temperature change and the proportion of cell-free to total EEA, suggesting seawater temperature and/or solar radiation as the most obvious abiotic mechanisms for the control of cell-free enzymatic activity. However, that was a field study of coastal waters, which includes the whole microbial community and many potential interactions and effects that can co-occur (e.g. production/consumption of free enzymes by microbes and variation in substrate concentration). Thus, to better understand the factors affecting marine free EEA we need to test the effect of environmental factors on free EEA under controlled conditions.

Here we isolated the free extracellular enzymes from a coastal site and specifically studied the effects of temperature and UVR on the activity of three cell-free extracellular enzyme groups: (i) alkaline phosphatase (APase), an enzyme used to acquire phosphorus from organic molecules; (ii) β-glucosidase (BGase), a glycolytic enzyme that targets carbohydrates groups; and (iii) leucine aminopeptidase (LAPase), an enzyme associated with the degradation of proteins. UVR treatments were hypothesized to reduce the activity of cell-free enzymes when compared to dark controls. Photodegradation with "high UVR dose" treatments (including the entire UV-B spectrum, 280 to 320 nm) were hypothesized to have a stronger degradative effect on cell-free enzymes than "low UVR dose" treatments (only including a fraction of the UV-B spectrum, 280 to 305 nm). This was based on the reported effects of UV-B on microbes and their metabolic rates including the total EEA (Herndl et al., 1993; Santos et al., 2012; Müller-Niklas et al., 1995; Demers, 2001). Compared to ambient temperatures (10 °C), cell-free enzymes exposed to high temperatures (15 °C) were hypothesized to be more active and vice versa, due to the general relationship between temperature and catalytic activity in enzymes (Daniel and Danson, 2010, 2013). Experiments carried out here are the first to directly test temperature effects on cell-free enzymes alone, and to directly test the effect of UVR on cell-free enzymes in the Southern Hemisphere and under in situ measured environmentally relevant UV irradiances.

2 Materials and methods

2.1 Study site, sampling, and experiment preparation

The experiments were conducted at the University of Otago's Portobello Marine Laboratory, situated on the Otago Harbour, Dunedin, New Zealand (45.8281° S, 170.6399° E). Otago Harbour is a tidal inlet which has an area of 46 km^2, consisting of two basins and with extensive sediment flats (Grove and Probert, 1999; Heath, 1975). The laboratory is based on the outer Otago Harbour, which has waters similar in composition to coastal seawater, owing to the short residence times for its waters exchanging with the open sea (Rainer, 1981; Grove and Probert, 1999). Samples were taken from the second metre of the water column off the marine laboratory's wharf that extends into a deep tidal channel. Prior to use, all sampling and laboratory equipment used were decontaminated using triplicate rinses of 18 MΩ × cm high-purity (Milli-Q™) water before and after soaking in 10 % hydrochloric acid for more than 6 h and oven-dried at 60 °C. To separate the cell-free extracellular enzymes from the total extracellular enzyme pool and the microbial community, samples were triple filtered through low-protein-binding 0.22 µm Acrodisc filters following published methods (Kim et al., 2007; Baltar et al., 2010). Then, 50 mL glass vials were filled with the 0.22 µm filtered seawater for use in experiments. Bacterial abundance was determined after both experiments by preserving samples in glutaraldehyde and processing them using SYBR Green nucleic acid stain with a BD Accuri C6 flow cytometer (BD Biosciences, USA). This was done to ensure that no significant bacterial growth occurred after filtering or during incubation. Bacterial abundance was reduced to less than 1 % of the prefiltered total and remained so during the 36 h incubations.

2.2 UVR experiments

To determine in situ UV irradiance and environmentally appropriate treatments for experiments, the attenuation of UVR was measured through the upper 2 m of the water column on site using a LI-COR LI1800UW spectroradiometer (LI-COR Biosciences, USA). The spectroradiometer was factory calibrated using NIST traceable standards. Once this was determined, artificial lighting was installed in a temperature-controlled room and set to the ambient seawater temperature of 10 °C. The lighting consisted of two FS20 UV-R lamps (General Electric, Schenectady, NY, USA) and a full spectrum Vita-Lite 72 lamp (Duro Test, Philadelphia, PA, USA), suspended above the samples. These lights were height-adjusted to yield an irradiance of 3.03 $W\,m^{-2}\,s^{-1}$, approximating UV irradiances measured in the field at 2 m depth (3.5 $W\,m^{-2}\,s^{-1}$). Schott WG and GG long-pass filters (15 × 15 cm) with nominal cutoffs (50 % transmission) in the UV-B (280 and 305 nm) were placed above the filtered cell-free enzyme seawater samples contained in glass vials, with either the high dose (<280 nm, 3.03 $W\,m^{-2}\,s^{-1}$, 130.8 kJ) or the low dose (<305 nm, 0.42 $W\,m^{-2}$, 18.1 kJ) of UVR. All light was blocked except that which passed directly through the long-pass filters onto the open glass vials to avoid any effect of the glass on the UVR dose. Controls were kept without light by wrapping the glass vials containing the filtered cell-free enzyme seawater samples in several layers of aluminium foil and were placed in the same temperature-controlled room. Readings of enzyme activity rates were taken from three replicates of each treatment at 12 and 36 h treatment time. UVR was not applied directly to the plate incubations as this can affect the fluorogenic substrate analogues used in the assays. Temperature inside the vials was also monitored to ascertain that the samples were constantly kept at the desired temperature.

2.3 Temperature experiments

For the temperature experiments we utilised a large graded heat block system (see Lamare et al., 2014, for design specifications). This heat block allowed for up to 15 replicate samples to be exposed to constant temperature treatments over time. The heat blocks were tested five times a day for 3 days in advance with blank samples to ensure the heat blocks were calibrated accurately; the variation in temperature was within 0.5 °C of the target temperatures (i.e. 5, 10, and 15 °C) in all measurements. These temperatures were selected because 5 to 15 °C is the annual range of temperature in the sampling site, and 10 °C was the in situ temperature at the time of sampling (unpublished data). All treatments were kept in the dark by wrapping the glass vials containing the filtered cell-free enzyme seawater samples in several layers of aluminium foil. Readings of enzyme activity rates were taken of three replicates of each treatment at 6, 12, 24, and 36 h. When incubating these samples, each was put into a separate incubator, which was set to the respective treatment temperature so as to avoid confounding the temperature treatments.

2.4 Extracellular enzymatic activity (EEA) assays

We used the method for assessing EEA rates based on the hydrolysis of fluorogenic substrate analogues developed by Hoppe (1983). The fluorogenic substrates 4-methylcoumarinyl-7-amide (MCA)-L-leucine-7-amido-4-methylcoumarin, 4-methylumbelliferyl (MUF)-β-D-glucoside, and MUF-phosphate were used to assess the leucine aminopeptidase, β-glucosidase, and alkaline phosphatase activities, respectively. Substrate concentrations of 100 µM were used for each enzyme based on pre-established kinetics tested in the lab. Although differences in UVR or temperature might affect the kinetic parameters, we decided to use the same concentration for all the enzymes (which was saturating at the in situ conditions) to allow for a better comparison and reduce confounding factors. The 96-well Falcon microplates were filled with six replicates of each of the three fluorogenic substrates (10 µL) and seawater (290 µL) to make up 300 µL reactions. Plates were read in a Spectramax M2 spectrofluorometer (Molecular Devices, USA), with excitation and emission wavelengths of 365 and 445 nm, both before and after 3 h incubations. All incubations were performed in the dark with UVR incubations set to the in situ seawater temperature, and temperature incubations set to each respective treatment temperature. Six samples without substrate addition served as blanks in each plate to determine the background fluorescence of the samples, which were used to correct the activity rates in the plate readings before and after incubation.

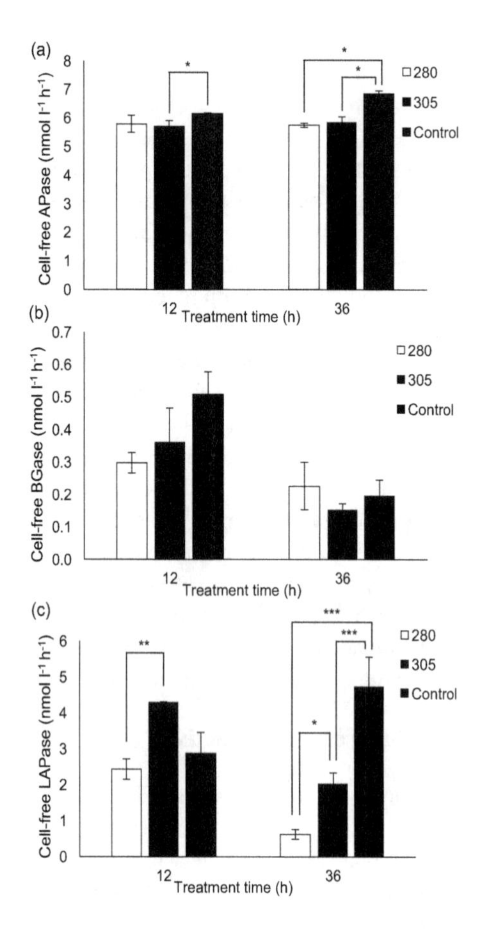

Figure 1. Results from UVR experiments showing the mean (±SE) cell-free extracellular enzyme activity for alkaline phosphatase **(a)**, beta-glucosidase **(b)**, and leucine aminopeptidase **(c)**, under a high dose (280 nm and above) and a low dose (305 nm and above) in comparison to dark controls. Asterisks above bars represent individual significant effects between treatments in post hoc Tukey test (* <0.05, ** <0.01, and *** <0.001) ($N = 3$).

2.5 Statistical analyses

In all analyses, parametric assumptions were first checked using the Shapiro–Wilk test for normality and the Levene's test for equal variance. Where appropriate, data were log-transformed to meet normality assumptions prior to analysis. Both experiments use a two-way analysis of variance (ANOVA) with an interaction term, with post hoc Tukey HSD tests run to assess the individual significant effects between treatments. All analyses were run in the R software environment (R development core team, Austria).

Table 1. Total, cell-free, and proportion of cell-free relative to total extracellular enzymatic activity (EEA) in situ for the seawater collected for the UVR and temperature experiments at the time of sampling.

	Total ($nmol\,L^{-1}\,h^{-1}$)	Cell-free ($nmol\,L^{-1}\,h^{-1}$)	% cell-free (%)
UVR experiment			
APase	75.4	70.3	93.3
BGase	2.3	2.2	96.7
LAPase	35.4	7.5	21.1
Temperature experiment			
APase	121.9	121.8	99.9
BGase	3.2	2.7	85.8
LAPase	33.1	9.9	30.0

3 Results and discussion

3.1 UVR experiments revealed photodegradation of cell-free enzymatic activities

The proportions of cell-free EEA in the seawater at the time of sampling were 93.3, 96.7, and 21.1 % for APase, BGase, and LAPase, respectively (Table 1). Overall, UVR significantly decreased cell-free APase when compared to dark controls ($p < 0.001$, $F_{2,12} = 15.85$, two-way ANOVA) (Fig. 1a). Individual significant effects between treatments in APase were seen as a significant decrease in activity in the low-dose treatment relative to the dark control at 12 h ($p < 0.05$, Tukey HSD), and between the dark control and both the high and low-dose treatments at the 36 h sampling point ($p < 0.05$, Tukey HSD). BGase cell-free activity was not significantly affected by UVR ($p = 0.53$, $F_{2,12} = 0.67$, two-way ANOVA). UVR had a significant overall effect on LAPase, decreasing the cell-free activity when compared to dark controls ($p < 0.01$, $F_{2,12} = 40.994$, two-way ANOVA) (Fig. 1c). Individual significant effects were seen in LAPase, showing a significant decrease in activity between the low and high dose at 12 h ($p < 0.01$, Tukey HSD) and showing a gradual decrease from high to low dose at 36 h ($p < 0.05$, Tukey HSD), and dark control to both low and high dose ($p < 0.001$, Tukey HSD). Changes observed in the controls of all the enzymes from 12 to 36 h were not statistically significant ($p > 0.05$, Tukey HSD).

Apart from the possibility that UVR treatments may have influenced the composition of the seawater substrate itself, these experiments revealed a significant reduction in cell-free EEA for both APase and LAPase in response to UVR, consistent with the predicted photodegradation, which was not evident for BGase. This was the first time that UVR has been demonstrated to reduce cell-free enzymatic activities at environmentally relevant intensities. The only previous study (Steen and Arnosti, 2011) did show a reduction in the cell-free EEA of APase and LAPase but only at artificially high UVR intensities where UV-B was 5–10 times more intense from artificial lamps in the lab than outdoors. Interestingly, they could not show significant UVR effects on BGase at any treatment level, which is consistent with the present study.

Both APase and LAPase showed the strongest effect from UVR at the 36 h sampling point, suggesting a UV-B dose-dependent response. LAPase also showed a gradual decrease in the effect between the low and high UVR treatments, which suggests the increase in UV-B irradiances also enhanced the degree of photodegradation. UV-B has been demonstrated to be a highly active part of the spectrum for degrading DNA in general (Sinha and Häder, 2002; Dahms and Lee, 2010), which is not only included in cells but is also an abundant component of the dissolved (extracellular) seawater fraction (Paul et al., 1987; Paul and David, 1989). Specific effects of UV-B on total extracellular enzymatic activities have been previously reported (Herndl et al., 1993; Santos et al., 2012; Demers, 2001; Müller-Niklas et al., 1995). However, it is important to distinguish these previous studies from the cell-free enzyme experiments performed here. Those previous studies tested the response of the entire microbial community, for total EEA, based on the assumption that UVR affects the organism (source of enzymes) directly. What is shown in this study is that UVR affects cell-free enzymes exclusively without the need to impact the source organism. The effects of UVR were different among the enzymes assessed, which may be of importance as some enzymes could be more impacted by UVR than others. For example, in this study, APase and LAPase were more affected by UVR than BGase, which could change the spectrum of extracellular enzyme activity in the surface of the ocean. The resulting higher BGase relative to APase or LAPase could potentially condition macromolecular DOC composition by hydrolysing relatively less proteins than carbohydrates in response to UV. In turn, it is conceivable that any change in the enzyme spectrum due to variability in UVR light could cause a loss of productivity (e.g. due to a decrease in the inorganic P made available through APase activities), as the nutrients made available by extracellular enzymes may not be in suitable ratios for the effective growth of microbes (Arnosti et al., 2014; Häder et al., 2007).

3.2 Temperature experiments revealed enhanced catalytic activity of cell-free enzymes

The proportion of cell-free EEA in the original seawater at the time of sampling was 99.9, 85.8, and 30.0 % for APase, BGase, and LAPase, respectively (Table 1). Temperature significantly increased cell-free APase at the high temperature of 15 °C when compared to the ambient control of 10 °C ($p < 0.01$, $F_{2,24} = 34.63$, two-way ANOVA) (Fig. 2a). APase activity was significantly increased, after 6 h, in the high relative to the low temperature ($p < 0.001$, Tukey HSD) and after 12 h between low and high temperature ($p < 0.001$, Tukey HSD), and control and high-treatments ($p < 0.05$, Tukey HSD). Cell-free BGase showed a similar pattern of increased activity in response to higher temperature but it was not significant (Fig. 2b). This lack of significant differences in cell-free BGase in response to temperature could be due to a relatively high variability in EEA among the high-temperature (15 °C) treatments. LAPase significantly decreased in the low-temperature treatment (5 °C), relative to the ambient control ($p < 0.01$, $F_{2,24} = 19.84$, two-way ANOVA) (Fig. 2c). LAPase cell-free activity significantly increased between the low and high-temperature treatments at the 6 h and 12 h time points ($p < 0.05$, Tukey HSD). The temperature effect was dependent on time, finding significant effects after 6 and 12 h, but not later for any of the studied enzymes.

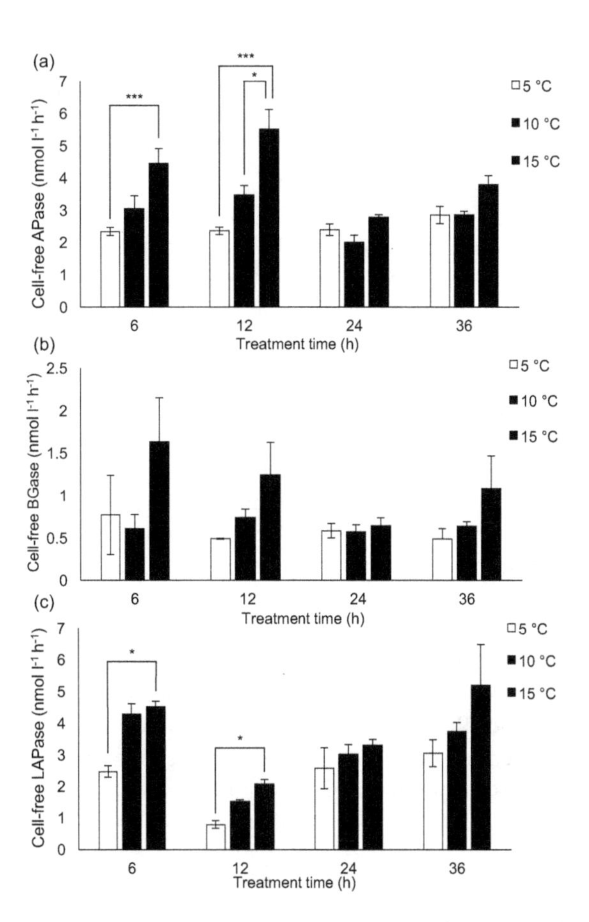

Figure 2. Results from temperature modification experiments showing the mean (±SE) cell-free extracellular enzyme activity for alkaline phosphatase **(a)**, beta-glucosidase **(b)**, and leucine aminopeptidase **(c)**, under high (15 °C) and low temperature (5 °C) treatments in comparison to ambient controls (10 °C). Asterisks above bars represent individual significant effects between treatments in post hoc Tukey test (* <0.05, ** <0.01, and *** <0.001) ($N = 3$).

The relationship found between temperature and cell-free enzyme activity is consistent with the general pattern of increased catalytic activity of enzymes in relation to temperature (Daniel and Danson, 2013). The positive relationship between temperature and the activity of cell-free enzymes observed in this study is contrary to the negative relationship between temperature and the proportion of cell-free relative to total EEA measured in a seasonal field study in the Baltic Sea (Baltar et al., 2016). However, it is important to take into consideration the fact that the study by Baltar et al. (2016) took place over a much longer timescale (1.5 years) and included the whole microbial community, whereas in this study different factors were teased apart by focusing only on the cell-free enzymes. This is supported by Baltar et al. (2016), where the proportion of cell-free relative to total EEA was significantly negatively correlated to prokaryotic heterotrophic production, suggesting that the low temperature preserves the constitutive activity of the cell-free enzymes (better than warm temperature) due to a reduction in the metabolism of heterotrophic microbes that would reduce the consumption/degradation of dissolved enzymes. The exclusion of heterotrophic microbes from our samples precluded this effect (i.e. heterotrophic degradation/consumption of free enzymes) of temperature from occurring and allowed us to tease apart the effect directly on the cell-free enzymatic activities. This also highlights the importance of scale when dealing with microbial oceanographic processes.

4 Conclusions

Overall, temperature and UVR were both demonstrated as potential control mechanisms for the activity of marine cell-free enzymes, providing a baseline for future research. This is the first report revealing the effects of photodegradation of cell-free enzymes at environmentally relevant levels of UVR and the effects of enhanced temperature on the catalytic activity of marine cell-free enzymes. Environmentally relevant UVR had a significant photodegradative effect that might be enzyme-specific (affecting APase and LAPase but not BGase) with the potential to alter not only the rates of cell-free EEA but also the spectrum of enzyme expression in the seawater. Alteration of the cell-free EEA spectrum from UVR variability could have ecological and biogeochemical implications like the conditioning of macromolecular DOM (i.e. affecting DOM composition by hydrolysing some DOM compounds more relative to others) and the change of the elemental ratio of some nutrients (e.g. affecting the availability of inorganic P due to a change in APase activity), with implications for productivity and nutrient cycling. Additionally, given the spatially and temporally variable UVR light regime (i.e. the 150 % increase in UV-B in polar regions during springtime ozone depletion; Smith et al., 1992) and the documented anthropogenic changes in ocean temperature (Chen et al., 2007), it is probable that the activity of cell-free enzymes and their contribution to organic matter remineralization might be affected in the future, if not already.

Competing interests. The authors declare that they have no conflict of interest.

Acknowledgements. We would like to thank the team of technicians out at Portobello Marine Laboratory, most notably Linda Groenewegen and Reuben Pooley. This research was supported by a University of Otago research grant and a Rutherford Discovery Fellowship (Royal Society of New Zealand) to Federico Baltar. We would like to acknowledge the support and insightful comments of the reviewers, which clearly helped improve the overall merit of the manuscript.

Edited by: Gerhard Herndl

References

Alderkamp, A. C., van Rijssel, M., and Bolhuis, H.: Characterization of marine bacteria and the activity of their enzyme systems involved in degradation of the algal storage glucan laminarin, FEMS Microbiol. Ecol., 59, 108–117, https://doi.org/10.1111/j.1574-6941.2006.00219.x, 2007.

Allison, S. D., Chao, Y., Farrara, J. D., Hatosy, S., and Martiny, A.: Fine-scale temporal variation in marine extracellular enzymes of coastal southern California, Front. Microbio., 3, 1–10, https://doi.org/10.3389/fmicb.2012.00301, 2012.

Amon, R. M. W. and Benner, R.: Bacterial utilization of different size classes of dissolved organic matter, Limnol. Oceanogr., 41, 41–51, https://doi.org/10.4319/lo.1996.41.1.0041, 1996.

Arnosti, C.: Microbial Extracellular Enzymes and the Marine Carbon Cycle, Annu. Rev. Mar. Sci., 3, 401–425, 2011.

Arnosti, C., Bell, C., Moorhead, D., Sinsabaugh, R., Steen, A., Stromberger, M., Wallenstein, M., and Weintraub, M.: Extracellular enzymes in terrestrial, freshwater, and marine environments: perspectives on system variability and common research needs, Biogeochemistry, 117, 5–21, https://doi.org/10.1007/s10533-013-9906-5, 2014.

Azam, F. and Cho, B.: Bacterial utilization of organic matter in the sea, Symposia of the Society for General Microbiology, Cambridge, in: Ecology of microbial communities, Cambridge University Press, Cambridge, 261–281, 1987.

Azam, F., Fenchel, T., Field, J., Gray, J., Meyer-Reil, L., and Thingstad, F.: The Ecological Role of Water-Column Microbes in the Sea, Mar. Ecol. Prog. Ser., 10, 257–263, https://doi.org/10.3354/meps010257, 1983.

Baltar, F., Arístegui, J., Gasol, J. M., Sintes, E., van Aken, H. M., and Herndl, G. J.: High dissolved extracellular enzymatic activity in the deep Central Atlantic Ocean, Aquat. Microb. Ecol., 58, 287–302, https://doi.org/10.3354/ame01377, 2010.

Baltar, F., Arístegui, J., Gasol, J., Yokokawa, T., and Herndl, G.: Bacterial Versus Archaeal Origin of Extracellular Enzymatic Activity in the Northeast Atlantic Deep Waters, Microb. Ecol., 65, 277–288, https://doi.org/10.1007/s00248-012-0126-7, 2013.

Baltar, F., Legrand, C., and Pinhassi, J.: Cell-free extracellular enzymatic activity is linked to seasonal temperature changes: a case study in the Baltic Sea, Biogeosciences, 13, 2815–2821, https://doi.org/10.5194/bg-13-2815-2016, 2016.

Benner, R. and Amon, R. M. W.: The Size-Reactivity Continuum of Major Bioelements in the Ocean, Annu. Rev. Mar. Sci., 7, 185–205, 2015.

Bochdansky, A. B., Puskaric, S., and Herndl, G.: Influence of zooplankton grazing on free dissolved enzymes in the sea, Mar. Ecol. Prog. Ser., 121, 53–63, https://doi.org/10.3354/meps121053, 1995.

Chen, Z., Marquis, M., Averyt, K. B., Tignor, M., and Miller, H.: Climate change 2007: the physical science basis. Contribution of Working Group I to the Fourth Assessment Report of the Intergovernmental Panel on Climate Change, Cambridge, Cambridge University, 2007.

Chróst, R. and Rai, H.: Ectoenzyme activity and bacterial secondary production in nutrient-impoverished and nutrient-enriched freshwater mesocosms, Microb. Ecol., 25, 131–150, https://doi.org/10.1007/BF00177191, 1993.

Dahms, H.-U. and Lee, J.-S.: UV radiation in marine ectotherms: Molecular effects and responses, Aquat. Toxicol., 97, 3–14, https://doi.org/10.1016/j.aquatox.2009.12.002, 2010.

Daniel, R. M. and Danson, M. J.: A new understanding of how temperature affects the catalytic activity of enzymes, Trends Biochem. Sci., 35, 584–591, https://doi.org/10.1016/j.tibs.2010.05.001, 2010.

Daniel, R. M. and Danson, M. J.: Temperature and the catalytic activity of enzymes: A fresh understanding, 587, 2738–1743, 2013.

Demers, S.: The Responses of a Natural Bacterioplankton Community to Different Levels of Ultraviolet-B Radiation: A Food Web Perspective, Microb. Ecol., 41, 56–68, 2001.

Duhamel, S., Dyhrman, S. T., and Karl, D. M.: Alkaline phosphatase activity and regulation in the North Pacific Subtropical Gyre, Limnol. Oceanogr., 55, 1414–1425, https://doi.org/10.4319/lo.2010.55.3.1414, 2010.

Grove, S. and Probert, P. K.: Sediment macrobenthos of upper Otago Harbour, New Zealand, New Zeal. J. Mar. Fresh., 33, 469–480, https://doi.org/10.1080/00288330.1999.9516892, 1999.

Häder, D. P., Kumar, H. D., Smith, R. C., and Worrest, R. C.: Effects of solar UV radiation on aquatic ecosystems and interactions with climate change, Photoch. Photobio. Sci., 6, 267–285, https://doi.org/10.1039/b700020k, 2007.

Heath, R. A.: Stability of some New Zealand coastal inlets, New Zeal. J. Mar. Fresh., 9, 449–457, https://doi.org/10.1080/00288330.1975.9515580, 1975.

Herndl, G. J., Müller-Niklas, G., and Frick, J.: Major role of ultraviolet-B in controlling bacterioplankton growth in the surface layer of the ocean, Nature, 361, 717–719, 1993.

Hoppe, H. G.: Significance of exoenzymatic activities in the ecology of brackish water: measurements by means of methylumbelliferyl-substrates, Mar. Ecol. Prog. Ser., 11, 299–308, https://doi.org/10.3354/meps011299, 1983.

Hoppe, H.-G.: Microbial extracellular enzyme activity: a new key parameter in aquatic ecology, in: Microbial enzymes in aquatic environments, Springer, 60–83, 1991.

Hoppe, H. G., Arnosti, C., Burns, R. G., and Dick, R. P.: Ecological significance of bacterial enzymes in the marine environment, edited by: Rijksuniversiteit, G., 73–97, 2002.

Kamer, M. and Rassoulzadegan, F.: Extracellular enzyme activity: Indications for high short-term variability in a coastal marine ecosystem, Microb. Ecol., 30, 143–156, https://doi.org/10.1007/BF00172570, 1995.

Kim, C., Nishimura, Y., and Nagata, T.: High potential activity of alkaline phosphatase in the benthic nepheloid layer of a large mesotrophic lake: implications for phosphorus regeneration in oxygenated hypolimnion, Aquat. Microb. Ecol., 49, 303–311, https://doi.org/10.3354/ame01137, 2007.

Lamare, M., Pecorino, D., Hardy, N., Liddy, M., Byrne, M., and Uthicke, S.: The thermal tolerance of crown-of-thorns (Acanthaster planci) embryos and bipinnaria larvae: implications for spatial and temporal variation in adult populations, Coral Reefs, 33, 207–219, https://doi.org/10.1007/s00338-013-1112-3, 2014.

Li, H., Veldhuis, M., and Post, A.: Alkaline phosphatase activities among planktonic communities in the northern Red Sea, Mar. Ecol. Prog. Ser., 173, 107–115, https://doi.org/10.3354/meps173107, 1998.

Müller-Niklas, G., Heissenberger, A., Puskaríc, S., and Herndl, G.: Ultraviolet-B radiation and bacterial metabolism in coastal waters, Aquat. Microb. Ecol., 9, 111–116, https://doi.org/10.3354/ame009111, 1995.

Paul, J. H. and David, A.W.: Production of extracellular nucleic acids by genetically altered bacteria in aquatic-environment microcosms, Escherichia coli, Pseudomonas aeroginosa, Pseudomonas cepacia, Bradyrhizobium japonicum, Appl. Environ. Microb., 55, 1865–1869, 1989.

Paul, J. H., Jeffrey, W. H., and DeFlaun, M. F.: Dynamics of extracellular DNA in the marine environment, Appl. Environ. Microb., 53, 170–179, 1987.

Rainer, S. F.: Soft-bottom benthic communities in Otago Harbour and Blueskin Bay, New Zealand, Blueskin Bay, New Zealand, Dept. of Scientific and Industrial Research, Wellington, 1981.

Rego, J. V., Billen, G., Fontigny, A., and Somville, M.: Free and attached proteolytic activity in water environments, Mar. Ecol. Prog. Ser, 21, 245–249, 1985.

Santos, A. L., Oliveira, V., Baptista, I. S., Henriques, I., Gomes, N. C. M., Almeida, A., Correia, A., and Cunha, A.: Effects of UV-B Radiation on the Structural and Physiological Diversity of Bacterioneuston and Bacterioplankton, Appl. Environ. Microb., 78, 517–535, 2012.

Sinha, R. P. and Häder, D.-P.: UV-induced DNA damage and repair: a review, Photoch. Photobio. Sci., 1, 225–236, 2002.

Smith, R. C., Prézelin, B. B., Baker, K. S., Bidigare, R. R., Boucher, N. P., Coley, T., Karentz, D., Macintyre, S., Matlick, H. A., Menzies, D., Ondrusek, M., Wan, Z., and Waters, K. J.: Ozone Depletion: Ultraviolet Radiation and Phytoplankton Biology in Antarctic Waters, Science, 255, 952–959, 1992.

Steen, A. D. and Arnosti, C.: Long lifetimes of β-glucosidase, leucine aminopeptidase, and phosphatase in Arctic seawater, Mar. Chem., 123, 127–132, https://doi.org/10.1016/j.marchem.2010.10.006, 2011.

Weiss, M. S., Abele, U., Weckesser, J., Welte, W., Schiltz, E., and Schulz, G. E.: Molecular Architecture and Electrostatic Properties of a Bacterial Porin, Science, 254, 1627–1630, 1991.

11

Alterations in microbial community composition with increasing fCO_2: A mesocosm study in the Eastern Baltic Sea

Katharine J. Crawfurd[1], **Santiago Alvarez-Fernandez**[2], **Kristina D. A. Mojica**[3], **Ulf Riebesell**[4], and **Corina P. D. Brussaard**[1,5]

[1]NIOZ Royal Netherlands Institute for Sea Research, Department of Marine Microbiology and Biogeochemistry and Utrecht University, PO Box 59, 1790 AB Den Burg, Texel, the Netherlands
[2]Alfred-Wegener-Institut Helmholtz-Zentrum für Polar- und Meeresforschung, Biologische Anstalt Helgoland, 27498 Helgoland, Germany
[3]Department of Botany and Plant Pathology, Cordley Hall 2082, Oregon State University, Corvallis, Oregon 97331-29052, USA
[4]GEOMAR Helmholtz Centre for Ocean Research Kiel, Biological Oceanography, Düsternbrooker Weg 20, 24105 Kiel, Germany
[5]Aquatic Microbiology, Institute for Biodiversity and Ecosystem Dynamics, University of Amsterdam, PO Box 94248, 1090 GE Amsterdam, the Netherlands

Correspondence to: Katharine J. Crawfurd (kate.crawfurd@gmail.com) and Corina P. D. Brussaard (corina.brussaard@nioz.nl)

Abstract. Ocean acidification resulting from the uptake of anthropogenic carbon dioxide (CO_2) by the ocean is considered a major threat to marine ecosystems. Here we examined the effects of ocean acidification on microbial community dynamics in the eastern Baltic Sea during the summer of 2012 when inorganic nitrogen and phosphorus were strongly depleted. Large-volume in situ mesocosms were employed to mimic present, future and far future CO_2 scenarios. All six groups of phytoplankton enumerated by flow cytometry ($< 20\,\mu m$ cell diameter) showed distinct trends in net growth and abundance with CO_2 enrichment. The picoeukaryotic phytoplankton groups Pico-I and Pico-II displayed enhanced abundances, whilst Pico-III, *Synechococcus* and the nanoeukaryotic phytoplankton groups were negatively affected by elevated fugacity of CO_2 (fCO_2). Specifically, the numerically dominant eukaryote, Pico-I, demonstrated increases in gross growth rate with increasing fCO_2 sufficient to double its abundance. The dynamics of the prokaryote community closely followed trends in total algal biomass despite differential effects of fCO_2 on algal groups. Similarly, viral abundances corresponded to prokaryotic host population dynamics. Viral lysis and grazing were both important in controlling microbial abundances. Overall our results point to a shift, with increasing fCO_2, towards a more regenerative system with production dominated by small picoeukaryotic phytoplankton.

1 Introduction

Marine phytoplankton are responsible for approximately half of global primary production (Field et al., 1998) with shelf sea communities contributing an average of 15–30 % (Kuliński and Pempkowiak, 2011). Since the industrial revolution, atmospheric carbon dioxide (CO_2) concentrations have increased by nearly 40 % due to anthropogenic emissions, primarily caused by the burning of fossil fuels and deforestation (Doney et al., 2009). Atmospheric CO_2 dissolves in the oceans where it forms carbonic acid that reduces seawater pH, which is a process commonly termed ocean acidification (OA). Currently, along with warming sea surface temperatures and changing light and nutrient conditions, marine ecosystems face unprecedented decreases in ocean pH (Doney et al., 2009; Gruber, 2011). Ocean acidification is considered one of the greatest current threats to marine ecosystems (Turley and Boot, 2010) and has been shown to alter phytoplankton primary production with the direction and magnitude of the responses dependent on community composition (e.g. Hein and Sand-Jensen, 1997; Tortell et al., 2002; Leonardos and Geider, 2005; Engel et al., 2008; Feng et al., 2009; Eberlein et al., 2017). Certain cyanobacteria, including diazotrophs, demonstrate stimulated growth under conditions of elevated CO_2 (Qiu and Gao, 2002; Barcelos e Ramos et al., 2007; Hutchins, et al., 2007; Dutkiewicz et al., 2015). However, no consistent trends have been found for *Synechococcus* (Schulz et al., 2017 and references therein). The responses of diatoms and coccolithophores also appear more variable (Dutkiewicz et al., 2015 and references therein), although coccolithophore calcification seems generally negatively impacted (Meyer and Riebesell, 2015; Riebesell et al., 2017). OA has also been reported to increase the abundances of small-sized photoautotrophic eukaryotes in mesocosm experiments (Engel et al., 2008; Meakin and Wyman, 2011; Brussaard et al., 2013; Schulz et al., 2017).

Recently, data regarding the effects of OA on taxa-specific phytoplankton growth rates were incorporated into a global ecosystem model. The results emphasized that elevated CO_2 concentrations can cause changes in community structure by altering the competitive fitness and thus the competition between phytoplankton groups (Dutkiewicz et al., 2015). Moreover, OA was found to have a greater impact on phytoplankton community size structure, function and biomass than either warming or reduced nutrient supply (Dutkiewicz et al., 2015). Many OA studies have been conducted using single species under controlled laboratory conditions and therefore cannot account for intrinsic community interactions that occur under natural conditions. Alternatively, larger-volume mesocosm experiments allow for OA manipulation of natural communities, and are more likely to capture and quantify the overall response of the natural ecosystems. To date, the majority of these experiments started under replete nutrient conditions or received nutrient additions (Paul et al., 2015 and references therein). Thus, limited data are available for oligotrophic conditions, which are present in $\sim 75\%$ of the world's oceans (Corno et al., 2007).

Whilst environmental factors, such as temperature, light, nutrient and CO_2 concentrations, regulate gross primary production, loss factors determine the fate of this photosynthetically fixed carbon. Grazing, sinking and viral lysis affect the cycling of elements in different manners, i.e. transferred to higher trophic levels through grazing, carbon sequestration in deep waters and sediments, and cellular content release by viral lysis (Wilhelm and Suttle, 1999; Brussaard et al., 2005). Released detrital and dissolved organic matter (DOM) is quickly utilized by heterotrophic bacteria, thereby stimulating activity within the microbial loop (Brussaard et al., 2008; Lønborg et al., 2013; Sheik et al., 2014; Middelboe and Lyck, 2002). Consequently, bacteria may be affected indirectly by OA through changes in the quality and/or quantity of DOM (Weinbauer et al., 2011). Viral lysis has been found to be as important as microzooplankton grazing to the mortality of natural bacterioplankton and phytoplankton (Weinbauer, 2004; Baudoux et al., 2006; Evans and Brussaard, 2012; Mojica et al., 2016). Thus far, most studies examining the effects of OA on microzooplankton abundance and/or grazing have found little or no direct effect (Suffrian et al., 2008; Rose et al., 2009; Aberle et al., 2013; Brussaard et al., 2013; Niehoff et al., 2013). To our knowledge, no viral lysis rates have been reported for natural phytoplankton communities under conditions of OA. A few studies have inferred rates based on changes in viral abundances under enhanced CO_2, but the results are inconsistent (Larsen et al., 2008; Brussaard et al., 2013). Therefore, the effect of OA on the relative share of these key loss processes is still understudied for most ecosystems.

Here we report on the temporal dynamics of microbes (phytoplankton, prokaryotes and viruses) under the influence of enhanced CO_2 concentrations in the low-salinity (around 5.7) Baltic Sea. Using large mesocosms with in situ light and temperature conditions, the pelagic ecosystem was exposed to a range of increasing CO_2 concentrations from ambient to future and far future concentrations. The study was performed during the summer in the Baltic Sea near Tvärminne when conditions were oligotrophic. Our data show that over the 43-day experiment, enhanced CO_2 concentrations elicited distinct shifts in the microbial community, most notably an increase in the net growth of small picoeukaryotic phytoplankton.

2 Materials and methods

2.1 Study site and experimental set-up

The present study was conducted in the Tvärminne Storfjärden (59°51.5′ N, 23°15.5′ E) between 14 June and 7 August 2012. Nine mesocosms, each enclosing $\sim 55\,m^3$ of water, were moored in a square arrangement at a site with a water depth of approximately 30 m. The mesocosms consisted of open-ended polyurethane bags 2 m in diameter and 18.5 m in length mounted onto floating frames covered at each end with a 3 mm mesh. Initially, the mesocosms were kept open for 5 days to allow for rinsing and water exchange while excluding large organisms from entering with the 3 mm mesh. During this time, the bags were positioned such that the tops were submerged 0.5 m below the water surface and the bottoms reached down to 17 m of depth in the water column. Photosynthetically active radiation (PAR) transparent plastic hoods (open on the side) prevented rain and bird droppings from entering the mesocosms, which would affect salinity and nutrients, respectively. Five days before the CO_2 treatment was to begin, the water column of the mesocosms was isolated from the influence of the surrounding water. To do

Table 1. The fCO_2 concentrations (µatm) averaged over the duration of the experiment (following CO_2 addition) and subsequent classification as low, intermediate or high. Mesocosms sampled for mortality assays are denoted by an asterisk. The symbols and colours are used throughout this paper and the corresponding articles in this issue.

Mesocosm	M1*	M5	M7	M6	M3*	M8
CO_2 level	Low	Low	Intermediate	Intermediate	High	High
Mean fCO_2 (µatm) days 1–43	365	368	497	821	1007	1231
Symbol						

so, the 3 mm mesh was removed and sediment traps (2 m long) were attached to close off the bottom of the mesocosms. The top ends of the bags were raised and secured to the frame 1.5 m above the water surface to prevent water from entering via wave action. The mesocosms were then bubbled with compressed air for 3.5 min to remove salinity gradients and ensure that the water body was fully homogeneous.

The present paper includes results from only six of the original mesocosms due to the unfortunate loss of three mesocosms, which were compromised by leakage. The mean fugacity of CO_2 (fCO_2) during the experiment, i.e. days 1–43, for the individual mesocosms were as follows: M1, 365 µatm; M3, 1007 µatm; M5, 368 µatm; M6, 821 µatm; M7, 497 µatm; M8, 1231 µatm (Table 1). The gradient of non-replicated fCO_2 in the present study (as opposed to a smaller number of replicated treatment levels) was selected as a balance between the necessary but manageable number of mesocosms and to minimize the impact of the high loss potential for the mesocosms to successfully address the underlying questions of the study (Schulz and Riebesell, 2013). Moreover, it maximizes the potential of identifying a threshold fCO_2 level concentration if present (by allowing for a larger number of treatment levels). Carbon dioxide manipulation was carried out in four steps and took place between days 0 and 4 until the target fCO_2 was reached. The initial fCO_2 was 240 µatm. For fCO_2 manipulations, 50 µm filtered natural seawater was saturated with CO_2 and then injected evenly throughout the depth of the mesocosms as described by Riebesell et al. (2013). Two mesocosms functioned as controls and were treated in a similar manner using only filtered seawater. On day 15, a supplementary fCO_2 addition was made to the top 7 m of mesocosms numbered 3, 6 and 8 to replace CO_2 lost due to outgassing (Paul et al., 2015; Spilling et al., 2016). Throughout this study we refer to fCO_2, which accounts for the nonideal behaviour of CO_2 gas and is considered the standard measurement required for gas exchange (Pfeil et al., 2013).

Initial nutrient concentrations were 0.05, 0.15, 6.2 and 0.2 $\mu mol L^{-1}$ for nitrate, phosphate, silicate and ammonium, respectively. Nutrient concentrations remained low for the duration of the experiment (Paul et al., 2015; this issue) and no nutrients were added. Salinity was relatively constant around 5.7. Temperature was more variable; on average temperature within the mesocosms (0–17 m) increased from $\sim 8\,°C$ to a maximum on day 15 of $\sim 15\,°C$ and then decreased again to $\sim 8\,°C$ by day 30. For further details of the experimental set-up, carbonate chemistry dynamics and nutrient concentrations throughout the experiment we refer to the general overview paper by Paul et al. (2015).

Collective sampling was performed every morning using depth-integrated water samplers (IWS; Hydro-Bios, Kiel). These sampling devices were gently lowered through the water column collecting ~ 5 L of water gradually between 0 and 10 m (top) or 0 and 17 m (whole water column). Water was collected from all mesocosms and the surrounding water. Subsamples were obtained for the enumeration of phytoplankton, prokaryotes and viruses. Samples for viral lysis and grazing experiments were taken from 5 m of depth using a gentle vacuum-driven pump system. Samples were protected against sunlight and warming by thick black plastic bags containing wet ice. Samples were processed at in situ temperature (representative of 5 m of depth) under dim light and handled using nitrile gloves. As viral lysis and grazing rates were determined from samples taken from 5 m of depth, the samples for microbial abundances reported here were taken from the top 10 m integrated samples.

The experimental period has been divided into four phases based on major physical and biological changes (Paul et al., 2015): Phase 0 before CO_2 addition (days −5 to 0), Phase I (days 1–16), Phase II (days 17–30) and Phase III (days 31–43). Throughout this paper, the data are presented using three colours (blue, grey and red), representing low (mesocosms M1 and M5), intermediate (M6 and M7) and high (M3 and M8) fCO_2 levels (Table 1).

2.2 Microbial abundances

Microbes were enumerated using a Becton Dickinson FACSCalibur flow cytometer (FCM) equipped with a 488 nm argon laser. The samples were stored on wet ice and in the dark until counting. The photoautotrophic cells ($< 20\,\mu m$) were counted directly using fresh seawater and were discriminated by their autofluorescent pigments (Marie et al., 1999). Six phytoplankton clusters were differentiated based on the bivariant plots of either chlorophyll (red autofluorescence) or phycoerythrin (orange autofluorescence for *Synechococcus* and Pico-III) against side scatter. The size of the different phytoplankton clusters was determined by gentle filtration through 25 mm diameter polycarbonate filters (Whatman) with a range of pore sizes (12, 10, 8, 5, 3, 2, 1 and 0.8 µm) according to Veldhuis and Kraay (2004). Average cell sizes for the different phytoplankton groups were 1, 1, 3, 2.9, 5.2 and 8.8 µm in diameter for the prokaryotic cyanobacteria *Synechococcus* spp. (SYN), picoeukaryotic phytoplankton I, II and III (Pico-I–III) and nanoeukaryotic phytoplankton I and II (Nano-I, Nano-II), respectively. Pico-III was discriminated from Pico-II (comparable average cell size) by a higher orange autofluorescence signature, potentially representing small-sized cryptophytes (Klaveness, 1989) or, alternatively, large single cells or microcolonies of *Synechococcus* (Haverkamp et al., 2009). The cyanobacterial species *Prochlorococcus* spp. were not observed during this experiment. Counts were converted to cellular carbon by assuming a spherical shape equivalent to the average cell diameters determined from size fractionations and applying conversion factors of 237 fg C μm^{-3} (Worden et al., 2004) and 196.5 fg C μm^{-3} (Garrison et al., 2000) for pico- and nano-sized plankton, respectively. Microbial net growth and loss rates were derived from exponential regressions of changes in the cell abundances over time.

Abundances of prokaryotes and viruses were determined from 0.5 % glutaraldehyde fixed, flash-frozen (−80 °C) samples according to Marie et al. (1999) and Brussaard (2004). The prokaryotes include heterotrophic bacteria, archaea and unicellular cyanobacteria, the latter accounting for a maximal 10 % of the total abundance in our samples as indicated by their autofluorescence. Thawed samples were diluted with sterile autoclaved Tris-EDTA buffer (10 mM Tris-HCl and 1 mM EDTA; pH 8.2; Mojica et al., 2014) and stained with the green fluorescent nucleic acid-specific dye SYBR Green I (Invitrogen Inc.) to a final concentration of the commercial stock of 1.0×10^{-4} (for prokaryotes) or 0.5×10^{-4} (for viruses). Virus samples were stained at 80 °C for 10 min and then allowed to cool for 5 min at room temperature in the dark. Prokaryotes were stained for 15 min at room temperature in the dark (Brussaard, 2004). Prokaryotes and viruses were discriminated in bivariate scatter plots of green fluorescence versus side scatter. Final counts were corrected for blanks prepared and analysed in a similar manner as the samples. Two groups of prokaryotes were identified by their stained nucleic acid fluorescence, referred here on as low (LNA) and high (HNA) fluorescence prokaryotes.

2.3 Viral lysis and grazing

Microzooplankton grazing and viral lysis of phytoplankton was determined using the modified dilution assay based on reducing grazing and viral lysis mortality pressure in a serial manner allowing for increased phytoplankton growth (over the incubation period) with dilution (Mojica et al., 2016). Two dilution series were created in clear 1.2 L polycarbonate bottles by gently mixing 200 µm sieved whole seawater with either 0.45 µm filtered seawater (i.e. microzooplankton grazers removed) or 30 kDa filtered seawater (i.e. grazers and viruses removed) to final dilutions of 20, 40, 70 and 100 %. The 0.45 µm filtrate was produced by gravity filtration of 200 µm mesh sieved seawater through a 0.45 µm Sartopore capsule filter. The 30 kDa ultrafiltrate was produced by tangential flow filtration of 200 µm pre-sieved seawater using a 30 kDa Vivaflow 200 PES membrane tangential flow cartridge (Vivascience). All treatments were performed in triplicate. Bottles were suspended next to the mesocosms in small cages at 5 m of depth for 24 h. Subsamples were taken at 0 and 24 h, and phytoplankton abundances of the grazing series (0.45 µm diluent) were enumerated by flow cytometry. Due to time constraints, the majority of the samples of the 30 kDa series were fixed with 1 % (final concentration) formaldehyde : hexamine solution (18 % v/v : 10 % w/v) for 30 min at 4 °C, flash-frozen in liquid nitrogen and stored at −80 °C until flow cytometry analysis in the home laboratory. Fixation had no significant effect (Student's t tests; p value > 0.05) as tested periodically against fresh samples. The modified dilution assay was only run for mesocosms 1 (low fCO_2) and 3 (high fCO_2) due to the logistics of handling times. Experiments were performed until day 31. Grazing rates and the combined rate of grazing and viral lysis were estimated from the slope of a regression of phytoplankton apparent growth versus dilution of the 0.45 µm and 30 kDa series, respectively. A significant difference between the two regression coefficients (as tested by analysis of covariance) indicated a significant viral lysis rate. Phytoplankton gross growth rate, in the absence of grazing and viral lysis, was derived from the y-intercept of the 30 kDa series regression. Similarly, significant differences between mesocosms M1 and M3 (low and high fCO_2) were determined through an analysis of covariance of the dilution series for the two mesocosms. A significance threshold of 0.05 was used, and significance is denoted throughout the paper by an asterisk (*). Occasionally, the regression of apparent growth rate versus fraction of natural water resulted in a positive slope (thus no reduction in mortality with dilution). In addition, very low phytoplankton abundances can also prohibit the statistical significance of results. Under such conditions dilution experiments were deemed unsuccessful (for limitations of the modified dilution method, see Baudoux et al., 2006; Kimmance and Brussaard, 2010; Stoecker et al., 2015).

Viral lysis of prokaryotes was determined according to the viral production assay (Wilhelm et al., 2002; Winget et al., 2005). After reduction of the natural virus concentration, new virus production by the natural bacterial community is sampled and tracked over time (24 h). Free viruses were reduced from a 300 mL sample of whole water by recirculation over a 0.2 µm pore size polyether sulfone membrane (PES) tangential flow filter (Vivaflow 50; Vivascience) at a filtrate expulsion rate of 40 mL min^{-1}. The concentrated sample was then reconstituted to the original volume using virus-free seawater. This process was repeated a total of three times to gradually wash away viruses. After the final reconstitution, 50 mL aliquots were distributed into six polycarbonate tubes. Mitomycin C (Sigma-Aldrich; final concentration 1 $\mu g\,mL^{-1}$; maintained at 4 °C), which induces lysogenic bacteria (Weinbauer and Suttle, 1996) was added to a second series of triplicate samples for each mesocosm. A third series of incubations with 0.2 µm filtered samples was used as a control for viral loss (e.g. viruses adhering to the tube walls) and showed no significant loss of free viruses during the incubations. At the start of the experiment, 1 mL subsamples were immediately removed from each tube and fixed as previously described for viral and bacterial abundance. The samples were dark incubated at in situ temperature and 1 mL subsamples were taken at 3, 6, 9, 12 and 24 h. Virus production was determined from linear regression of viral abundance over time. Viral production due to induction of lysogeny was calculated as the difference between production in the unamended samples and the production of samples to which mitomycin C was added. Although mortality experiments were initially planned to be employed for mesocosms 1, 2 and 3 representing low, mid and high fCO_2 conditions, mesocosm 2 was compromised due to leakage. Additionally, due to logistical reasons assays were only performed until day 21.

To determine grazing rates on prokaryotes, fluorescently labelled bacteria (FLBs) were prepared from enriched natural bacterial assemblages (originating from the North Sea) labelled with 5-([4,6-dichlorotriazin-2-yl]amino)fluorescein (DTAF 36565; Sigma-Aldrich; 40 $\mu g\,mL^{-1}$) according to Sherr et al. (1993). Frozen ampoules of FLB (1–5 % of total bacterial abundance) were added to triplicate 1 L incubation bottles containing whole water gently passed through 200 µm mesh. Samples of 20 mL were taken immediately after addition (0 h) and the headspace was removed by gently squeezing air from the bottle. The 1 L bottles were incubated on a slow turning wheel (1 rpm) at in situ light and temperature conditions (representative of 5 m of depth) for 24 h. Sampling was repeated after 24 h. All samples were fixed to a 1 % final concentration of gluteraldehyde (0.2 µm filtered; 25 % EM-grade), stained (in the dark for 30 min at 4 °C) with 4',6-diamidino-2-phenylindole dihydrochloride (DAPI) solution (0.2 µm filtered; Acrodisc® 25 mm syringe filters; Pall Life Sciences; 2 $\mu g\,mL^{-1}$ final concentration; Sherr et al., 1993) and filtered onto 25 mm, 0.2 µm black polycarbonate filters (GE Healthcare Life Sciences). Filters were then mounted on microscopic slides and stored at −20 °C until analysis. FLBs present on a $\sim 0.75\,mm^2$ area were counted using a Zeiss Axioplan 2 microscope. Grazing (μd^{-1}) was measured according to $N_{T24} = N_{T0} \cdot e^{-\mu t}$, where N_{T24} and N_{T0} are the number of FLBs present at 24 and 0 h, respectively.

2.4 Statistics

Non-metric multidimensional scaling (NMDS) was used to follow microbial community development in each mesocosm over the experimental period. NMDS is an ordination technique which represents the dissimilarities obtained from an abundance data matrix in a two-dimensional space (Legendre and Legendre, 1998). In this case, the data matrix was comprised of abundance data for each phytoplankton group in each mesocosm for every day of sampling. The treatment effect was assessed by an analysis of similarity (ANOSIM; Clarke, 1993) and inspection of the NMDS biplot. ANOSIM compares the mean of ranked dissimilarities in mesocosms between fCO_2 treatments (low: 1, 5, 7; high: 6, 3, 8) to the mean of ranked dissimilarities within treatments per phase. The NMDS plots allowed divergence periods in the development and community composition between treatments to be visually assessed (period 1 from days 3–13 and period 2 from days 16–24). The net growth rates of each of the different microbial groups were calculated for these identified divergence periods. Relationships between net growth rates and peak cell abundances with fCO_2 were evaluated by linear regression against the average fCO_2 per mesocosm during each period or peak day. A generalized linear model was used to test the relationship between prokaryote abundance and carbon biomass with an ARMA correlation structure of order 3 to account for temporal autocorrelation. The model fulfilled all assumptions, such as homoscedasticity and avoiding autocorrelation of the residuals (Zuur et al., 2007). A significance threshold of $p \leq 0.05$ was used, and significance is denoted by an asterisk (*). All analyses were performed using the statistical software program R with the packages nlme (Pinheiro et al., 2017) and vegan (Oksanen et al., 2017; R core Team, 2017). Where averages of low and high mesocosm abundance data are reported, the values represent the average of mesocosms 1, 5 and 7 (mean fCO_2 365–497 µatm) and 6, 3 and 8 (821–1231 µatm).

3 Results

3.1 Total phytoplankton dynamics in response to CO_2 enrichment

During Phase 0, low variability in phytoplankton abundances in the different mesocosms ($1.5 \pm 0.05 \times 10^5\,mL^{-1}$) indicated good replicability of initial conditions prior to CO_2 manipulation (Fig. 1). This was further supported by the high similarity between microbial communities in the different mesocosms as indicated by the tight clustering of points in the NMDS plot during this period (Fig. 2). During Phase 0, the phytoplankton community (< 20 µm) was dominated by pico-sized autotrophs, with the prokaryotic cyanobacteria *Synechococcus* (SYN) and Pico-I accounting for 69 and 27 % of total phytoplankton abundance, respectively. After CO_2 addition, there were two primary peaks in phytoplankton, which occurred on day 4 in Phase I and day 24 in Phase II (Fig. 1a). The phytoplankton community became significantly different over time in the different treatments (ANOSIM, $p = 0.01$; Fig. 2). Two periods

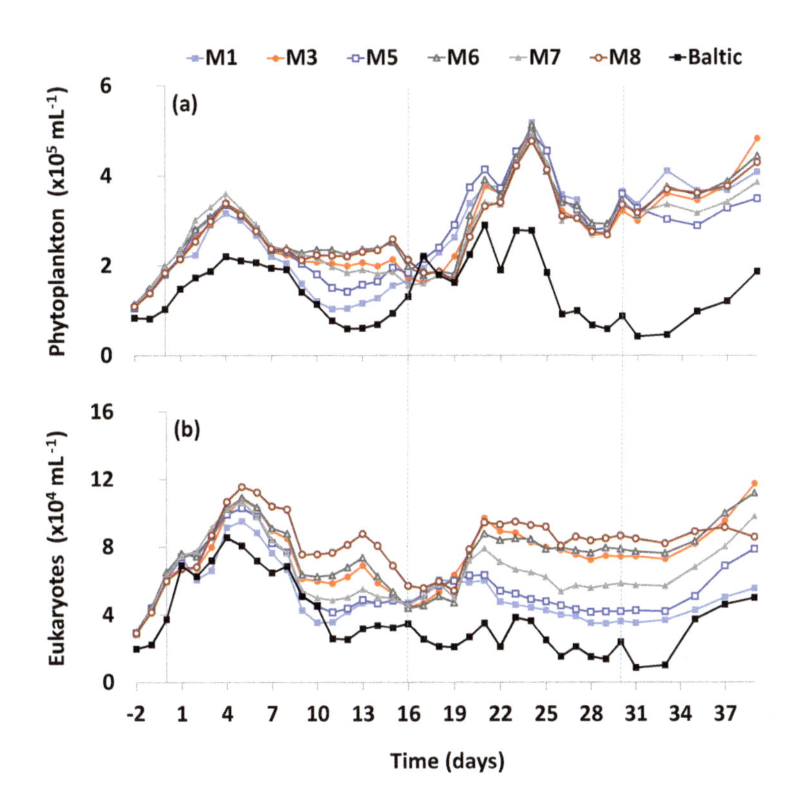

Figure 1. **(a)** Time series plot of depth-integrated (0.3–10 m) total phytoplankton abundance (< 20 μm) and **(b)** total eukaryotic phytoplankton abundance for each mesocosm and the surrounding waters (Baltic). Dotted lines indicate the end of Phase I and the end of Phase II. Colours and symbols represent the different mesocosms and are consistent throughout the paper. Mean fCO_2 during the experiment (days 1–43): M1, 365 μatm; M3, 1007 μatm; M5, 368 μatm; M6, 821 μatm; M7, 497 μatm; M8, 1231 μatm.

were identified based on their divergence (Fig. 2). The first (NMDS-based period 1) followed the initial peak in abundance (days 3–13) with the highest abundances occurring in the elevated CO_2 mesocosms (Fig. 1a). During the second period (NMDS-based period 2; days 16–24), abundances were higher in the low fCO_2 mesocosms (Fig. 1a). In general the NMDS plot shows that throughout the experiment, mesocosm M1 followed the same basic trajectory as mesocosms M5 and M7, whilst mesocosm M3 followed M6 and M8 (Fig. 2). Thus, the two mesocosms (representing high and low fCO_2 treatments) deviated from each other during Phase I and were clearly separated during Phases II and III (Fig. 2).

Phytoplankton abundances in the surrounding water started to differ from the mesocosms during Phase 0 (on average 44 % lower), which was primarily due to lower abundances of SYN. This effect was seen from day −1 prior to CO_2 addition but following bubbling with compressed air (day −5). On day 15, a deep mixing event occurred as a result of storm conditions (with consequent alterations in temperature and salinity). As a result phytoplankton abundances in the surrounding open water diverged more strongly from the mesocosms but remained similar in their dynamics (Fig. 3). Microbial abundances in the 0–17 m samples were slightly lower but showed very similar dynamics to those in the 0–10 m samples (Fig. S1 in the Supplement).

3.1.1 Synechococcus

The prokaryotic cyanobacteria *Synechococcus* (SYN) accounted for the majority of total abundance, i.e. 74 % averaged across all mesocosms over the experimental period. Abundances of SYN showed distinct variability between the different CO_2 treatments,starting on day 7, with the low CO_2 mesocosms exhibiting nearly 20 % lower abundances between days 11 and 15 compared to high fCO_2 mesocosms (Fig. 3a). SYN net growth rates during days 3–13 (NMDS-based period 1) were positively correlated with CO_2 ($p = 0.10$, $R^2 = 0.53$; Table 2, Fig. S2a). One explanation for higher net growth rates at elevated CO_2 could be the significantly ($p < 0.05$) higher grazing rate in the low fCO_2 mesocosm M1 ($0.56\,d^{-1}$) compared to the high fCO_2 M3 ($0.27\,d^{-1}$) as measured on day 10 (Fig. 4a). After day 16, SYN abundances increased in all mesocosms, and during this period (days 16–24) net growth rates had a significant negative correlation with fCO_2 ($p = 0.05$, $R^2 = 0.63$; Figs. 3a and S3a, Table 2). Consequently, the net increase in SYN abundances during this period was on average 20 % higher at low fCO_2 compared to high fCO_2. This corresponded

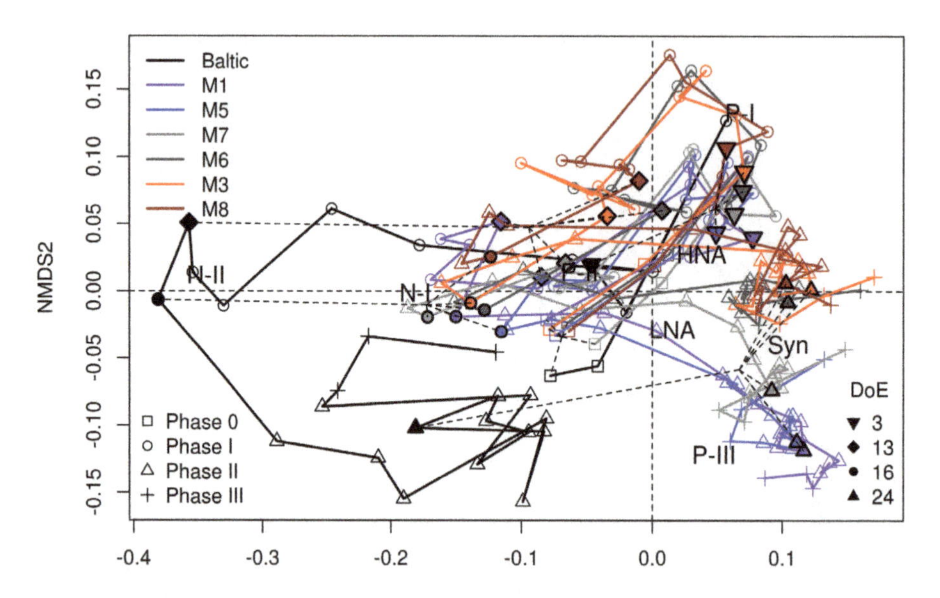

Figure 2. Non-metric multidimensional scaling (NMDS) ordination plot of microbial community development in each mesocosm and the surrounding waters (Baltic) over the experimental period. Phases are indicated by different open symbols. Days of experiment (DoE) when communities separate (3, 13, 16 and 24) are indicated by different closed symbols. Phytoplankton groups are denoted as SYN (Syn), Pico-I (P-I), Pico-II (P-II), Pico-III (P-III), Nano-I (N-I), Nano-II (N-II), low NA prokaryotes (LNA) and high NA prokaryotes (HNA).

to higher total loss rates in high fCO_2 treatments measured on day 17 (0.33 vs. 0.17 d^{-1} for M3 and M1, respectively; Fig. 4a). The higher net growth most likely led to the peak in SYN abundance observed on day 24 (maximum 4.7×10^5 mL^{-1}), which was negatively correlated with fCO_2 ($p = 0.01$, $R^2 = 0.80$; Table 3, Fig. S4a). After this period (days 24–28), SYN abundances declined at comparable rates in the different mesocosms irrespective of fCO_2 (Fig. 3a). Abundances in the low fCO_2 mesocosms remained higher into Phase III (Fig. 3a). SYN abundances in the surrounding water were generally lower than in the mesocosms, with the exception of days 17–21.

3.1.2 Picoeukaryotes

In contrast to the prokaryotic photoautotrophs, the eukaryotic phytoplankton community showed a strong positive response to elevated fCO_2 (Fig. 1b). Pico-I was the numerically dominant group of eukaryotic phytoplankton, accounting for an average 21–26 % of total phytoplankton abundances. Net growth rates leading up to the first peak in abundance (from days 1 to 5) had a strong positive correlation with fCO_2 ($p < 0.01$, $R^2 = 0.90$; Figs. 3b and S5a, Table 3). Accordingly, the peak on day 5 (maximum 1.1×10^5 mL^{-1}; Fig. 3b) was also correlated positively with fCO_2 ($p = 0.01$, $R^2 = 0.81$; Table 3, Fig. S4b). During Phase I from days 3 to 13 (i.e. NMDS-based period 1), net growth rates of Pico-I remained positively correlated with CO_2 concentration ($p = 0.01$, $R^2 = 0.80$; Table 2, Fig. S2b). However, during this period there was also a decline in abundance (days 5–9; $p < 0.01$, $R^2 = 0.89$; Table 3, Fig. S5b) with 23 % more

Table 2. The fit (R^2) and significance (p value) of linear regressions applied to assess the relationship between net growth rate and temporally averaged fCO_2 for the different microbial groups distinguished by flow cytometry. The results presented are for two periods distinguished from NMDS analysis: NMDS-based period 1 (days 3–13) and 2 (days 16–24). A significance level of $p \leq 0.05$ was taken and significant results are shown in bold.

Phytoplankton group	NMDS period 1 (days 3–13)		NMDS period 2 (days 16–24)	
	p	R^2	p	R^2
SYN	0.10	0.53	**0.05**	**0.63**
Pico-I	**0.01**	**0.80**	**0.05**	**0.64**
Pico-II	0.52	0.11	0.10	0.52
Pico-III	**0.04**	**0.67**	**<0.01**	**0.91**
Nano-I	**0.01**	**0.79**	0.26	0.30
Nano-II	0.20	0.36	0.06	0.61
HNA	**0.05**	**0.64**	0.89	0.00
LNA	**<0.01**	**0.95**	**0.02**	**0.76**

cells lost in the low fCO_2 mesocosms. Accordingly, following this period, gross growth rate was significantly higher in the high fCO_2 mesocosm M3 compared to the low fCO_2 mesocosm M1 (day 10, $p < 0.05$; Fig. 4b). Pico-I abundances in the surrounding open water started to deviate from the mesocosms after day 10 and were on average around half that of the low fCO_2 mesocosms (Fig. 3b). Following a brief increase (occurring between days 11 and 13) correlated with fCO_2 ($p < 0.01$, $R^2 = 0.94$; Table 3, Fig. S4c), abundances declined sharply between days 13 and 16 (Fig. 3b), coincid-

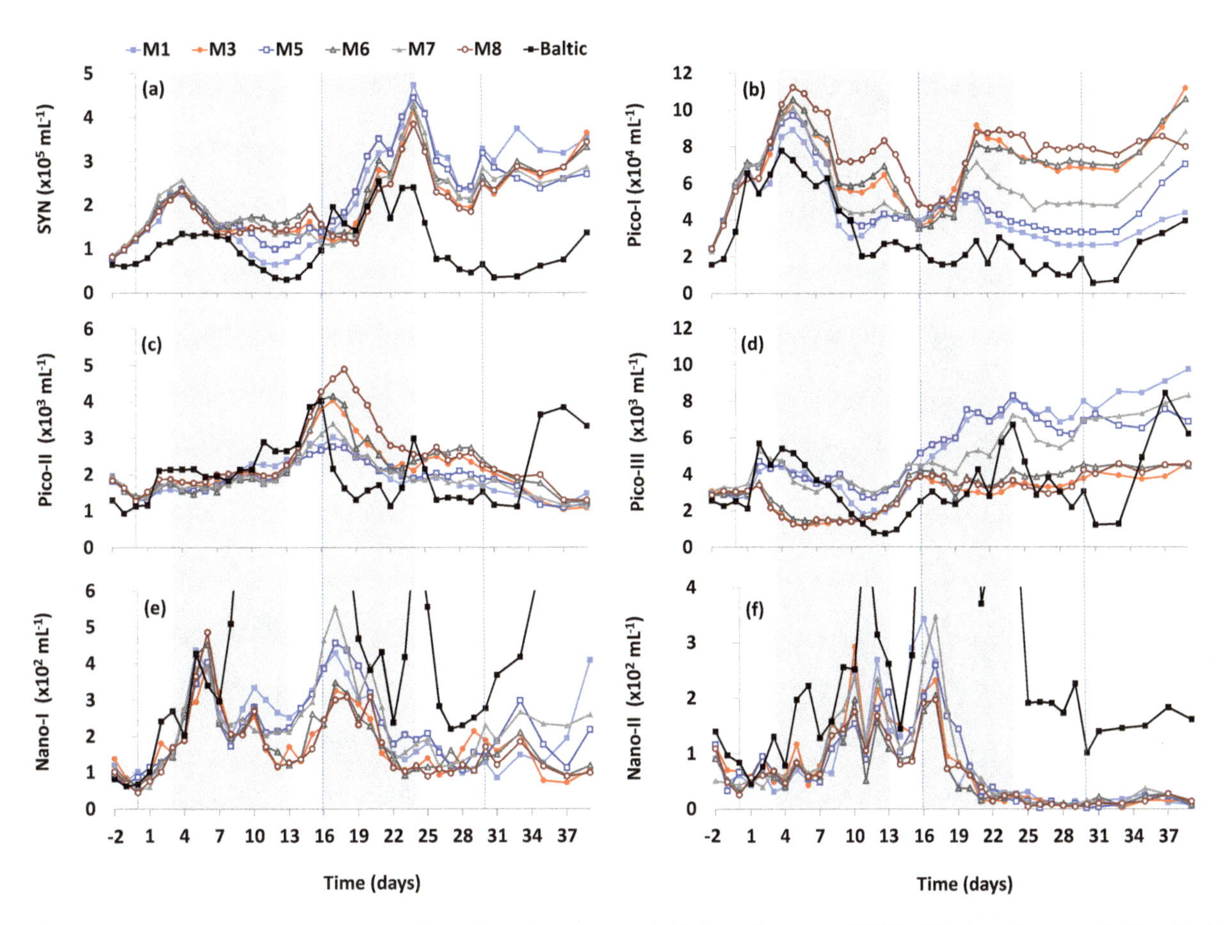

Figure 3. Time series plot of depth-integrated (0.3–10 m) abundances of **(a)** *Synechococcus* (SYN), **(b)** picoeukaryotes I (Pico-I), **(c)** picoeukaryotes II (Pico-II), **(d)** picoeukaryotes III (Pico-III), **(e)** nanoeukaryotes I (Nano-I) and **(f)** nanoeukaryotes II (Nano-II) distinguished by flow cytometric analysis of the microbial community in each mesocosm and the surrounding waters (Baltic). Dotted lines indicate the end of Phase I and the end of Phase II; grey areas indicate NMDS-based periods 1 and 2 during which net growth rates were analysed.

Table 3. The fit (R^2) and significance (p value) of linear regressions used to relate peak abundances and net growth rate with temporally averaged fCO_2 for the different microbial groups distinguished by flow cytometry during specific periods of interest. A significance level of $p \leq 0.05$ was taken and significant results are shown in the table below.

	Peak abundance		Net growth rate	
	p	R^2	p	R^2
SYN day 24	0.01	0.80	–	–
Pico-I day 5	0.01	0.81	–	–
Pico-I day 13	< 0.01	0.94	–	–
Pico-I day 21	0.01	0.84	–	–
Pico-II day 17	< 0.01	0.93	–	–
Pico-III day 24	< 0.01	0.91	–	–
Nano-I day 17	0.04	0.67	–	–
Pico-I days 1–5	–	–	< 0.01	0.90
Pico-I days 5–9	–	–	< 0.01	0.89
Pico-II days 12–17	–	–	0.01	0.82

ing with a significantly higher total mortality rate in the high fCO_2 mesocosm M3 (day 13; Fig. 4b). Viral lysis was a substantial loss factor relative to grazing for this group, comprising an average 45 and 70 % of total losses in M1 and M3, respectively (Table S1). During NMDS-based period 2, net growth rates of Pico-I were significantly higher at high fCO_2 ($p = 0.05$, $R^2 = 0.64$; Table 2, Fig. S3b). By day 21, abundances in the high fCO_2 mesocosms were (on average) ~2-fold higher than at low fCO_2 (maximum abundances 8.7×10^4 and 5.9×10^4 mL^{-1} for high and low fCO_2 mesocosms; $p = 0.01$, $R^2 = 0.84$; Table 3, Fig. S4d). Standing stock of Pico-I remained high in the elevated fCO_2 mesocosms for the remainder of the experiment (7.9×10^4 vs. 4.3×10^4 mL^{-1} on average for high and low fCO_2 mesocosms, respectively; Fig. 3b). Additionally, gross growth rates during this final period were relatively low (0.14 and 0.16 d^{-1} in M1 and M3, respectively) and comparable to total loss rates (averaging 0.13 and 0.10 d^{-1} over days 25–31 for M1 and M3, respectively; Fig. 4b).

Another picoeukaryote group, Pico-II, slowly increased in abundance until day 13, when it increased more rapidly

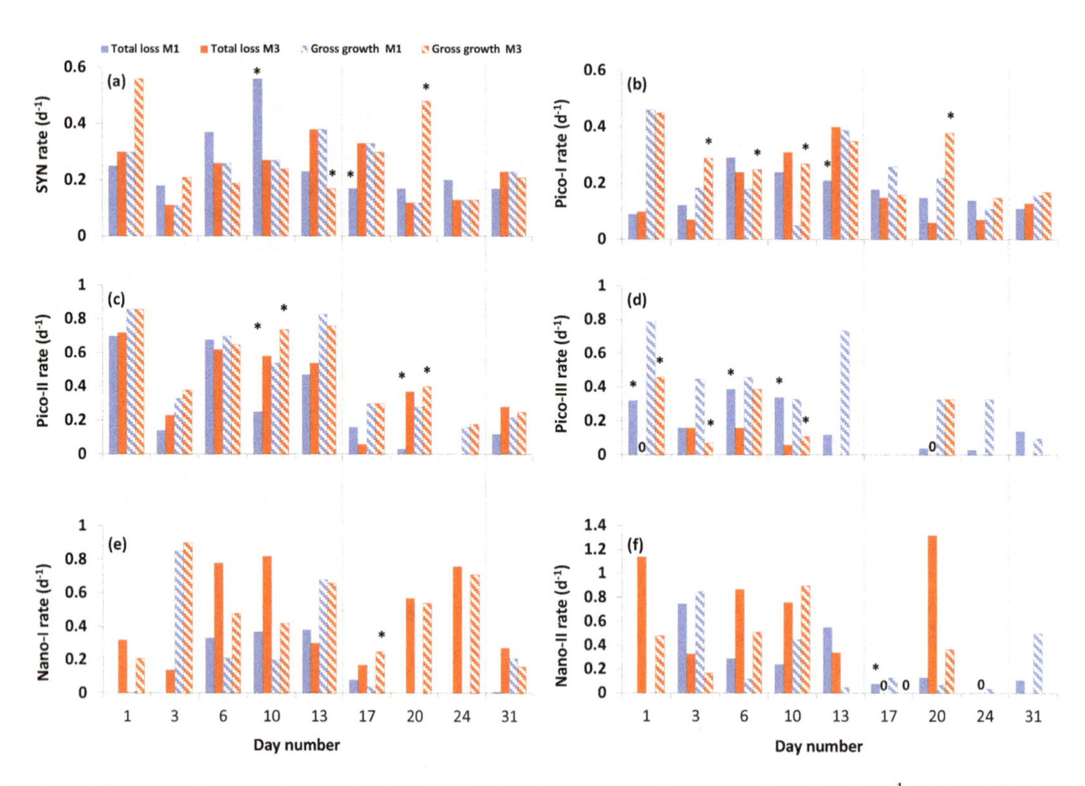

Figure 4. Total mortality rates (i.e. grazing and lysis; solid bars) and gross growth rates (striped bars) d^{-1} of the different phytoplankton groups in mesocosms M1 (blue) and M3 (red) on the day indicated: **(a)** *Synechococcus* (SYN), **(b)** picoeukaryotes I (Pico-I), **(c)** picoeukaryotes II (Pico-II), **(d)** picoeukaryotes III (Pico-III), **(e)** nanoeukaryotes I (Nano-I) and **(f)** nanoeukaryotes II (Nano-II). Significant ($p \leq 0.05$) differences between mesocosms are indicated by an asterisk above the relevant bar (either total loss or gross growth). A coloured zero indicates that a rate of zero was measured in the mesocosm of the corresponding colour; the absence of a bar or zero indicates a failed experiment. Dotted lines indicate the end of Phase I and the end of Phase II.

(Fig. 3c). Gross growth rates measured during Phase I were high (0.69 and 0.72 d^{-1} on average in the low and high fCO_2 mesocosms M1 and M3, respectively; Fig. 4c) and comparable to loss processes (0.46 and 0.58 d^{-1}), indicative of a relatively high turnover rate of production. Overall net growth rates during days 3–13 (NMDS-based period 1) did not correlate with CO_2 ($p = 0.52$, $R^2 = 0.11$; Table 2, Fig. S2c). However, during periods of rapid increases in net growth, abundances were positively correlated with CO_2 concentration (days 12–17; $p = 0.01$, $R^2 = 0.82$; Table 3, Fig. S5c). Accordingly, the peak in abundances of Pico-II on day 17 displayed a distinct positive correlation with fCO_2 ($p < 0.01$, $R^2 = 0.93$; Table 3, Fig. S4e) with maximum abundances of 4.6×10^3 and 3.4×10^3 mL^{-1} for the high and low fCO_2 mesocosms, respectively (Fig. 3c). In M8 (the highest fCO_2 mesocosm), abundances increased for an extra day with the peak occurring on day 18, resulting in an average of 23 % higher abundances. During the decline in the Pico-II peak (days 16–24), net growth rates were negatively correlated with fCO_2 ($p = 0.10$, $R^2 = 0.52$; Table 2, Fig. S3c). Moreover, the rate of decline was faster for the high fCO_2 mesocosms during days 18–21 ($p < 0.01$, $R^2 = 0.85$). The Pico-II abundances in the surrounding water were comparable to the mesocosms during Phases 0 and I, lower during Phase II and higher during Phase III (Fig. 3c).

Pico-III exhibited a short initial increase in abundances in the low fCO_2 treatments, resulting in nearly 2-fold higher abundances at low fCO_2 by day 3 compared to the high fCO_2 treatment (Fig. 3d). After this initial period, net growth rates of this group had a significant positive correlation with fCO_2 (days 3–13; $p = 0.04$, $R^2 = 0.67$; Table 2, Fig. S2d). In general, during Phase I gross growth ($p < 0.01$; days 1, 3, 10; Fig. 4d) and total mortality ($p < 0.05$; days 1, 6, 10; Fig. 4d) were significantly higher in the low fCO_2 mesocosm M1 compared to the high fCO_2 mesocosm M3, resulting in low net growth rates. During Phase II (days 16–24; NMDS-based period 2) the opposite occurred; i.e. net growth rates were negatively correlated with fCO_2 ($p < 0.01$, $R^2 = 0.86$; Table 2, Fig. S3d). Maximum Pico-III abundances (day 24: 4.2×10^3 and 8.3×10^3 mL^{-1} for high and low fCO_2) had a strong negative correlation with fCO_2 ($p < 0.01$, $R^2 = 0.91$; Table 3, Fig. S4f). Pico-III abundances remained noticeably higher in the low fCO_2 mesocosms during Phases II and III (on average 80 %; Fig. 3d). Unfortunately, almost half of the mortality assays in this second half of the experiment failed (see Sect. 2), but the successful assays suggest that losses were minor (< 0.15 d^{-1};

Fig. 4d) and primarily due to grazing, as no significant viral lysis was detected (Table S1).

3.1.3 Nanoeukaryotes

Nano-I showed maximum abundances ($4.3\pm0.4\times10^2$ mL^{-1}) on day 6 (except M1, which peaked on day 5) independent of fCO_2 ($p = 0.23$, $R^2 = 0.33$; Fig. 3e). There was, however, a negative correlation of net growth rate with fCO_2 during days 3–13 (NMDS-based period 1; $p = 0.01$, $R^2 = 0.79$; Table 2, Fig. S2e). A second major peak in abundance of Nano-I occurred on day 17, with markedly higher numbers in the low fCO_2 mesocosms (4.1×10^2 mL^{-1} compared to 2.4×10^2 mL^{-1} in high fCO_2 mesocosms; $p = 0.04$, $R^2 = 0.67$; Figs. 3e and S4g, Table 3). Total loss rates in the high fCO_2 mesocosm M3 on days 6 and 10 were 2.3-fold higher compared to the low fCO_2 mesocosm M1 (Fig. 4e), which may help to explain this discrepancy in total abundance between low and high fCO_2 mesocosms. Viral lysis accounted for up to 98 % of total losses in the high fCO_2 mesocosm M3 during this period, whilst in M1 viral lysis was only detected on day 13 (Table S1 in the Supplement). Peak abundances (around 5.0×10^2 mL^{-1}) were much lower compared to those in the surrounding waters (max $\sim 2.4 \times 10^3$ mL^{-1}; Figs. 3e and S6a). During Phase II, Nano-I abundances in the surrounding waters displayed rather erratic dynamics compared to those of the mesocosms but converged during certain periods (e.g. days 19–22). No significant relationship was found between net loss rates and fCO_2 for the second NMDS-based period ($p = 0.26$, $R^2 = 0.30$; Table 2, Fig. S3e). At the end of Phase II, abundances were similar in all mesocosms but diverged again during Phase III (days 31–39) due primarily to a negative effect of CO_2 on Nano-I abundances, as depicted in the average 36 % reduction in Nano-I.

The temporal dynamics of Nano-II, the least abundant phytoplankton group analysed in our study, displayed the largest variability (Fig. 3f), perhaps due to the spread of this cluster in flow cytographs (which may indicate that this group represents several different phytoplankton species). No significant relationship was found between net growth rate and fCO_2 for this group for the two NMDS-based periods (Table 2, Figs. S2f and S3f) nor with the peak in abundances on day 17 ($p = 0.13$, $R^2 = 0.46$; Fig. S4h). Moreover, no consistent trend was detected in mortality rates (Fig. 4f). Similar to Nano-I, abundances in the surrounding water were often higher than in the mesocosms (maximum 3.5×10^2 mL^{-1} vs. 1.1×10^4 mL^{-1}, respectively; Figs. 3f and S6b).

3.1.4 Algal carbon biomass

The mean combined biomass of Pico-I and Pico-II showed a strong positive correlation with fCO_2 throughout the experiment ($p < 0.05$, $R^2 = 0.95$; Fig. 5a), an effect already noticeable by day 2. Their biomass in the high fCO_2 mesocosms was, on average 11 % higher than in the low fCO_2 mesocosms between days 10 and 20 and 20 % higher between days 20 and 39. Conversely, the remaining algal groups showed an average 10 % reduction in carbon biomass at enhanced fCO_2 (days 3–39, the sum of SYN, Pico-III, Nano-I and II; $p < 0.01$; Fig. 5b). The most notable response was found for the biomass of Pico-III, which showed an immediate negative response to CO_2 addition (Fig. S7a) and remained on average 29 % lower throughout the study period (days 2–39). For Nano-I and Nano-II the lower carbon biomass only became apparent during the end of Phase I and the beginning of Phase II (days 14–20; Fig. S7b). Due to its small cell size, the numerically dominant SYN accounted for an average of 40 % of total carbon biomass.

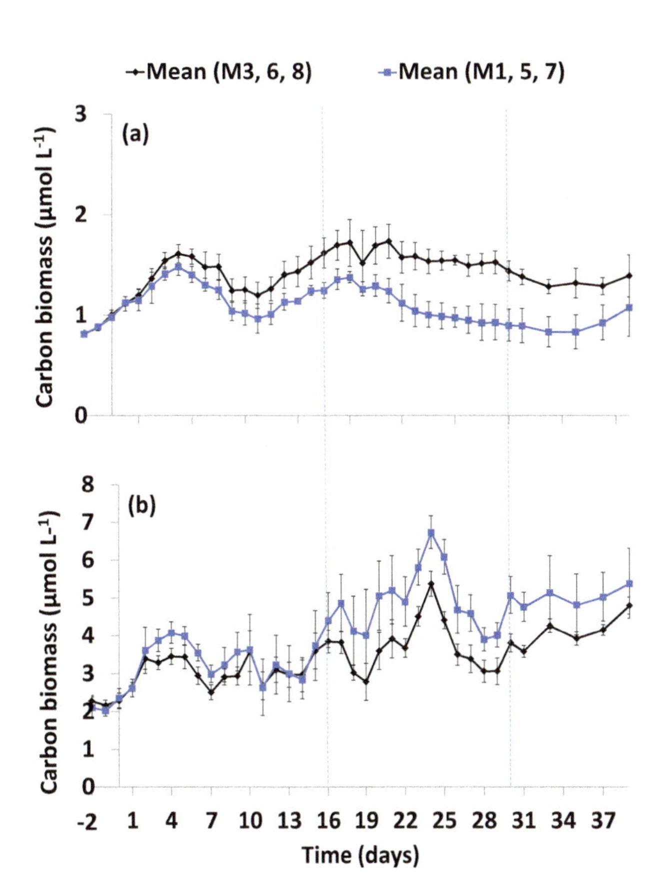

Figure 5. Time series plot of the mean phytoplankton carbon biomass in high fCO_2 (M3, M6, M8; red) and low fCO_2 (M1, M5, M7; blue) mesocosms of **(a)** Pico-I and Pico-II combined and **(b)** SYN, Pico III, Nano-I and Nano-II combined. Error bars represent 1 standard deviation from the mean. Carbon biomass is calculated assuming a spherical diameter equivalent to the mean average cell diameters for each group and conversion factors of 237 fg C µm^{-3} (Worden et al., 2004) and 196.5 fg C µm^{-3} (Garrison et al., 2000) for pico- and nano-sized plankton, respectively. Dotted lines indicate the end of Phase I and the end of Phase II.

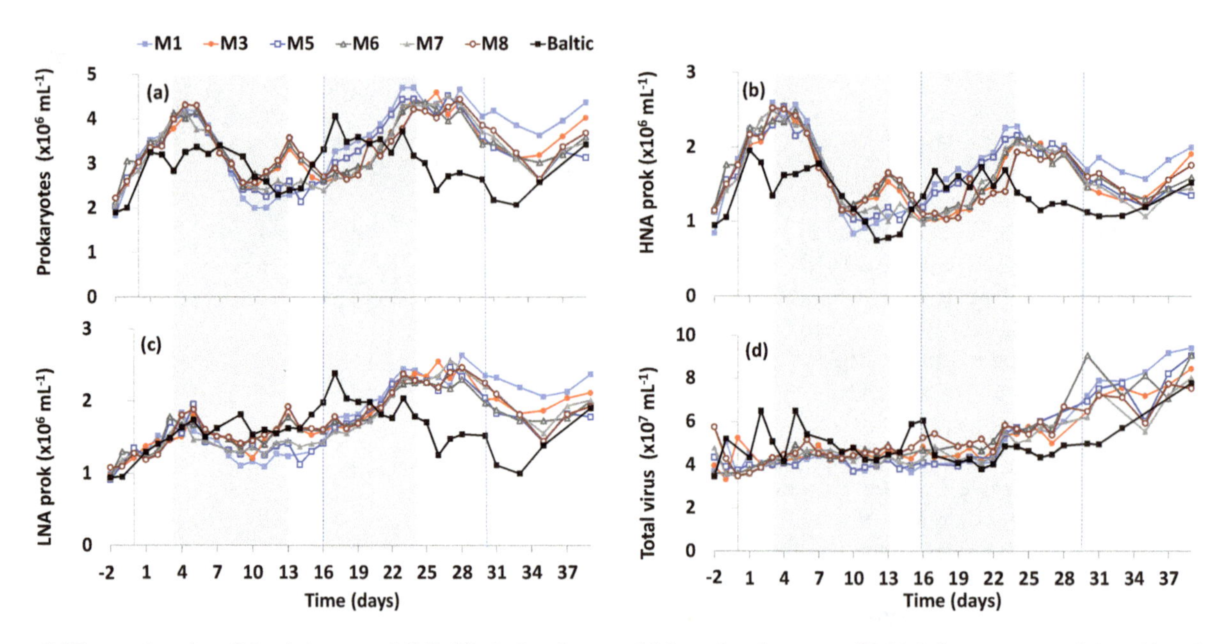

Figure 6. Time series plot of depth-integrated (0.3–10 m) abundances of **(a)** total prokaryotes, **(b)** high fluorescent nucleic acid prokaryote population (HNA), **(c)** low fluorescent nucleic acid prokaryote population (LNA) and **(d)** total virus. Dotted lines indicate the end of Phase I and the end of Phase II; grey areas indicate NMDS-based periods during which net growth rates were analysed.

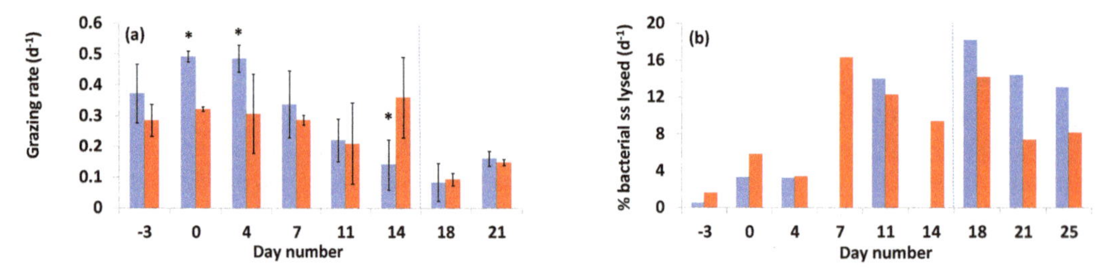

Figure 7. Prokaryote mortality rates: **(a)** total grazing (d^{-1}) and **(b)** viral lysis rates as % of prokaryote standing stock in mesocosms M1 (low fCO_2; blue) and M3 (high fCO_2; red). Grazing rates were determined from fluorescently labelled prey, and viral lysis rates from viral production assays. Error bars represent 1 standard deviation of triplicate assays. Significant ($p \leq 0.05$) differences between mesocosms are indicated by an asterisk. Dotted lines indicate the end of Phase I.

3.2 Prokaryote and virus population dynamics

Prokaryote abundance in the mesocosms was positively related to total algal biomass independent of treatment ($p < 0.05$, $R^2 = 0.33$; Fig. 8) and generally followed total algal biomass (Fig. S7c). The initial increase in total prokaryote abundances occurred during the first few days following the closure of the mesocosms (Fig. 6a). This was primarily due to increases in the HNA prokaryote group (Fig. 6b), which displayed higher net growth rates ($0.22\,d^{-1}$) compared to the LNA prokaryotes ($0.14\,d^{-1}$ on days −3 to 3; Fig. 6c). A similar, albeit somewhat lower, increase was also recorded in the surrounding waters (Fig. 6a). The decline in the first peak in prokaryote abundances coincided with the decay in phytoplankton abundance and biomass (Figs. 1a and S7c). Concurrently the share of viral lysis increased, representing 37–39 % of total mortality on day 11 (Fig. 7b). No measurable rates of lysogeny were found for the prokaryotic community during the experimental period (all phases). From days 10 to 15 prokaryote dynamics (total, HNA and LNA) became noticeably affected by CO_2 concentration with a significant positive correlation between net growth and fCO_2 during Phase I (days 3–13; NMDS-based period 1; Table 2, Fig. S2g and h). In the higher fCO_2 mesocosms, the decline in prokaryote abundance occurring between days 13 and 16 (Fig. 6a) was largely (70 %) due to decreasing HNA prokaryote numbers (Fig. 6b). The grazing was 1.6-fold higher in the high fCO_2 mesocosm M3 compared to M1 (0.36 ± 0.13 and $0.14 \pm 0.08\,d^{-1}$ on day 14; Fig. 7a). At the same time, viral abundance increased in the high fCO_2 mesocosms (Fig. 6d).

During Phase II, prokaryote abundances increased steadily until day 24 (for both HNA and LNA), corresponding to increased algal biomass (Figs. 6 and S7c) and lowered grazing rates (Fig. 7a). Specifically, during days 16–24 (NMDS-

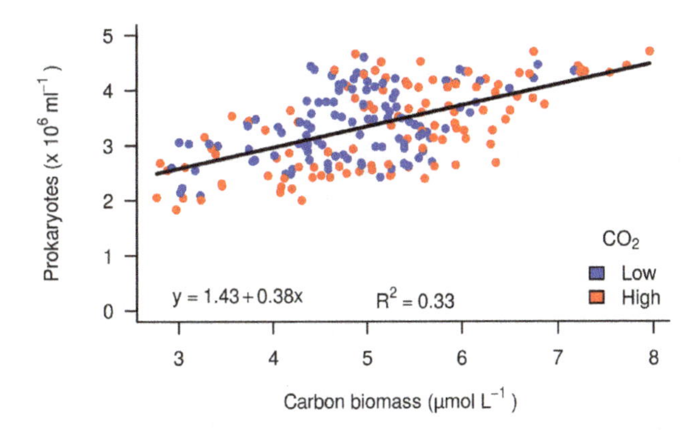

Figure 8. Correlation between total carbon biomass (µmol L^{-1}) and total prokaryote abundance in low fCO_2 mesocosms (M1, M5, M7; blue) and high fCO_2 mesocosms (M3, M6, M8; red) throughout the experiment (days −2 to 39).

based period 2), the HNA prokaryotes showed an average 10 % higher abundances in the low compared to the high fCO_2 mesocosms (Fig. 6b). However, a significant negative correlation of net growth rates and fCO_2 was only found for LNA (Table 2, Fig. S3g and h). No significant differences in loss rates between M1 and M3 were found during Phase II ($p = 0.22$ and $p = 0.46$ on days 18 and 21, respectively; Fig. 7). Halfway through Phase II (day 24), the prokaryote abundance in the surrounding water levelled off (Fig. 6a). Prokaryote abundance ultimately declined during days 28–35 (Fig. 6a), and the net growth of LNA was again negatively correlated with enhanced CO_2 ($p = 0.02$, $R^2 = 0.76$; Table 2, Fig. S3g). Unfortunately, no experimental data on grazing and lysis of prokaryotes are present after day 25. However, viral abundances increased steadily at $2.2 \times 10^6\ d^{-1}$ concomitant with a decline in prokaryote abundance (Fig. 6a and d). There was no significant correlation between viral abundances and fCO_2 during Phases II and III ($p = 0.36$, $R^2 = 0.21$).

4 Discussion

In most experimental mesocosm studies, nutrients have been added to stimulate phytoplankton growth (Schulz et al., 2017); therefore limited data exists for oligotrophic phytoplankton communities. In this study, we describe the impact of increased fCO_2 on the brackish Baltic Sea microbial community during summer (nutrient depleted; Paul et al., 2015). Small-sized phytoplankton numerically dominated the autotrophic community, in particular SYN and Pico-I (both about 1 µm in cell diameter). Our results demonstrate variable effects of fCO_2 manipulation on temporal phytoplankton dynamics, dependent on phytoplankton group. In particular, Pico-I and Pico-II showed significant positive responses, whilst the abundances of Pico-III, SYN and Nano-I were negatively influenced by elevated fCO_2. The impact of OA on the different groups was, at times, a direct consequence of alterations in gross growth rate, whilst overall phytoplankton population dynamics could be explained by the combination of growth and losses. OA effects on community composition in these systems may have consequences on both the food web and biogeochemical cycling.

4.1 Comparison with surrounding waters

During Phase 0, the microbial assemblage showed good replicability among all mesocosms; however, they had already began to deviate from the community in the surrounding waters. This was most likely a consequence of water movement altering the physical conditions and biological composition of the surrounding water body. The dynamic nature of water movement in this region has been shown to alter the entire phytoplankton community several times over within a few months due to fluctuations in nutrient supply, advection, replacement or mixing of water masses and water temperature (Lips and Lips, 2010). Alternatively, the effects of enclosure and the techniques (bubbling) used to ensure a homogenous water column may have stimulated SYN within the mesocosms, which has been found to occur in several mesocosm experiments (Paulino et al., 2008; Gazeau et al., 2017). By Phases II and III, the microbial abundances within the mesocosms were distinctly different from the surrounding waters, with generally fewer SYN and Pico-I and more Nano-I and Nano-II. Our statistical analysis shows that during this time, there was little similarity between the surrounding waters and mesocosms regardless of the CO_2 treatment level. Thus, the deviations during this time were most likely due to an upwelling event in the archipelago (days 17–30; Paul et al., 2015). Cold, nutrient-rich deep water has been shown to upwell during summer with a profound positive influence on ecosystem productivity (Nômmann et al., 1991; Lehmann and Myrberg, 2008). A relaxation from nutrient limitation in vertically stratified waters disproportionately favours larger-sized phytoplankton due to their higher nutrient requirements and lower capacity to compete at low concentrations dictated by their lower surface to volume ratio (Raven, 1998; Veldhuis et al., 2005). Inside the mesocosms, which were isolated from upwelled nutrients, picoeukaryotes dominated similar to a stratified water column. Following this upwelling event, the pH of the surrounding waters dropped from 8.3 to 7.8, a level comparable to the highest CO_2 treatment (M8) on day 32 (Paul et al., 2015). This suggests that other factors contributed to the observed differences between mesocosms and the surrounding water than can be accounted for by CO_2 concentration alone, e.g. nutrients. Alternatively, the magnitude and source of mortality occurring in the surrounding water may have been altered compared to within the mesocosms after such an upwelling event. Although the grazer community in the surrounding waters was not studied during this campaign, it is likely that the grazing community was completely restructured during the upwelling event (Uitto et al., 1997). It is nonetheless noteworthy that the phytoplankton groups with distinct responses to CO_2 enrichment (either positive or negative) in the low (ambient) fCO_2 mesocosms diverged from those in the surrounding water before the upwelling event occurred.

4.2 Phytoplankton dynamics

Synechococcus showed significantly lower net growth rates and peak abundances at higher fCO_2. Both in laboratory and mesocosm experiments, *Synechococcus* has been reported to have diverse responses to CO_2 with approximately equal accounts of positive (Lu et al., 2006; Schulz et al., 2017), negative (Paulino et al., 2008; Hopkins et al., 2010; Traving et al., 2014,) and insignificant changes (Fu et al., 2007; Lu et al., 2006) in net growth rate with fCO_2. This variable response is probably due, at least in part, to the broad physiological and genetic diversity of this species. In the Gulf of Finland alone, 46 different strains of *Synechococcus* were isolated in July 2004 (Haverkamp et al., 2009). Direct effects on physiology have been implied from laboratory studies. One isolate, a phycoerythrin-rich strain of *Synechococcus* WH7803 (Traving et al., 2014), elicited a negative physiological effect on the growth rate from increased CO_2. This was most likely a consequence of higher sensitivity to the lower pH (Traving et al., 2014) and the cellular cost of maintaining pH homeostasis or, conversely, a direct effect on protein export. Additionally, Lu et al. (2006) reported increased growth rates in a cultured phycocyanin-rich but not a phycoerythrin-rich strain of *Synechococcus*, suggesting that pigments may play some part in defining the direct physiological response within *Synechococcus*. In addition, within natural communities (Paulino et al., 2008; Hopkins et al., 2010; Schulz et al., 2017) variability can also arise from indirect effects, such altering competition with other picoplankton (Paulino et al., 2008). The delay and dampened effect of fCO_2 on SYN abundances within our study was more likely due to indirect effects arising from alterations in food web dynamics than to direct impacts on the physiology of this species. Specifically, significant differences in grazing rates of SYN between M1 and M3 (days 10 and 17, no significant lysis detected) could be responsible for the differing dynamics between the mesocosms at the end of Phase I and the beginning of Phase II.

The gross growth rates of Pico-I were significantly higher ($p < 0.05$) at high fCO_2 compared to the low CO_2 concentrations during the first 10 days of Phase I. Moreover, no differences were detected in the measured loss rates, demonstrating that increases in Pico-I were the due to increases in growth alone. The stimulation of Pico-I by elevated fCO_2 may be due to a stronger reliance on diffusive CO_2 entry compared to larger cells. Model simulations reveal that whilst near-cell CO_2 and pH conditions are close to those of the bulk water for cells $< 5\,\mu m$ in diameter, they diverge as cell diameters increase (Flynn et al., 2012). This is due to the size-dependent thickness of the diffusive boundary layer, which determines the diffusional transport across the boundary layer and to the cell surface (Wolf-Gladrow and Riebesell, 1997; Flynn et al., 2012). It is suggested that larger cells may be more able to cope with fCO_2 variability as their carbon acquisition is more geared towards handling low CO_2 concentrations in their diffusive boundary layer, e.g. by means of active carbon acquisition and bicarbonate utilization (Wolf-Gladrow and Riebesell, 1997; Flynn et al., 2012). Moreover, as the Baltic Sea experiences particularly large seasonal fluctuations in pH and fCO_2 (Jansson et al., 2013) due to the low buffering capacity of the waters, phytoplankton here are expected to have a higher degree of physiological plasticity. Our results agree with previous mesocosm studies, which reported enhanced abundances of picoeukaryotic phytoplankton (Brussaard et al., 2013; Davidson et al, 2016; Schulz et al., 2017), particularly the prasinophyte *Micromonas pusilla* at higher fCO_2 (Engel et al., 2008; Meakin and Wyman, 2011). Furthermore, Schaum et al. (2012) found that 16 ecotypes of *Ostreococcus tauri* (another prasinophyte similar in size to Pico-I) increased in growth rate by 1.4–1.7-fold at 1000 compared to 400 µatm fCO_2. All ecotypes increased their photosynthetic rates, and those with the most plasticity (those most able to vary their photosynthetic rate in response to changes in fCO_2) were more likely to increase in frequency within the community. It is possible that Pico-I cells are adapted to a highly variable carbonate system regime and are able to increase their photosynthetic rate when additional CO_2 is available. This ability would allow them to out-compete other phytoplankton (e.g. nanoeukaryotes in this study) in an environment when nutrients are scarce.

The net growth rates and peak abundances of Pico-II were also positively affected by fCO_2. Gross growth rates were significantly higher at high fCO_2 on only two occasions (days 10 and 20) and were accompanied by high total mortality rates. Pigment analysis suggests that both Pico-I and Pico-II are chlorophytes (Paul et al., 2015) and as such may share a common evolutionary history (Schulz et al., 2017); thus Pico-II may be stimulated by fCO_2 in a similar manner to Pico I. Chlorophytes are found in high numbers at this site throughout the year (Kuosa, 1991), suggesting the ecological relevance of Pico-I and Pico-II in this ecosystem. In addition, Pico-II bloomed exactly when Pico-I declined, which may suggest potential competitive exclusion.

Pico-III showed the most distinct and immediate response to CO_2 addition. The significant reduction in gross growth rates observed during Phase I suggests a direct negative effect of CO_2 on the physiology of these cells. For this group, the lower gross growth rates were matched by lower total mortality rates with increased fCO_2. Although the mean cell size of Pico-III and Pico-II were comparable (2.9 and 2.5 µm, respectively), they showed opposing responses to fCO_2 enrichment (lower Pico-III abundances at high fCO_2). These differences may arise from taxonomic differences between the two groups. Pico-III displayed relatively high phycoerythrin orange autofluorescence, likely representing small-sized cryptophytes (Klaveness, 1989), although rod-shaped *Synechococcus* up to 2.9 µm in length (isolated from this region; Haverkamp et al., 2009) or *Synechococcus* microcolonies (often only two cells in the Baltic; Motwani and Gorokhove, 2013) cannot be excluded. In agreement with Pico-III response to CO_2 enrichment, Hopkins et al. (2010) reported reduced abundances of small cryptophytes under increased CO_2 in a mesocosm study in a Norwegian fjord near Bergen.

Lastly, the two nanoeukaryotic phytoplankton groups also displayed a negative response to fCO_2 enrichment; Nano-II was the least defined, most likely due to a high taxonomic diversity in this group. Nano-I started to display lower abundances at high fCO_2 during Phase I (after day 10), which was likely the result of greater differences between gross growth and total mortality (compared to low fCO_2). Alternatively, enhanced nutrient competition due to increased abundances of SYN and Pico-I (and later also Pico-II) at elevated fCO_2 may also have contributed to the dampened response of Nano-I in the high fCO_2 mesocosms. The overall decline in Nano-I during Phase II and the sustained low abundances during Phase III may well have been the result of grazing by the increased mesozooplankton abundances during Phase II (Lischka et al., 2017).

4.3 Microbial loop

The strong association of prokaryote abundance with algal biomass, which was present throughout the experiment, suggests that the effect of CO_2 was an indirect consequence of alterations in the availability of phytoplankton carbon. Others have reported a tight coupling of autotrophic and heterotrophic communities at this location, with an estimated 35 % of the total net primary production being utilized directly by bacteria or heterotrophic flagellates (Kuosa and Kivi, 1989), suggesting a highly efficient microbial loop in this ecosystem. In addition to phytoplankton exudation, viral lysis may also contribute to the dissolved organic carbon pool (Wilhelm and Suttle, 1999; Brussaard et al., 2005; Lønborg et al., 2013). We calculated that viral lysis of phytoplankton between days 9 and 13 resulted in the release of 1.3 and 13.1 $ng\,C\,mL^{-1}$ for M1 and M3, respectively. Assuming a bacterial growth efficiency of 30 % and cellular carbon conversion of 7 $fg\,C\,cell^{-1}$ (Hornick et al., 2017), we estimate that the organic carbon required to support bacterial dynamics during this period (taking into account the net growth and loss rates) was 2.9 and 11.5 $ng\,C\,mL^{-1}$ in the low and high fCO_2 mesocosms M1 and M3, respectively. These results suggest that viral lysis of phytoplankton was an important source of organic carbon for the bacterial community. Our results are consistent with bacterial–phytoplankton coupling during this eastern Baltic Sea mesocosm study (Hornick et al., 2017) and agree with earlier work on summer carbon flow in the northern Baltic Sea showing that prokaryotic growth was largely supported by recycled carbon (Uitto et al., 1997). The average net growth rates of the prokaryotes during the first period of increase in Phases 0 and I ($0.2\,d^{-1}$) were comparable to rates reported for this region (Kuosa, 1991). In order to sustain the concomitant daily mortality (between 0.3 and $0.5\,d^{-1}$) measured during our study, prokaryotic gross growth rates must have been close to one doubling a day (0.5–$0.7\,d^{-1}$). During Phase I, grazing was the dominant loss factor of the prokaryotic community, although there was also evidence that viral lysis was occurring. Bermúdez et al. (2016) reported the highest biomass of protozoans around day 15. This was predominantly the heterotrophic choanoflagellate *Calliacantha natans*, which selectively feeds on particles < 1 µm in diameter (Marchant and Scott, 1993; Hornick et al., 2017). Indeed, an earlier study in this area showed that heterotrophic nanoflagellates were the dominant grazers of bacteria responsible for the ingestion of approximately 53 % of bacterial production compared to only 11 % being grazing by ciliates (Uitto et al., 1997). During the first half of Phase II, grazing was reduced and likely contributed to the steady increase in prokaryote abundances. Specifically, a negative relationship between the abundances of HNA prokaryotes and fCO_2 was detected and corresponded to reduced bacterial production and respiration at higher fCO_2 (Hornick et al., 2017; Spilling et al., 2016). Although CO_2 enrichment may not directly affect bacterial growth, a co-occurring global rise in temperature can increase enzyme activities, affecting bacterial production and respiration rates (Piontek et al., 2009; Wohlers et al., 2009; Wohlers-Zöllner et al., 2011). The enhanced bacterial remineralization of organic matter may stimulate autotrophic production by the small-sized phytoplankton (Riebesell et al., 2009; Riebesell and Tortell, 2011; Engel et al., 2013), intensifying the selection of small cell sizes.

Mean viral abundances were higher under CO_2 enrichment towards the end of Phase I and into Phase II, which is expected under conditions of increased phytoplankton and prokaryote biomass. The estimated average viral burst size obtained from this increase in total viral abundance and concurrent decline in bacterial abundances was about 30, which is comparable to published values (Parada et al., 2006; Wommack and Colwell, 2000). Viral lysis rates of prokaryotes were measured until day 25 and indicated that during days 18–25 an average 10–15 % of the total prokaryote population was lysed per day. Moreover, the concurrent steady increase in viral abundances during Phase III indicates that viral lysis of the prokaryotes remained important. Thus, the combined impact of increased viral mortality together with reduced production (Hornick et al., 2017) ultimately led to the decline in prokaryote abundance (this study). Lysogeny did not appear to be an important life strategy of viruses during our campaign. Direct effects of higher fCO_2 on viruses are not expected, as marine virus isolates are quite stable (both in terms of particle decay and loss of infectivity) over the range of pH in the present study (Danovaro et al., 2011; Mojica and Brussaard, 2014). The few studies which have inferred viral lysis rates based on changes in viral abundances show reduced abundances of algal viruses (e.g. *Emiliania huxleyi*) under enhanced CO_2 (Larsen et al., 2008), while mesocosm results by Brussaard et al. (2013) indicated a stronger impact of viruses on bacterial abundance dynamics with CO_2 enrichment.

5 Conclusions

Due to the low buffering capacity of the Baltic Sea and the paucity of data regarding OA impact in nutrient-limited waters, the results presented here are pertinent to increasing our understanding of how projected rises in $f\mathrm{CO_2}$ will affect the microbial communities in this region. Our study provides evidence that cell size, taxonomy and sensitivity to loss can all play a role in the outcome of CO_2 enrichment. Physiological constraints of cell size favour nutrient uptake by small cells under conditions of reduced nutrients, and our results show that these effects can be further exacerbated by OA. Gross growth rates along with the complementary mortality rates allowed for a more comprehensive understanding of the phytoplankton population dynamics and thus perception of how microbial food web dynamics can influence the response of the autotrophic and heterotrophic components of the community. Our results further suggest that alterations in CO_2 concentrations are expected to affect prokaryote communities (mainly) indirectly through alterations in phytoplankton biomass, productivity and viral lysis. Overall, the combination of growth and losses (grazing and viral lysis) could explain the microbial population dynamics observed in this study. It is noteworthy to mention a recent study in the oligotrophic northeastern Atlantic Ocean, which reported a shift from a grazing-dominated to a viral-lysis-dominated phytoplankton community with strengthening vertical stratification (shoaling the mixed layer depth and enhancing nutrient limitation; Mojica et al., 2016). Thus, we highly recommend that future research on OA combine mesocosm studies focusing on changes in microbial community composition and activity with experiments aimed at understanding the effects of OA on food web dynamics, i.e. partitioning mortality between grazing and viral lysis (Brussaard et al., 2008).

Author contributions. Design and overall coordination of research by CB. Organization and performance of analyses in the field by KC. Data analysis by KC, CB and SA-F. Design and coordination of the overall KOSMOS mesocosm project by UR. All authors contributed to the writing of the paper (KC, KM and CB are lead authors).

Competing interests. The authors declare that they have no conflict of interest.

Special issue statement. This article is part of the special issue "Effects of rising CO_2 on a Baltic Sea plankton community: ecological and biogeochemical impacts". It is not associated with a conference.

Acknowledgements. This project was funded through grants to Corina P. D. Brussaard by the Darwin project, the Netherlands Institute for Sea Research (NIOZ) and the EU project MESOAQUA (grant agreement number 228224). We thank the KOSMOS project organizers and team, in particular Andrea Ludwig, the staff of the Tvärminne Zoological Station and the diving team. We give special thanks to Anna Noordeloos, Kirsten Kooijman and Richard Doggen for their technical assistance during this campaign. We also gratefully acknowledge the captain and crew of R/V *Alkor* for their work transporting, deploying and recovering the mesocosms. The collaborative mesocosm campaign was funded by BMBF projects BIOACID II (FKZ 03F06550) and SOPRAN phase II (FKZ 03F0611).

Edited by: Eric Achterberg

References

Aberle, N., Schulz, K. G., Stuhr, A., Malzahn, A. M., Ludwig, A., and Riebesell, U.: High tolerance of microzooplankton to ocean acidification in an Arctic coastal plankton community, Biogeosciences, 10, 1471–1481, https://doi.org/10.5194/bg-10-1471-2013, 2013.

Barcelos e Ramos, J., Biswas, H., Schulz, K. G., LaRoche, J., and Riebesell, U.: Effect of rising atmospheric carbon dioxide on the marine nitrogen fixer *Trichodesmium*, Global Biogeochem. Cy., 21, GB2028, https://doi.org/10.1029/2006GB002898, 2007.

Baudoux, A. C., Noordeloos, A. A. M., Veldhuis, M. J. W., and Brussaard, C. P. D.: Virally induced mortality of *Phaeocystis globosa* during two spring blooms in temperate coastal waters, Aquat. Microb. Ecol., 44, 207–217, https://doi.org/10.3354/ame044207, 2006.

Bermúdez, R., Winder, M., Stuhr, A., Almén, A.-K., Engström-Öst, J., and Riebesell, U.: Effect of ocean acidification on the structure and fatty acid composition of a natural plankton community in the Baltic Sea, Biogeosciences, 13, 6625–6635, https://doi.org/10.5194/bg-13-6625-2016, 2016.

Brussaard, C. P. D.: Optimization of Procedures for Counting Viruses by Flow Cytometry, Appl. Environ. Microb., 70, 1506–1513, https://doi.org/10.1128/AEM.70.3.1506-1513.2004, 2004.

Brussaard, C. P. D., Kuipers, B., and Veldhuis, M. J. W.: A mesocosm study of *Phaeocystis globosa* population dynamics: I. Regulatory role of viruses in bloom control, Harmful Algae, 4, 859–874, https://doi.org/10.1016/j.hal.2004.12.015, 2005.

Brussaard, C. P. D., Wilhelm, S. W., Thingstad, F., Weinbauer, M. G., Bratbak, G., Heldal, M., Kimmance, S. A., Middelboe, M., Nagasaki, K., Paul, J. H., Schroeder, D. C., Suttle, C. A., Vaqué, D., and Wommack, K. E.: Global-scale processes with a nanoscale drive: the role of marine viruses, ISME J., 2, 575–578, https://doi.org/10.1038/ismej.2008.31, 2008.

Brussaard, C. P. D., Noordeloos, A. A. M., Witte, H., Collenteur, M. C. J., Schulz, K., Ludwig, A., and Riebesell, U.: Arctic microbial community dynamics influenced by elevated CO_2 levels, Biogeosciences, 10, 719–731, https://doi.org/10.5194/bg-10-719-2013, 2013.

Clarke, K. R.: Non-parametric multivariate analyses of changes in community structure, Aust. J. Ecol., 18, 117–143, 1993.

Corno, G., Karl, D., Church, M., Letelier, R., Lukas, R., Bidigare, R., and Abbott, M.: Impact of climate forcing on ecosystem processes in the North Pacific Subtropical Gyre, J. Geophys. Res., 112, C04021, https://doi.org/10.1029/2006JC003730, 2007.

Danovaro, R., Corinaldesi, C., Dell'Anno, A., Fuhrman, J. A., Middelburg, J. J., Noble, R. T., and Suttle, C. A.: Marine viruses and global climate change, FEMS Microbiol. Rev., 35, 993–1034, 2011.

Davidson, A., McKinlay, J., Westwood, K., Thomson, P., Van Den Enden, R., De Salas, M., Wright, S., Johnson, R., and Berry, K.: Enhanced CO_2 concentrations change the structure of Antarctic marine microbial communities, Mar. Ecol.-Prog. Ser., 552, 93–113, https://doi.org/10.3354/meps11742, 2016.

Doney, S. C., Fabry, V. J., Feely, R. A., and Kleypas, J. A.: Ocean acidification: The other CO_2 problem, Annual Review of Marine Science, 1, 169–192, 2009.

Dutkiewicz, S., Morris, J. J., Follows, M. J., Scott, J., Levitan, O., Dyhrman, S. T., and Berman-Frank, I.: Impact of ocean acidification on the structure of future phytoplankton communities, Nature Climate Change, 5, 1002–1006, 2015.

Eberlein, T., Wohlrab, S., Rost, B., John, U., Bach, L. T., Riebesell, U., and Van de Waal, D. B.: Effects of ocean acidification on primary production in a coastal North Sea phytoplankton community, PLoS ONE, 12, e0172594, https://doi.org/10.1371/journal.pone.0172594, 2017.

Engel, A., Schulz, K. G., Riebesell, U., Bellerby, R., Delille, B., and Schartau, M.: Effects of CO_2 on particle size distribution and phytoplankton abundance during a mesocosm bloom experiment (PeECE II), Biogeosciences, 5, 509–521, https://doi.org/10.5194/bg-5-509-2008, 2008.

Engel, A., Borchard, C., Piontek, J., Schulz, K. G., Riebesell, U., and Bellerby, R.: CO_2 increases ^{14}C primary production in an Arctic plankton community, Biogeosciences, 10, 1291–1308, https://doi.org/10.5194/bg-10-1291-2013, 2013.

Evans, C. and Brussaard, C. P. D.: Regional variation in lytic and lysogenic viral infection in the Southern Ocean and its contribution to biogeochemical cycling, Appl. Environ. Microb., 78, 6741–6748, 2012.

Feng, Y., Leblanc, K., Rose, J. M., Hare, C. E., Zhang, Y., Lee, P. A., Wilhelm, S. W., DiTullio, G. R., Rowe, J. M., Sun, J., Nemcek, N., Gueguen, C., Passow, U., Benner, I., Hutchins, D. A., and Brown, C.: Effects of increased pCO_2 and temperature on the North Atlantic spring bloom. I. The phytoplankton community and biogeochemical response, Mar. Ecol.-Prog. Ser., 388, 13–25, https://doi.org/10.3354/meps08133, 2009.

Field, C. B., Behrenfeld, M. J., Randerson, J. T., and Falkowski, P.: Primary Production of the Biosphere: Integrating Terrestrial and Oceanic Components, Science, 281, 237–240, https://doi.org/10.1126/science.281.5374.237, 1998.

Flynn, K. J., Blackford, J. C., Baird, M. E., Raven, J. A., Clark, D. R., Beardall, J., Brownlee, C., Fabian, H., and Wheeler, G. L.: Letter. Changes in pH at the exterior surface of plankton with ocean acidification, Nature Climate Change, 2, 510–513, https://doi.org/10.1038/nclimate1696, 2012.

Fu, F. X., Warner, M. E., Zhang, Y., Feng, Y., and Hutchins, D. A.: Effects of increased temperature and CO_2 on photosynthesis, growth, and elemental ratios in marine *Synechococcus* and *Prochlorococcus* (Cyanobacteria), J. Phycol., 43, 485–496, https://doi.org/10.1111/j.1529-8817.2007.00355.x, 2007.

Garrison, D. L., Gowing, M. M., Hughes, M. P., Campbell, L., Caron, D. A., Dennett, M. R., Shalapyonok, A., Olson, J. A., Landry, M. R., Brown, S. L., Liu, H-B., Azam, F., Steward, G. F., Ducklow, H. W., and Smith, D. C.: Microbial food web structure in the Arabian Sea: A US JGOFS study, Deep-Sea Res. Pt. II, 47, 1387–1422, https://doi.org/10.1016/S0967-0645(99)00148-4, 2000.

Gazeau, F., Sallon, A., Pitta, P., Tsiola, A., Maugendre, L., Giani, M., Celussi, M., Pedrotti, M. L., Marro, S., and Guieu, C.: Limited impact of ocean acidification on phytoplankton community structure and carbon export in an oligotrophic environment: results from two short-term mesocosm studies in the Mediterranean Sea, Estuar. Coast. Shelf S., 186, 72–88, https://doi.org/10.1016/j.ecss.2016.11.016, 2017.

Gruber, N.: Warming up, turning sour, losing breath: Ocean biogeochemistry under global change, Philos. T. Roy. Soc. A, 369, 1980–1996, 2011.

Haverkamp, T., Schouten, D., Doeleman, M., Wollenzien, U., Huisman, J., and Stal, L.: Colorful microdiversity of *Synechococcus* strains (picocyanobacteria) isolated from the Baltic Sea, ISME J., 3, 397–408, 2009.

Hein, M. and Sand-Jensen, K.: CO_2 increases oceanic primary production, Nature, 388, 526–527, 1997.

Hopkins, F. E., Turner, S. M., Nightingale, P. D., Steinke, M., Bakker, D., and Liss, P. S.: Ocean acidification and marine trace gas emissions, P. Natl. Acad. Sci. USA, 107, 760–765, https://doi.org/10.1073/pnas.0907163107, 2010.

Hornick, T., Bach, L. T., Crawfurd, K. J., Spilling, K., Achterberg, E. P., Woodhouse, J. N., Schulz, K. G., Brussaard, C. P. D., Riebesell, U., and Grossart, H.-P.: Ocean acidification impacts bacteria-phytoplankton coupling at low-nutrient conditions, Biogeosciences, 14, 1–15, https://doi.org/10.5194/bg-14-1-2017, 2017.

Hutchins, D. A., Fe, F-X., Zhang, Y., Warner, M. E., Feng, Y., Portune, K., Bernhardt, P. W., and Mulholland, M. R.: CO_2 control of *Trichodesmium* N_2 fixation, photosynthesis, growth rates, and elemental ratios: implications for past, present, and future ocean biogeochemistry, Limnol. Oceanogr., 52, 1293–1304, 2007.

Jansson, A., Norkko, J., and Norkko, A.: Effects of Reduced pH on *Macoma balthica* Larvae from a System with Naturally Fluctuating pH-Dynamics, PLoS One, 8, e68198, https://doi.org/10.1371/journal.pone.0068198, 2013.

Kimmance, S. A. and Brussaard, C. P. D.: Estimation of viral-induced phytoplankton mortality using the modified dilution method, in: Manual of Aquatic Viral Ecology, edited by: Wilhelm, S. W., Weinbauer, M., and Suttle, C. A., ASLO, 65–73, 2010.

Klaveness, D.: Biology and ecology of the Cryptophyceae: status and challenges, Biological Oceanography, 6, 257–270, 1989.

Kuliński, K. and Pempkowiak, J.: The carbon budget of the Baltic Sea, Biogeosciences, 8, 3219–3230, https://doi.org/10.5194/bg-8-3219-2011, 2011.

Kuosa, H.: Picoplanktonic algae in the northern Baltic Sea: seasonal dynamics and flagellate grazing, Mar. Ecol.-Prog. Ser., 73, 269–276, https://doi.org/10.3354/meps073269, 1991.

Kuosa, H. and Kivi, K.: Bacteria and heterotrophic flagellates in the pelagic carbon cycle in the northern Baltic Sea, Mar. Ecol.-Prog. Ser., 53, 93–100, https://doi.org/10.3354/meps053093, 1989.

Larsen, J. B., Larsen, A., Thyrhaug, R., Bratbak, G., and Sandaa, R.-A.: Response of marine viral populations to a nutrient induced phytoplankton bloom at different pCO_2 levels, Biogeosciences, 5, 523–533, https://doi.org/10.5194/bg-5-523-2008, 2008.

Legendre, P. and Legendre, L.: Numerical Ecology, Elsevier, Amsterdam, New York, 1998.

Lehmann, A. and Myrberg, K.: Upwelling in the Baltic Sea – A review, J. Marine Syst., 74, S3–S12, https://doi.org/10.1016/j.jmarsys.2008.02.010, 2008.

Leonardos, N. and Geider, R. J.: Elevated atmospheric carbon dioxide increases organic carbon fixation by *Emiliania huxleyi* (Haptophyta), under nutrient-limited high-light conditions, J. Phycol., 41, 1196–1203, https://doi.org/10.1111/j.1529-8817.2005.00152.x, 2005.

Lips, I. and Lips, U.: Phytoplankton dynamics affected by the coastal upwelling events in the Gulf of Finland in July–August 2006, J. Plankton Res., 32, 1269–1282, 2010.

Lischka, S., Bach, L. T., Schulz, K.-G., and Riebesell, U.: Ciliate and mesozooplankton community response to increasing CO_2 levels in the Baltic Sea: insights from a large-scale mesocosm experiment, Biogeosciences, 14, 447–466, https://doi.org/10.5194/bg-14-447-2017, 2017.

Lønborg, C., Middelboe, M., and Brussaard, C. P. D.: Viral lysis of Micromonas pusilla: Impacts on dissolved organic matter production and composition, Biogeochemistry, 116, 231–240, https://doi.org/10.1007/s10533-013-9853-1, 2013.

Lu, Z., Jiao, N., and Zhang, H.: Physiological changes in marine picocyanobacterial *Synechococcus* strains exposed to elevated CO_2 partial pressure, Mar. Biol. Res., 2, 424–430, https://doi.org/10.1080/17451000601055419, 2006.

Marchant, H. J. and Scott, F. J: Uptake of sub-micrometre particles and dissolved organic material by Antarctic choanoflagellates, Mar. Ecol.-Prog. Ser., 92, 59–64, 1993.

Marie, D., Brussaard, C. P. D., Thyrhaug, R., Bratbak, G., and Vaulot, D.: Enumeration of marine viruses in culture and natural samples by flow cytometry, Appl. Environ. Microbiol., 65, 45–52, 1999.

Meakin, N. G. and Wyman, M.: Rapid shifts in picoeukaryote community structure in response to ocean acidification, ISME J., 5, 1397–405, https://doi.org/10.1038/ismej.2011.18, 2011.

Meyer, J. and Riebesell, U.: Reviews and Syntheses: Responses of coccolithophores to ocean acidification: a meta-analysis, Biogeosciences, 12, 1671–1682, https://doi.org/10.5194/bg-12-1671-2015, 2015.

Middelboe, M. and Lyck, P. G.: Regeneration of dissolved organic matter by viral lysis in marine microbial communities, Aquat. Microb. Ecol., 27, 187–194, https://doi.org/10.3354/ame027187, 2002.

Mojica, K. D. A. and Brussaard, C. P. D.: Factors affecting virus dynamics and microbial host–virus interactions in marine environments, FEMS Microbiol. Ecol., 89, 495–515, https://doi.org/10.1111/1574-6941.12343, 2014.

Mojica, K. D. A., Evans, C., and Brussaard, C. P. D.: Flow cytometric enumeration of marine viral populations at low abundances, Aquat. Microb. Ecol., 71, 203–209, https://doi.org/10.3354/ame01672, 2014.

Mojica, K. D. A., Huisman, J., Wilhelm, S. W., and Brussaard, C. P. D.: Latitudinal variation in virus-induced mortality of phytoplankton across the North Atlantic Ocean, ISME J., 10, 500–513, https://doi.org/10.1038/ismej.2015.130, 2016.

Motwani, N. H. and Gorokhova, E.: Mesozooplankton grazing on picocyanobacteria in the Baltic Sea as inferred from molecular diet analysis, PLoS ONE, 8, e79230, https://doi.org/10.1371/journal.pone.0079230, 2013.

Niehoff, B., Schmithüsen, T., Knüppel, N., Daase, M., Czerny, J., and Boxhammer, T.: Mesozooplankton community development at elevated CO_2 concentrations: results from a mesocosm experiment in an Arctic fjord, Biogeosciences, 10, 1391–1406, https://doi.org/10.5194/bg-10-1391-2013, 2013.

Nõmmann, S., Sildam, J., Nõges, T., and Kahru, M.: Plankton distribution during a coastal upwelling event off Hiiumaa, Baltic Sea: impact of short-term flow field variability, Cont. Shelf Res., 11, 95–108, 1991.

Oksanen, J., Blanchet, F. G., Friendly, M., Kindt, R., Legendre, P., McGlinn, D., Minchin, P., R., O'Hara, R. B., L. Simpson, G. L., Solymos, P., Stevens, M. H. H., Szoecs, E., and Wagner, H.: vegan: Community Ecology Package, R package version 2.4-3, available at: https://CRAN.R-project.org/package=vegan, 2017.

Parada, V., Herndl, G., and Weinbauer, M.: Viral burst size of heterotrophic prokaryotes in aquatic systems, J. Mar. Biol. Assoc. UK, 86, 613–621, https://doi.org/10.1017/S002531540601352X, 2006.

Paul, A. J., Bach, L. T., Schulz, K.-G., Boxhammer, T., Czerny, J., Achterberg, E. P., Hellemann, D., Trense, Y., Nausch, M., Sswat, M., and Riebesell, U.: Effect of elevated CO_2 on organic matter pools and fluxes in a summer Baltic Sea plankton community, Biogeosciences, 12, 6181–6203, https://doi.org/10.5194/bg-12-6181-2015, 2015.

Paulino, A. I., Egge, J. K., and Larsen, A.: Effects of increased atmospheric CO_2 on small and intermediate sized osmotrophs during a nutrient induced phytoplankton bloom, Biogeosciences, 5, 739–748, https://doi.org/10.5194/bg-5-739-2008, 2008.

Pfeil, B., Olsen, A., Bakker, D. C. E., Hankin, S., Koyuk, H., Kozyr, A., Malczyk, J., Manke, A., Metzl, N., Sabine, C. L., Akl, J., Alin, S. R., Bates, N., Bellerby, R. G. J., Borges, A., Boutin, J., Brown, P. J., Cai, W.-J., Chavez, F. P., Chen, A., Cosca, C., Fassbender, A. J., Feely, R. A., González-Dávila, M., Goyet, C., Hales, B., Hardman-Mountford, N., Heinze, C., Hood, M., Hoppema, M., Hunt, C. W., Hydes, D., Ishii, M., Johannessen, T., Jones, S. D., Key, R. M., Körtzinger, A., Landschützer, P., Lauvset, S. K., Lefèvre, N., Lenton, A., Lourantou, A., Merlivat, L., Midorikawa, T., Mintrop, L., Miyazaki, C., Murata, A., Nakadate, A., Nakano, Y., Nakaoka, S., Nojiri, Y., Omar, A. M., Padin, X. A., Park, G.-H., Paterson, K., Perez, F. F., Pierrot, D., Poisson, A., Ríos, A. F., Santana-Casiano, J. M., Salisbury, J., Sarma, V. V. S. S., Schlitzer, R., Schneider, B., Schuster, U., Sieger, R., Skjelvan, I., Steinhoff, T., Suzuki, T., Takahashi, T., Tedesco, K., Telszewski, M., Thomas, H., Tilbrook, B., Tjiputra, J., Vandemark, D., Veness, T., Wanninkhof, R., Watson, A. J., Weiss, R., Wong, C. S., and Yoshikawa-Inoue, H.: A uniform, quality controlled Surface Ocean CO_2 Atlas (SOCAT), Earth Syst. Sci. Data, 5, 125–143, https://doi.org/10.5194/essd-5-125-2013, 2013.

Pinheiro, J., Bates, D., DebRoy, S., Sarkar, D., and R Core Team: nlme: Linear and Nonlinear Mixed Effects Models, R package version 3.1-131, available at: https://CRAN.Rproject.org/package=nlme, 2017.

Piontek, J., Händel, N., Langer, G., Wohlers, J., Riebesell, U., and Engel, A.: Effects of rising temperature on the formation and microbial degradation of marine diatom aggregates, Aquat. Microb. Ecol., 54, 305–318, https://doi.org/10.3354/ame01273, 2009.

Qiu, B. and Gao, K.: Effects of CO_2 enrichment on the bloom-forming cyanobacterium *Microcystis aeruginosa* (Cyanophyceae): Physiological responses and relationships with the availability of dissolved inorganic carbon, J. Phycol., 38, 721–729, https://doi.org/10.1046/j.1529-8817.2002.01180.x, 2002.

R Core Team: R: A language and environment for statistical computing, R Foundation for Statistical Computing, Vienna, Austria, available at: https://www.R-project.org/, 2017.

Raven, J. A.: The twelfth Tansley Lecture. Small is beautiful: the picophytoplankton, Funct. Ecol., 12, 503–513, https://doi.org/10.1046/j.1365-2435.1998.00233.x, 1998.

Riebesell, U., Körtzinger, A., and Oschlies, A.: Sensitivities of marine carbon fluxes to ocean change, P. Natl. Acad. Sci. USA, 106, 20602–20609, https://doi.org/10.1073/pnas.0813291106, 2009.

Riebesell, U. and Tortell, P. D.: Effects of ocean acidification on pelagic organisms and ecosystems, in: Ocean acidification, edited by: Gattuso, J.-P. and Hansson, L., Oxford University Press, Oxford, UK, 99–121, 2011.

Riebesell, U., Czerny, J., von Bröckel, K., Boxhammer, T., Büdenbender, J., Deckelnick, M., Fischer, M., Hoffmann, D., Krug, S. A., Lentz, U., Ludwig, A., Muche, R., and Schulz, K. G.: Technical Note: A mobile sea-going mesocosm system – new opportunities for ocean change research, Biogeosciences, 10, 1835–1847, https://doi.org/10.5194/bg-10-1835-2013, 2013.

Riebesell, U., Bach, L. T., Bellerby, R. G. J., Bermudez, R., Boxhammer, T., Czerny, J., Larsen, A., Ludwig, A., and Schulz, K. G.: Ocean acidification impairs competitive fitness of a predominant pelagic calcifier, Nat. Geosci., 10, 19–24, https://doi.org/10.1038/ngeo2854, 2017.

Rose, J. M., Feng, Y., Gobler, C. J., Gutierrez, R., Harel, C. E., Leblancl, K., and Hutchins, D. A.: Effects of increased pCO_2 and temperature on the North Atlantic spring bloom. II. Microzooplankton abundance and grazing, Mar. Ecol.-Prog. Ser., 388, 27–40, https://doi.org/10.3354/meps08134, 2009.

Schaum, E., Rost, B., Millar, A. J., and Collins, S.: Variation in plastic responses of a globally distributed picoplankton species to ocean acidification, Nature Climate Change, 3, 298–302, https://doi.org/10.1038/nclimate1774, 2012.

Schulz, K. G. and Riebesell, U.: Diurnal changes in seawater carbonate chemistry speciation at increasing atmospheric carbon dioxide, Mar. Biol., 160, 1889–1899, https://doi.org/10.1007/s00227-012-1965-y, 2013.

Schulz, K. G., Bach, L. T., Bellerby, R. G. J., Bermúdez, R., Büdenbender, J., Boxhammer, T., Czerny, J., Engel, A., Ludwig, A., Meyerhöfer, M., Larsen, A., Paul, A. J., Sswat, M., and Riebesell, U.: Phytoplankton Blooms at Increasing Levels of Atmospheric Carbon Dioxide: Experimental Evidence for Negative Effects on Prymnesiophytes and Positive on Small Picoeukaryotes, Frontiers in Marine Science, 4, 64, https://doi.org/10.3389/fmars.2017.00064, 2017.

Sheik, A. R., Brussaard, C. P. D., Lavik, G., Lam, P., Musat, N., Krupke, A., Littmann, S., Strous, M., and Kuypers, M. M. M.: Responses of the coastal bacterial community to viral infection of the algae Phaeocystis globosa, ISME J., 8, 212–25, https://doi.org/10.1038/ismej.2013.135, 2014.

Sherr, E. B., Caron, D. A., and Sherr, B. F.: Staining of heterotrophic protists for visualization via epifluorescence microscopy, in: Current Methods in Aquatic Microbial Ecology, edited by: Sherr, B., Sherr, E., and Kemp, J. C. P., Lewis Publ., NY, 213–228, 1993.

Spilling, K., Paul, A. J., Virkkala, N., Hastings, T., Lischka, S., Stuhr, A., Bermúdez, R., Czerny, J., Boxhammer, T., Schulz, K. G., Ludwig, A., and Riebesell, U.: Ocean acidification decreases plankton respiration: evidence from a mesocosm experiment, Biogeosciences, 13, 4707–4719, https://doi.org/10.5194/bg-13-4707-2016, 2016.

Stoecker, D. K., Nejstgaard, J., Madhusoodhanan, R., and Larsen, A.: Underestimation of microzooplankton grazing in dilution experiments due to inhibition of phytoplankton growth, Limnol. Oceanogr., 60, 1426–1438, https://doi.org/10.1002/lno.10106, 2015.

Suffrian, K., Simonelli, P., Nejstgaard, J. C., Putzeys, S., Carotenuto, Y., and Antia, A. N.: Microzooplankton grazing and phytoplankton growth in marine mesocosms with increased CO_2 levels, Biogeosciences, 5, 1145–1156, https://doi.org/10.5194/bg-5-1145-2008, 2008.

Tortell, P. D., DiTullio, G. R., Sigman, D. M., and Morel, F. M. M.: CO_2 effects on taxonomic composition and nutrient utilization in an Equatorial Pacific phytoplankton assemblage, Mar. Ecol.-Prog. Ser., 236, 37–43, 2002.

Traving, S. J., Clokie, M. R. J., and Middelboe, M.: Increased acidification has a profound effect on the interactions between the cyanobacterium *Synechococcus* sp. WH7803 and its viruses, FEMS Microbiol. Ecol., 87, 133–141, https://doi.org/10.1111/1574-6941.12199, 2014.

Turley, C. and Boot, K.: UNEP emerging issues: Environmental consequences of ocean acidification: A threat to food security, United Nations Environment Programme, 2010.

Uitto, A., Heiskanen, A., Lignell, R., Autio, R., and Pajuniemi, R.: Summer dynamics of the coastal planktonic food web in the northern Baltic Sea, Mar. Ecol.-Prog. Ser., 151, 27–41, https://doi.org/10.3354/meps151027, 1997.

Veldhuis, M. J. W. and Kraay, G. W.: Phytoplankton in the subtropical Atlantic Ocean: Towards a better assessment of biomass and composition, Deep-Sea Res. Pt I, 51, 507–530, https://doi.org/10.1016/j.dsr.2003.12.002, 2004.

Veldhuis, M. J. W., Timmermans, K. R., Croot, P., and Van Der Wagt, B.: Picophytoplankton; A comparative study of their biochemical composition and photosynthetic properties, J. Sea Res., 53, 7–24, https://doi.org/10.1016/j.seares.2004.01.006, 2005.

Weinbauer, M. G.: Ecology of prokaryotic viruses, FEMS Microbiol. Rev., 28, 127–181, https://doi.org/10.1016/j.femsre.2003.08.001, 2004.

Weinbauer, M. G. and Suttle, C. A.: Potential significance of lysogeny to bacteriophage production and bacterial mortality in coastal waters of the Gulf of Mexico, Appl. Environ. Microb., 62, 4374–4380, 1996.

Weinbauer, M. G., Mari, X., and Gattuso, J.-P.: Effect of ocean acidification on the diversity and activity of heterotrophic marine microorganisms, in: Ocean Acidification, edited by: Gattuso, J.-P.

and Hansson, L., Oxford University Press, Oxford, UK, 83–98, 2011.

Wilhelm, S. W. and Suttle, C. A.: Viruses and Nutrient Cycles in the Sea aquatic food webs, BioScience, 49, 781–788, https://doi.org/10.2307/1313569, 1999.

Wilhelm, S. W., Brigden, S. M., and Suttle, C. A.: A dilution technique for the direct measurement of viral production: A comparison in stratified and tidally mixed coastal waters, Microb. Ecol., 43, 168–173, https://doi.org/10.1007/s00248-001-1021-9, 2002.

Winget, D. M., Williamson, K. E., Helton, R. R., and Wommack, K. E.: Tangential flow diafiltration: An improved technique for estimation of virioplankton production, Aquat. Microb. Ecol., 41, 221–232, https://doi.org/10.3354/ame041221, 2005.

Wohlers, J., Engel, A., Zöllner, E., Breithaupt, P., Jürgens, K., Hoppe, H.-G., Sommer, U., and Riebesell, U.: Changes in biogenic carbon flow in response to sea surface warming, P. Natl. Acad. Sci. USA, 106, 7067–7072, https://doi.org/10.1073/pnas.0812743106, 2009.

Wohlers-Zöllner, J., Breithaupt, P., Walther, K., Jürgens, K., and Riebesell, U.: Temperature and nutrient stoichiometry interactively modulate organic matter cycling in a pelagic algal-bacterial community, Limnol. Oceanogr., 56, 599–610, https://doi.org/10.4319/lo.2011.56.2.0599, 2011.

Wolf-Gladrow, D. and Riebesell, U.: Diffusion and reactions in the vicinity of plankton: A refined model for inorganic carbon transport, Mar. Chem., 59, 17–34, https://doi.org/10.1016/S0304-4203(97)00069-8, 1997.

Wommack, K. E. and Colwell, R. R.: Virioplankton: viruses in aquatic ecosystems, Microbiol. Mol. Biol. R., 64, 69–114, 2000.

Worden, A. Z., Nolan, J. K., and Palenik, B.: Assessingthe dynamics and ecology of marine picophytoplankton: The importance of the eukaryotic component, Limnol. Oceanogr., 49, 168–179, https://doi.org/10.4319/lo.2004.49.1.0168, 2004.

Zuur, A. F., Ieno, E. N., and Smith, G. M.: Analysing Ecological Data, in: Statistics for Biology and Health, Springer, New York, https://doi.org/10.1007/978-0-387-45972-1, 2007.

12

Effects of changes in nutrient loading and composition on hypoxia dynamics and internal nutrient cycling of a stratified coastal lagoon

Yafei Zhu[1,2]**, Andrew McCowan**[2]**, and Perran L. M. Cook**[1]

[1]Water Studies Centre, Monash University, Clayton 3800, Australia
[2]Water Technology Pty Ltd, 15 Business Park Drive, Notting Hill 3168, Australia

Correspondence to: Yafei Zhu (yafei.zhu@monash.edu)

Abstract. The effects of changes in catchment nutrient loading and composition on the phytoplankton dynamics, development of hypoxia and internal nutrient dynamics in a stratified coastal lagoon system (the Gippsland Lakes) were investigated using a 3-D coupled hydrodynamic biogeochemical water quality model. The study showed that primary production was equally sensitive to changed dissolved inorganic and particulate organic nitrogen loads, highlighting the need for a better understanding of particulate organic matter bioavailability. Stratification and sediment carbon enrichment were the main drivers for the hypoxia and subsequent sediment phosphorus release in Lake King. High primary production stimulated by large nitrogen loading brought on by a winter flood contributed almost all the sediment carbon deposition (as opposed to catchment loads), which was ultimately responsible for summer bottom-water hypoxia. Interestingly, internal recycling of phosphorus was more sensitive to changed nitrogen loads than total phosphorus loads, highlighting the potential importance of nitrogen loads exerting a control over systems that become phosphorus limited (such as during summer nitrogen-fixing blooms of cyanobacteria). Therefore, the current study highlighted the need to reduce both total nitrogen and total phosphorus for water quality improvement in estuarine systems.

1 Introduction

Excessive anthropogenic nutrient loading, particularly nitrogen, has led to widespread hypoxia and other ecological damages in estuarine and coastal areas (Howarth et al., 2011). About half of the known hypoxic events have been caused by eutrophication (Diaz and Rosenberg, 2008). High primary production as a result of eutrophication can lead to hypoxia or anoxia in poorly mixed bottom water and subsequently enhance the recycling of both nitrogen and phosphorus, which again can reinforce eutrophication (Correll, 1998). This has been found in many stratified estuarine systems around the world, including the Baltic Sea (Vahtera et al., 2007) and the Black Sea (Capet et al., 2016), the Neuse River Estuary (Paerl et al., 1995), and the Gippsland Lakes (Scicluna et al., 2015). The magnitude of sediment phosphorus release is related to severity of bottom-water dissolved oxygen (DO) depletion as well as the duration of hypoxia and/or anoxia. For example, Conley et al. (2002) found that the annual change in dissolved inorganic phosphorus (DIP) in the Baltic Sea was proportional to the area covered by hypoxic water rather than the catchment phosphorus load.

Although some researchers argued that a reduction in both nitrogen and phosphorus were important to improve hypoxia in areas such as the Gulf of Mexico (Rabalais et al., 2007) and the Baltic Sea (Vahtera et al., 2007), others considered that nitrogen should be the primary factor driving marine coastal eutrophication (Diaz, 2001; Hagy et al., 2004; Howarth and Marino, 2006) and thus hypoxia. Regardless of this controversy, the global river export of phosphorus to the coastal ocean has already decreased significantly as a result of advances in wastewater treatment technology since the start of the 21st century; however, nitrogen export still remained high (Howarth et al., 2011). The form and composition of nitrogen export, i.e. dissolved inorganic nitrogen (DIN) and particulate organic nitrogen (PON), can also have a significant impact on receiving coastal waters, as they have different residence times and bioavailability. Seitzinger et al. (2002) showed that total global PON and DIN export by rivers in 1990 were similar, but the DIN : PON ratios varied from region to region. Generally speaking, the DIN : PON ratio was much higher in areas with larger population, indicating that anthropogenic activities had a larger influence on DIN export compared to PON export. The global DIN input to coastal systems was predicted to increase by more than 120 % to 47 million t N year^{-1} by 2050 compared to the level in 1990 (Seitzinger et al., 2002). The relative importance of internally generated (primary production stimulated by nitrogen export) and externally supplied organic matter in hypoxia has only been previously studied by a few researchers (Paerl et al., 1998; Turner et al., 2007). Paerl et al. (1998) showed that hypoxia in estuaries can be stimulated either by internally generated or externally supplied organic matter depending on the meteorological and hydrological conditions. Most importantly, there was a lack of understanding of how different forms of nitrogen (i.e. DIN and PON) in the catchment load can influence the dynamics of hypoxia and the sediment phosphorus cycle in estuarine systems.

Coupled hydrodynamic and biogeochemical models are now increasingly sophisticated and can capture complex biogeochemical feedbacks. There have been a number of successful applications of these models in studying the effects of changes in anthropogenic nutrient loading on the water quality dynamics in estuarine waters (Kiirikki et al., 2001; Webster et al., 2001; Neumann et al., 2002; Pitkänen et al., 2007; Skerratt et al., 2013). However, all these studies primarily focused on the effectiveness of alternative management scenarios for estuarine systems. None of these studies have addressed the sensitivity of hypoxia dynamics and internal nutrient cycling to different forms of nitrogen, phosphorus and organic carbon inputs, which has important scientific and management implications for estuarine water quality.

In this study, we utilised a 3-D coupled hydrodynamic–biogeochemical model to evaluate (1) the sensitivity of phytoplankton and hypoxia dynamics in the Gippsland Lakes to the change in composition of anthropogenic nutrient loading and (2) the consequent impact on internal nutrient dynamics.

2 Method

2.1 Study site

The Gippsland Lakes, located in the southeast of Australia, are the largest estuarine coastal lagoon system in Australia (Fig. 1). The system consists of three main lakes with a total surface area of about 360 km^2. The depths vary from less than 4 m deep in Lake Wellington to 5–10 m deep in Lake King. Lake Wellington and Lake Victoria are linked by the McLennan Strait, which is a narrow channel about 10 km long, 80 m wide and up to 11 m deep. The lakes are connected to the ocean through an artificial entrance at Lakes Entrance constructed in 1889. The Gippsland Lakes has a catchment area of more than 20 000 km^2 and mainly receives freshwater inflow from six major rivers as shown in Fig. 1.

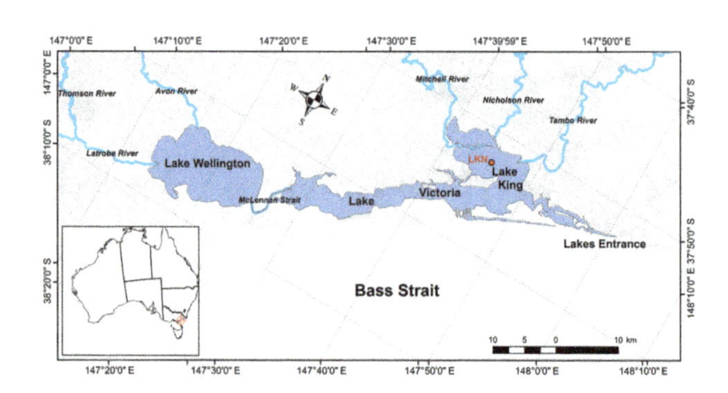

Figure 1. Gippsland Lakes, major tributaries and the location of Lake King North (LKN).

The Gippsland Lakes suffer recurring blooms of toxic nitrogen-fixing cyanobacteria in summers following floods in winter and spring. Together with stratification, high carbon delivery to the sediment in winter and spring following flood-induced diatom and dinoflagellate blooms caused depleted bottom-water oxygen in summer and a subsequent large release of phosphorus from the sediment in the central basins of Lake King and Victoria. Between July and August 2011, the Gippsland Lakes experienced two consecutive floods. In the following summer, a large toxic cyanobacteria bloom occurred in the lakes that persisted from mid-November 2011 to the end of January 2012. It was found that sediment phosphorus release as a result of depleted bottom DO rather than catchment load supplied most of the phosphorus to support the development of the bloom (Zhu et al., 2016).

2.2 Model description

The coupled model used in the current study was developed by Zhu et al. (2016) using DHI's MIKE3 FM and ECO Lab. The model consists of two components, the hydrodynamic model and the biogeochemical model. The hydrodynamic model simulated the transport and turbulent mixing in the water column. The horizontal domain of the hydrodynamic model was discretised as triangular and quadrilateral elements with element areas ranging from 1500 m^2 near the entrance channel and 3 km^2 in Bass Strait. The model had a total of 33 fixed z and varying sigma layers for the vertical discretisation, with height from less than 0.5 m close to the surface to 5 m down to the bottom. Smagorinsky formulations for the horizontal and the standard κ-ϵ model for the vertical have been used to simulate the transport and turbulent mixing in the water column. A scaled eddy formulation was used for the horizontal and vertical dispersion processes, which made the dispersion coefficients directly related to the eddy viscosity calculated by the turbulence models. With the same wind forcing data and model domain, a spectral wave model was also developed for the Gippsland Lakes, using

DHI's MIKE 21 spectral wave module. The wave model results were used to calculate wind–wave shear stress. Since there were not any measured wave data inside the lakes, the default wave input parameters have been used.

The water quality model contains 41 state variables describing the chemical, biological and ecological processes occurring in the water column and sediment compartments. The model included three groups of phytoplankton, which were N-fixing cyanobacteria, vertically migrating dinoflagellates and fast-growing diatoms. One group of grazers was included and configured to avoid grazing on cyanobacteria. The mortality of phytoplankton together with the catchment input was the major source of organic matter. The organic matter was represented by particulate organic carbon (POC), PON, and particulate organic phosphorus (POP) and was further divided into labile and refractory fractions. To simplify, the dissolved organic carbon and nutrients and the hydrolysis process (conversion from particulate to dissolved organic carbon and nutrients) have not been modelled explicitly. Instead, the model has been configured in the way that mineralisation of particulate organic carbon and nutrients took place without going through hydrolysis first. An accumulation of organic matter layer in the bottom water was formed due to settling. Some of the accumulated organic matter would deposit in the sediment and some would return to the water column by resuspension. The rates of deposition and resuspension were calculated based on the modelled local shear stress and the critical shear stress defined for deposition and resuspension. Burial was active when the total thickness of sediment organic matter exceeded 20 cm. The thickness was calculated using a density of 16.959 kg C m^{-3} assuming a porosity of 0.8. The buried sediment organic matter was removed from simulations.

It has previously been shown that internal phosphorus recycling is a key process within the Gippsland Lakes, requiring a refined implementation into water quality models (Webster et al., 2001). The present model overcame previous limitations by implementing sorption and desorption of sediment phosphate and bioirrigation into the model, enabling an accurate simulation of sediment phosphorus dynamics. The sorption and desorption of sediment phosphate were modelled explicitly based on the penetration depth of oxygen and nitrate and the sediment iron concentration, which was a spatially varying constant estimated by using the data collected from previous studies. The impact of bioirrigation was modelled by introducing a scaling factor that was used to adjust the diffusion rates of oxygen and inorganic nutrients at the sediment–water interface. The scaling factor was a function of temperature, DO and labile organic matter. In addition, the model also included a simple cohesive sediment transport module with two bed layers that took into account the salinity and shear stress due to wind–wave and current interactions.

The initial conditions, especially the sediment nutrient storage, could have a large impact on the model simulation. The initial condition of organic carbon, nitrogen, and phosphorus and inorganic nitrogen in the sediment was estimated by iteratively simulating the model for a year, and the concentration at the end of a simulation was used as the initial condition for successive simulations. This was repeated until the sediment nutrient inventory did not change substantially by the end of the simulation. A spatially varying sediment iron-bound phosphate distribution was estimated based on previous field studies.

The total sediment DIN, iron-bound phosphate, and labile POC, PON, and POP in the Gippsland Lakes were approximately 118 t N, 3234 t P, 2861 t C, 238 t N, and 34 t P, respectively. The coupled model was calibrated and validated for the period between May 2010 and July 2012 and it has reproduced the hydrodynamic and biogeochemical conditions in the lakes well and successfully replicated the 2011–2012 summer cyanobacterial bloom (Zhu et al., 2016).

2.3 Nutrient scenarios

Catchment nutrient load data were obtained from the Water Measurement Information System (previously was known as Victorian Water Resources Data Warehouse), which is managed by the Department of Environment, Land, Water and Planning (DELWP). The nutrient data consisted of various constituent concentrations including total nitrogen (TN), nitrate, nitrite, ammonia, total Kjeldahl nitrogen (TKN), total phosphorus (TP) and DIP. The concentrations of the total inorganic and particulate organic nutrients were first calculated using the raw data, and the particulate organic nutrients were then further divided into labile and refractory fractions. There were no measured data for the riverine carbon input; thus, the catchment organic carbon load was estimated using the organic nitrogen load, assuming a C : N weight ratio of 10 (Meybeck, 1982), which also agrees with previous studies that showed that the sediment C : N weight ratio was around 8 to 13 in the Gippsland Lakes (Longmore, 2000; Holland et al., 2013). It can be assumed that organic matter with a C : N ratio close to the Redfield ratio, 5.7 (on a mass basis), should be labile. The Redfield C : N ratio was close to 60 % of that of the estimated catchment C : N ratio. Therefore, it was assumed that 60 % of the catchment organic nutrient loads were labile and the rest were refractory and not bioavailable over the timescale of water residence time. Therefore, a very low mineralisation rate (0.005 day^{-1}) was used for the refractory portion (Zhu et al., 2016).

On average, the western rivers (Latrobe, Thomson and Avon rivers) and eastern rivers (Mitchell, Tambo and Nicholson rivers) each supplied approximately 52 and 48 % of the riverine freshwater inflows to the lake system. However, the western rivers contributed up to approximately 70 % of catchment nutrient loads between May 2010 and July 2012 (Table 1). The majority of these nutrients were delivered during the wet season and were associated with major floods (Fig. 2). The DIN : PON ratio for the eastern rivers was only

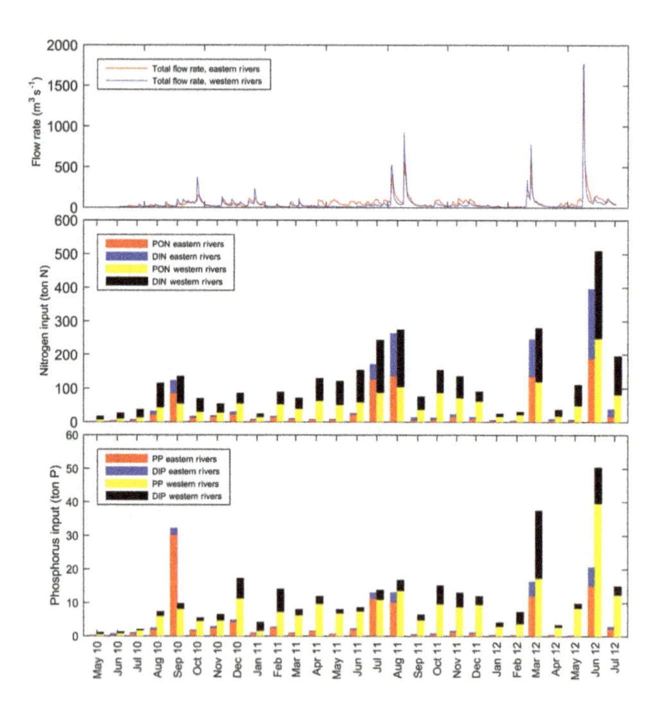

Figure 2. River flow rates and monthly catchment nutrient input.

Table 1. Nutrient loads between May 2010 and July 2012.

	Eastern rivers	Western rivers
DIN (tN)	411.92	1403.76
PON[a] (tN)	681.58	1197.09
DIP (tP)	18.45	77.77
PP[a,b] (tP)	89.82	171.18
POC (tC)	6815.82	11 970.9

[a] The refractory faction has been excluded.
[b] PP is particulate P, which is TP – DIP.

0.6 compared to 1.17 for the western rivers, consistent with more intense human activities in the western catchment. In other words, the percentages of riverine bioavailable nitrogen (DIN + labile PON) delivered to the Gippsland Lakes in the form of labile PON were around 63 and 46 % for the eastern and western rivers, respectively. These numbers were very close to the ones reported for the other major rivers around the world, including the Mississippi (40 %), Amazon (62 %), and Yangtze (45 %) rivers (Mayer et al., 1998). This confirmed that 60 % of the particulate nutrients being labile was a reasonable assumption to describe the characteristics of catchment organic nutrient loads. The TN : TP ratios for both eastern and western rivers were around 10 (on a mass basis), which was close to the Redfield ratio of 7.23 (on a mass basis). While the DIN : DIP ratios were 22.3 and 18 (on a mass basis) for the eastern rivers and western rivers, respectively.

For the current study, we used the calibrated model as the base case and simulated a number of nutrient load scenarios for the same period. There were five sets of scenarios with adjusted loads for DIN, PON, TN (DIN+PON), TP (DIP + particulate P), or TN and TP. In all the scenarios, POC values were set to vary at the same proportion as PON. Since POC load was estimated based on PON load, the C : N ratio in the model was also used to define if particulate matter was labile or refractory. Each set of scenarios had eight simulations that decreased and increased the load by 25, 50, 75 and 100 %. To ensure the consistency of the comparisons and analysis, all scenarios had the same initial conditions including the sediment nutrient inventory, which was derived from the base case. The response of bottom-water DO and sediment processes to different nutrient scenarios over the 2-year simulation period between May 2010 and July 2012 was analysed and discussed, with the focus on the central basins of northern Lake King where the most severe hypoxia, highest sediment DIP fluxes and cyanobacterial blooms were located.

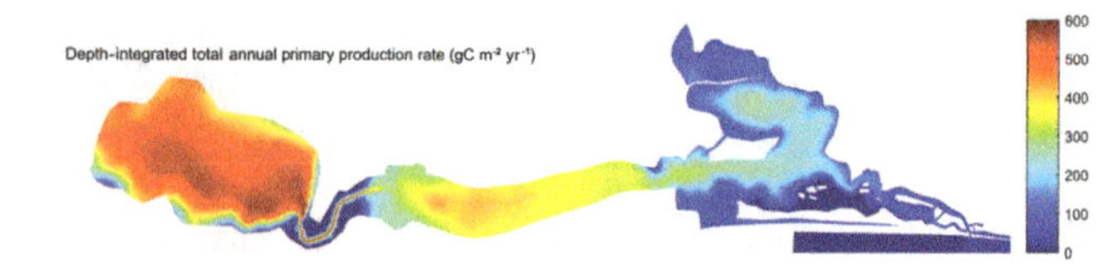

Figure 3. Modelled depth-integrated total annual primary production for the base case.

3 Results

3.1 Primary production

The annual total primary production (TPP) rate in Lake Wellington could reach as high as $600\,\mathrm{g\,C\,m^2\,year^{-1}}$, about $350\,\mathrm{g\,C\,m^2\,year^{-1}}$ in Lake Victoria followed by $250\,\mathrm{g\,C\,m^2\,year^{-1}}$ in Lake King (Fig. 3). The spatial variation in primary production rate was caused by a number of factors, mainly the higher nutrient loads from the western rivers and longer residence time for Lake Wellington. As a result, Lake King only contributed about 12 % of the TPP in the lakes. The total catchment POC load was only 7.5 % of the TPP for the simulation period and was expected to have a minor impact on the sediment biogeochemistry in the lakes. As expected, the TPP in Lake King and the catchment nutrient load had a positive correlation (Fig. 4). TPP was most sensitive to changes in TN + TP loads and least sensitive to reductions in PON loads and increases in TP loads. TPP was very sensitive to reductions in TP loads at reductions > 25 %. The TN reduction scenarios displayed an intermediate response. TPP was more sensitive to reductions in TN than TP until reduction exceeded 75 % when TPP became more sensitive to TP.

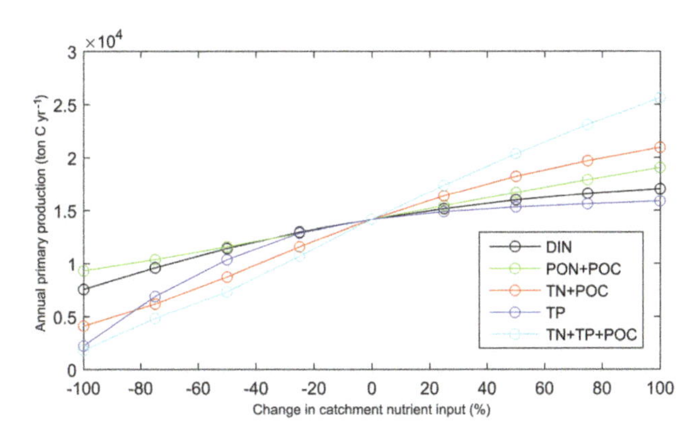

Figure 4. Modelled total primary production in Lake King between May 2010 and July 2012: DIN, change in dissolved inorganic nitrogen load; PON + POC, change in particulate organic nitrogen and carbon loads; TN + POC, change in total nitrogen and particulate organic carbon loads; TP, change in total phosphorus load; TN + TP + POC, change in total nitrogen, total phosphorus and particulate organic carbon loads.

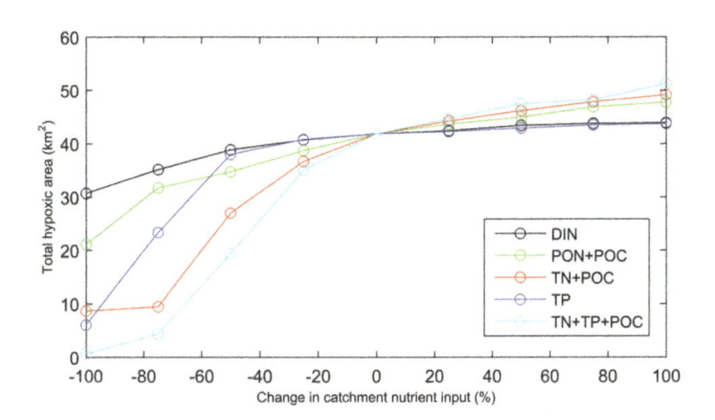

Figure 5. Modelled total area experiencing hypoxia in Lake King.

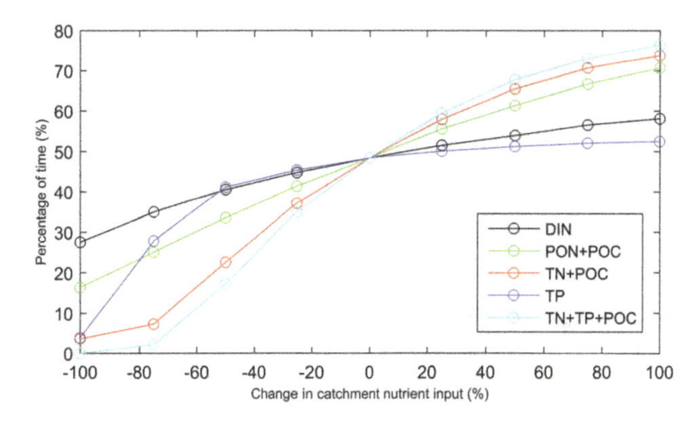

Figure 6. Occurrence of hypoxia as a percentage of time at LKN.

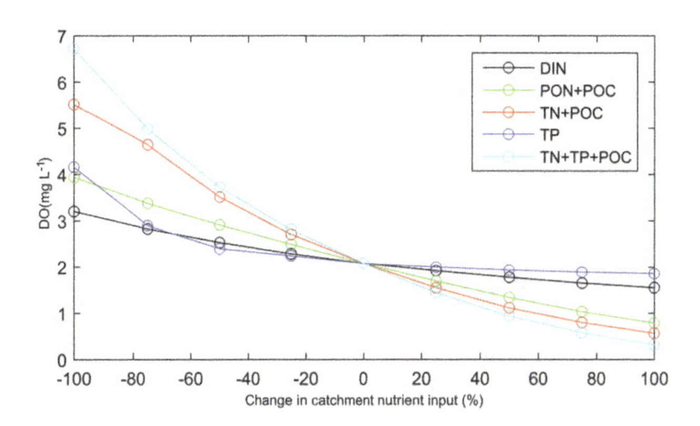

Figure 7. Median bottom-water DO concentration.

3.2 Bottom-water oxygen

The total area in Lake King covered by hypoxic bottom water was about $40\,km^2$ for the base case (Fig. 5). Any further increase in catchment DIN or TP load had no obvious impact on the size of the total hypoxic area (the total area with 24 h averaged bottom-water DO concentration $< 2\,mg\,L^{-1}$ that occurred at least once), while slight increases were seen for the other three scenarios. This was because TPP did not increase much when either DIN or TP increased. Complete removal of catchment DIN and PON loads would result in 25 and 50 % reductions in the total area covered by hypoxic bottom water. The decrease in hypoxic area was insignificant when the TP load was reduced by 50 % but was followed by an accelerating decline if TP was further reduced. Conversely, the hypoxic area decreased more steadily when TN was reduced and the magnitude of decrease in the hypoxic area was most obvious when both TN and TP were reduced. The most severe and persistent hypoxia and/or anoxia was found in the northern Lake King basin. Therefore, the statistics of the time series bottom DO concentration from Lake King North (LKN, location shown in Fig. 1) were extracted and analysed. It was found that the bottom-water DO concentration at LKN would still decrease to close to $0\,mg\,L^{-1}$ even if either catchment TN or TP were completely removed. It would require complete removal of both catchment TN and TP input to eliminate hypoxia in the northern Lake King basin. For the base case, the bottom DO concentration was below the $2\,mg\,L^{-1}$ threshold for almost 43 % of the time (Fig. 6) and the median bottom DO concentration was just slightly above $2\,mg\,L^{-1}$ (Fig. 7). Compared to DIN and TP, doubling the PON load would result in much more frequent hypoxia at LKN, and the occurrence of hypoxia could substantially increase by almost 70 % and could also lead to a 73 % reduction in median bottom DO concentration. This was likely because the POC was also adjusted with PON. The median DO concentration would significantly increase to $5.5\,mg\,L^{-1}$ if catchment TN was completely removed. A 100 % reduction in TP load would improve the median bottom DO concentrations to about $4\,mg\,L^{-1}$ and reduction in PON would have a similar effect. Nonetheless, DIN tended to have a relatively smaller impact on the bottom DO concentrations at LKN.

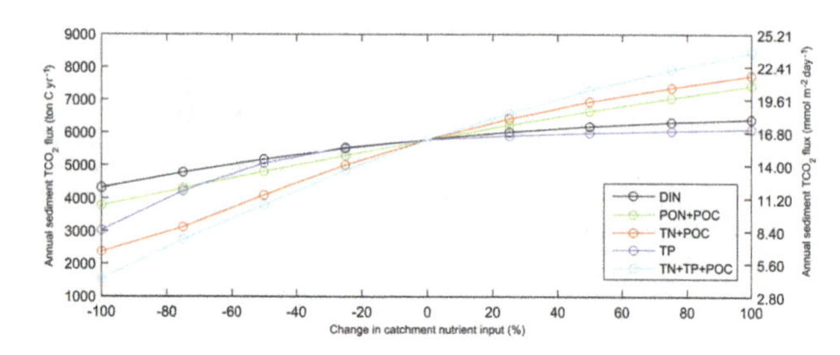

Figure 8. Annual sediment TCO_2 flux from Lake King.

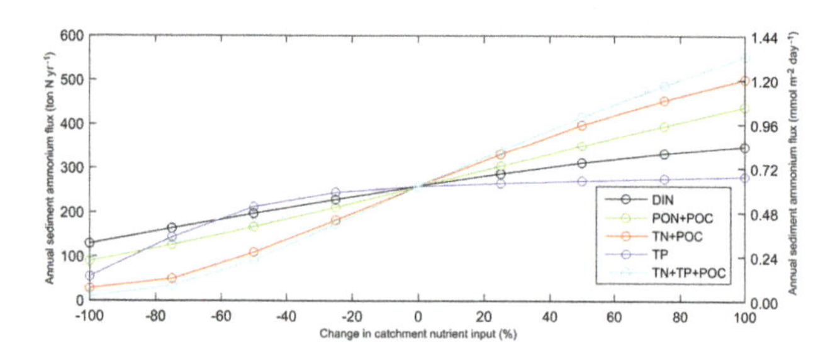

Figure 9. Annual sediment ammonia flux from Lake King.

3.3 Sediment nutrient fluxes

The total CO_2 (TCO_2) flux was a good indication of how much labile organic matter is deposited on the lake bed (Fig. 8). To compare the effect of nutrient reductions on TCO_2 fluxes, the effect of the initial sediment nutrient condition was taken into account by subtracting the TCO_2 flux and TPP for the simulation with no catchment nutrient input from all the model results. The TCO_2 flux and TPP were highly correlated and the TCO_2 flux was approximately equivalent to 8.5 % (calculated by linear regression: $R^2 = 0.97$, $n = 41$) of the TPP across the entire lake system. The ratio was much higher at Lake King and increased to 33 % ($R^2 = 0.93$, $n = 41$). This indicated that the deposition rate in Lake King was much higher than in the rest of the lakes and/or the deposited organic matter could have come from the other parts of the lakes. Increased PON loads resulted in a greater increase in TCO_2 flux than DIN or TP. The sediment ammonia fluxes followed very similar trends to the TCO_2 fluxes, whereas ammonia fluxes were more dependent on PON load (Fig. 9). The average annual nitrogen removal rate through denitrification was about 240 t N $year^{-1}$ in the sediment of Lake King for the base case. However, the nitrogen removal rate in Lake King for all the scenarios only varied marginally from the base case by around ±15 %, except for the simulation of no anthropogenic nutrient load, which had a slightly higher reduction in denitrification rate at approximately 35 %. As a result, the denitrification efficiency in the sediment of Lake King had a negative correlation with the catchment nutrient load (Fig. 10).

Denitrification efficiency is a commonly used measure of the efficiency of nitrogen removal from sediments. It has

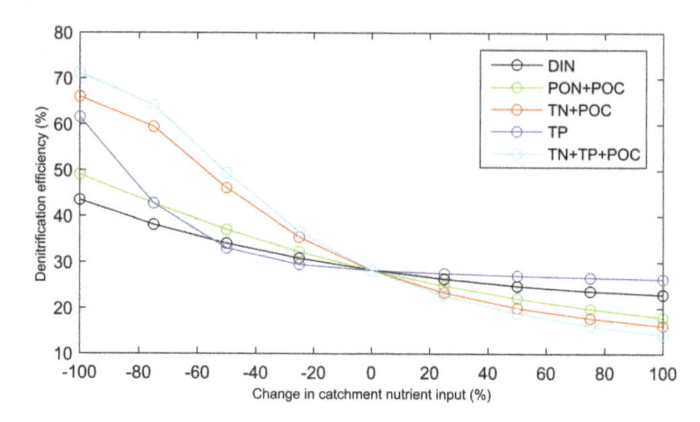

Figure 10. Annual average denitrification efficiency in Lake King.

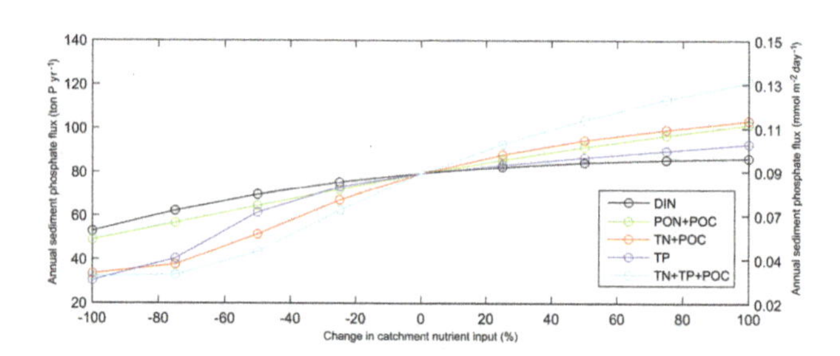

Figure 11. Annual sediment phosphate flux from Lake King.

been defined as the percentage of inorganic nitrogen released from the sediment as dinitrogen gas ($g N m^2 year^{-1}$) and can be calculated by $[N_2/(FNH + FN_3 + N_2) \times 100\,\%]$ (Eyre and Ferguson, 2009). FNH and FN_3 are the sediment ammonia and nitrate fluxes in $g N m^{-2} year^{-1}$. Interestingly, sediment phosphate flux was more sensitive to the change in TN rather than TP loads (Fig. 11). This was because the majority of the phosphate fluxes were a consequence of desorption processes under hypoxic and/or anoxic conditions. The results also showed that even if the catchment nutrient load was completely removed, it would not stop sediment phosphate release, and a 60 % reduction was predicted. One explanation would be that the desorption process still took place as stratification prevented bottom-water oxygen replenishment after the flood event.

4 Discussion

Through observations and modelling, we have previously documented the seasonal dynamics of phytoplankton and nutrient cycling in the Gippsland Lakes (Cook and Holland, 2012; Cook et al., 2010). High winter inflows carried nitrogen into the Gippsland Lakes, which stimulated phytoplankton production. Inputs of organic matter from internal production and the catchment led to hypoxia throughout spring and summer, which then caused phosphorus release from the sediment. We now discuss the sensitivity of this conceptual model to changes in external nutrient loading rates.

4.1 Primary production

Outside the summer cyanobacterial blooms, the lakes are typically nitrogen limited (Holland et al., 2012), and we therefore expected a strong sensitivity of primary production to nitrogen loading rates. Surprisingly, the model showed that primary production was equally sensitive to inorganic and particulate nitrogen loading and that there were two distinct mechanisms by which these two nitrogen forms were trapped within the lakes. Particulate nitrogen can settle down to the sediment while only a negligible portion of the inorganic nitrogen can be transported to the sediment by diffusion unless converted to particulate form by photosynthesis. To calculate how much PON or DIN could have been retained in the lakes after the floods in July and August 2011, we used the model to simulate the transport of PON and DIN, excluding all the biological and chemical processes. The results from this simulation showed that 79 % of flood-introduced PON was retained in the lakes by the end of August. Of the retained labile PON, 95 % reached the sediment where roughly 90 % was recycled over the following 3 months. The importance of PON as a nitrogen source within aquatic systems depends strongly on the degradation kinetics. In this study, we estimated the degradation kinetics of PON based on a surprisingly small pool of literature (Cerco and Cole, 1994; Robson and Hamilton, 2004; DHI Water & Environment, 2012; Deltares, 2013). We therefore suggest that further studies need to be undertaken to better understand the degradation kinetics of PON and the factors that control this such as land use, which may generate PON of different degradability.

Conversely, without any biogeochemical processes such as phytoplankton uptake, only 32 % of the DIN remained in the lakes after the flood. In fact, all the simulations except for the TP reduction scenario showed that around 70 % (calculated by linear regression: $R^2 = 0.99$, $n = 37$) of TN contributed by the flood event still remained in the lakes by the end of August 2011. It is suggested that the majority of the DIN was assimilated by phytoplankton and settled to the sediment during the high flow period between July and August 2011. TN exported to the ocean as a percentage of the total catchment nitrogen input increased by 2.3, 8.1, 24 and 50 % correspondingly when TP was reduced by 25, 50, 75 and 100 %, implying that the N : P ratio in flood waters can be an important factor controlling the residence time of flood-introduced TN in the lakes. This is also reflected in Fig. 9, which shows a sharp decline in sediment ammonia flux when catchment TP was reduced, highlighting the importance of biogeochemical factors in the residence time of nutrients in estuaries (Church, 1986). Under the same hydrodynamic conditions, nitrogen in particulate form has a longer residence time as it can settle down to the bottom of the lake but DIN can be assimilated by phytoplankton and converted to PON before being washed out of the lakes. The rate and efficiency of the conversion from DIN to PON ultimately determine the residence time of DIN in the system.

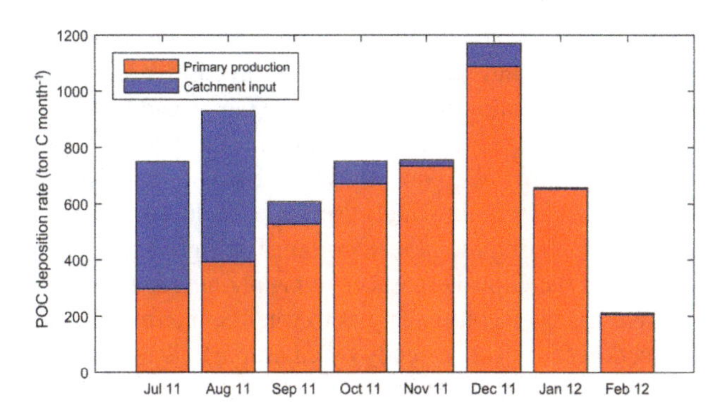

Figure 12. Sediment POC deposition rate at Lake King contributed by primary production and catchment input.

4.2 Hypoxia

Seasonal hypoxia is controlled by both stratification and inputs of organic carbon. The hypoxia observed in the Gippsland Lakes coincided with the recent transition to higher flow following the Millennium drought, which ended in 2010. Boesch et al. (2001) also reported the extended hypoxia in the Chesapeake Bay in the 1970s, which coincided with a transition from drought to wet years. Large freshwater inflows do not only enhance stratification but also increase catchment carbon and nutrient loads. Similar to many other estuary systems, such as the Baltic Sea (Conley et al., 2002) and the Black Sea (Capet et al., 2013), the suppressed oxygen replenishment due to stratification and high oxygen consumption from the mineralisation of deposited organic carbon were the main causes of the hypoxia in Lake King. Stratification in Lake King could last up to several months before the water column became well-mixed again. Lake King also had a higher deposition rate compared to the other parts of the lakes and the majority of the POC deposition was found to be in the deeper basin in northern Lake King. This was because a semi-closed circulation pattern was formed in this area as a result of the interaction between the outgoing river flow and the incoming tidal flow from the ocean, resulting in a lot of POC being trapped in the area unless there was a large flood or storm surge. Another important reason for POC retention in the Lake King basin was that the bottom shear stress in this area was generally low and the 90th percentile shear velocity was only $0.34\,\mathrm{cm\,s^{-1}}$, which was lower than the reported critical shear velocity (0.4–$0.8\,\mathrm{cm\,s^{-1}}$) for the resuspension of phytoplankton-derived organic matter (Beaulieu, 2003). The normal current or wave conditions did not exert enough force to resuspend the sediment in this area once deposited. Furthermore, increases in nutrient load did not proportionally increase the total area subject to hypoxia in Lake King. This was because the areas subject to high detritus deposition were determined by the hydrodynamics and wave conditions, and these depositional areas remained the same for all nutrient scenarios. Furthermore, unlike the Baltic Sea where low DO water could move upwards from individual hypoxic basins and become connected to form a larger hypoxic region (Conley et al., 2009a), the Gippsland Lakes are a much shallower system and the bottom water in areas with depths < 3–4 m was frequently ventilated with wind-driven upwelling and incoming tidal currents, even during the stratified period, highlighting the importance of high organic matter loads in the Gippsland Lakes in maintaining hypoxia.

There has been controversy as to whether internal primary production stimulated directly by anthropogenic nutrients or external catchment organic carbon inputs caused hypoxia in estuaries such as the Gulf of Mexico (Boesch et al., 2009). Previous studies have suggested that carbon either derived from algal blooms or the catchment could result in estuarine hypoxia depending on the hydrological and meteorological conditions (Paerl et al., 1998). To compare the relative importance of the catchment carbon and primary production to the development of hypoxia in Lake King and sediment phosphorus flux, a mass balance calculation was undertaken to calculate the amount of labile POC deposited in the sediment of Lake King. The mass balance calculation was carried out for the period between July 2011 and January 2012 and this was when severe hypoxia and high sediment P flux were observed (Fig. 12). As the model in the current study was not able to trace the origin of the POC, the mass balance calculation was only an approximation and was based on the difference between the base case and the simulation of complete removal of the catchment TN and TP. For a given period, the change in sediment labile POC (ΔC) is approximately equal to the sum of settled labile POC contributed by catchment input and primary production, subtracted from the sediment CO_2 flux due to mineralisation of labile POC. If the conversion from labile to refractory POC can be excluded from the calculation, then sediment TCO_2 flux for the zero catchment nutrient simulation can be assumed to be completely contributed by the mineralisation of refractory POC in the sediment (Cmr). The sediment CO_2 flux due to mineralisation of labile POC is equal to sediment TCO_2 flux minus Cmr. Since the primary production for the zero catchment TP scenario between July 2011 and January 2012 was only equivalent to 0.38 % of that for the base case, it can be assumed that all the sediment POC accumulation resulted from catchment input, which is given by $\Delta C + TCO_2 - \text{Cmr}$. By applying the same approach to the base case, the total POC deposited in the sediment from both catchment and internal generation can be calculated.

The 2011 winter flood brought approximately 4500 t of labile organic carbon into the lakes and about 22 % of this load settled in Lake King between July and August. For this period, catchment carbon contributed almost half of the sediment TCO_2 flux in Lake King. However, bottom-water hypoxia did not develop in Lake King until mid-October, by which time most of the sediment labile POC derived from the catchment was mineralised based on the results of the zero-catchment TP simulation. The elevated temperature and higher light levels during November resulted in higher primary production, which consequently enhanced the sediment carbon enrichment and increased the mineralisation rate of sediment POC. Severe bottom-water hypoxia developed and lasted through towards the end of the simulation. In addition, the catchment POC only contributed less than 7 % to the sediment TCO_2 flux between September 2011 and January 2012. This number is very close to the catchment POC to TPP ratio of 8.3 % for the entire 2-year simulation period. Therefore, bottom-water oxygen depletion in Lake King was primarily related to the planktonic organic matter stimulated by nutrient fluxes from the catchment.

4.3 Internal nutrient dynamics

Internal nutrient recycling can be a critical supply of nutrients to algal growth and it is therefore important to consider the sensitivity of these processes to changes in external loading. Consistent with previous studies (Mulholland et al., 2008; Gardner and McCarthy, 2009), denitrification efficiency increased with reduced nitrogen loading rates, which reduced sediment hypoxia and sediment organic carbon mineralisation rates. Interestingly, at high reductions in phosphorus loading, there was also a large increase in denitrification efficiency, which resulted from the already noted transition to P limitation, meaning less organic nitrogen input to the sediment.

We have previously shown that the primary source of phosphorus fuelling summer blooms of N-fixing cyanobacteria was from the sediment (Scicluna et al., 2015; Zhu et al., 2016). This sediment release of phosphorus was induced by hypoxia, which was in turn driven by internal primary production. As already discussed, this primary production was driven by nitrogen during the winter and spring months and one would therefore expect sediment phosphorus release to be sensitive to nitrogen loads. Importantly, the model showed that internal phosphorus release from the sediment was more sensitive to TN loading than it was to total phosphorus loading at load reductions < 80 %. Conversely, increases in total phosphorus loads were expected to have a minimal effect on internal phosphorus recycling. There has previously been strong debate as to the importance of nitrogen versus phosphorus in coastal systems and it has been argued that the key focus of eutrophication management is to control phosphorus as it is the nutrient that ultimately limits productivity (Schindler et al., 2008). Using a mechanistic approach, the present study highlights that both N and P reductions are required to reduce internal recycling of phosphorus and that phosphorus load reductions alone are likely to be ineffective (Paerl, 2009; Conley et al., 2009b).

4.4 Management implications

Compared to DIN, PON had a slightly larger impact on the bottom DO concentration in Lake King most likely because POC load was related to PON load. The results showed that LKN was more susceptible to TN loading to a certain extent when compared to TP. However, initial input of catchment phosphorus was essential to stimulate primary production, which contributed the majority of the carbon enrichment in the sediment. To eliminate hypoxia in the Gippsland Lakes within the timescale (~ 2 years) of the current model simulation would require a complete removal of catchment nutrient input. Many studies have demonstrated that the effectiveness of external nutrient reduction could be compromised by the sediment supply (Rossi and Premazzi, 1991; Istvánovics et al., 2002; Søndergaard et al., 2003; Jeppesen et al., 2005; Wu et al., 2017). In reality, it is very difficult to reduce nutrients sufficiently and for long enough. It has previously been estimated that the feasible reductions in TN and TP were only 25 % (Ladson, 2012) and 20 % (Roberts et al., 2012), respectively, for the Gippsland Lakes. The results showed that even if the feasible reduction target was achieved, the immediate improvement in the water quality in the lakes was marginal.

For the Gippsland Lakes, the majority of the catchment nutrient flux was non-point source introduced by the flood, making it difficult to manage. The reduction in dissolved inorganic nutrients in flood waters would be particularly more challenging. However, the current study has shown that the water quality in the lakes was also largely influenced by particulate nitrogen and phosphorus, each of which comprised about 60 and 80 % of the total catchment loads, respectively, between 2005 and 2011. Therefore, erosion control is the key to reduce particulate nutrient load from a non-point source and to improve the water quality in the Gippsland Lakes. Vegetated buffers, particularly riparian forests, have been considered to be the simplest but the most effective management option to reduce agricultural non-point source pollutants (Phillips, 1989), especially for nutrient in particulate forms. Vegetated buffers have formed an important part of the water quality improvement strategies for many coastal and estuarine systems such as the Gulf of Mexico (Mitsch et al., 2001) and Chesapeake Bay (Lowrance et al., 1997). Zhang et al. (2010) analysed more than 50 published studies on the performance of vegetated buffers on nutrient removal and found that the median removal efficiency was 68 % for nitrogen and 72 % for phosphorus. However, the performance of vegetated buffers can be compromised on a catchment scale and the removal efficiency can be reduced dramatically to less than 20 % (Verstraeten et al., 2006), due to subsurface hydrological pathways, breakthrough surface runoff or bypass through roads (Mainstone and Parr, 2002). Other measures such as modification in agricultural practices should also be considered. Carefully designed and properly managed vegetated buffers can be a part of an integrated non-point source control strategy for estuarine water quality improvement.

5 Conclusion

Hypoxia and associated sediment phosphorus release in Lake King were predominantly driven by stratification and sediment carbon enrichment. Primary production stimulated by nitrogen loads rather than catchment organic carbon input contributed the majority of the carbon enrichment and was therefore responsible for the depletion of bottom-water DO in summer. Although a significant amount of phosphorus was stored in the sediment, it would only be released under low bottom-water DO conditions in which a large quantity of POC settled in the sediment, which was ultimately driven by nitrogen loading. In addition, the residence time of flood-introduced DIN could be largely influenced by a number of factors including the availability of phosphorus in flood water. It was found that DIN introduced by floods could be converted to PON by photosynthesis quickly enough to prevent being flushed out of the lakes. The current study demonstrated that it is important to reduce both TN and TP in hypoxia mitigation in estuarine systems.

Competing interests. The authors declare that they have no conflict of interest.

Acknowledgements. This work was supported by the Australian research council grant LP140100087 to PLMC and the Victorian Department of Environment, Land, Water and Planning. The authors also thank DHI for provision of licenses for MIKE21 SW, MIKE3 FM and ECO Lab.

Edited by: Caroline P. Slomp

References

Beaulieu, S. E.: Resuspension of phytodetritus from the sea floor: A laboratory flume study, Limnol. Oceanogr., 48, 1235–1244, 2003.

Boesch, D. F., Brinsfield, R. B., and Magnien, R. E.: Chesapeake bay eutrophication, J. Environ. Qual., 30, 303–320, 2001.

Boesch, D. F., Boynton, W. R., Crowder, L. B., Diaz, R. J., Howarth, R. W., Mee, L. D., Nixon, S. W., Rabalais, N. N., Rosenberg, R., and Sanders, J. G.: Nutrient enrichment drives Gulf of Mexico hypoxia, Eos, Transactions American Geophysical Union, 90, 117–118, 2009.

Capet, A., Beckers, J.-M., and Grégoire, M.: Drivers, mechanisms and long-term variability of seasonal hypoxia on the Black Sea northwestern shelf – is there any recovery after eutrophication?, Biogeosciences, 10, 3943–3962, https://doi.org/10.5194/bg-10-3943-2013, 2013.

Capet, A., Stanev, E. V., Beckers, J.-M., Murray, J. W., and Grégoire, M.: Decline of the Black Sea oxygen inventory, Biogeosciences, 13, 1287–1297, https://doi.org/10.5194/bg-13-1287-2016, 2016.

Cerco, C. F. and Cole, T. M.: Three-Dimensional Eutrophication Model of Chesapeake Bay, Volume 1: Main Report, DTIC Document, 1994.

Church, T. M.: Biogeochemical factors influencing the residence time of microconstituents in a large tidal estuary, Delaware Bay, Mar. Chem., 18, 393–406, 1986.

Conley, D. J., Humborg, C., Rahm, L., Savchuk, O. P., and Wulff, F.: Hypoxia in the Baltic Sea and basin-scale changes in phosphorus biogeochemistry, Environ. Sci. Technol., 36, 5315–5320, 2002.

Conley, D. J., Björck, S., Bonsdorff, E., Carstensen, J., Destouni, G., Gustafsson, B. G., Hietanen, S., Kortekaas, M., Kuosa, H., and Markus Meier, H.: Hypoxia-related processes in the Baltic Sea, Environ. Sci. Technol., 43, 3412–3420, 2009a.

Conley, D. J., Paerl, H. W., Howarth, R. W., Boesch, D. F., Seitzinger, S. P., Havens, K. E., Lancelot, C., and Likens, G. E.: Controlling eutrophication: nitrogen and phosphorus, Science, 323, 1014–1015, 2009b.

Cook, P. L. M. and Holland, D. P.: Long term nutrient loads and phytoplankton dynamics in a large temperate Australian lagoon system affected by recurring blooms of *Nodularia spumigena*, Biogeochemistry, 107, 261–274, 2012.

Cook, P. L. M., Holland, D. P., and Longmore, A. R.: Effect of a flood event on the dynamics of phytoplankton and biogeochemistry in a large temperate Australian lagoon, Limnol. Oceanogr., 55, 1123–1133, 2010.

Correll, D. L.: The role of phosphorus in the eutrophication of receiving waters: A review, J. Environ. Qual., 27, 261–266, 1998.

Deltares: D-Water Quality Processes Technical Reference Manual, Delft Hydraulics, the Netherlands, 2013.

DHI Water & Environment: DHI Eutrophication Model 1, ECO Lab Template, DHI Water & Environment, Denmark, 2012.

Diaz, R. J.: Overview of hypoxia around the world, J. Environ. Qual., 30, 275–281, 2001.

Diaz, R. J. and Rosenberg, R.: Spreading dead zones and consequences for marine ecosystems, Science, 321, 926–929, 2008.

Eyre, B. D. and Ferguson, A. J.: Denitrification efficiency for defining critical loads of carbon in shallow coastal ecosystems, Hydrobiologia, 629, 137–146, 2009.

Gardner, W. S. and McCarthy, M. J.: Nitrogen dynamics at the sediment–water interface in shallow, sub-tropical Florida Bay: why denitrification efficiency may decrease with increased eutrophication, Biogeochemistry, 95, 185–198, 2009.

Hagy, J. D., Boynton, W. R., Keefe, C. W., and Wood, K. V.: Hypoxia in Chesapeake Bay, 1950–2001: long-term change in relation to nutrient loading and river flow, Estuaries, 27, 634–658, 2004.

Holland, D. P., Van Erp, I. C., Beardall, J., and Cook, P. L. M.: Environmental controls on the growth of the nitrogen-fixing cyanobacterium Nodularia spumigena Mertens in a temperate lagoon system in South-Eastern Australia, Mar. Ecol.-Prog. Ser., 461, 47–57, 2012.

Holland, D., Jennings, M., Beardall, J., Gell, P., Doan, P., Mills, K., Briles, C., Zawadzki, A., and Cook, P.: Two hundred years of blue-green algae blooms in the Gippsland Lakes, Water Studies Centre, Moans University, Melbourne, Australia, 2013.

Howarth, R. W. and Marino, R.: Nitrogen as the limiting nutrient for eutrophication in coastal marine ecosystems: evolving views over three decades, Limnol. Oceanogr., 51, 364–376, 2006.

Howarth, R., Chan, F., Conley, D. J., Garnier, J., Doney, S. C., Marino, R., and Billen, G.: Coupled biogeochemical cycles: eutrophication and hypoxia in temperate estuaries and coastal marine ecosystems, Front. Ecol. Environ., 9, 18–26, 2011.

Istvánovics, V., Somlyódy, L., and Clement, A.: Cyanobacteria-mediated internal eutrophication in shallow Lake Balaton after load reduction, Water Res., 36, 3314–3322, 2002.

Jeppesen, E., Søndergaard, M., Jensen, J. P., Havens, K. E., Anneville, O., Carvalho, L., Coveney, M. F., Deneke, R., Dokulil, M. T., and Foy, B.: Lake responses to reduced nutrient loading–an analysis of contemporary long-term data from 35 case studies, Freshwater Biol., 50, 1747–1771, 2005.

Kiirikki, M., Inkala, A., Kuosa, H., Pitkänen, H., Kuusisto, M., and Sarkkula, J.: Evaluating the effects of nutrient load reductions on the biomass of toxic nitrogen-fixing cyanobacteria in the Gulf of Finland, Baltic Sea, Boreal Environ. Res., 6, 131–146, 2001.

Ladson, A.: Importance of catchment-sourced nitrogen loads as a factor in determining the health of the Gippsland Lakes, Gippsland Lakes and Catchments Task Force, Melbourne, Australia, 2012.

Longmore, A. R.: Gippsland Lakes sediment nutrient inventory, Marine and Freshwater Resources Institute, Queenscliff, Australia, 2000.

Lowrance, R., Altier, L. S., Newbold, J. D., Schnabel, R. R., Groffman, P. M., Denver, J. M., Correll, D. L., Gilliam, J. W., Robinson, J. L., and Brinsfield, R. B.: Water quality functions of riparian forest buffers in Chesapeake Bay watersheds, Environ. Manage., 21, 687–712, 1997.

Mainstone, C. P. and Parr, W.: Phosphorus in rivers – ecology and management, Sci. Total Environ., 282, 25–47, 2002.

Mayer, L. M., Keil, R. G., Macko, S. A., Joye, S. B., Ruttenberg, K. C., and Aller, R. C.: Importance of suspended particulates in riverine delivery of bioavailable nitrogen to coastal zones, Global Biogeochem. Cy., 12, 573–579, 1998.

Meybeck, M.: Carbon, nitrogen, and phosphorus transport by world rivers, Am. J. Sci., 282, 401–450, 1982.

Mitsch, W. J., Day Jr., J. W., Gilliam, J. W., Groffman, P. M., Hey, D. L., Randall, G. W., and Wang, N.: Reducing Nitrogen Loading to the Gulf of Mexico from the Mississippi River Basin: Strategies to Counter a Persistent Ecological Problem: Ecotechnology – the use of natural ecosystems to solve environmental problems – should be a part of efforts to shrink the zone of hypoxia in the Gulf of Mexico, BioScience, 51, 373–388, 2001.

Mulholland, P. J., Helton, A. M., Poole, G. C., Hall, R. O., Hamilton, S. K., Peterson, B. J., Tank, J. L., Ashkenas, L. R., Cooper, L. W., and Dahm, C. N.: Stream denitrification across biomes and its response to anthropogenic nitrate loading, Nature, 452, 202–205, 2008.

Neumann, T., Fennel, W., and Kremp, C.: Experimental simulations with an ecosystem model of the Baltic Sea: A nutrient load reduction experiment, Global Biogeochem. Cycles, 16, 7-1–7-19, https://doi.org/10.1029/2001GB001450, 2002.

Paerl, H.: Controlling Eutrophication along the Freshwater–Marine Continuum: Dual Nutrient (N and P) Reductions are Essential, Estuar. Coast., 32, 593–601, 2009.

Paerl, H. W., Mallin, M. A., Donahue, C. A., Go, M., and Peierls, B. L.: Nitrogen loading sources and eutrophication of the Neuse River Estuary, North Carolina: Direct and indirect roles of atmospheric deposition, Water Resources Research Institute of the University of North Carolina, 1995.

Paerl, H. W., Pinckney, J. L., Fear, J. M., and Peierls, B. L.: Ecosystem responses to internal and watershed organic matter loading: consequences for hypoxia in the eutrophying Neuse River Estuary, North Carolina, USA, Mar. Ecol.-Prog. Ser., 166, 17–25, 1998.

Phillips, J. D.: Nonpoint source pollution control effectiveness of riparian forests along a coastal plain river, J. Hydrol., 110, 221–237, 1989.

Pitkänen, H., Kiirikki, M., Savchuk, O. P., Räike, A., Korpinen, P., and Wulff, F.: Searching efficient protection strategies for the eutrophied Gulf of Finland: the combined use of 1D and 3D modeling in assessing long-term state scenarios with high spatial resolution, AMBIO, 36, 272–279, 2007.

Rabalais, N. N., Turner, R., Gupta, B. S., Boesch, D., Chapman, P., and Murrell, M.: Hypoxia in the northern Gulf of Mexico: Does the science support the plan to reduce, mitigate, and control hypoxia?, Estuar. Coast., 30, 753–772, 2007.

Roberts, A. M., Pannell, D. J., Doole, G., and Vigiak, O.: Agricultural land management strategies to reduce phosphorus loads in the Gippsland Lakes, Australia, Agr. Syst., 106, 11–22, 2012.

Robson, B. J. and Hamilton, D. P.: Three-dimensional modelling of a Microcystis bloom event in the Swan River estuary, Western Australia, Ecolog. Model., 174, 203–222, 2004.

Rossi, G. and Premazzi, G.: Delay in lake recovery caused by internal loading, Water Res., 25, 567–575, 1991.

Schindler, D. W., Hecky, R., Findlay, D., Stainton, M., Parker, B., Paterson, M., Beaty, K., Lyng, M., and Kasian, S.: Eutrophication of lakes cannot be controlled by reducing nitrogen input: results of a 37-year whole-ecosystem experiment, P. Natl. Acad. Sci. USA, 105, 11254–11258, 2008.

Scicluna, T. R., Woodland, R. J., Zhu, Y. F., Grace, M. R., and Cook, P. L. M.: Deep dynamic pools of phosphorus in the sediment of a temperate lagoon with recurring blooms of diazotrophic cyanobacteria, Limnol. Oceanogr., 60, 2185–2196, 2015.

Seitzinger, S., Kroeze, C., Bouwman, A., Caraco, N., Dentener, F., and Styles, R.: Global patterns of dissolved inorganic and particulate nitrogen inputs to coastal systems: Recent conditions and future projections, Estuaries, 25, 640–655, 2002.

Skerratt, J., Wild-Allen, K., Rizwi, F., Whitehead, J., and Coughanowr, C.: Use of a high resolution 3D fully coupled hydrodynamic, sediment and biogeochemical model to understand estuarine nutrient dynamics under various water quality scenarios, Ocean Coast. Manage., 83, 52–66, 2013.

Søndergaard, M., Jensen, J. P., and Jeppesen, E.: Role of sediment and internal loading of phosphorus in shallow lakes, Hydrobiologia, 506, 135–145, 2003.

Turner, R., Rabalais, N., Alexander, R., McIsaac, G., and Howarth, R.: Characterization of nutrient, organic carbon, and sediment loads and concentrations from the Mississippi River into the Northern Gulf of Mexico, Estuar. Coast., 30, 773–790, 2007.

Vahtera, E., Conley, D. J., Gustafsson, B. G., Kuosa, H., Pitkänen, H., Savchuk, O. P., Tamminen, T., Viitasalo, M., Voss, M., and Wasmund, N.: Internal ecosystem feedbacks enhance nitrogen-fixing cyanobacteria blooms and complicate management in the Baltic Sea, AMBIO, 36, 186–194, 2007.

Verstraeten, G., Poesen, J., Gillijns, K., and Govers, G.: The use of riparian vegetated filter strips to reduce river sediment loads: an overestimated control measure?, Hydrol. Process., 20, 4259–4267, 2006.

Webster, I. T., Parslow, J. S., Grayson, R. B., Molloy, R. P., Andrewartha, J., Sakov, P., Tan, K. S., Walker, S. J., and Wallace, B. B.: Assessing Options for Improving Water Quality and Ecological Function, Gippsland Lakes Environmental Study, Gippsland Coastal Board, Melbourne, Australia, 2001.

Wu, Z., Liu, Y., Liang, Z., Wu, S., and Guo, H.: Internal cycling, not external loading, decides the nutrient limitation in eutrophic lake: A dynamic model with temporal Bayesian hierarchical inference, Water Res., 116, 231–240, 2017.

Zhang, X., Liu, X., Zhang, M., Dahlgren, R. A., and Eitzel, M.: A review of vegetated buffers and a meta-analysis of their mitigation efficacy in reducing nonpoint source pollution, J. Environ. Qual., 39, 76–84, 2010.

Zhu, Y., Hipsey, M. R., McCowan, A., Beardall, J., and Cook, P. L. M.: The role of bioirrigation in sediment phosphorus dynamics and blooms of toxic cyanobacteria in a temperate lagoon, Environ. Modell. Softw., 86, 277–304, 2016.

13

Source, composition, and environmental implication of neutral carbohydrates in sediment cores of subtropical reservoirs, South China

Dandan Duan[1,2], Dainan Zhang[1,2], Yu Yang[1], Jingfu Wang[3], Jing'an Chen[3], and Yong Ran[1]

[1]State Key Laboratory of Organic Geochemistry, Guangzhou Institute of Geochemistry, Chinese Academy of Sciences, Guangzhou, Guangdong 510640, China
[2]University of Chinese Academy of Sciences, Beijing 100049, China
[3]State Key Laboratory of Environmental Geochemistry, Institute of Geochemistry, Chinese Academy of Sciences, Guiyang 55002, China

Correspondence to: Yong Ran (yran@gig.ac.cn)

Abstract. Neutral monosaccharides, algal organic matter (AOM), and carbon stable isotope ratios in three sediment cores of various trophic reservoirs in South China were determined by high-performance anion-exchange chromatography, Rock-Eval pyrolysis, and Finnigan Delta Plus XL mass spectrometry, respectively. The carbon isotopic compositions were corrected for the Suess effect. The concentrations of total neutral carbohydrates (TCHO) range from 0.51 to 6.4 $mg\,g^{-1}$ at mesotrophic reservoirs, and from 0.83 to 2.56 $mg\,g^{-1}$ at an oligotrophic reservoir. Monosaccharide compositions and diagnostic parameters indicate a predominant contribution of phytoplankton in each of the three cores, which is consistent with the results inferred from the corrected carbon isotopic data and C/N ratios. The sedimentary neutral carbohydrates are likely to be structural polysaccharides and/or preserved in sediment minerals, which are resistant to degradation in the sediments. Moreover, the monosaccharide contents are highly related to the carbon isotopic data, algal productivity estimated from the hydrogen index, and increasing mean air temperature during the past 60 years. The nutrient input, however, is not a key factor affecting the primary productivity in the three reservoirs. The above evidence demonstrates that some of the resistant monosaccharides have been significantly elevated by climate change, even in low-latitude regions.

1 Introduction

Carbohydrates are the most abundant compounds in the biosphere, and are present in the natural environment as both structural and storage compounds of aquatic and terrestrial organisms. They comprise about 20–40 wt% of plankton (Parsons et al., 1984), more than 40 wt% of bacteria (Moers et al., 1993), and more than 75 wt% of vascular plants (Moers et al., 1993). Due to their high biological reactivity and availability, carbohydrates are preferentially utilized by heterotrophic organisms (e.g., bacteria and fungi) in water columns (Hernes et al., 1996; Khodse et al., 2007), leading to the preservation of some refractory structural carbohydrates in sediments (Cowie and Hedges, 1994; Burdige et al., 2000; Jensen et al., 2005; He et al., 2010). Moreover, the compositional signature of structural carbohydrates depends more on planktonic sources than the diagenetic pathway (Hernes et al., 1996). Some structural fractions could be used for elucidating sources, deposition processes, and diagenetic fates of organic matter (OM) in aquatic environments (Cowie and Hedges, 1984; Moers et al., 1990; Hicks et al., 1994; Meyers, 1997; Unger et al., 2005; Aufdenkampe et al., 2007; Skoog et al., 2008; Khodse and Bhosle, 2012; Panagiotopoulos et al., 2012).

Carbon isotope analyses in sedimentary OM offer another important tool for reconstructing the history of nutrient loading and eutrophication in lacustrine sediments (Schelske and Hodell, 1991, 1995). Phytoplankton preferentially remove dissolved $^{12}CO_2$ from epilimnetic water and deplete ^{12}C in the remaining dissolved inorganic carbon (Hodell and Schelske, 1998). As supplies of $^{12}CO_2$ become diminished, phytoplankton discriminate less against ^{13}C and sinking OM, incorporating more $^{13}CO_2$. Therefore, increased or decreased productivity can be reflected by enriched or depleted values of $\delta^{13}C$ in OM from the underlying sediments. However, during recent years, the $\delta^{13}C$ values in atmospheric CO_2, water column, and sedimentary OM have been significantly diminished by the Suess effect (Schelske and Hodell, 1995), which is defined as the change in the abundance of carbon isotopes (^{14}C, ^{13}C, ^{12}C) in natural OM reservoirs due to anthropogenic activities (e.g., fossil fuel combustion) (Keeling, 1979). Thus, the Suess effect needs to be considered when applying $\delta^{13}C$ in lacustrine sediments as a proxy for aquatic productivity. Although O'Reilly et al. (2005) had not corrected for the Suess effect in the heterotrophic Lake Tanganyika in Africa, Verburg (2007) found that the corrected $\delta^{13}C$ values were used as a productivity proxy. In the Pearl River Delta (PRD), the development of industrialization and urbanization could enhance the high emission of CO_2 during recent years. Hence, the $\delta^{13}C$ values of reservoir sediments in the PRD need to be corrected for the Suess effect.

Several studies have shown that climate warming plays a significant role in algal productivity by using the S2 proxy in the Arctic lakes during recent decades (Outridge et al., 2005, 2007; Stern et al., 2009; Carrie et al., 2010). However, Kirk et al. (2011) investigated 14 Canadian Arctic and sub-Arctic lakes and found that the relationship between the S2 proxy and climate warming was irrelevant. In addition, only limited investigations have focused on the impact of global change on aquatic productivity in subtropical lakes (Hambright et al., 1994; Smol et al., 2005). Therefore, the above observations call for more investigations on the effects of trophic levels, early diagenesis, and sources of organic matter on the relationship between algal productivity and climate warming.

For this investigation, subtropical reservoirs in rural areas with different trophic states were chosen. Our purpose is to assess the source and diagenetic state of the carbohydrates, and their relationship with algal productivity in sediment cores by using carbohydrate compositions combined with Rock-Eval parameters, carbon isotopic composition, and elemental C/N ratios. In addition, trace metal data (Cu and Zn) cited from our previous paper (Duan et al., 2014) are used to help in understanding the source of carbohydrates. Moreover, neutral carbohydrates and recorded temperature data were statistically analyzed to explore the effects of climate warming on the historical variations of neutral carbohydrates in subtropical regions over the last 6 decades.

2 Materials and methods

2.1 Study area and sample collection

Zengtang reservoir (ZT), Lian'an reservoir (LA), and Xinfengjiang reservoir (XFJ) with different depths and trophic states were chosen for this investigation. The reservoirs are mainly supplied by rainfall and are far away from the industrial center. Aquaculture and dredging activities are forbidden in the investigated reservoirs. Detailed descriptions of the study sites are shown in the previous literature (Duan et al., 2015). In brief, ZT is a shallow, polymictic reservoir with a mesotrophic level, whereas LA is a deep and mesotrophic reservoir. XFJ is a deep and oligotrophic reservoir. Both LA and XFJ are monomictic reservoirs. The most abundant algal species in the ZT, LA, and XFJ reservoirs are green algae, cyanophyta, and diatoms.

Undisturbed sediment cores were collected from the central part of the studied lakes using a 6 cm diameter gravity corer with a Plexiglass liner in 2010 and 2011. The water depths for the sampling sites at the ZT, LA, and XFJ reservoirs are 3, 17, and 36 m, respectively. The core liners were put down slowly in order to avoid disturbance. The sediment cores were sliced into 2 cm thick intervals using extrusion equipment. It is noted that the top four slices of the ZT core were merged into two intervals (0–4 and 5–8 cm) due to the insufficient amount of samples for neutral sugar analysis. All subsamples were immediately placed in plastic bags, sealed, and stored at a low temperature (0–10 °C), and then were quickly transported to the laboratory, where they were freeze-dried and stored until further analysis.

2.2 Physicochemical properties in water

Vertical and temporal variation of chlorophyll *a*, dissolved oxygen, and temperature in the water column of the LA reservoir were recorded by a CTD-90M probe (Sea & Sun Technology, Germany) in increment mode, which enables us to carry out a great number of profile records in the field (Fig. S1 in the Supplement). For the XFJ reservoir, the physicochemical record was conducted only in March 2014. The lack of data from the ZT reservoir is due to its reconstruction after the sediment core sampling.

2.3 Rate of sedimentation

Each sliced sample was analyzed for ^{210}Pb and ^{137}Cs radiometric dating (Duan et al., 2015). Briefly, the activities of ^{210}Pb and ^{137}Cs were measured by a S-100 Multi Channel Spectrometer (Canberra, USA) with a PIPS Si detector and a GCW3022 H-P Ge coaxial detector, respectively. Excess ^{210}Pb activities were measured by subtracting the average ^{210}Pb activities of deeper layers in sediment cores and the constant rates of supply dating models (CRS) were used for the calculation of chronology and sedimentation rates.

2.4 Rock-Eval analysis

All of the samples were analyzed by Rock-Eval 6 (Vinci Technologies, France). The detailed procedures were reported in the previous literature (Duan et al., 2014, 2015). Briefly, bulk sediment was firstly pyrolyzed in an inert, O_2-free oven (100–650 °C) and secondly combusted in an oxidation oven (400–850 °C). Several parameters such as S1, S2, S3, residue carbon (RC) and total organic carbon (TOC) can be generated by a flame ionization detector (FID) and infrared spectroscopy (IR). S1 and S2 represent the fractions of hydrocarbons (HCs) released during the pyrolysis step, where S3 was derived from the fractions of CO and CO_2 released during the two procedures. The residue carbon after combustion defined as RC and the sum of the generated organic fractions is TOC. The hydrogen index (HI) and oxygen index (OI) are calculated by normalizing the contents of S2 and S3 to TOC.

2.5 Stable carbon isotopic analysis

Samples were initially decarbonated by the moderate HCl solution, and then the stable carbon isotopic composition was measured by Finnigan Delta Plus XL mass spectrometry. The $\delta^{13}C$ values (‰) were given by the equation below:

$$\delta^{13}C\,(‰) = (R_{sample}/R_{standard} - 1) \times 1000,$$

where R is the $^{13}C/^{12}C$ ratio and the standard is V-Pee Dee Belemnite. Black carbon (product ID GBW04408, National Research Center for Certified Reference Materials, China) was used as the reference standard for the determination of accuracy and precision. The precision of $\delta^{13}C$ for the replicates was $< 0.19‰ \pm 0.12$ ($n = 40$).

Measured values of $\delta^{13}C$ were corrected for the Suess effect with the following polynomial equation (Schelske and Hodell, 1995), where t is time (in years):

$$\delta^{13}C\,(‰) = -4577.8 + 7.3430t - 3.9213 \times 10^{-3}t^2 + 6.9812 \times 10^{-7}t^3.$$

The calculated time-dependent depletion in $\delta^{13}C$ induced by fossil fuel combustion since 1840 was subtracted from the measured $\delta^{13}C$ for each dated sediment section.

2.6 Neutral sugar analysis

Sediment samples (about 5 mg) from the three reservoirs were weighted and hydrolyzed in glass ampules with 12 M H_2SO_4 for 2 h at room temperature. After 9 mL of Milli-UV + water were added (1.2 M H_2SO_4, final concentration of acid), the ampules were flame-sealed and the samples were stirred and hydrolyzed in a 100 °C water bath for 3 h. The hydrolysis was terminated by placing the ampules in an ice bath for 5 min. Then, the deoxyribose was added as the internal standard (Philben et al., 2015). Before instrumental analysis, the samples were run through a mixed bed of anion (AG 2-X8, 20–50 meshes, Bio-Rad) and cation (AG 50W-X8, 100–200 meshes, Bio-Rad) exchange resins (Philben et al., 2015). Self-absorbed AG11 A8 resin was utilized to remove the acid. The volume of resin needed for complete neutralization depended on the amount of acid used for hydrolysis (Philben et al., 2015). After purification with a mixture of cation and anion exchange resins, neutral sugars were isocratically separated with 25 mM NaOH on a PA 1 column in a Dionex 500 ion chromatography system, which was equipped with a pulsed amperometric detector (PAD, model ED40) (Philben et al., 2015). The detector setting was based on Skoog and Benner (1997). Chromatographic data were recorded with a personal computer equipped with Hewlett Packard Chemstation software (Skoog and Benner, 1997; Kaiser and Benner, 2000).

For every 10 analyses, a blank sample and a duplicate sample were analyzed to check accuracy and precision. No neutral sugars were detected in the blanks. Only glucose, galactose, mannose, rhamnose, fucose, and xylose were detected and analyzed in samples due to the loss of ribose in the process of acid hydrolysis. The recovery of the monosaccharides in the sediments ranged from 73 to 95 %. The analytical precision of duplicate samples performed on different days was within ±3 % for glucose and ±5 % for other sugars. In this study, the total neutral carbohydrates (TCHO) are defined as the sum of all identified monosaccharides.

3 Results

3.1 Physicochemical properties of water

As shown in Fig. S1, chlorophyll *a* concentrations in the water column of LA are higher in spring and summer than in fall and winter, which is consistent with the seasonal distribution of the contents of dissolved oxygen. The vertical profiles of chlorophyll *a* and dissolved oxygen also show similar patterns; depleted dissolved oxygen in the hypolimnion accompanies a low content of chlorophyll *a* throughout the year, suggesting the productivity significantly contributes to dissolved oxygen contents. In general, the bottom sediments in LA are mostly under anaerobic conditions. Water temperature is higher in summer and fall, resulting in thermal stratification in the water column. During winter, the lake mixes from top to bottom due to the decrease in temperature (so-called autumn overturn). Therefore, the relatively high contents of nutrients can be transported by the water flow to the upper depths, resulting in the increase in nutrients and productivity in the entire water column.

The concentration of chlorophyll *a* (average: 0.94 $\mu g\,L^{-1}$) in the water column of XFJ is much lower than that in LA. The oxygen contents vary from 9 $mg\,L^{-1}$ at a depth of 1 m to 8.54 $mg\,L^{-1}$ at a depth of 36 m (Fig. S1), suggesting the bottom sediments are under aerobic conditions.

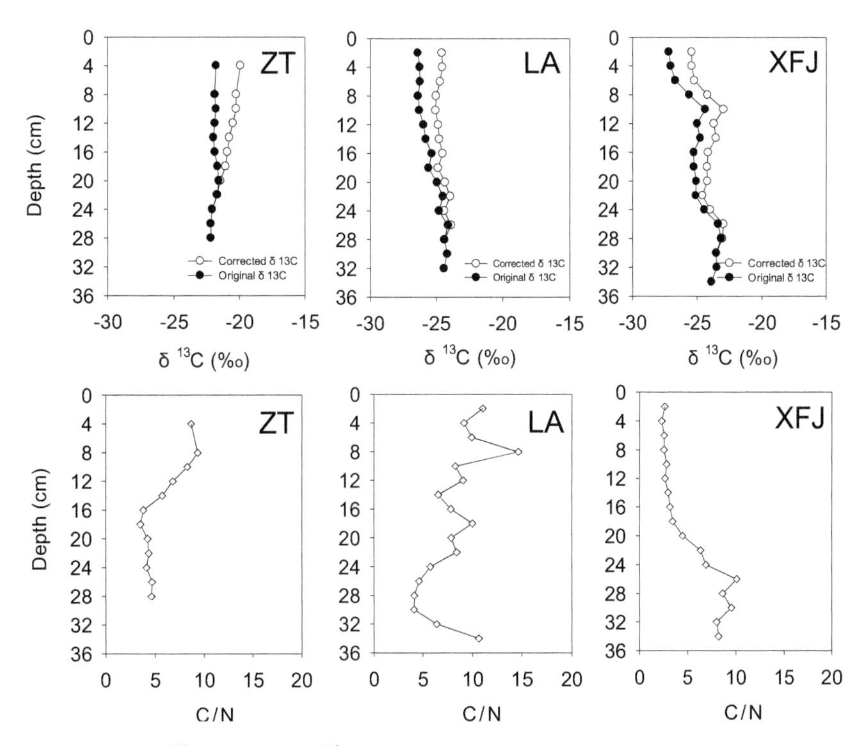

Figure 1. Vertical variations of original $\delta^{13}C$, corrected $\delta^{13}C$, and C/N in sediment cores of three reservoirs.

3.2 Characteristics of OM

The data from Rock-Eval pyrolysis are listed in Table S1 in the Supplement, which were reported previously (Duan et al., 2015). As shown in Table S1, pyrolytic parameters S1 and S2 represent the fractions of HCs released during the pyrolysis step, where S3 was derived from the fractions of CO and CO_2 released during the pyrolysis and oxidation procedures. The HI is calculated by normalizing the contents of S2 to TOC (Duan et al., 2015). TOC and HI are in the ranges of 0.78–2.98 and 114–231 mg HC g^{-1} TOC, respectively, in the ZT core; in the ranges of 0.88–4.31 and 151–229 mg HC g^{-1} TOC, respectively, in the LA core; and in the ranges of 0.47–1.76 and 141–196 mg HC g^{-1} TOC, respectively, in the XFJ core. In general, the HI values are enriched in the surface layers of all the sediment cores.

$\delta^{13}C$ values (‰) range from −22.2 to −21.6 ‰ in the ZT core, from −26.4 to −24.1 ‰ in the LA core, and from −27.2 to −23.1 ‰ in the XFJ core, with average values of −21.9, −25.4, and −24.9 ‰, respectively (Fig. 1). After the correction for the Suess effect, $\delta^{13}C$ values (‰) vary from −21.7 to −19.9 ‰ in the ZT core, from −25.1 to −23.9 ‰ in the LA core, and from −25.5 to −22.9 ‰ in the XFJ core, with average values of −20.8, −24.6, and −24.1 ‰, respectively (Fig. 1).

Elemental C/N ratios vary from 3.51 to 9.34 in the ZT core, from 4.12 to 14.7 in the LA core, and from 2.3 to 10.1 in the XFJ core, with mean values of 5.56, 8.15, and 5.14, respectively (Fig. 1).

The detailed results from ^{210}Pb and ^{137}Cs radiometric dating were reported previously (Duan et al., 2015). The mass accumulation rates (MARs) were cited and displayed in Table S1. As the ZT, LA, and XFJ reservoirs had been small lakes before the dam construction, the chronological records for the ZT, LA, and XFJ cores are longer than the timescales of the dam construction.

3.3 Neutral sugar data

The concentrations of seven monosaccharides (glucose, galactose, mannose, arabinose, rhamnose, fucose, and xylose) are presented in Figs. 2 and S2, and/or listed in Table S2. Glucose (0.2–2.34 mg g^{-1}) is the most abundant sugar in all the reservoir sediments, followed by galactose (0.09–1.2 mg g^{-1}), mannose (0.03–0.92 mg g^{-1}), and xylose (0.01–0.95 mg g^{-1}). Concentrations of arabinose (0.05–0.77 mg g^{-1}), rhamnose (0.04–0.67 mg g^{-1}), and fucose (0.02–0.43 mg g^{-1}) are relatively low in the reservoir sediments. The TCHO concentrations at the ZT, LA, and XFJ reservoirs range from 1.94 to 5.36 mg g^{-1}, from 0.51 to 6.4 mg g^{-1}, and from 0.83 to 2.56 mg g^{-1}, respectively, and show decreasing downcore trends in all the sediment cores. Carbohydrate yield (%) is the molar concentration of monosaccharide carbon normalized by TOC. Carbohydrate yield (%) ranges from 7.08 to 10.9 % in the ZT core, from 2.31 to 13.53 % in the LA core, and from 1.93 to 12.52 % in the XFJ core, with average values of 8.68, 7.79, and 7.33 %, respectively (Fig. 3).

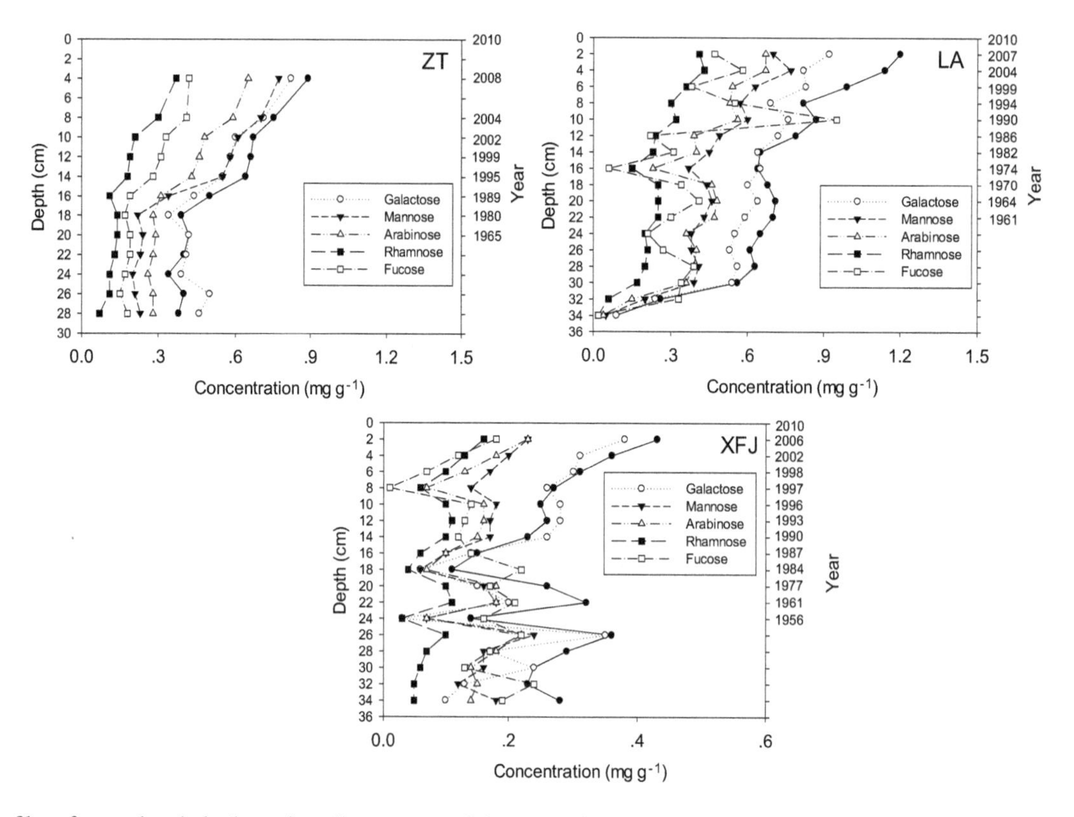

Figure 2. Profiles of neutral carbohydrates in sediment cores of the reservoirs.

Compositions of neutral sugars in the three sediment cores are calculated based on their concentrations. Glucose (21.1–27.6 mol %) is the most abundant monosaccharide, followed by mannose (14–21.5 mol %), galactose (14.7–18.7 mol %), arabinose (10.7–17.1 mol %), rhamnose (11–12.9 mol %), xylose (8.8–10.1 mol %), and fucose (3.2–7.2 mol %) in the ZT sediments. Glucose (22.2–37.4 mol %) is also the most abundant monosaccharide, followed by galactose (12–20 mol %), mannose (11.1–20.5 mol %), arabinose (10.7–14.4 mol %), rhamnose (6.9–11.9 mol %), xylose (5.09–18.32 mol %), and fucose (3.3–7.74 mol %) in the LA sediments. For the XFJ core, glucose (18.5–48 mol %) is still the most abundant monosaccharide, followed by galactose (12.3–3.7 mol %), mannose (3.8–23 mol %), arabinose (8.3–14.5 mol %), rhamnose (6–11.2 mol %), xylose (0.75–21.6 mol %), and fucose (2.9–8.2 mol %) in the sediment core.

3.4 Meteorological records of the studied areas

The 5-year moving average temperature (T_5) was calculated from the reported database (Duan et al., 2015). The mean air temperature in the Guangzhou area has increased by about 1.5 °C since 1960, and the mean air temperature in the Heyuan area increased by about 1.52 °C between 1957 and 2004. Therefore, the above data suggest a significant trend in climate warming in the investigated areas during the last 6 decades (Duan et al., 2015). The annual hours of daylight in Guangzhou and Heyuan on the timescale of 60 years have been obtained from the China Meteorological Data Sharing Service System (CMDSSS). The annual hours of daylight in both areas are somehow variable and show a progressively decreasing trend from 1950 to 2010 (Fig. S6).

4 Discussion

4.1 OM characteristics in sediment cores

The OM fractions derived from the Rock-Eval analysis could provide the source and early diagenetic information of OM in the reservoir cores. The S1, S2, S3, TOC, and HI show significant decreasing downcore trends in the ZT and LA cores, suggesting that the sedimentary OM has been affected either by autochthonous inputs or by extensive degradation (Duan et al., 2015). For the XFJ core, the TOC as well as the other pyrolytic parameters (except the HI proxy) show increasing trends with depth, suggesting that the degradation and oxidation of OM and/or terrestrial inputs of the OM are the primary factors affecting the variation of OM.

Carbon isotope analyses offer an important tool for identifying the sources of OM in lacustrine sediments. Different primary producers have distinctive carbon isotope compositions. The average $\delta^{13}C$ values of C3 plants are around −27 to −26 ‰, whereas the C4 plants have average $\delta^{13}C$ values

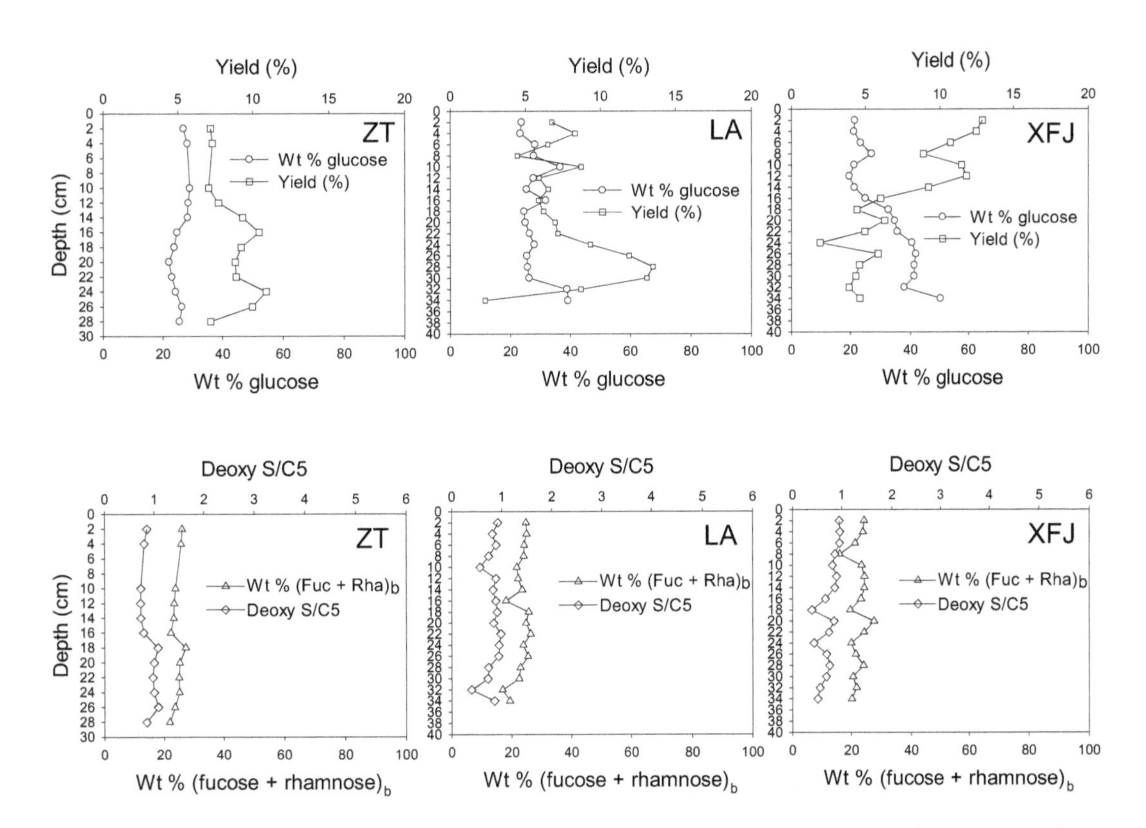

Figure 3. Vertical profiles of yield (%), deoxy S / C5, wt % glucose and wt % (fucose + rhamnose)$_b$ in the ZT, LA, and XFJ reservoirs.

of −14 to −13 ‰. Although the $\delta^{13}C$ values in phytoplankton are in a broad range of −17 to −45 ‰ (Boschker et al., 1995), phytoplankton can be identified by the combination of other proxies (e.g., elemental C/N ratios). All the corrected $\delta^{13}C$ values of sedimentary OM in the three reservoirs vary from −25.5 to −19.9 ‰ (Fig. 1), which are in the range of phytoplankton and C3 plants. However, their corresponding C/N ratios are relatively lower than those for higher plants (> 12) (Fig. 1), suggesting the predominant contribution of phytoplankton in the OM of reservoirs. It is noted that the very low C/N ratios in the XFJ core are likely to be related to inorganic N in minerals. As the TOC contents are quite low in XFJ, their inorganic N contents will affect the C/N ratios. The $\delta^{13}C$ values in ZT sediments are more enriched (average: −20.8 ‰) than in LA (average: −24.6 ‰) or XFJ (average: −24.1 ‰) sediments, which may be attributed to high phytoplankton productivity (chlorophyll $a = 90.7\ \mu g\,L^{-1}$), anaerobic sediments with high rates of methanogenesis, and lack of terrestrial carbon inputs in shallow water bodies. High phytoplankton can enhance isotopic fractionation and result in enrichment of ^{13}C in dissolved inorganic carbon. The removal of CH_4 (^{13}C light) by intensive methanogenesis also leads to the accumulation of ^{13}C-heavy OM in sediments (Gu et al., 2004).

After correcting for the Suess effect, the OM in sediments becomes more enriched in ^{13}C from the bottom to the top of the ZT core (Fig. 1), reflecting a progressive increase in historical productivity, which is consistent with the vertical variations of TOC, C/N, and HI. Similar observations were also found in the LA core from a depth of 16 cm to the surface layer. Therefore, both the ZT and LA reservoirs undergo significant increases in primary productivity during the recent years. As shown in Fig. 1, the correction for the Suess effect can result in opposite conclusions regarding the aquatic productivity, based on uncorrected $\delta^{13}C$ values for ZT and LA. Similar results were also observed in Lake Brunnsviken (Routh et al., 2004), Lake Eric (Schelske and Hodell, 1995), and the deep Lake Tahoe (Chandra et al., 2005), suggesting the importance, in terms of productivity, of the correction for the Suess effect in the recent $\delta^{13}C$ values for lacustrine sediment cores.

For the XFJ reservoir, the corrected $\delta^{13}C$ values increase from depths of 22 to 10 cm and then decrease abruptly to the surface layers, which may be related to the biodegradation of OM in aerobic sediments and/or a great amount of recent terrestrial loading. However, values of C/N ratios in the upper layers of the XFJ core are very low (C/N ≈ 3) and indicate a contribution of algal origin in the sediments (Fig. 1). Therefore, the decrease in corrected $\delta^{13}C$ at XFJ is mainly due to the biodegradation of OM under aerobic conditions. The preferential degradation of more ^{13}C-enriched organic compounds (e.g., carbohydrates and proteins) would lead to a decrease in the $\delta^{13}C$ values in the residue OM (Lehmann et al., 2002).

4.2 Monosaccharide compositions in sediment cores

As shown in Table. S2, the TCHO concentrations in the ZT and LA cores show a significant decrease in the downcore sediments, which are similar to the vertical profiles of S2, TOC, C/N, corrected δ^{13}C, and HI (Tables S1 and S2). For the XFJ core, the TCHO and HI still decline in the downcore sediments, as do those in the ZT and LA cores (Tables S1 and S2). In general, the contents of TCHO (0.51–6.4 $mg\,g^{-1}$) in the three reservoirs are similar to those of a sediment core in the eutrophic French Aydat lake (1.19–4.58 $mg\,g^{-1}$) (Ogier et al., 2001), which was also enriched with neutral sugars at the surface layers of the sediment core.

Monosaccharide compositions were calculated for investigating the applicability of neutral sugars as productivity proxies. The compositions of neutral sugars in ZT and LA cores show that glucose is the most abundant sugar in these two reservoir sediments, while galactose, mannose, and arabinose are relatively more abundant than rhamnose, xylose, and fucose (Figs. 2 and S2), which is similar to the monosaccharide composition in phytoplankton (Hamilton and Hedges, 1988). Moreover, the relative abundances of monosaccharides do not vary much in the ZT and LA cores, except for the apparent changes in glucose and xylose at depths of 10, 16, 32, and 34 cm in the LA core. For the XFJ reservoir, the monosaccharide composition in the upper layers (0–16 cm) of the sediment core also indicates a dominant origin of algal carbohydrates. However, xylose significantly increases at a few depths (XFJ core: at 18, 24, and 32 cm depths) (Fig. 2). Furthermore, glucose shows an increasing trend and is correlated with the abundant xylose, especially between the depths of 18 and 34 cm. The pattern of carbohydrate composition in these samples (18–34 cm) is not in agreement with the post-depositional process of diagenesis as observed in previous studies (Hamilton and Hedges, 1988; Hedges et al., 1994). They found that the diagenesis process often led to a decrease in glucose along with a corresponding increase in bacteria-derived deoxy sugars (rhamnose and fucose) in sediment cores. Therefore, these outliers (18–34 cm) might be related to increasing vascular plant input or hydrological variation at XFJ, as discussed in the following paragraph.

4.3 Source of neutral carbohydrates

Molecular-level diagnostic parameters have often been used to differentiate microbial, planktonic, and terrestrial sources (Cowie and Hedges, 1984; Guggenberger et al., 1994). They could be used as potential proxies for productivity in lakes and reservoirs. Diagnostic carbohydrate parameters and their results are presented in Figs. S3 and S4.

The ratios of mannose to xylose could indicate the OM sources derived from phytoplankton, bacteria, gymnosperm, and angiosperm tissues (Cowie and Hedges, 1984). As shown in Fig. S4, the values of the mannose / xylose ratios in most of the sediments at ZT and LA range from 1.51 to 2.70 and 1.49 to 3.50, respectively, except at a depth of 8 cm (1.46) in the ZT core and at the depths of 4 (1.18), 8 (1.05), 10 (0.67), 16 (8.60), and 28–32 cm (0.6–1.33) in the LA core. Thus, most of the samples can be identified as a phytoplankton source (1.5–3.5), suggesting the important contribution of algal organic matter (AOM) in these two reservoirs. However, most of the mannose / xylose ratios are in the range of 0.23–0.87 for gymnosperm tissues (< 1) at depths deeper than 16 cm in the XFJ sediments, which indicates the presence of terrestrial OM derived from angiosperm leaves and grasses.

The above conclusion is also confirmed by the %xylose$_b$ parameters ("b" represents a value on a glucose-free base) plotted in Figs. 2 and S4. %xylose$_b$ is a useful biomarker to differentiate the type of terrestrial input (Cowie and Hedges, 1984). In most of the samples at ZT and LA, %xylose$_b$ is in the ranges of 9.34–11.8 and 7.10–15.8, respectively, except at the depths of 10 (23.5), 16 (3.02), and 32 cm (26.4) in the LA core. The low %xylose$_b$ values (6.2–17.0) indicate the primary phytoplankton origin of neutral sugars at ZT and LA. The high values at the depths of 10 and 32 cm in the LA core might indicate important terrestrial input. Further evidence is also obtained from % (arabinose + galactose)$_b$ plotted in Fig. S4 and from % (fucose + rhamnose)$_b$ plotted in Fig. S3. The results from the %xylose$_b$ vs. % (fucose + rhamnose)$_b$ plots suggest a phytoplankton origin in reservoir sediments, which is consistent with that reported in the literature (Boschker et al., 1995). As for the % (arabinose + galactose)$_b$ ratios, their values are mostly in the range of 30.7–44.4 in the sediment cores at ZT, LA, and XFJ, indicating that the sedimentary OM samples are largely derived from phytoplankton with the ratios of 22–47. Only a few high values at a depth of 16 cm (48) in the LA core and at depths of 8 cm (50.6) and 34 cm (48.9) in the XFJ core are likely to indicate an additional origin from non-woody angiosperm tissues and grasses, as demonstrated by Cowie and Hedges (1994). The above result also implies a different origin for neutral sugars in the upper layers of the XFJ core (0–16 cm) than in the lower layers ($>$ 18 cm), which have been increasingly affected by terrestrial input.

In order to support the above conclusions of the sources of OM derived from neutral sugars in the reservoirs, monosaccharide concentrations and heavy metal data are compared (Tables S3 and S4). It is found that almost all of the monosaccharides (except xylose at LA and XFJ) are significantly related to heavy metals (e.g., Zn and Cu) in the sediment cores of ZT and LA, and the upper layers of the XFJ core (0–16 cm) (Table S4). However, only galactose, mannose, and fucose are positively correlated with Zn and Cu in the core of XFJ (0–34 cm), although the lower layers are increasingly affected by allochthonous input, as discussed in the above paragraphs. As Zn and Cu are essential nutrients for phytoplankton growth, these relationships provide additional evidence for the important contribution of AOM to carbohydrates in the investigated sediments. However, the Pb contents in the sediments of these reservoirs are very low, suggesting that there is no or little industry contamination in the investigated areas.

The ZT reservoir is a shallow reservoir and has a relatively higher trophic level. The LA reservoir is deeper and has a longer water residence time and a medium trophic level. Both the ZT and LA reservoirs are dominated by autochthonous green algae and diatoms, which contribute to the majority of carbohydrates in their sediments. However, the XFJ reservoir receives a relatively high input of higher plants from the surrounding runoff. Therefore, the different sources of carbohydrates are related not only to input of algae and plants, but also to historical changes in hydrological conditions and nutrient levels, etc.

4.4 Diagenesis of neutral carbohydrates

Carbohydrates are useful not only in identifying the sources of OM, but also in evaluating early diagenetic processes occurring in the post-depositional environment. The four parameters, deoxysugar/pentose (deoxy S / C5) ratio, glucose content (mol % or wt %), % (fucose + rhamnose)$_b$, and carbohydrate yield (%) in the sediment cores are often used to evaluate diagenetic changes in OM (Cowie and Hedges, 1984; Ittekkot and Arain, 1986; Opsahl and Benner, 1999; Benner and Opsahl, 2001; Kaiser and Benner, 2009).

Glucose content is an important factor used to assess the degradation state of OM. Glucose accounts for 58 to 90 % of the carbohydrates in fresh plankton and terrestrial tissues (Cowie and Hedges, 1984; Opsahl and Benner, 1999; da Cunha et al., 2002). Hernes et al. (1996) proposed that relative mol % glucose in particulate OM could be used as a diagenetic indicator of organic material in the equatorial Pacific region. In this investigation, wt% glucose in the sediments ranges from 22 to 29.1 % at ZT, from 23.3 to 39.2 % at LA, and from 19.6 to 50.3 % at XFJ (Fig. 3), suggesting that neutral sugars are biodegraded in the sedimentary OM. This conclusion is also confirmed by the carbohydrate yields (%), which usually represent 30–40 % of TOC in fresh tissues of plant and phytoplankton but less than 9 % in sediments (Cowie and Hedges, 1984; Opsahl and Benner, 1999). Carbohydrate yields range from 1.93 to 13.53 % in the sediments of the three reservoirs (Fig. 3). It is also suggested that neutral sugars degrade significantly in the investigated sediments. These results are consistent with the general observation from previous studies (Ogier et al., 2001). In general, the carbohydrates in the reservoir sediments are extensively transformed and degraded. However, the stability of their compositions was observed in the downcore sediments. Whether the carbohydrate compounds are degraded mainly in the water column or in the sediment core will be discussed below (Fig. 3).

Keil et al. (1998) found that the % wt (fucose+rhamnose)$_b$ values could reflect the diagenesis process of neutral carbohydrates. This index was elevated as the sediment particle sizes decreased, suggesting that smaller size fractions showed a higher degree of degradation. Their observation was also consistent with other diagenetic indices such as lignin and non-protein biomarkers (Keil et al., 1998). In this study, the values of % wt (fucose + rhamnose)$_b$ increase slightly, but the values of wt % glucose do not change significantly in each of the ZT, LA, and XFJ downcore sediments (Fig. 3), suggesting that the process of degradation occurs mainly during the settling period rather than after deposition. Further evidence in support of this conclusion can be obtained from the ratio of deoxy sugars (e.g., rhamnose and fucose) to C5 (e.g., arabinose and xylose) (deoxy S / C5). The deoxy S / C5 ratios increase slightly in the downcore sediments of ZT, LA, and XFJ (Fig. 3). Therefore, although sinking organic matter-rich particles and their carbohydrates suffer from intensive oxidation and degradation in the water column during their transit to bottom sediments, some fractions are selectively preserved in the sediment cores, and remained almost unchanged during post-deposition, as observed before (Cowie and Hedges, 1984; Moers et al., 1990; Hicks et al., 1994).

Carbohydrates consist of storage polymers and cell membranes of phytoplankton. In general, glucose is bound mainly in an unbranched, starchlike β-1,3-glucan storage polymer (Handa, 1969), whereas mannose, galactose, xylose, fucose, and rhamnose are characteristically more abundant in the cell walls (Cowie et al., 1992). In addition, a portion of carbohydrates is highly likely to be preserved by sedimentary clay minerals. Hence, the neutral carbohydrates in the sediment cores could be resistant to microbial degradation.

In support of the above conclusion, the k values of deoxy S / C5 were calculated using a "multi-G" model (Wang et al., 1998) to evaluate neutral sugar degradation. The k value is 0.0025 yr^{-1} for ZT, 0.0021 yr^{-1} for LA, and 0.0025 yr^{-1} for XFJ. Thus, the k values of deoxy S / C5 indicate that the decomposition of 95 % neutral sugar in reservoir sediments will take thousands of years, which is similar to the results of TCHO in the ocean sediments (Wang et al., 1998) and the degradation of OM in the lacustrine sediments (Li et al., 2013).

As delineated above, the variations of OM and neutral sugars are related to trophic states, OM sources, and hydrological changes in different depositional environments. Moreover, the downcore OM profiles in some of the sediment cores investigated in other aquatic environments have not exhibited decreasing trends (Kirk et al., 2011; Meyers, 1997), which are not consistent with the traditional degradation model and mechanism. Hence, more works are needed for investigating sources, fractionations, and biodegradation products of carbohydrates in different aquatic environments.

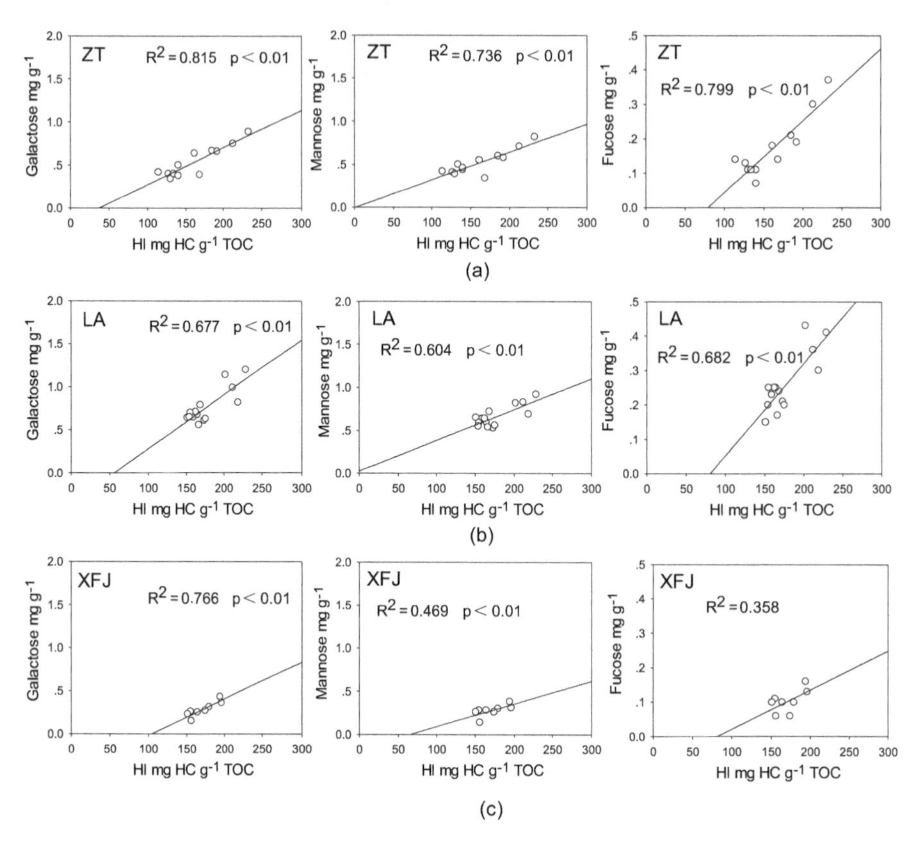

Figure 4. Relationship of HI with monosaccharide compounds in the ZT, LA, and XFJ reservoirs (for LA, the sample at 8–10 cm depths and at 30–34 cm depths were excluded; for XFJ, samples below 16 cm were excluded). The HI data are cited from Duan et al. (2015).

4.5 Effects of climate change on primary productivity and carbohydrates

As demonstrated above, a considerable amount of resistant structural carbohydrates could be preserved in the sediments. Moreover, HI has been widely utilized as a useful indicator of primary productivity during recent years (Gasse et al., 1991; Stein et al., 2006; Bechtel and Schubert, 2009). As shown in Fig. 4 and Table S4, the HI values in the ZT and LA cores are positively correlated with the concentrations of galactose, mannose, fucose, and arabinose, which are usually dominant in 89 cell walls of planktonic algae (Ittekkot and Arain, 1986; Hamilton and Hedges, 1988; D'souza et al., 2003; Hecky et al., 1973; Haug and Myklestad, 1976). For the XFJ core, significant correlations are also found between the concentrations of some monosaccharides and the HI values, except for rhamnose, fucose, and xylose. It is noted that a few samples at depths of 8–10 cm and 30–34 cm in the LA core and at depths of 18–34 cm in the XFJ core are excluded due to the inputs of allochthonous OM to the sediment cores. Therefore, monosaccharides (e.g., galactose and mannose) can be used for the reconstruction of historical productivity in the subtropical reservoirs.

As shown in Table S4, the contents of glucose, galactose, arabinose, mannose, fucose, and rhamnose are also positively correlated with S2 and HI values in the ZT, LA, and XFJ sediment cores. However, each of them shows weak correlations with the diagenetic parameters of neutral sugars such as % (fucose + rhamnose)^{b} and deoxy S / C5. Moreover, the xylose concentrations representing terrestrial inputs do not show significant correlations with the productivity proxies such as S2 and HI, and with any of the other monosaccharide concentrations at LA and XFJ. Thus, the above evidence supports the finding that the increasing neutral sugars in the reservoir sediments are mainly attributed to algal productivity rather than the degradation of neutral sugars during post-diagenesis.

In order to understand the effects of climate change on the historical variations of primary productivity, the T_5 values over 60 years are compared with each of the carbohydrate profiles in the three reservoirs (Fig. 5 and Table S4). The monosaccharide profiles at ZT and LA show a good correlation with T_5 during the past 60 years (for ZT, galactose: T_5, $R^2 = 0.824$, $p < 0.01$; mannose: T_5, $R^2 = 0.824$, $p < 0.01$; fucose: T_5, $R^2 = 0.805$, $p < 0.01$; for LA, galactose: T_5, $R^2 = 0.885$, $p < 0.01$; mannose: T_5, $R^2 = 0.699$, $p < 0.01$; fucose: T_5, $R^2 = 0.883$, $p < 0.01$;), suggesting that an increase in temperature enhances the deposition of carbohydrates in sediments (Fig. 5 and Table S4). Moreover, total nitrogen (TN) and total phosphorus (TP) concentrations show

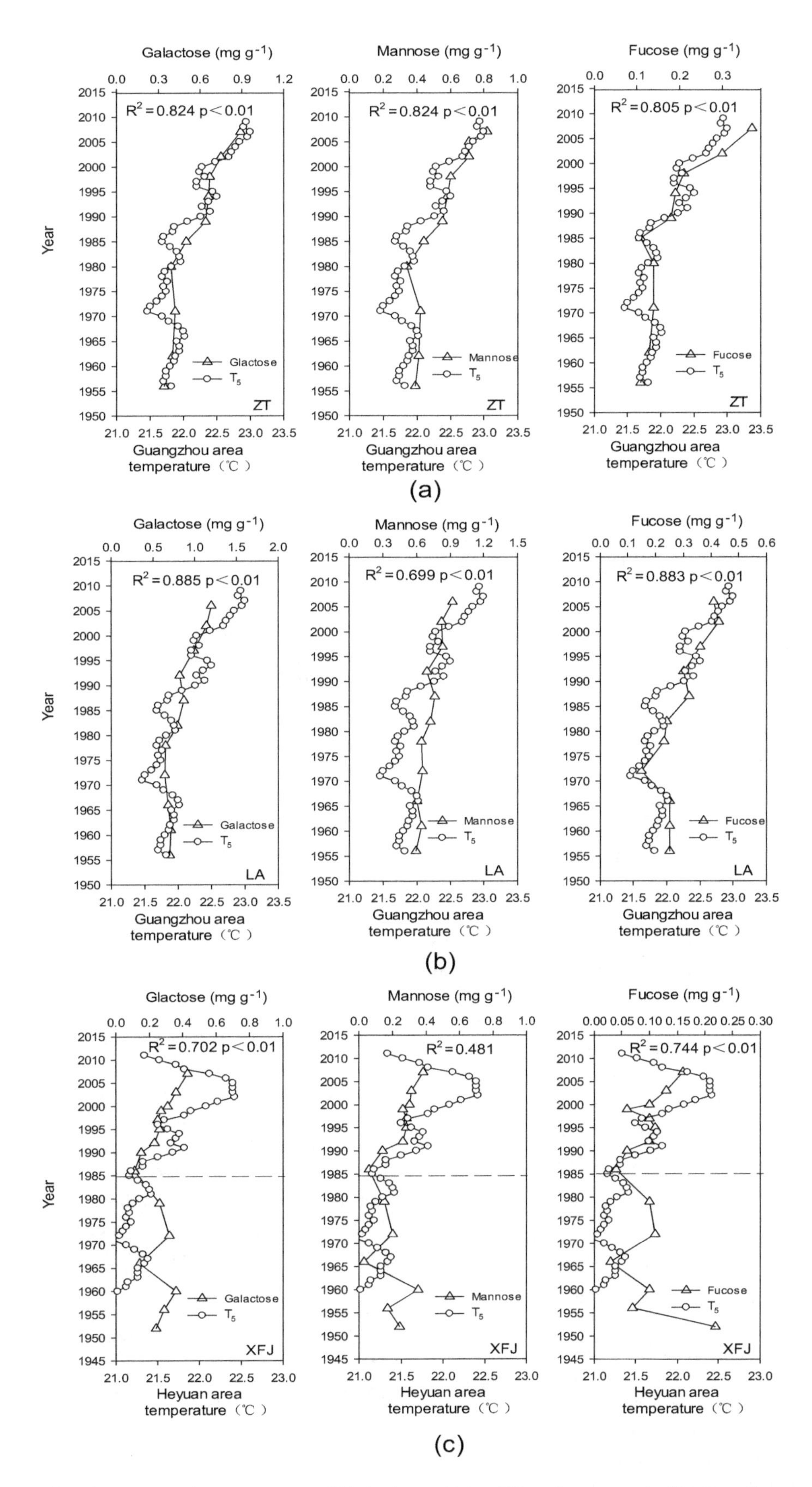

Figure 5. Temporal profiles of 5-year moving temperature (T_5), hydrogen index (HI) and monosaccharides in sediment cores from the three reservoirs.

weaker correlations with T_5 than carbohydrates for the sediment cores of ZT and LA (Table S5; Duan et al., 2015). The TN and TP concentrations can be used to reflect the historical inputs of nutrients in the ZT and LA reservoirs. Furthermore, the TP and TN concentrations at ZT and LA remained at a low level (mostly TP $< 0.1\,\mathrm{mg\,g^{-1}}$, TN $< 0.4\,\%$) during the past 6 decades and are far lower than those in sediments of other eutrophic reservoirs (Duan et al., 2015). Further evidence can be found from principal component analyses (PCAs) in Fig. S5. The T_5, HI, and monosaccharides are in the first principal component and account for 76.5 and 67.3 % of the total variance in the LA and XFJ reservoirs, whereas the second principal component of TP and TN accounts for 8.36 and 11.7 % of the total variance, respectively. Hence, the nutrient input is not the key factor affecting carbohydrates in the three reservoirs. In addition, the factor of light can also be excluded by the records of the annual hours of daylight, which show progressively decreasing trends from 1950 to 2010 at Guangzhou and Heyuan (Fig. S6). In conclusion, the increase in carbohydrates in the sediments at ZT and LA correlates very well with the increase in temperature (Fig. 5). For the XFJ reservoir, the profiles of monosaccharides also show a positive correlation with the temperature variations (for XFJ, galactose: T_5, $R^2 = 0.702$, $p < 0.01$; fucose: T_5, $R^2 = 0.744$, $p < 0.01$;), but show no relationship with TN or TP concentrations (Fig. 5, Table S4). Although the primary productivity is lower in the XFJ reservoir than in the ZT or LA reservoirs, it is still significantly affected by the increasing temperature. Although the TOC and S2 values have declined with increasing temperature in the XFJ core profile due to the phosphorus-limited trophic level, the monosaccharide contents still increase (Fig. 5). Therefore, the above results suggest that some monosaccharides derived from algae cell walls in the ZT, LA, and XFJ reservoirs are strongly associated with climate warming in the subtropical area.

As shown in Table S4, positive correlations among corrected δ^{13}C values, HI, monosaccharide contents, and T_5 are significant in the ZT core. However, the corrected δ^{13}C in the LA and XFJ cores are not correlated with other production parameters and T_5. This may be related to the impacts of organic matter degradation, terrestrial inputs, and human activities on the carbon isotopic composition (δ^{13}C) in mesotrophic LA and oligotrophic XFJ reservoirs. Moreover, the HI values and monosaccharide data show the same changing trends, and each of them is positively correlated with five T_5 in each of the three reservoirs. Thus, the HI parameters and monosaccharide contents are more reliable for reconstructing historical productivity in subtropical reservoirs. Therefore, the combined uses of these parameters are strongly recommended, which can help us to better understand the historical change in productivity in the subtropical reservoirs. It is challenging to find the appropriate indicators of primary production in aquatic ecosystems. More works are needed to investigate specific fractions of organic matter as productivity proxies. Meanwhile, multiple biomarker proxies are also employed to trace the source of biological productivity, and to rule out the impact of human activities. Moreover, compound-specific isotope ratios and biodegradation products of biomarkers such as neutral sugars and lipids could provide more information on algal organic matter in aquatic ecosystems. Furthermore, the modeling of the relationships between air temperature and algal organic matter parameters is worth being exploited and established.

5 Conclusions

The source, composition, and diagenesis of carbohydrates in the three sediment cores were investigated by conducting acid hydrolysis coupling with high-performance liquid chromatography (HPLC) with pulsed amperometric detection (PAD). Glucose, mannose, and galactose are the most abundant monosaccharides. Monosaccharide composition and diagnostic parameters (mannose / xylose ratio, arabinose plus galactose, xylose) indicate a predominant contribution of phytoplankton in the whole sediment cores of the ZT and LA reservoirs and in the upper layers of the XFJ core (0–16 cm). The carbohydrates are partially degraded during the settling. It was found that the degradation proxies (S2 / RC, % wt (fucose + rhamnose)$_b$, and deoxy S / C5) did not vary much in each of the downcore sediments.

The corrected δ^{13}C values can be used to reflect the historical changes in productivity. Elevated productivity derived from corrected δ^{13}C values was observed in the upper layers of the ZT and LA reservoirs, which is consistent with the increasing monosaccharides (e.g., galactose, mannose, fucose, and arabinose) and HI. But the corrected δ^{13}C values abruptly decrease with very low C/N ratios, which may indicate a different OM source and biodegradation in the downcore XFJ sediments. Moreover, strong positive correlations between some of the monosaccharides and HI were observed in both the mesotrophic ZT and LA reservoirs and in the oligotrophic XFJ reservoir, suggesting that the monosaccharides derived from algae cell walls and/or preserved in the mineral matrix are related to primary productivity in the studied subtropical reservoirs. Furthermore, increasing levels of carbohydrates in the three reservoir cores show significant relationships with T_5 during the last 60 years. Elevated temperatures lead to increasing levels of carbohydrates in the sediment profiles. Therefore, this investigation provides important evidence for the effect of climate change on the aquatic ecosystems in the low-latitude region. To further develop the productivity indicators of carbohydrates, more investigations are needed to understand their fractionation, biodegradation products, and compound-specific isotope ratios, and to improve their detection of other monosaccharides such as ribose.

Author contributions. JC and YR designed the experiments and DD, DZ, JW, and YY carried them out. DD prepared the manuscript with contributions from all co-authors.

Competing interests. The authors declare that they have no conflict of interest.

Acknowledgements. This study was supported by a project of the National Natural Science Foundation of China (41473103), a key project of NNSFC-Guangdong (U1201235), and a project of the Earmarked Foundation of the State Key Laboratory (SKLOG2016A06). We thank Ronald Benner and Michael Philben at the University of South Carolina for help with the analysis of neutral sugars conducted in Benner's laboratory. This is contribution no. IS-2434 from GIGCAS.

Edited by: Gerhard Herndl

References

Aufdenkampe, A. K., Mayorga, E., Hedges, J. I., Llerena, C., Quay, P. D., Gudeman, J., Krusche, A. V., and Richey, J. E.: Organic matter in the Peruvian headwaters of the Amazon: compositional evolution from the Andes to the lowland Amazon mainstem, Org. Geochem., 38, 337–364, 2007.

Bechtel, A. and Schubert, C. J.: A biogeochemical study of sediments from the eutrophic Lake Lugano and the oligotrophic Lake Brienz, Switzerland, Org. Geochem., 40, 1100–1114, 2009.

Benner, R. and Opsahl, S.: Molecular indicators of the sources and transformations of dissolved organic matter in the Mississippi river plume, Org. Geochem., 32, 597–611, 2001.

Boschker, H. T. S., Dekkers, E. M. J., Pel, R., and Cappenberg, T. E.: Sources of organic carbon in the littoral of Lake Gooimeer as indicated by stable carbon isotope and carbohydrate compositions, Biogeochemistry, 29, 89–105, 1995.

Burdige, D., Skoog, A., and Gardner, K.: Dissolved and particulate carbohydrates in contrasting marine sediments, Geochim. Cosmochim. Ac., 64, 1029–1041, 2000.

Carrie, J., Wang, F., Sanei, H., Macdonald, R. W., Outridge, P. M., and Stern, G. A.: Increasing contaminant burdens in an Arctic fish, Burbot (Lota lota), in a warming climate, Environ. Sci. Technol., 44, 316–322, 2010.

Chandra, S., Jake Vander Zanden, M., Heyvaert, A. C., Richards, B. C., Allen, B. C., and Goldman, C. R.: The effects of cultural eutrophication on the coupling between pelagic primary producers and benthic consumers, Limnol. Oceanogr., 50, 1368–1376, 2005.

Cowie, G. L. and Hedges, J. I.: Determination of neutral sugars in plankton, sediments, and wood by capillary gas chromatography of equilibrated isomeric mixtures, Anal. Chem., 56, 497–504, 1984.

Cowie, G. L. and Hedges, J. I.: Biochemical indicators of diagenetic alteration in natural organic matter mixtures, Nature, 369, 304–307, 1994.

Cowie, G. L., Hedges, J. I., and Calvert, S. E.: Sources and relative reactivities of amino acids, neutral sugars, and lignin in an intermittently anoxic marine environment, Geochim. Cosmochim. Ac., 56, 1963–1978, 1992.

da Cunha, L. C., Serve, L., and Blazi, J. L.: Neutral sugars as biomarkers in the particulate organic matter of a French Mediterranean river, Org. Geochem., 33, 953–964, 2002.

D'souza, F., Garg, A., and Bhosle, N. B.: Biogeochemical characteristics of sedimenting particles in Dona Paula Bay, India, Estuar. Coast. Shelf S., 58, 311–320, 2003.

Duan, D., Ran, Y., Cheng, H. F., Chen, J. A., and Wan, G. J.: Contamination trends of trace metals and coupling with algal productivity in sediment cores in Pearl River Delta, South China, Chemosphere, 103, 35–43, 2014.

Duan, D. D., Huang, Y. D., Cheng, H. F., and Ran, Y.: Relationship of polycyclic aromatic hydrocarbons with algae-derived organic matter in sediment cores from a subtropical region, J. Geophys. Res.-Biogeosci., 120, 2243–2255, 2015.

Gasse, F., Arnold, M., Fontes, J. C., Fort, M., Gibert, E., Huc, A., Li, B. Y., Li, Y. F., Liu, Q., Melieres, F., Van Campo, E., Wang, F. B., and Zhang, Q. S.: A 13,000-year climate record from western Tibet, Nature, 353, 742–745, 1991.

Gu, B., Schelske, C. L., and Hodell, D. A.: Extreme ^{13}C enrichments in a shallow hypereutrophic lake: Implications for carbon cycling, Limnol. Oceanogr., 49, 1152–1159, 2004.

Guggenberger, G., Christensen, B. T., and Zech, W.: Land-use effects on the composition of organic matter in particle-size separates of soil: I. Lignin and carbohydrate signature, Eur. J. Soil Sci., 45, 449–458, 1994.

Hambright, K. D., Gophen, M., and Serruya, S.: Influence of Long-Term Climatic Changes on the Stratification of a Subtropical, Warm Monomictic Lake, Limnol. Oceanogr., 39, 1233–1242, 1994.

Hamilton, S. E. and Hedges, J. I.: The comparative geochemistries of lignins and carbohydrates in an anoxic fjord, Geochim. Cosmochim. Ac., 52, 129–142, 1988.

Handa, N.: Carbohydrate metabolism in the marine diatom skeletonema costatum, Mar. Biol., 4, 208–214, 1969.

Haug, A. and Myklestad, S.: Polysaccharides of marine diatoms with special reference to Chaetoceros species, Mar. Biol., 34, 217–222, 1976.

He, B., Dai, M., Huang, W., Liu, Q., Chen, H., and Xu, L.: Sources and accumulation of organic carbon in the Pearl River Estuary surface sediment as indicated by elemental, stable carbon isotopic, and carbohydrate compositions, Biogeosciences, 7, 3343–3362, https://doi.org/10.5194/bg-7-3343-2010, 2010.

Hecky, R. E., Mopper, K., Kilham, P., and Degens, E. T.: The amino acid and sugar composition of diatom cell-walls, Mar. Biol., 19, 323–331, 1973.

Hedges, J. I., Cowie, G. L., Richey, J. E., Quay, P. D., Benner, R., Strom, M., and Forsberg, B. R.: Origins and processing of organic-matter in the Amazon River as indicated by carbohydrates and amino-acids, Limnol. Oceanogr., 39, 743–761, 1994.

Hernes, P. J., Hedges, J. I., Peterson, M. L., Wakeham, S. G., and Lee, C.: Neutral carbohydrate geochemistry of particulate material in the central equatorial Pacific, Deep-Sea Res. Pt. II, 43, 1181–1204, 1996.

Hicks, R. E., Owen, C. J., and Aas, P.: Deposition, resuspension, and decomposition of particulate organic matter in the sediments of

Lake Itasca, Minnesota, USA, Hydrobiologia, 284, 79–91, 1994.
Ittekkot, V. and Arain, R.: Nature of particulate organic matter in the river Indus, Pakistan, Geochim. Cosmochim. Ac., 50, 1643–1653, 1986.
Jensen, M. M., Holmer, M., and Thamdrup, B.: Composition and diagenesis of neutral carbohydrates in sediments of the Baltic-North Sea transition, Geochim. Cosmochim. Ac., 69, 4085–4099, 2005.
Kaiser, K. and Benner, R.: Determination of amino sugars in environmental samples with high salt content by high-performance anion-exchange chromatography and pulsed amperometric detection, Anal. Chem., 72, 2566–2572, 2000.
Kaiser, K. and Benner, R.: Biochemical composition and size distribution of organic matter at the Pacific and Atlantic time-series stations, Mar. Chem., 113, 63–77, 2009.
Keeling, C. D.: The Suess effect: 13 carbon-14 carbon interrelations, Environ. Int., 2, 229–300, 1979.
Keil, R. G., Tsamakis, E., Giddings, J. C., and Hedges, J. I.: Biochemical distributions (amino acids, neutral sugars, and lignin phenols) among size-classes of modern marine sediments from the Washington coast region, Geochim. Cosmochim. Ac., 62, 1347–1364, 1998.
Khodse, V. B. and Bhosle, N. B.: Nature and sources of suspended particulate organic matter in a tropical estuary during the monsoon and pre-monsoon: insights from stable isotopes (δ^{13}C POC, δ^{15}N TPN) and carbohydrate signature compounds, Mar. Chem., 145, 16–28, 2012.
Khodse, V. B., Fernandes, L., Gopalkrishna, V., Bhosle, N. B., Fernandes, V., Matondkar, S., and Bhushan, R.: Distribution and seasonal variation of concentrations of particulate carbohydrates and uronic acids in the northern Indian Ocean, Mar. Chem., 103, 327–346, 2007.
Kirk, J. L., Muir, D. C., Antoniades, D., Douglas, M. S., Evans, M. S., Jackson, T. A., Kling, H., Lamoureux, S., Lim, D. S. S., Pienitz, R., Smol, J. P., Stewart, K., Wang, X. W., and Yang, F.: Climate change and mercury accumulation in Canadian high and subarctic lakes, Environ. Sci. Technol., 45, 964–970, 2011.
Lehmann, M. F., Bernasconi, S. M., Barbieri, A., and McKenzie, J. A.: Preservation of organic matter and alteration of its carbon and nitrogen isotope composition during simulated and in situ early sedimentary diagenesis, Geochim. Cosmochim. Ac., 66, 3573–3584, 2002.
Li, H., Minor, E. C., and Zigah, P. K.: Diagenetic changes in Lake Superior sediments as seen from FTIR and 2D correlation spectroscopy, Org. Geochem., 58, 125–136, 2013.
Meyers, P. A.: Organic geochemical proxies of paleoceanographic, paleolimnologic, and paleoclimatic processes, Org. Geochem., 27, 213–250, 1997.
Philben, M., Holmquist, J., MacDonald, G., Duan, D., Kaiser, K., and Benner, R.: Temperature, oxygen, and vegetation controls on decomposition in a James Bay peatland, Global Biogeochem. Cy., 29, 729–743, 2015.
Moers, M., Baas, M., De Leeuw, J., and Schenck, P.: Analysis of neutral monosaccharides in marine sediments from the equatorial eastern Atlantic, Org. Geochem., 15, 367–373, 1990.
Moers, M., Jones, D., Eakin, P., Fallick, A., Griffiths, H., and Larter, S.: Carbohydrate diagenesis in hypersaline environments: application of GC-IRMS to the stable isotope analysis of derivatized saccharides from surficial and buried sediments, Org. Geochem., 20, 927–933, 1993.
Ogier, S., Disnar, J. R., Albéric, P., and Bourdier, G.: Neutral carbohydrate geochemistry of particulate material (trap and core sediments) in an eutrophic lake (Aydat, France), Org. Geochem., 32, 151–162, 2001.
Opsahl, S. and Benner, R.: Characterization of carbohydrates during early diagenesis of five vascular plant tissues, Org. Geochem., 30, 83–94, 1999.
Outridge, P., Stern, G., Hamilton, P., Percival, J., McNeely, R., and Lockhart, W.: Trace metal profiles in the varved sediment of an Arctic lake, Geochim. Cosmochim. Ac., 69, 4881–4894, 2005.
Outridge, P., Sanei, H., Stern, G., Hamilton, P., and Goodarzi, F.: Evidence for control of mercury accumulation rates in Canadian High Arctic lake sediments by variations of aquatic primary productivity, Environ. Sci. Technol., 41, 5259–5265, 2007.
O'Reilly, C. M., Dettman, D. L., and Cohen, A. S.: Paleolimnological investigations of anthropogenic environmental change in Lake Tanganyika: VI. Geochemical indicators, J. Paleolimnol., 34, 85–91, 2005.
Panagiotopoulos, C., Sempéré, R., Para, J., Raimbault, P., Rabouille, C., and Charrière, B.: The composition and flux of particulate and dissolved carbohydrates from the Rhone River into the Mediterranean Sea, Biogeosciences, 9, 1827–1844, https://doi.org/10.5194/bg-9-1827-2012, 2012.
Parsons, T. R., Takahashi, M., and Hargrave, B.: Biological Oceanographic Processes, Pergamon Press, Oxford, New York, 1984.
Routh, J., Meyers, P. A., Gustafsson, O., Baskaran, M., Hallberg, R., and Schöldström, A.: Sedimentary geochemical record of human-induced environmental changes in the Lake Brunnsviken watershed, Sweden, Limnol. Oceanogr., 49, 1560–1569, 2004.
Schelske, C. L. and Hodell, D. A.: Recent changes in productivity and climate of Lake Ontario detected by isotopic analysis of sediments, Limnol. Oceanogr., 36, 961–975, 1991.
Schelske, C. L. and Hodell, D. A.: Using carbon isotopes of bulk sedimentary organic matter to reconstruct the history of nutrient loading and eutrophication in Lake Erie, Limnol. Oceanogr., 40, 918–929, 1995.
Skoog, A. and Benner, R.: Aldoses in various size fractions of marine organic matter: Implications for carbon cycling, Limnol. Oceanogr., 42, 1803–1813, 1997.
Skoog, A., Alldredge, A., Passow, U., Dunne, J., and Murray, J.: Neutral aldoses as source indicators for marine snow, Mar. Chem., 108, 195–206, 2008.
Smol, J. P., Wolfe, A. P., Birks, H. J. B., Douglas, M. S., Jones, V. J., Korhola, A., Pienitz, R., Rühland, K., Sorvari, S., and Antoniades, D.: Climate-driven regime shifts in the biological communities of Arctic lakes, Proc. Natl. Acad. Sci. USA, 102, 4397–4402, 2005.
Stein, R., Boucsein, B., and Meyer, H.: Anoxia and high primary production in the Paleogene central Arctic Ocean: First detailed records from Lomonosov Ridge, Geophys. Res. Lett., 33, L18606, https://doi.org/10.1029/2006GL026776, 2006.
Stern, G., Sanei, H., Roach, P., Delaronde, J., and Outridge, P.: Historical interrelated variations of mercury and aquatic organic matter in lake sediment cores from a subarctic lake in Yukon, Canada: further evidence toward the algal-mercury scavenging hypothesis, Environ. Sci. Technol., 43, 7684–7690, 2009.

Unger, D., Ittekkot, V., Schäfer, P., and Tiemann, J.: Biogeochemistry of particulate organic matter from the Bay of Bengal as discernible from hydrolysable neutral carbohydrates and amino acids, Mar. Chem., 96, 155–184, 2005.

Verburg, P.: The need to correct for the Suess effect in the application of $\delta^{13}C$ in sediment of autotrophic Lake Tanganyika, as a productivity proxy in the Anthropocene, J. Paleolimnol., 37, 591–602, 2007.

Wang, X. C., Druffel, E. R., Griffin, S., Lee, C., and Kashgarian, M.: Radiocarbon studies of organic compound classes in plankton and sediment of the northeastern Pacific Ocean, Geochim. Cosmochim. Ac., 62, 1365–1378, 1998.

14

Modelled estimates of spatial variability of iron stress in the Atlantic sector of the Southern Ocean

Thomas J. Ryan-Keogh[1,2], **Sandy J. Thomalla**[1], **Thato N. Mtshali**[1], **and Hazel Little**[2]

[1]Southern Ocean Carbon and Climate Observatory, Natural Resources and Environment, CSIR, Rosebank, Cape Town 7700, South Africa
[2]Department of Oceanography, University of Cape Town, Rondebosch, Cape Town 7701, South Africa

Correspondence to: Thomas J. Ryan-Keogh (thomas.ryan-keogh@uct.ac.za)

Abstract. The Atlantic sector of the Southern Ocean is characterized by markedly different frontal zones with specific seasonal and sub-seasonal dynamics. Demonstrated here is the effect of iron on the potential maximum productivity rates of the phytoplankton community. A series of iron addition productivity versus irradiance (*PE*) experiments utilizing a unique experimental design that allowed for 24 h incubations were performed within the austral summer of 2015/16 to determine the photosynthetic parameters α^B, P^B_{max} and E_k. Mean values for each photosynthetic parameter under iron-replete conditions were 1.46 ± 0.55 (µg (µg Chl *a*)$^{-1}$ h^{-1} (µM photons m^{-2} s^{-1})$^{-1}$) for α^B, 72.55 ± 27.97 (µg (µg Chl *a*)$^{-1}$ h^{-1}) for P^B_{max} and 50.84 ± 11.89 (µM photons m^{-2} s^{-1}) for E_k, whereas mean values under the control conditions were 1.25 ± 0.92 (µg (µg Chl *a*)$^{-1}$ h^{-1} (µM photons m^{-2} s^{-1})$^{-1}$) for α^B, 62.44 ± 36.96 (µg (µg Chl *a*)$^{-1}$ h^{-1}) for P^B_{max} and 55.81 ± 19.60 (µM photons m^{-2} s^{-1}) for E_k. There were no clear spatial patterns in either the absolute values or the absolute differences between the treatments at the experimental locations. When these parameters are integrated into a standard depth-integrated primary production model across a latitudinal transect, the effect of iron addition shows higher levels of primary production south of 50° S, with very little difference observed in the subantarctic and polar frontal zone. These results emphasize the need for better parameterization of photosynthetic parameters in biogeochemical models around sensitivities in their response to iron supply. Future biogeochemical models will need to consider the combined and individual effects of iron and light to better resolve the natural background in primary production and predict its response under a changing climate.

1 Introduction

Phytoplankton primary production (PP) in the Southern Ocean is a key contributor to global atmospheric CO_2 drawdown, responsible for 30–40 % of global anthropogenic carbon uptake (Khatiwala et al., 2009; Mikaloff Fletcher et al., 2006; Schlitzer, 2002). High nutrient availability fuels this phytoplankton production, but growth is ultimately constrained by the lack of availability of the micronutrient iron (Fe) (de Baar et al., 1990; Martin et al., 1990). This leads to high levels of macronutrients that remain unutilized by phytoplankton growth in what is known as a high-nutrient, low-chlorophyll (HNLC) conditions. Maximum primary productivity rates of the Southern Ocean are also limited by light availability due to low incident solar angles, persistent cloud cover and deep mixed layers that curtail production and subsequently affect the efficiency of the biological carbon pump. Under future climate change scenarios, altered upwelling and mixed layer stratification (Boyd et al., 2001; Boyd and Doney, 2002), changes in sea ice cover (Close and Goosse, 2013; de Lavergne et al., 2014; Montes-Hugo et al., 2008; Zhang, 2007) and food-web dynamics (Dubischar and Bathmann, 1997; Moore et al., 2013; Pakhomov and Froneman, 2004; Smetacek et al., 2004) will alter both the nutrient and light supply, strongly impacting primary production rates. As such, it is important that we understand the sensitivity of phytoplankton production to light and micronutrient availability so that we may improve our predictive capability of the response of the Southern Ocean carbon pump to a changing climate.

Iron plays a critical role in modulating PP due to the high requirements of the photosynthetic apparatus, photosystems I and II (Quigg et al., 2003; Raven, 1990; Strzepek and Harrison, 2004; Twining and Baines, 2013). Light availability can further increase the demand for iron, as low irradiance levels increase requirements associated with the synthesis of additional photosynthetic units to increase potential light absorption (Maldonado et al., 1999; Raven, 1990; Strzepek et al., 2012; Sunda and Huntsman, 1997). Iron is also required to activate both nitrate and nitrite reductase (de Baar et al., 2005), which facilitate the assimilation of nitrate and nitrite and their subsequent intracellular reduction to ammonium. In HNLC regions, such as the Southern Ocean, nitrate uptake rates (ρNO_3^-) have also frequently been reported as becoming iron-limited (Cochlan, 2008; Lucas et al., 2007; Moore et al., 2013; Price et al., 1994). However, it has also been demonstrated that iron limitation rather than inhibiting nitrate reductase activity results in a bottleneck further downstream due to a reduction in photosynthetically derived reductant (Milligan and Harrison, 2000). This would lead to an excretion of excess nitrate back into the water column that would further contribute to HNLC conditions such as those present in the Southern Ocean.

Estimating PP in the oceans towards an improved understanding of the effects of iron and light limitation requires an understanding of the relationship between photosynthesis (P) and irradiance (E) (Behrenfeld and Falkowski, 1997b; Dower and Lucas, 1993; Platt et al., 2007). *PE* responses are derived from an equation by Platt et al. (1980), where the responses are parameterized as a function of irradiance. The parameters derived include P^{B}_{max}, the biomass-specific rate of photosynthesis at saturating irradiances; α^{B}, the irradiance-limited biomass-specific initial slope; and E_k, the irradiance at which saturation is initiated. The response of these parameters can be not only a function of temperature (Behrenfeld and Falkowski, 1997b) but also as a change in the quantum efficiency of photosynthesis, usually as the result of changes in iron availability. In previous iron fertilization experiments a doubling of α^{B} has been reported (Hiscock et al., 2008), yet this response is not consistent across Southern Ocean waters (Feng et al., 2010; Hopkinson et al., 2007; Moore et al., 2007; Smith Jr. and Donaldson, 2015). Given their relative importance within PP models (Behrenfeld and Falkowski, 1997a, b; Sathyendranath and Platt, 2007), a greater understanding of the drivers of the variability within these photosynthetic parameters is therefore required, particularly if we are to accurately quantify and constrain PP in the Southern Ocean to examine seasonal and interannual variability and trends.

The Atlantic sector of the Southern Ocean is composed of a series of circumpolar fronts that are characterized by large geostrophic velocities (Nowlin and Klinck, 1986; Orsi et al., 1995). The fronts constrain water masses with distinct physical and chemical properties that define different oceanographic zones. These spatial zones display not only zonal variability with the fronts but also display important seasonal contrasts (Thomalla et al., 2011), with differing bloom initiation dates and temporal extent of bloom duration. Whilst the bloom initiation dates can in part be explained by day length and sea ice cover further polewards, the differences in the extent and duration of blooms between the zones requires an alternative and more nuanced explanation. One theory that has been postulated is that the supply mechanisms of iron to the mixed layer following the spring bloom vary between zones (Thomalla et al., 2011). Weak diapycnal inputs and a heavy reliance on iron recycling was suggested by Tagliabue et al. (2014) to match approximate phytoplankton utilization within the pelagic zones. An alternative theory that postulates the importance of summer storms may also be pivotal in understanding the seasonal dynamics of phytoplankton primary productivity (Nicholson et al., 2016; Swart et al., 2015; Thomalla et al., 2015), with respect to the sustained bloom observed in the sub-Antarctic Zone (SAZ). Here, summer storms are said to periodically deepen the mixed layer to below the ferricline followed by rapid shoaling during quiescent periods that balances the supply of light and iron in the upper oceans favouring phytoplankton growth that culminates in a sustained summer bloom (Swart et al., 2015). Regardless of the mechanisms at play, an understanding of when and where iron concentrations and supply mechanisms limits potential phytoplankton growth and productivity is needed to better understand the drivers that determine the characteristics of the Southern Ocean seasonal cycle.

To this end, a research cruise was conducted in the austral summer of 2015/16 as part of the third multidisciplinary Southern Ocean Seasonal Cycle Experiment (SOSCEx III), which aimed to identify and understand the physical and chemical controls on the seasonal cycle of the biological carbon pump. As part of this study, shipboard nutrient addition *PE* experiments were performed to determine the extent of iron limitation upon phytoplankton primary production.

2 Materials and methods

2.1 Oceanographic sampling

The samples and data presented here were obtained during the 55th South African National Antarctic Expedition (3 December 2015 to 11 February 2016) on board the *S.A. Agulhas II* to the Atlantic sector of the Southern Ocean as part of SOSCEx III (Swart et al., 2012). During the cruise, six nutrient addition *PE* long-term experiments were performed within the Atlantic sector of the Southern Ocean (Fig. 1) to determine the extent to which relief from iron limitation could alter the maximal primary productivity rates of the phytoplankton community. Uncontaminated whole seawater was collected from 30 to 50 m depth using Teflon-lined,

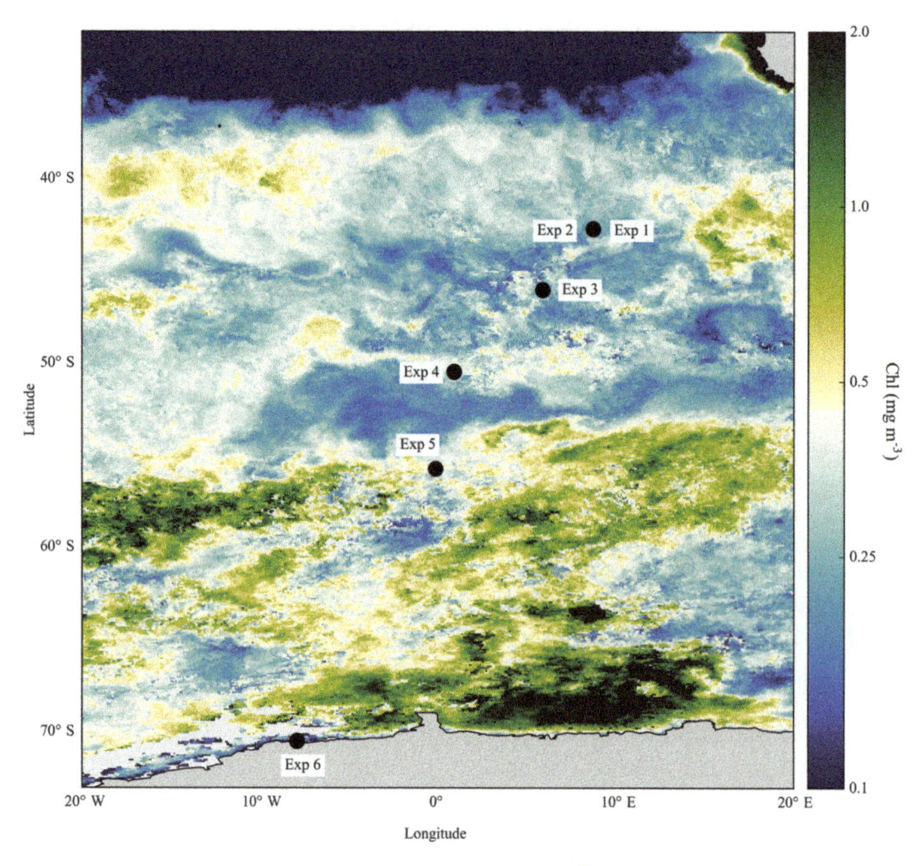

Figure 1. Composite map of MODIS (8 days, 9 km) derived chlorophyll ($mg\,m^{-3}$) from November 2015 to March 2016 for the Atlantic sector of the Southern Ocean with locations of the nutrient addition productivity versus irradiance (*PE*) experiments.

external closure 12 L Go-Flo samplers deployed on a trace metal clean CTD rosette system.

2.2 *PE* experimental setup

Phytoplankton productivity was measured by the incorporation of ^{13}C stable isotopes in response to an increasing light gradient. Inside a trace metal clean laboratory class-100 container, bulk trace metal clean seawater was decanted unscreened into an acid-washed 50 L LDPE carboy (Thermo Scientific) to ensure homogenization; this was then redistributed into acid-cleaned 1.0 L polycarbonate bottles (Nalgene). All experimental conditions were conducted and carried out following trace metal clean standards and conditions. Sample manipulations were conducted under a laminar flow hood. All bottles were inoculated with ^{13}C ($10\,\mu M\,NaH_2^{13}CO_3\,100\,mL^{-1}$) spikes to achieve an enrichment of $\sim 5\,\%$; 11 bottles received the addition of $FeCl_3$ (+2.0 nM, "Fe"), whereas 11 bottles received the ^{13}C spikes alone ("Control"). The bottles were incubated in screened (LEE Filters) LDPE boxes within light- and temperature-controlled incubators. Experimental temperature was set to mimic the in situ sample collection temperature. Irradiances were measured within the screened boxes using a handheld 4π PAR sensor (Biospherical Instruments) and ranged from 0–400 μM photons $m^{-2}\,s^{-1}$. Bottles tops were covered with parafilm and double-bagged with clear polyethylene bags to minimize contamination risks during the incubation. Due to physical constraints, the experiments were not conducted as triplicates, and as such evaluation of the precision/error within experiments is not possible.

Experiments were incubated for 24 h, after which the samples were vacuum filtered through a pre-combusted (400 °C for 24 h) GF/F filter. Samples were acid-fumed with concentrated HCl for 24 h to remove inorganic carbon before being dried in an oven at 40 °C for 24 h. The isotopic composition of all samples were determined by mass spectrometry on a Flash EA 1112 series elemental analyser (Thermo Finnigan). Carbon uptake rates ($\mu M\,C\,h^{-1}$) were calculated from the equation of Dugdale and Wilkerson (1986), utilizing in situ determinations of dissolved inorganic carbon (DIC). The uptake rates normalized to the chlorophyll *a* (Chl) concentration, were used to calculate the maximal light-saturated Chl-specific photosynthetic fixation rates (P^B_{max}), the light-limited slope (α^B) and the photoacclimation parameter (E_k). The curves and parameters were generated using a non-linear least squares fit to the equation of Platt et al. (1980).

2.3 Chlorophyll *a* and nutrient analysis

Samples for Chl analysis, 250 mL, were filtered onto GF/F filters and then extracted into 90 % acetone for 24 h in the dark at −20 °C, followed by analysis with a fluorometer (TD70; Turner Designs) (Welschmeyer, 1994). Macronutrient samples were drawn into 50 mL diluvials and stored at −20 °C until analysis on land. Nitrate + nitrite and silicate were measured using a Lachat flow injection analyser (Egan, 2008; Wolters, 2002), whilst nitrite and phosphate were determined manually by colorimetric method as specified by Grasshoff et al. (1983). Dissolved iron samples (DFe) were carefully collected in acid-washed 125 mL LDPE bottles, acidified with 30 % HCl Suprapur to pH ∼ 1.7 (using $2\,\mathrm{mL\,L^{-1}}$ criteria) and stored at room temperature until analysis on land at UniBrest in France using the chemiluminescence–flow injection analyser (CL-FIA) method (Obata et al., 1993). Accuracy and precision of the method was verified by analysis of in-house internal standards and SAFe reference seawater samples (Johnson et al., 2007); the limits of detection were on the order of 10 pM.

2.4 Phytoplankton photosynthetic physiology

Variable Chl fluorescence was measured using a Chelsea Scientific Instruments FastOcean fast repetition rate fluorometer (FRRf) integrated with a FastAct laboratory system. Samples were acclimated in dark bottles at in situ temperatures, and FRRf measurements were blank-corrected using carefully prepared 0.2 µm filtrates for all samples (Cullen and Davis, 2003). Protocols for FRRf measurements consisted of the following: 100 × 2 µs saturation flashlets with a 2 µs interval, followed by 25 × 1 µs relaxation flashlets with an interval of 84 µs with a sequence interval of 100 ms. Sequences were repeated 32 times, resulting in an acquisition length of 3.2 s. The power of the excitation LED (λ450) was adjusted between samples to saturate the observed fluorescence transients within a given range of $R\sigma_{\mathrm{PSII}}$. $R\sigma_{\mathrm{PSII}}$, the probability of a reaction centre being closed during the first flashlet, is optimized between 0.042 and 0.064 per the manufacturer specifications. By adopting this approach, it ensures the best signal-to-noise ratio in the recovered parameters whilst accommodating significant variations in the photophysiology of the phytoplankton community without having to adjust the protocol. Data from the FRRf were analysed to derive fluorescence parameters as defined in Baker et al. (2001) and Roháček (2002) by fitting transients to the model of Kolber et al. (1998).

2.5 Pigment analysis and CHEMTAX

Pigment samples were collected by filtering 0.5–2.0 L of water onto GF/F filters. Filters were frozen and stored at −80 °C until analysis in Villefranche, France, on an Agilent Technologies HPLC 1200. Filters were extracted in 100 % methanol, disrupted by sonication, clarified by filtration and analysed by HPLC following the methods of Ras et al. (2008). Limits of detection were on the order of $0.1\,\mathrm{ng\,L^{-1}}$. Pigment composition data were standardized through root square transformation before cluster analysis utilizing multi-dimensional scaling where similar samples appear together and dissimilar samples do not. Samples were grouped and analysed in CHEMTAX (Mackey et al., 1996) using the pigment ratios from Gibberd et al. (2013). Multiple iterations of pigment ratios were used to reduce uncertainty in the taxonomic abundance as described in Gibberd et al. (2013), with the solution that had the smallest residual used for the estimated taxonomic abundance.

2.6 Particle size analysis

The size distribution of the particle population was measured by running 40 mL of water sample through a 100 µm aperture on a Beckman Coulter multisizer (20 runs at 2.0 mL per run), binning the size counts into 400 bins between 2 and 60 µm. Data were subsequently analysed utilizing custom MATLAB scripts to calculate the effective diameter of particles within the sample following Hansen and Travis (1974).

2.7 Depth-integrated production

Water column primary production rates were calculated according to Platt et al. (1980) and Platt and Sathyendranath (1993) as in Thomalla et al. (2015), where

$$\mathrm{PP}_0 = P_{\max} \times \left(1 - e^{\left(\frac{-\alpha \times E_0^{\mathrm{m}} \times 0.5}{P_{\max}}\right)}\right). \quad (1)$$

PP_0 ($\mathrm{mg\,C\,m^{-2}\,d^{-1}}$) is the primary production at the surface, $P_{\max}$ the maximal light-saturated photosynthetic fixation rate, α the light-limited slope and E_0^{m} is daily PAR at the surface, calculated by assuming maximum PAR at midday, zero PAR at sunrise and sunset, a constant gradient of light between time steps and extrapolating the measured PAR (from an above-water Biospherical 4π PAR sensor at the time of the station into an isosceles triangle; see also Thomalla et al., 2015).

$$E_*^{\mathrm{m}} = \frac{E_0^{\mathrm{m}}}{E_{\mathrm{k}}} \quad (2)$$

The results were generalized by calculating $E_*^{\mathrm{m}}(2)$, the dimensionless daily surface irradiance, while primary productivity over the entire water column $\mathrm{PP_{wc}}$ ($\mathrm{mg\,C\,m^{-2}\,d^{-1}}$) was calculated with the following Eq. (3). The dimensionless function $f(E_*^{\mathrm{m}})$ for daily primary productivity was solved analytically by Platt et al. (1980). Rates were calculated for both the iron addition and control treatments, allowing the difference between the integrated rates to be solved.

$$\mathrm{PP_{wc}} = \mathrm{PP}_0 \times \frac{f(E_*^{\mathrm{m}})}{k_{\mathrm{d}}} \quad (3)$$

Table 1. Locations for *PE* experiments conducted during the cruise along with details for the initial chemical, physiological and physical setup conditions.

Experiment	1	2	3	4	5	6
Initiation date	08/12/2015	05/01/2016	07/01/2016	08/01/2016	09/01/2016	26/01/2016
Latitude (° S)	−42.69	−42.69	−45.99	−50.45	−55.70	−70.44
Longitude (° E/W)	08.74	08.74	05.93	01.04	−00.00	−07.82
Collection depth (m)	30	35	35	35	50	35
Sunrise–sunset	03:30–18:30	04:00–19:00	04:00–19:00	04:00–19:00	04:00–19:00	00:00–00:00[a]
Chl ($\mu g\,L^{-1}$)	0.97	0.84	0.89	2.30	1.15	1.49
Nitrate (μM)	7.21	10.20	15.83	21.07	17.02	23.81
Silicate (μM)	0.86	0.72	0.09	3.76	30.83	48.81
Phosphate (μM)	0.88	0.76	0.95	1.28	1.11	0.94
DFe (nM)	0.16	0.17	0.07	0.03	0.05	0.10
F_v/F_m	0.19	0.30	0.35	0.30	0.35	0.37
σ_{PSII} (nm^{-2})	14.79	6.45	5.50	5.59	5.37	3.89
MLD (m)	33.77	56.96	108.42	70.11	42.89	40.80
Salinity	33.87	33.70	33.88	33.80	33.73	33.72
Temp. (°C)	10.80	10.44	6.72	3.17	−1.42	−1.51
Average daytime PAR (μM photons $m^{-2}\,s^{-1}$)[b]	1055.31	787.35	289.18	524.41	769.87	673.62
Euphotic depth (m)	72.79	75.10	52.95	47.92	69.13	78.07

[a] 24 h day length; [b] see Sect. 2.7 for details.

K_d was initially calculated as the slope of the natural log of in situ PAR with depth from CTD profiles. When in situ PAR with depth was not available, K_d was also calculated from in situ surface Chl concentrations with the following Eq. (4) (Morel, 1988; Morel et al., 2007). Co-located calculations utilizing in situ PAR versus chlorophyll-derived K_d demonstrated on average a 40 % higher K_d when calculated with chlorophyll.

$$K_d = 0.0166 + 0.0773 \times [\mathrm{Chl}]^{0.6715} \quad (4)$$

2.8 Ancillary physical data

Temperature and salinity profiles were obtained from a Sea-Bird CTD mounted on the rosette system. The mixed layer depth (MLD) was calculated following de Boyer Montégut et al. (2004), which identifies the MLD as the depth where the temperature differs from the temperature at 10 m by more than 0.2 °C ($\Delta T_{10\,m} = 0.2$ °C). The position of the fronts was determined using sea surface height (SSH) data from maps of absolute dynamic topography (MADT) according to Swart et al. (2010).

3 Results

3.1 Oceanographic context

The experimental setup locations covered a wide range of pelagic zones from the SAZ to the marginal ice zone (MIZ), each with different physical, chemical and biological properties (see Table 1). Chl concentrations between experiment initiation locations varied between 0.84 and 2.30 $\mu g\,L^{-1}$, peaking just south of the polar front at ~ 50° S. Initial temperatures displayed a characteristic decrease from 10.80 °C at the most northerly location to −1.51 °C at the MIZ, whereas there were no distinct differences in salinity ranging from 33.70 to 33.88. Macronutrient concentrations all increased polewards, with peaks of 28.15, 1.34 and 48.81 μM for nitrate, phosphate and silicate respectively. Dissolved iron concentrations decreased polewards from a maximum of 0.17 nM in the SAZ to minimum values of 0.03 and 0.05 nM at 50 and 55° S respectively, before increasing again in the MIZ to 0.10 nM.

Phytoplankton photophysiology, F_v/F_m, increased polewards from a minimum of 0.19 to a maximum of 0.37, whereas σ_{PSII}, the effective absorption cross section of PSII, decreased polewards from 14.79 to 3.89 nm^{-2}. The effective diameter of the phytoplankton population, a relative measure of size, increased polewards from a minimum of 4.29 ± 0.35 μm in the SAZ to a maximum of 8.59 ± 0.68 μm in the MIZ. Estimated taxonomic abundance through HPLC analysis and CHEMTAX determined that the dominant groups at all stations were either diatoms, haptophytes or a mix of the two. Haptophytes were the dominant group (> 68 % of total Chl) in the SAZ during experiments 1 and 2, with diatoms becoming dominant (> 70 % of total Chl) from experiment 4 onwards.

MLDs were highly variable and ranged from ~ 34 m at experiment 1 to ~ 108 m at experiment 3. The MLD was typically deeper than the experimental setup depth (average difference of ~ 15 m) at all experiments except for experiment

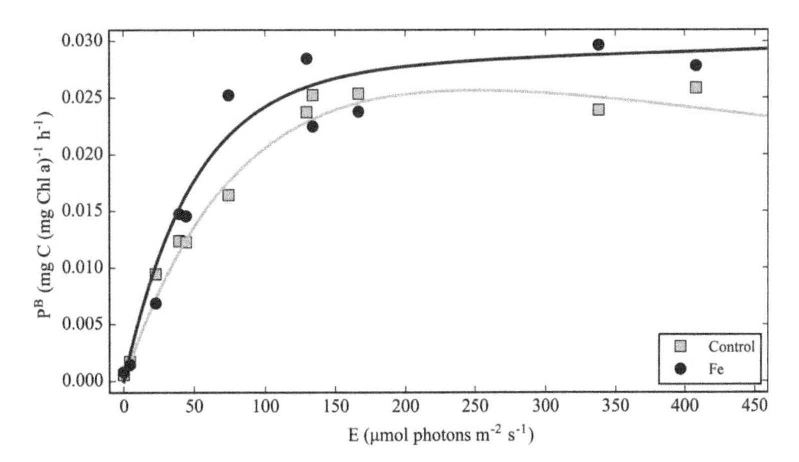

Figure 2. An example of a *PE* curve of productivity (mg C (mg Chl *a*)$^{-1}$ h^{-1}) versus irradiance (µmol photons m^{-2} s^{-1}), with (Fe) and without (Control) the addition of iron; the lines represent a non-linear least squares fit to the equation of Platt et al. (1980).

Table 2. Summary of *PE* parameters, α^B (mg (mg Chl *a*)$^{-1}$ h^{-1} (µmol photons m^{-2} s^{-1})$^{-1}$), P^B_{max} (mg (mg Chl *a*)$^{-1}$ h^{-1}) and E_k (µmol photons m^{-2} s^{-1}), for the ρC nutrient addition experiments.

	Experiment	1	2	3	4	5	6
	$\alpha^B_{(Fe)}$ ($\times 10^{-3}$)	1.73	2.23	1.23	1.56	1.43	0.56
	$\alpha^B_{(Control)}$ ($\times 10^{-3}$)	2.43	2.16	1.19	1.21	0.13	0.37
ρC	$P^B_{max\,(Fe)}$ ($\times 10^{-2}$)	10.67	9.30	8.46	6.22	6.04	2.86
	$P^B_{max(Control)}$ ($\times 10^{-2}$)	9.23	9.14	9.48	5.99	1.06	2.56
	$E_{k\,(Fe)}$	61.52	41.72	68.59	39.80	42.29	51.12
	$E_{k\,(Control)}$	38.03	42.40	79.77	49.46	83.21	69.37

5, where the collection depth was 7 m below the MLD. The CTD density profile at experiment 5 was indicative of two mixed layers present, with the experiment performed above the deeper of the mixed layers (~ 56 m). Experiments 1 and 2 that were set up in the same location in the SAZ but 28 days apart had markedly different setup conditions: a 41 % increase in the nitrate concentration from 7.21 to 10.20 µM, a 2-fold increase in F_v/F_m from 0.19 to 0.35 with a concurrent 56 % decrease in σ_{PSII} from 14.79 to 6.45 nm^{-2} and a deepening of the MLD from ~ 34 to ~ 57 m.

The light environment within the water column at each location was determined by calculating the percentage light depth as a function of the vertical attenuation coefficient of irradiance (K_d). The percentage light depths of the experiments ranged between 3.46 and 14.78 %. The 1 % light depth, which typically coincides with the compensation light depth i.e. the depth where rates of production equate to rates of respiration, is consistently below the MLD, except for experiment 4, where it was 22 m above the mixed layer.

3.2 *PE* parameters

PE curves for carbon uptake (ρC) (Figs. 2 and S1 in the Supplement), summarized in Table 2, display consistent results with greater values of α^B and P^B_{max} with the addition of iron compared to the control treatments (Figs. S2–S3). The *PE* curves for the control treatments did not display any significant outliers (r^2 => 95 %), we can assume that contamination levels were minimal, as no measurements of DFe in the sample bottles were collected. The values derived here fall within the range previously reported for iron addition experiments in the Southern Ocean (Hiscock et al., 2008; Hopkinson et al., 2007; Moore et al., 2007; Smith Jr. and Donaldson, 2015). Maximum values of α^B (mg C (mg Chl *a*)$^{-1}$ h^{-1} (µmol photons m^{-2} s^{-1})$^{-1}$) for ρC were 2.23 × 10^{-3} from experiment 2 Fe treatment and 2.43 × 10^{-3} from experiment 1 control treatment, with minimum values of 0.13 × 10^{-3} from experiment 5 control treatment and 0.56 × 10^{-3} from experiment 6 Fe treatment. P^B_{max} (mg C (mg Chl *a*)$^{-1}$ h^{-1}) values peaked in experiment 1 Fe treatment, with a minimum value of 1.06 × 10^{-2} in experiment 5 control treatment. E_k (µ mol photons m^{-2} s^{-1}) peaked at 79.77, with minimum values in experiment 1 control treatment. Despite the substantial differences in setup conditions for experiments 1 and 2 in the SAZ, occupied twice over the space of 28 days, there were no significant differences in the responses of the *PE* parameters to Fe. Due to constraints in light levels for the incubator setup, light levels that may result in photoinhibi-

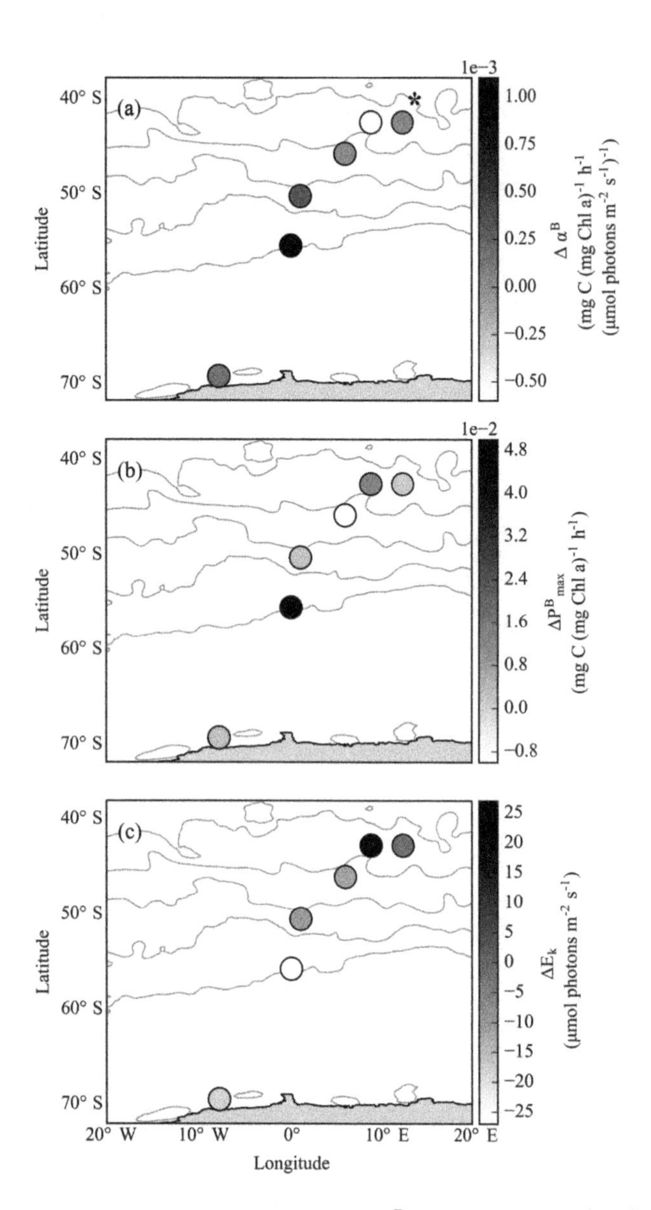

Figure 3. Experimental values of **(a)** $\Delta\alpha^{B}$ (mg C (mg Chl *a*)$^{-1}$ h^{-1} (μmol photons m^{-2} s^{-1})$^{-1}$), **(b)** ΔP^{B}_{max} (mg C (mg Chl *a*)$^{-1}$ h^{-1}) and **(c)** ΔE_{k} (μmol photons m^{-2} s^{-1}) for experiments set up in the Atlantic sector of the Southern Ocean. Ocean fronts, indicated by grey lines, were determined from MADT from the CLS/AVISO product (Rio et al., 2011) and their position averaged over 5 months (November 2015 to March 2016). * Position of experiment 3 moved 2.5° eastwards for presentation purposes.

tion (> 400 μ mol photons m^{-2} s^{-1}) were not achieved and as such no measurements of β were determined.

To better understand the effects of iron limitation on the *PE* parameters, the absolute differences (Fig. 3) of α^{B}, P^{B}_{max}, and E_{k} between the iron treatments and control treatments were calculated. $\Delta\alpha^{B}$ ranged from -6.94×10^{-4} to 1.30×10^{-3}, with minimum and maximum percentage differences of −40.04 and 91.12 % respectively. ΔP^{B}_{max} ranged between 4.98×10^{-2} and -1.02×10^{-2}, with minimum and maximum percentage differences of −12.10 and 82.52 %; the greatest value for ΔE_{k} was −40.92 for experiment 5. Maximal values of all differences were consistently found in experiment 5, which was set up just south of the Southern Boundary Front (Fig. 3).

Potential drivers of variability within the photosynthetic parameters were determined through a Pearson's linear correlation coefficient matrix (Fig. 4), revealing significant negative and positive relationships with sea surface temperature (SST), salinity, nitrate and silicate concentrations; photosynthetic physiology parameters (F_{v}/F_{m} and σ_{PSII}); and measures of the community structure, effective diameter and ratio of diatoms to haptophytes. There were no significant relationships with either dissolved iron concentrations or chlorophyll concentrations. Other parameters that did not show any relationships were excluded from the matrix include MLD, the light environment (in situ PAR and 1 % light depth) and phosphate concentrations. α^{B} for the control treatments displayed the greatest number of relationships with SST, nitrate concentrations, community structure variables and F_{v}/F_{m}. The relative differences in all the parameters showed strong positive correlations with SST and salinity ($p < 0.05$). A principal component analysis (PCA) was carried out on the data with the variables' PCA projection on the factor plane represented in Fig. S4 in the Supplement. The sum of the first two PCs explained 76.74 % of the total variance. The factor plane representation splits the variables, both experimental and initial conditions, into the four different quadrants. The grouping of the variables within each quadrant agree with the positive correlations determined within the correlation coefficient matrix, whereas variables in opposite quadrants agree with the negative correlations.

3.3 Primary production

Depth-integrated primary production (PP_{wc}) was calculated at each experimental location and displayed a wide range of variability with and without iron (Fig. 5). On average PP_{wc} was higher in the iron addition treatments (Fig. 5a), with an average of 387.32 ± 207.18 (mg C m^{-2} d^{-1}) for iron addition and an average of 315.37 ± 229.37 (mg C m^{-2} d^{-1}) for the control. The maximum absolute differences in PP_{wc} (ΔPP_{wc}, Fig. 5b) of 228.82 mc C m^{-2} s^{-1} was found in experiments 5 at ~ 55° S near the Southern Boundary Front, with very little difference observed in ΔPP_{wc} at experiments 3 and 4.

The responses of Fe addition to primary production from the six experiments were extrapolated onto broader spatial and temporal scales, whereby underway measurements of Chl were converted into K_{d} using Eq. (4). This, when combined with underway measurements of surface PAR, allowed us to look at latitudinal gradients in primary production (as per Eqs. 1, 2 and 3). As the *PE* parameters displayed strong linear correlations with latitude, (α $R^{2} = 0.73$ and 0.66, P_{max} $R^{2} = 0.91$ and 0.68 for Fe and Control respectively), a linear

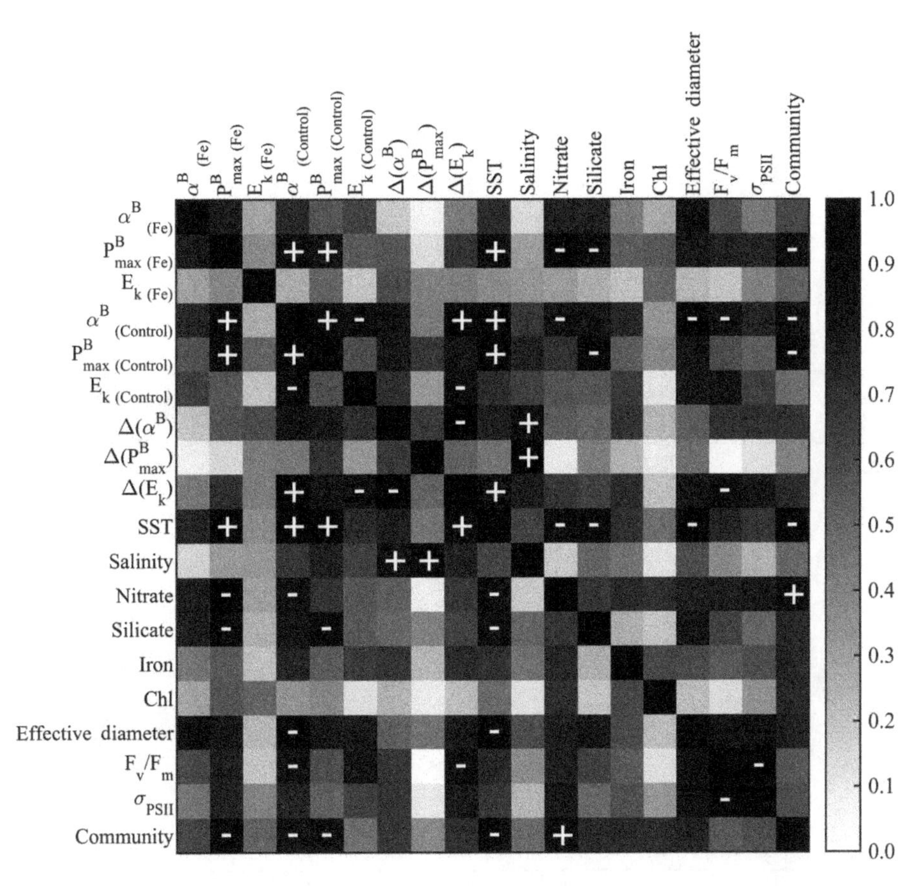

Figure 4. Matrix of Pearson's linear correlation coefficients between the photosynthetic parameters determined experimentally and in situ variables measured, including α^B, P^B_{max} and E_k from the both Fe and control treatments; the relative differences; sea surface temperature (SST); salinity, nitrate, silicate and dissolved iron concentration; Chl concentration; effective diameter; F_v/F_m; σ_{PSII}; and community composition (ratio of diatoms to haptophytes). The strength of the linear relationship associated between each pair of variables is indicated by the colour of the square, with the negative and positive correlations denoted by "−" and "+" within all squares where significant ($p<0.05$).

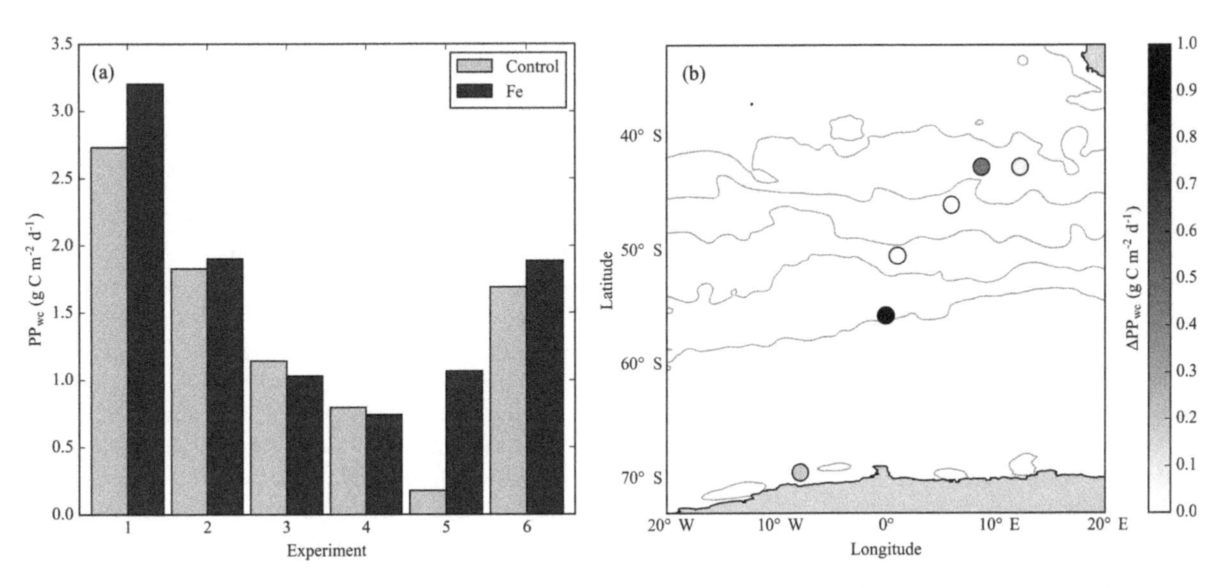

Figure 5. Modelled outputs of primary production utilizing experimentally derived photosynthetic parameters. **(a)** Depth-integrated primary production (PP_{wc}) (mg C m^{-2} d^{-1}) and **(b)** ΔPP_{wc} (mg C m^{-2} d^{-1}). Ocean fronts, indicated by grey lines, are displayed as in Fig. 3.

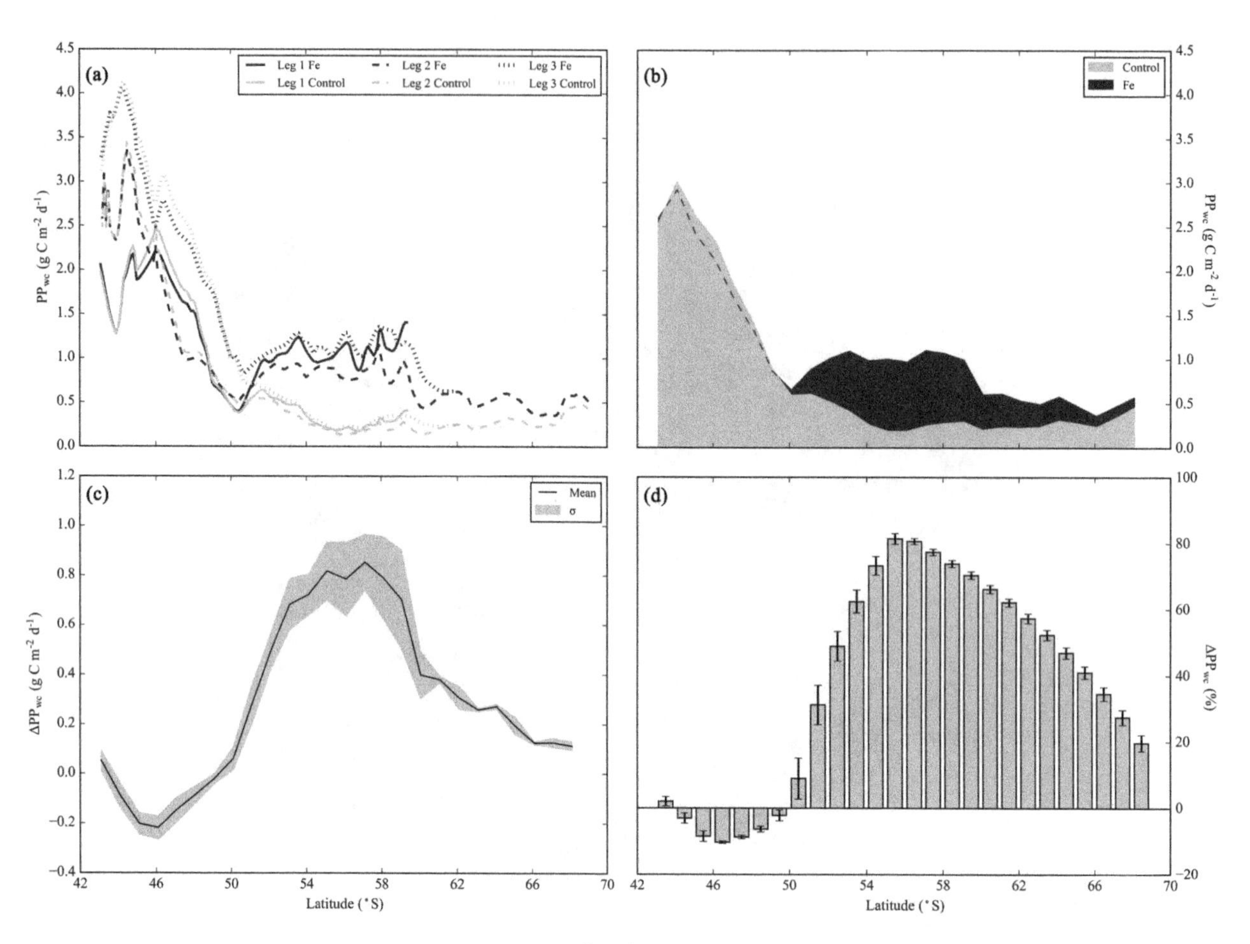

Figure 6. Depth-integrated primary production (PP_{wc}) ($mg\,C\,m^{-2}\,d^{-1}$) for each transect (Leg 1–3) **(a)** interpolated along the transect line utilizing linearly interpolated values for α and P_{max} as determined from the Fe and Control treatments. **(b)** Mean PP_{wc} ($mg\,C\,m^{-2}\,d^{-1}$) with ±standard deviation (σ). **(c)** The mean absolute differences in PP_{wc} (ΔPP_{wc}) with ±standard deviation between the Fe and Control treatments. **(d)** ΔPP_{wc} represented as the mean percentage difference with ±standard deviations.

interpolation was applied to P_{max} and α, extrapolating the values from six points to a 0.1° resolution along the cruise track. The interpolated values of P_{max} and α were combined with underway measurements of K_d and PAR to calculate PP_{wc} with and without Fe addition for the three different occupations of the same transect line (Fig. 6a). A high degree of variability was revealed between occupations in the SAZ and polar frontal zone (PFZ) but no clear differences between the iron and control treatments. Variability in the SAZ and PFZ appears to be temporally driven, with higher values of PP_{wc} found in the third occupation of the transect line later in the summer season. Differences in PP_{wc} between the two treatments become evident south of 50° S (Fig. 6a and b), with all three iron treatment occupations being $\sim 0.5\,g\,C\,m^{-2}\,d^{-1}$ higher than their control treatment counterparts. The differences between the control and Fe treatments were calculated for each transect, which when combined allowed for the calculation of an average absolute difference in primary productivity (ΔPP_{wc}, Fig. 6c). ΔPP_{wc} is slightly negative within the SAZ and PFZ, before sharply increasing to a maximum difference of $0.85\,g\,C\,m^{-2}\,d^{-1}$ at 58° S. ΔPP_{wc} begins to decrease with increasing latitude before reaching an average difference of $0.11\,g\,C\,m^{-2}\,d^{-1}$ in the MIZ. Representing these differences in PP_{wc} as a percentage difference (Fig. 6d) shows that within the SAZ, PFZ and MIZ the differences are ±10–20 %, whereas within the Antarctic zone (55–65° S) the differences between the treatments can be as much as 80 %.

Given the limitations of our data set (which requires the use of interpolated values of P_{max} and α) together with the weight we place on the conversion of these parameters to PP (with chlorophyll and PAR), it is important that we understand the sensitivity of the PP model to variability in the different input parameters. To test this, we performed a series of sensitivity tests to determine which components present the greatest influence on the final PP values. The sensitivity tests were divided into the three components of the equation: K_d derived from chlorophyll (Fig. S5), surface PAR (Fig. S6) and the photosynthetic parameters (P_{max} and α) (Fig. S7). For consistency, the range of variation for each parameter was calculated and used as a factor to alter each component.

The mean range of variability for K_d was 84.33 %, surface PAR was 68.73 %, and α and P_{max} were 82.85 and 83.01 % respectively. If K_d values are increased by 84.33 % this results in a 29.61 % decrease in ΔPP_{wc}, whereas a decrease of K_d results in an increase in ΔPP_{wc} of 59.17 %. Increasing surface PAR resulted in an increase in ΔPP_{wc} of 3.50 %, whilst decreasing PAR corresponded to a decrease of 8.06 %. The largest differences in ΔPP_{wc} were generated when P_{max} was altered by 83.01 %, in accordance with the range of variability, resulting in an increase of 42.97 % and a decrease of 80.92 % in ΔPP_{wc} (for an increase and decrease in P_{max} respectively). The other *PE* parameter, α, did not result in the same level of changes in ΔPP_{wc} and only increased by 4.01 % and decreased by 12.22 % for an increase and decrease in α by 82.85 % respectively.

4 Discussion

Phytoplankton biomass in the Southern Ocean is potentially limited in their extent and magnitude predominantly by the availability of the micronutrient iron (Blain et al., 2007; Boyd et al., 2000; Pollard et al., 2009). This conclusion is based on the combination of two factors: the high iron requirements for photosynthetic proteins (Quigg et al., 2003; Raven, 1990; Strzepek and Harrison, 2004; Twining and Baines, 2013) and the lack of supply sources of iron to the Southern Ocean (Duce and Tindale, 1991; Tagliabue et al., 2014). The result of this is an environment that displays high degrees of spatial and temporal variability in primary production in response to highly variable iron supply mechanisms that result in chlorophyll patchiness (Fig. 1) and a complex seasonality (Thomalla et al., 2011). Iron limitation is potentially strongest during the summer months, when light levels are not considered limiting (Boyd et al., 2010) and the spring bloom is expected to have utilized the bulk of the winter iron resupply. In the austral summer of 2015/2016 a series of iron addition photosynthesis versus irradiance experiments were performed in the Atlantic sector of the Southern Ocean to determine the extent to which iron availability was limiting maximal rates of primary productivity.

The addition of iron appeared to stimulate increased productivity to varying degrees (Figs. 2, 3b, and S1–S3) with average P_{max} and α values being higher for an iron-replete system (12.75 ± 6.95 and 0.25 ± 0.14) compared to a control system (11.17 ± 8.23 and 0.22 ± 0.19), suggestive that iron is indeed a micronutrient-limiting phytoplankton production in this region. Similar responses have been reported by Hiscock et al. (2008) under conditions of sub-saturating light conditions, where the addition of iron can result in a doubling of photosynthetic rates. However, a nutrient addition *PE* experiment in the Ross Sea demonstrated no significant increases in α^B or P^B_{max} (Smith Jr. and Donaldson, 2015). One potential reason for this is the length of their incubation period, which was only 2 h and may not have been enough for the phytoplankton to incorporate the iron into their photosynthetic proteins and produce higher productivity rates. Indeed, nutrient addition experiments performed under similar conditions were shown to require 24 h to see any significant differences in initial changes in photophysiology (Browning et al., 2014; Ryan-Keogh et al., 2017; Ryan-Keogh et al., 2013) with changes in biomass only being reported after 48 h. This shortcoming highlights the attraction of the unique experimental design utilized here, which allows for 24 h Fe addition and control incubations at varying light levels and constant temperature. However, it should be noted that a time length of 24 h may not be sufficient to complete alleviate the iron-mediated photosynthetic response and as such these results may only reflect initial responses rather than longer-term community-level responses to relief from iron limitation. It should be noted, however, that light acclimation can between 2 and 6 h and as such be reflected in the potential iron demand, a lower demand at higher irradiances (Strzepek et al., 2012). Such incidences would impact the observed differences between *PE* parameters in control versus Fe addition experiments. However, since the light range of the experiments (0–400) fall below the maximum light intensities measured in situ (Table 1), acclimation responses are unlikely to dominate and, if occurring, would indeed result in an underestimation of the differences between control and addition experiments. The experimental design of 24 h, whilst suitable for investigating iron limitation, means that results are not truly representative of in situ photosynthetic parameters and should not be interpreted as such.

Potential factors that are known to be associated with iron-induced enhanced primary productivity include temperature, macronutrient concentrations, Chl, MLD, light history and community composition. A Pearson's linear correlation matrix (Fig. 4) was carried out on an array of variables to examine the influence of key physical, chemical and biological factors on the variability of photosynthetic parameters in this study. Significant relationships were found with SST, salinity and macronutrient concentrations, which show strong latitudinal gradients. A proxy for the community structure that utilized the ratio of the two dominant groupings (diatoms and haptophytes) also indicated strong significant relationships with the *PE* parameters, which is potentially driven by Si availability controlling community structure. Indeed, it has been demonstrated that in the SAZ, where haptophytes dominated during this study, there is evidence for Fe-Si co-limitation. In a study by Hutchins et al. (2001) it was demonstrated that the addition of both Fe and Si resulted in the greatest responses in chlorophyll and the photosynthetic parameters. The relationship here may not be driven by Fe availability on the *PE* parameters but rather community-level limitation. No significant relationships were, however, found between *PE* parameters and iron or Chl concentrations. The lack of significant relationships could be due to the small range of variability observed in these parameters; for example, Chl concentrations at all stations were typically low (0.84–$2.30\,\mu g\,L^{-1}$) when compared to the range of chlorophyll concentrations measured throughout the entire cruise (0.01–$11.25\,\mu g\,L^{-1}$). The lack of a relationship with dissolved iron concentrations highlights how this proxy is not necessarily a good indicator of iron stress, as any limiting nutrient would be expected to be severely depleted by biological uptake with a resultant ambient concentration that would remain close to zero despite possible event scale supply (Ryan-Keogh et al., 2017).

The photosynthetic parameters derived here are important components in a suite of models that derive estimates of phytoplankton primary production (Behrenfeld and Falkowski, 1997a, b; Sathyendranath and Platt, 2007). Different primary production models inherently consist of certain biases towards modelling the photosynthetic parameters, whereas others have excluded them entirely from the computation of primary productivity rates. Hiscock et al. (2008) demonstrated that the variables in the Behrenfeld and Falkowski (1997b) standard depth-integrated model (DIM) exerted considerably different forcing mechanisms on the final primary productivity rates. In the case of this DIM, phytoplankton biomass was the dominant variable that could result in 3 orders of magnitude changes in primary production, compared to only a 40-fold change when altering the photosynthetic parameter P^{B}_{opt} (i.e. P^{B}_{max}). This highlights the need to understand the sensitivity of different PP models to variability within their input parameters.

Results from the production model applied here (Eqs. 1–3) show a general decrease with latitude in depth-integrated primary production (PP_{wc}), with significant differences between treatments (t test, $p<0.05$). One station near the Southern Boundary exhibited the greatest differences in ΔPP_{wc} with a value of $0.89\,g\,C\,m^{-2}\,d^{-1}$ (Fig. 5b), with the lowest observed ΔPP_{wc} of $0.11\,g\,C\,m^{-2}\,d^{-1}$ south of the polar front. The low sampling frequency of the experiments both spatially and temporally (six experiments spanning 2 months and the entire latitudinal extent of the Southern Ocean) together with the diverse range of initial setup conditions (Table 1) make it difficult to interpret the causal relationships observed within each experiment with any certainty. Instead, the information from these experiments was maximized through an alternate approach that utilized the range of variability in *PE* parameters in control versus iron addition experiments to gain a broader spatial interpretation of the response of phytoplankton production to iron addition.

A linear interpolation of the *PE* parameters (P_{max} and α) with latitude, together with underway measurements of PAR and K_d (derived from surface Chl) allow for the generation of high-resolution rates of PP_{wc} with and without Fe addition for three occupations of the cruise transect (Fig. 6a). Within the SAZ and PFZ there was a high degree of variability between the three occupations, with higher PP_{wc} values later in the growing season (Fig. 6a). However, there were no clear differences between the iron and control treatments in any of the occupations. This may not reflect a lack of iron limitation in the SAZ, as it has been demonstrated previously that there is ecological and physiological iron limitation (Coale et al., 2004), with longer experiments demonstrating increases in P_{max} and α following iron addition (Hutchins et al., 2001). However, south of 50° S there were no differences observed as the growing season progressed with similar PPwc values across the three occupations of the cruise transect, but there was a clear difference between the iron and control treatments (Fig. 6b and c). Here, a maximum percentage difference of $\sim 80\,\%$ (Fig. 6d) was observed between control and iron-replete conditions, with ΔPP_{wc} peaking at $0.85\,g\,C\,m^{-2}\,yr^{-1}$ at 55° S. Differences between iron addition and control systems begin to decline within the MIZ (Fig. 6c). These results suggest that there are potential differences in iron availability and supply within different zones of the Southern Ocean, which agrees with previous studies which postulated that the bloom extent and duration within the SAZ could potentially be driven by enhanced iron supply through storm–eddy interaction (Nicholson et al., 2016) while in the MIZ addition iron is supplied through melting ice (Gao et al., 2003; Grotti et al., 2005; Sedwick and DiTullio, 1997). The Fe addition test performed here demonstrates the sensitivity of waters south of 50° S to Fe availability. If models do not consider this sensitivity then the degree of error for PP models can be as high as 80 %. It must be noted that the transects will not only reflect latitudinal gradients but also contain a seasonal signal as the cruise spanned 2 months across the austral summer. A seasonal shift in community structure of haptophytes increasing their dominance beyond the SAZ into the PFZ was evident from underway measurements of community structure (data not shown), indicative of seasonal Si limitation for this region (Boyd et al., 2010). Moreover, the complex seasonality of this region represents shifts between varying co-limitations that will be represented not only in the *PE* parameters measured but also in the additional components utilized to calculate PP_{wc}.

From these results, it became clear that higher values of P_{max} and α because of iron addition were significantly influencing the model outputs of primary production. However, the extent to which changes in the *PE* parameters were responsible for the latitudinal trend in ΔPP_{wc} versus changes in ancillary parameters (e.g. Chl, PAR) is unclear. To test our interpretation of the variability in PP_{wc} being a direct response to Fe availability through changes in the *PE* parameters, a series of sensitivity analyses were performed which showed that PAR and α exerted very little influence (Figs. S6 and S7). Biomass (Chl), as represented through K_d, did exert a large influence on PP_{wc} (up to 59 %, Fig. S5), but this influence could be overestimated due to potential errors in the calculation of K_d (Morel et al., 2007). However the greatest influence was P_{max} (up to 81 %, Fig. S7). As such, we can conclude that the primary driver of the latitudinal trend in ΔPP_{wc} is the result of changes in the maximum photosynthetic capacity (P_{max}) to iron addition; however, regions along the transect may be experiencing seasonal co-limitation of Fe and Si, particularly during the third transect conducted during late summer.

The photosynthetic parameters P_{max} and α remain difficult to fully parameterize due to interacting effects of iron, light availability, temperature and community structure, yet these parameters remain critical components of different biogeochemical models. Our results show that if models fail to capture the interacting effects of iron and other parameters on primary productivity, then the degree of error across vast extents of the Southern Ocean can be significant (as much as 80 %). On the other hand, any model that can correctly account for variability in these parameters will better reproduce the natural background levels of primary productivity and the seasonal cycle for application to iron-limited areas of the ocean including the Subarctic Pacific and the Southern Ocean.

Competing interests. The authors declare that they have no conflict of interest.

Acknowledgements. We would like to thank the South African National Antarctic Programme (SANAP) and the captain and crew of the *S.A. Agulhas II* for their professional support throughout the cruise. Ryan Cloete and Ryan Miltz were involved in experimental setup; Natasha van Horsten and Warren Joubert performed the DIC determinations for calculation of carbon assimilation. This work was undertaken and supported through the CSIR's Southern Ocean Carbon and Climate Observatory (SOCCO) Programme (http://socco.org.za/). This work was supported by CSIR's Parliamentary Grant funding and an NRF SANAP grant (SNA14073184298).

Edited by: Gerhard Herndl

References

Baker, N. R., Oxborough, K., Lawson, T., and Morrison, J. I. L.: High resolution imaging of photosynthetic activities of tissues, cells and chloroplasts in leaves, J. Exp. Bot., 52, 615–621, https://doi.org/10.1093/jxb/52.356.615, 2001.

Behrenfeld, M. J. and Falkowski, P. G.: A consumer's guide to phytoplankton primary productivity models, Limnol. Oceanogr., 42, 1479–1491, https://doi.org/10.4319/lo.1997.42.7.1479, 1997a.

Behrenfeld, M. J. and Falkowski, P. G.: Photosynthetic rates derived from satellite-based chlorophyll concentration, Limnol. Oceanogr., 42, 1–20, https://doi.org/10.4319/lo.1997.42.1.0001, 1997b.

Blain, S., Queguiner, B., Armand, L., Belviso, S., Bombled, B., Bopp, L., Bowie, A., Brunet, C., Brussaard, C., Carlotti, F., Christaki, U., Corbiere, A., Durand, I., Ebersbach, F., Fuda, J. L., Garcia, N., Gerringa, L., Griffiths, B., Guigue, C., Guillerm, C., Jacquet, S., Jeandel, C., Laan, P., Lefevre, D., Lo Monaco, C., Malits, A., Mosseri, J., Obernosterer, I., Park, Y. H., Picheral, M., Pondaven, P., Remenyi, T., Sandroni, V., Sarthou, G., Savoye, N., Scouarnec, L., Souhaut, M., Thuiller, D., Timmermans, K., Trull, T., Uitz, J., van Beek, P., Veldhuis, M., Vincent, D., Viollier, E., Vong, L., and Wagener, T.: Effect of natural iron fertilization on carbon sequestration in the Southern Ocean, Nature, 446, 1070–1074, https://doi.org/10.1038/nature05700, 2007.

Boyd, P. W. and Doney, S. C.: Modelling regional responses by marine pelagic ecosystems to global climate change, Geophys. Res. Lett., 29, 53-1–53-4, https://doi.org/10.1029/2001GL014130, 2002.

Boyd, P. W., Watson, A. J., Law, C. S., Abraham, E. R., Trull, T., Murdoch, R., Bakker, D. C. E., Bowie, A. R., Buesseler, K. O., Chang, H., Charette, M., Croot, P., Downing, K., Frew, R., Gall, M., Hadfield, M., Hall, J., Harvey, M., Jameson, G., LaRoche, J., Liddicoat, M., Ling, R., Maldonado, M. T., McKay, R. M., Nodder, S., Pickmere, S., Pridmore, R., Rintoul, S., Safi, K., Sutton, P., Strzepek, R., Tanneberger, K., Turner, S., Waite, A., and Zeldis, J.: A mesoscale phytoplankton bloom in the polar Southern Ocean stimulated by iron fertilization, Nature, 407, 695–702, https://doi.org/10.1038/35037500, 2000.

Boyd, P. W., Crossely, A. C., DiTullio, G. R., Griffiths, F. B., Hutchins, D. A., Queguiner, B., Sedwick, P. N., and Trull, T. W.: Control of phytoplankton growth by iron supply and irradiance in the subantarctic Southern Ocean: Experimental results from the SAZ Project, J. Geophys. Res., 106, 31573–31583, https://doi.org/10.1029/2000JC000348, 2001.

Boyd, P. W., Strzepek, R., Fu, F. X., and Hutchins, D. A.: Environmental control of open-ocean phytoplankton groups: Now and in the future, Limnol. Oceanogr., 55, 1353–1376, https://doi.org/10.4319/lo.2010.55.3.1353, 2010.

Browning, T. J., Bouman, H. A., Moore, C. M., Schlosser, C., Tarran, G. A., Woodward, E. M. S., and Henderson, G. M.: Nutrient regimes control phytoplankton ecophysiology in the South Atlantic, Biogeosciences, 11, 463–479, https://doi.org/10.5194/bg-11-463-2014, 2014.

Close, S. E. and Goosse, H.: Entrainment-driven modulation of Southern Ocean mixed layer properties and sea ice variability in CMIP5 models, J. Geophys. Res.-Oceans, 118, 2811–2827, https://doi.org/10.1002/jgrc.20226, 2013.

Coale, K. H., Johnson, K. S., Chavez, F. P., Buesseler, K. O., Barber, R. T., Brzezinski, M. A., Cochlan, W. P., Millero, F. J., Falkowski, P. G., Bauer, J. E., Wanninkhof, R. H., Kudela, R. M., Altabet, M. A., Hales, B. E., Takahashi, T., Landry, M. R., Bidigare, R. R., Wang, X., Chase, Z., Strutton, P. G., Friederich, G. E., Gorbunov, M. Y., Lance, V. P., Hilting, A. K., Hiscock, M. R., Demarest, M., Hiscock, W. T., Sullivan, K. F., Tanner, S. J., Gordon, R. M., Hunter, C. N., Elrod, V. A., Fitzwater, S. E., Jones, J. L., Tozzi, S., Koblizek, M., Roberts, A. E., Herndon, J., Brewster, J., Ladizinsky, N., Smith, G., Cooper, D., Timothy, D., Brown, S. L., Selph, K. E., Sheridan, C. C., Twining, B. S., and Johnson, Z. I.: Southern Ocean Iron Enrichment Experiment: Carbon Cycling in High- and Low-Si Waters, Science, 304, 408–414, 2004.

Cochlan, W. P.: Nitrogen Uptake in the Southern Ocean, in: Nitrogen in the Marine Environment, Elsevier, Amsterdam, the Netherlands, 2008.

Cullen, J. J. and Davis, R. F.: The blank can make a big difference in oceanographic measurements, Limnology and Oceanography Bulletin, 12, 29–35, https://doi.org/10.1002/lob.200312229, 2003.

de Baar, H. J. W., Buma, A. G. J., Nolting, R. F., Cadee, G. C., Jacques, G., and Treguer, P. J.: On iron limitation of the Southern Ocean: Experimental observations in the Weddell and Scotia Seas, Mar. Ecol.-Prog. Ser., 65, 105–122, 1990.

de Baar, H. J. W., Boyd, P. W., Coale, K. H., Landry, M. R., Tsuda, A., Assmy, P., Bakker, D. C. E., Bozec, Y., Barber, R. T., Brezinski, M. A., Buesseler, K. O., Boyé, M., Croot, P. L., Gervais, F., Gorbunov, M. Y., Harrison, P. J., Hiscock, W. T., Laan, P., Lancelot, C., Law, C. S., Levasseur, M., Marchetti, A., Millero, F. J., Nishioka, J., Nojiri, Y., van Oijen, T., Riebesell, U., Rijkenberg, M. J. A., Saito, H., Takeda, S., Timmermans, K. R., Veldhuis, M. J. W., Waite, A. M., and Wong, C. S.: Synthesis of iron fertilization experiments: From the iron age in the age of enlightenment, J. Geophys. Res., 110, C09S16, https://doi.org/10.1029/2004JC002601, 2005.

de Boyer Montégut, C., Madec, G., Fischer, A. S., Lazar, A., and Iudicone, D.: Mixed layer depth over the global ocean: an examination of profile data and a profile-based climatology, J. Geophys. Res., 109, C12003, https://doi.org/10.1029/2004JC002378, 2004.

de Lavergne, C., Palter, J. B., Galbraith, E. D., Bernardello, R., and Marinov, I.: Cessation of deep convection in the open Southern Ocean under anthropogenic climate change, Nature Climate Change, 4, 278–282, https://doi.org/10.1038/nclimate2132, 2014.

Dower, K. M. and Lucas, M. I.: Photosynthesis-irradiance relationships and production associated with a warm-core ring shed from the Agulhas Retroflection south of Africa, Mar. Ecol.-Prog. Ser., 95, 141–154, https://doi.org/10.3354/meps095141, 1993.

Dubischar, C. D. and Bathmann, U. V.: Grazing impacts of copepods and salps on phytoplankton in the Atlantic sector of the Southern Ocean, Deep-Sea Res. Pt. II, 44, 415–433, https://doi.org/10.1016/S0967-0645(96)00064-1, 1997.

Duce, R. A. and Tindale, N. W.: Atmospheric Transport of Iron and Its Deposition in the Ocean, Limnol. Oceanogr., 36, 1715–1726, https://doi.org/10.4319/lo.1991.36.8.1715, 1991.

Dugdale, R. C. and Wilkerson, F. P.: The use of 15N to measure nitrogen uptake in eutrophic oceans; experimental considerations, Limnol. Oceanogr., 31, 673–689, https://doi.org/10.4319/lo.1986.31.4.0673, 1986.

Egan, L.: QuickChem Method 31-107-04-1-C – Nitrate and/or Nitrite in brackish or seawater, Lachat Instruments, Colorado, USA, 2008.

Feng, Y., Hare, C. E., Rose, J. M., Handy, S. M., DiTullio, G. R., Lee, P. A., Smith, W. O., Jr., Peloquin, J., Tozzi, S., Sun, J., Zhang, Y., Dunbar, R. B., Long, M. C., Sohst, B., Lohan, M., and Hutchins, D. A.: Interactive effects of iron, irradiance and CO_2 on Ross Sea phytoplankton, Deep-Sea Res. Pt. I, 57, 368–383, https://doi.org/10.1016/j.dsr.2009.10.013, 2010.

Gao, Y., Fan, S.-M., and Sarmiento, J. L.: Aeolian iron input to the ocean through precipitation scavenging: A modeling perspective and its implication for natural iron fertilization in the ocean, J. Geophys. Res., 108, 4221, https://doi.org/10.1029/2002JD002420, 2003.

Gibberd, M.-J., Kean, E., Barlow, R., Thomalla, S., and Lucas, M.: Phytoplankton chemotaxonomy in the Atlantic sector of the Southern Ocean during late summer 2009, Deep-Sea Res. Pt. I, 78, 70–78, https://doi.org/10.1016/j.dsr.2013.04.007, 2013.

Grasshoff, K., Ehrhardt, M., and Kremling, K.: Methods of seawater analysis, Verlag Chemie, Wienheim, Germany, 1983.

Grotti, M., Soggia, F., Ianni, C., and Frache, R.: Trace metals distributions in coastal sea ice of Terra Nova Bay, Ross Sea, Antarctica, Antarct. Sci., 17, 289–300, 2005.

Hansen, J. E. and Travis, L. D.: Light scattering in planetary atmospheres, Space Sci. Rev., 16, 527–610, https://doi.org/10.1007/BF00168069, 1974.

Hiscock, M. R., Lance, V. P., Apprill, A. M., Johnson, Z., Bidigare, R. R., Mitchell, B. G., Smith, W. O. J., and Barber, R. T.: Photosynthetic maximum quantum yield increases are an essential component of Southern Ocean phytoplankton iron response, P. Natl. Acad. Sci. USA, 105, 4775–4780, https://doi.org/10.1073/pnas.0705006105, 2008.

Hopkinson, B. M., Mitchell, B. G., Reynolds, R. A., Wang, H., Selph, K. E., Measures, C. I., Hewes, C. D., Holm-Hansen, O., and Barbeau, K. A.: Iron limitation across chlorophyll gradients in the southern Drake Passage: Phytoplankton responses to iron addition and photosynthetic indicators of iron stress, Limnol. Oceanogr., 52, 2540–2554, https://doi.org/10.4319/lo.2007.52.6.2540, 2007.

Hutchins, D. A., Sedwick, P. N., DiTullio, G. R., Boyd, P. W., Quéguiner, B., Griffiths, F. B., and Crossely, C.: Control of phytoplankton growth by iron and silicic acid availability in the subantarctic Southern Ocean: Experimental results from the SAZ Project, J. Geophys. Res., 106, 31559–31572, https://doi.org/10.1029/2000JC000333, 2001.

Johnson, K. S., Elrod, V. A., Fitzwater, S. E., Plant, J., Boyle, E., Bergquist, B., Bruland, K. W., Aguilar-Islas, A. M., Buck, K., Lohan, M. C., Smith, G. J., Sohst, B. M., Coale, K. H., Gordon, M., Tanner, S., Measures, C. I., Moffett, J., Barbeau, K. A., King, A., Bowie, A. R., Chase, Z., Cullen, J. J., Laan, P., Landing, W., Mendez, J., Milne, A., Obata, H., Doi, T., Ossiander, L., Sarthou, G., Sedwick, P. N., Van den Berg, S., Laglera-Baquer, L., Wu, J.-F., and Cai, Y.: Developing standards for dissolved iron in seawater, Eos, Transactions American Geophysical Union, 88, 131–132, https://doi.org/10.1029/2007EO110003, 2007.

Khatiwala, S., Primeua, F., and Hall, T.: Reconstruction of the history of anthropogenic CO_2 concentrations in the ocean, Nature, 462, 346–349, https://doi.org/10.1038/nature08526, 2009.

Kolber, Z. S., Prášil, O., and Falkowski, P. G.: Measurements of variable chlorophyll fluorescence using fast repetition rate techniques: defining methodology and experimental protocols, Biochim. Biophys. Acta, 1367, 88–106, https://doi.org/10.1016/S0005-2728(98)00135-2, 1998.

Lucas, M., Seeyave, S., Sanders, R., Moore, C. M., Williamson, R., and Stinchcombe, M.: Nitrogen uptake responses to a naturally Fe-fertilised phytoplankton bloom during the 2004/2005 CROZEX study, Deep-Sea Res. Pt. II, 54, 2138–2173, https://doi.org/10.4319/lo.2007.52.6.2540, 2007.

Mackey, M. D., Mackey, D. J., Higgins, H. W., and Wright, S. W.: CHEMTAX – a program for estimating class abundances from chemical markers: application to HPLC measurements of phytoplankton, Mar. Ecol.-Prog. Ser., 144, 265–283, 10.3354/meps144265, 1996.

Maldonado, M. T., Boyd, P. W., Harrison, P. J., and Price, N. M.: Co-limitation of phytoplankton growth by light and Fe during winter in the NE subarctic Pacific Ocean, Deep-Sea Res. Pt. II, 46, 2475–2485, https://doi.org/10.1016/S0967-0645(99)00072-7, 1999.

Martin, J. H., Gordon, R. M., and Fitzwater, S. E.: Iron in Antarctic waters, Nature, 345, 156–158, https://doi.org/10.1038/345156a0, 1990.

Mikaloff Fletcher, S. E., Gruber, N., Jacobson, A. R., Doney, S. C., Dutkiewicz, S., Gerber, M., Follows, M., Joos, F., Lindsay, K., Menemenlis, D., Mouchet, A., Müller, S. A., and Sarmiento, J. L.: Inverse estimates of anthropogenic CO_2 uptake, transport, and storage by the ocean, Global Biogeochem. Cy., 20, GB2002, https://doi.org/10.1029/2005GB002530, 2006.

Milligan, A. J. and Harrison, P. J.: Effects of non-steady-state iron limitation on nitrogen assimilatory enzymes in the marine diatom Thalassiosira weissflogii (Bacillariophyceae), J. Phycol., 36, 78–86, https://doi.org/10.1046/j.1529-8817.2000.99013.x, 2000.

Montes-Hugo, M., Doney, S. C., Ducklow, H. W., Fraser, W., Martinson, D., Stammerjohn, S. E., and Schofield, O.: Recent changes in phytoplankton communities associated with rapid regional climate change along the Western Antarctic Peninsula, Science, 323, 1470–1473, https://doi.org/10.1126/science.1164533, 2008.

Moore, C. M., Seeyave, S., Hickman, A. E., Allen, J. T., Lucas, M. I., Planquette, H., Pollard, R. T., and Poulton, A. J.: Iron-light interactions during the CROZet natural iron bloom and EXport experiment (CROZEX) I: Phytoplankton growth and photophysiology, Deep-Sea Res. Pt. II, 54, 2045–2065, https://doi.org/10.1016/j.dsr2.2007.06.011, 2007.

Moore, C. M., Mills, M. M., Arrigo, K. R., Berman-Frank, I., Bopp, L., Boyd, P. W., Galbraith, E. D., Geider, R. J., Guieu, C., Jaccard, S. L., Jickells, T. D., La Roche, J., Lenton, T. M., Mahowald, N. M., Marañón, E., Marinov, I., Moore, J. K., Nakatsuka, T., Oschlies, A., Saito, M. A., Thingstad, T. F., Tsuda, A., and Ulloa, O.: Processes and patterns of oceanic nutrient limitation, Nat. Geosci., 6, 701–710, https://doi.org/10.1038/NGEO1765, 2013.

Morel, A.: Optical modelling of the upper ocean in relation to its biogenous matter content (case 1 waters), J. Geophys. Res., 93, 10749–10768, https://doi.org/10.1029/JC093iC09p10749, 1988.

Morel, A., Huot, Y., Gentili, B., Werdell, P. J., Hooker, S. B., and Franz, B. A.: Examining the consistency of products derived from various ocean color sensors in open ocean (Case 1) waters in the perspective of a multi-sensor approach, Remote Sens. Environ., 111, 69–88, https://doi.org/10.1016/j.rse.2007.03.012, 2007.

Nicholson, S.-A., Lévy, M., Llort, J., Swart, S., and Monteiro, P. M. S.: Investigating into the impact of storms on sustaining summer primary productivity in the Sub-Antarctic Ocean, Geophys. Res. Lett., 43, 9192–9199, https://doi.org/10.1002/2016GL069973, 2016.

Nowlin, W. D. and Klinck, J. M.: The physics of the Antarctic Circumpolar Current, Revi. Geophys., 24, 469–491, https://doi.org/10.1029/RG024i003p00469, 1986.

Obata, H., Karatani, H., and Nakayama, E.: Automated determination of iron in seawater by chelating resin concentration and chemiluminescence detection, Anal. Chem., 65, 1524–1528, https://doi.org/10.1021/ac00059a007, 1993.

Orsi, A. H., Whitworth III, T. W., and Nowlin, W. D.: On the meridional extent and front of the Antarctic Circumpolar Current, Deep-Sea Res. Pt. I, 42, 641–673, https://doi.org/10.1016/0967-0637(95)00021-W, 1995.

Pakhomov, E. A. and Froneman, P. W.: Zooplankton dynamics in the eastern Atlantic sector of the Southern Ocean during austral summer 1997/1998, Deep-Sea Res. Pt. II, 51, 2599–2616, https://doi.org/10.1016/j.dsr2.2000.11.001, 2004.

Platt, T. and Sathyendranath, S.: Estimators of primary production for interpretation of remotely-sensed data on ocean colour, J. Geophys. Res., 98, 14561–14576, https://doi.org/10.1029/93JC01001, 1993.

Platt, T., Gallegos, C. L., and Harrison, W. G.: Photoinhibition of photosynthesis in natural assemblages of marine phytoplankton, J. Mar. Res., 38, 687–701, 1980.

Platt, T., Sathyendranath, S., and Fuentes-Yaco, C.: Biological oceanography and fisheries management: perspective after 10 years, ICES J. Mar. Sci., 64, 863–869, https://doi.org/10.1093/icesjms/fsm072, 2007.

Pollard, R. T., Salter, I., Sanders, R. J., Lucas, M. I., Moore, C. M., Mills, R. A., Statham, P. J., Allen, J. T., Baker, A. R., Bakker, D. C. E., Charette, M. A., Fielding, S., Fones, G. R., French, M., Hickman, A. E., Holland, R. J., Hughes, J. A., Jickells, T. D., Lampitt, R. S., Morris, P. J., Nedelec, F. H., Nielsdottir, M., Planquette, H., Popova, E. E., Poulton, A. J., Read, J. F., Seeyave, S., Smith, T., Stinchcombe, M., Taylor, S., Thomalla, S., Venables, H. J., Williamson, R., and Zubkov, M. V.: Southern Ocean deep-water carbon export enhanced by natural iron fertilization, Nature, 457, 577–581, https://doi.org/10.1038/nature07716, 2009.

Price, N. M., Ahner, B. A., and Morel, F. M. M.: The equatorial Pacific Ocean: Grazer-controlled phytoplankton populations in an iron-limited ecosystem, Limnol. Oceanogr., 39, 520–534, https://doi.org/10.4319/lo.1994.39.3.0520, 1994.

Quigg, A., Finkel, Z. V., Irwin, A. J., Rosenthal, Y., Ho, T.-Y., Reinfelder, J. R., Schofield, O., Morel, F. M. M., and Falkowski, P. G.: The evolutionary influence of elemental stoichiometry in marine phytoplankton, Nature, 425, 291–294, 2003.

Ras, J., Claustre, H., and Uitz, J.: Spatial variability of phytoplankton pigment distributions in the Subtropical South Pacific Ocean: comparison between in situ and predicted data, Biogeosciences, 5, 353–369, https://doi.org/10.5194/bg-5-353-2008, 2008.

Raven, J. A.: Predictions of Mn and Fe use efficiencies of phototrophic growth as a function of light availability for growth and C assimilation pathway, New Phytol., 116, 1–18, https://doi.org/10.1111/j.1469-8137.1990.tb00505.x, 1990.

Rio, M. H., Guinehut, S., and Larnicol, G.: New CNES-CLS09 global mean dynamic topography computed from the combination of GRACE data, altimetry, and in situ measurements, J. Geophys. Res., 116, C07018, https://doi.org/10.1029/2010JC006505, 2011.

Roháček, K.: Chlorophyll Fluorescence Parameters: The Definitions, Photosynthetic Meaning, and Mutual Relationships, Photosynthetica, 40, 13–29, https://doi.org/10.1023/A:1020125719386, 2002.

Ryan-Keogh, T. J., Macey, A. I., Nielsdóttir, M., Lucas, M. I., Steigenberger, S. S., Stinchcombe, M. C., Achterberg, E. P., Bibby, T. S., and Moore, C. M.: Spatial and temporal development of phytoplankton iron stress in relation to bloom dynamics in the high-latitude North Atlantic Ocean, Limnol. Oceanogr., 58, 533–545, https://doi.org/10.4319/lo.2013.58.2.0533, 2013.

Ryan-Keogh, T. J., DeLizo, L. M., Smith, W. O., Jr., Sedwick, P. N., McGillicuddy Jr., D. J., Moore, C. M., and Bibby, T. S.:

Temporal progression of photosynthetic strategy by phytoplankton in the Ross Sea, Antarctica, J. Marine Syst., 166, 87–96, https://doi.org/10.1016/j.jmarsys.2016.08.014, 2017.

Sathyendranath, S. and Platt, T.: Spectral effects in bio-optical control on the ocean system, Oceanologia, 49, 5–39, 2007.

Schlitzer, R.: Carbon export fluxes in the Southern Ocean: results from inverse modeling and comparion with satellite-based estimates, Deep-Sea Res. Pt. II, 49, 1623–1644, https://doi.org/10.4319/lo.1994.39.3.0520, 2002.

Sedwick, P. N. and DiTullio, G. R.: Regulation of algal blooms in Antarctic shelf waters by the release of iron from melting sea ice, Geophys. Res. Lett., 24, 2515–2518, https://doi.org/10.1029/97GL02596, 1997.

Smetacek, V., Assmy, P., and Henjes, J.: The role of grazing in structuring Southern Ocean pelagic ecosystems and biogeochemical cycles, Antarct. Sci., 16, 541–558, https://doi.org/10.1017/S0954102004002317, 2004.

Smith Jr., W. O. and Donaldson, K.: Photosynthesis-irradiance responses in the Ross Sea, Antarctica: a meta-analysis, Biogeosciences, 12, 3567–3577, https://doi.org/10.5194/bg-12-3567-2015, 2015.

Strzepek, R. F. and Harrison, P. J.: Photosynthetic architecture differs in coastal and oceanic diatoms, Nature, 431, 689–692, https://doi.org/10.1038/nature02954, 2004.

Strzepek, R. F., Hunter, K. A., Frew, R. D., Harrison, P. J., and Boyd, P. W.: Iron-light interactions differ in Southern Ocean phytoplankton, Limnol. Oceanogr., 57, 1182–1200, https://doi.org/10.4319/lo.2012.57.4.1182, 2012.

Sunda, W. G. and Huntsman, S. A.: Interrelated influence of iron, light and cell size on marine phytoplankton growth, Nature, 390, 389–392, https://doi.org/10.1038/37093, 1997.

Swart, S., Speich, S., Ansorge, I. J., and Lutjeharms, J. R. E.: An altimetry-based gravest empirical mode south of Africa: 1. Development and validation, J. Geophys. Res., 115, C03002, https://doi.org/10.1029/2009JC005299, 2010.

Swart, S., Chang, N., Fauchereau, N., Joubert, W., Lucas, M., Mtshali, T., Roychoudhury, A., Tagliabue, A., Thomalla, S., Waldron, H., and Monteiro, P. M. S.: Southern Ocean Seasonal Cycle Experiment 2012: Seasonal scale climate and carbon cycle links, S. Afr. J. Sci., 108, 11–13, https://doi.org/10.4102/sajs.v108i3/4.1089, 2012.

Swart, S., Thomalla, S. J., and Monteiro, P. M. S.: The seasonal cycle of mixed layer dynamics and phytoplankton biomass in the Sub-Antarctic Zone: A high-resolution glider experiment, J. Marine Syst., 147, 103–115, https://doi.org/10.1016/j.jmarsys.2014.06.002, 2015.

Tagliabue, A., Sallée, J.-B., Bowie, A. R., Lévy, M., Swart, S., and Boyd, P. W.: Surface-water iron supplies in the Southern Ocean sustained by deep winter mixing, Nat. Geosci., 7, 314–320, https://doi.org/10.1038/ngeo2101, 2014.

Thomalla, S. J., Fauchereau, N., Swart, S., and Monteiro, P. M. S.: Regional scale characteristics of the seasonal cycle of chlorophyll in the Southern Ocean, Biogeosciences, 8, 2849–2866, https://doi.org/10.5194/bg-8-2849-2011, 2011.

Thomalla, S. J., Racault, M.-F., Swart, S., and Monteiro, P. M. S.: High-resolution view of the spring bloom initiation and net community production in the Subantarctic Southern Ocean using glider data, ICES J. Mar. Sci., 72, 1999–2020, https://doi.org/10.1093/icesjms/fsv105, 2015.

Twining, B. S. and Baines, S. B.: The trace metal composition of marine phytoplankton, Annual Review of Marine Science, 5, 191–215, https://doi.org/10.1146/annurev-marine-121211-172322, 2013.

Welschmeyer, N. A.: Fluorometric analysis of chlorophyll *a* in the presence of chlorophyll *b* and pheopigments, Limnol. Oceanogr., 39, 1985–1992, https://doi.org/10.4319/lo.1994.39.8.1985, 1994.

Wolters, M.: Quickchem Method 31-114-27-1-D – Silicate in Brackish or Seawater, Lachat Instruments, Colorado, USA, 2002.

Zhang, J. L.: Increasing Antarctic sea ice under warming atmospheric and oceanic conditions, J. Climate, 20, 2515–2529, https://doi.org/10.1175/jcli4136.1, 2007.

15

Does denitrification occur within porous carbonate sand grains?

Perran Louis Miall Cook[1], **Adam John Kessler**[1], **and Bradley David Eyre**[2]

[1]Water Studies Centre, School of Chemistry, Monash University, Clayton, Australia
[2]Centre for Coastal Biogeochemistry, Southern Cross University, Lismore, Australia

Correspondence to: Perran Louis Miall Cook (perran.cook@monash.edu)

Abstract. Permeable carbonate sands form a major habitat type on coral reefs and play a major role in organic matter recycling. Nitrogen cycling within these sediments is likely to play a major role in coral reef productivity, yet it remains poorly studied. Here, we used flow-through reactors and stirred reactors to quantify potential rates of denitrification and the dependence of denitrification on oxygen concentrations in permeable carbonate sands at three sites on Heron Island, Australia. Our results showed that potential rates of denitrification fell within the range of 2–28 $\mu mol\,L^{-1}$ sediment h^{-1} and were very low compared to oxygen consumption rates, consistent with previous studies of silicate sands. Denitrification was observed to commence at porewater oxygen concentrations as high as 50 µM in stirred reactor experiments on the coarse sediment fraction (2–10 mm) and at oxygen concentrations of 10–20 µM in flow-through and stirred reactor experiments at a site with a median sediment grain size of 0.9 mm. No denitrification was detected in sediments under oxic conditions from another site with finer sediment (median grain size: 0.7 mm). We interpret these results as confirmation that denitrification may occur within anoxic microniches present within porous carbonate sand grains. The occurrence of such microniches has the potential to enhance denitrification rates within carbonate sediments; however further work is required to elucidate the extent and ecological significance of this effect.

1 Introduction

Nitrogen is typically regarded as one of the key nutrients limiting production in the coastal environment (Howarth and Marino, 2006). Coral reefs are examples of highly oligotrophic environments that are coming under increasing threat from increased nutrient loads (De'ath et al., 2012). Denitrification is a key remedial process in the nitrogen cycle, leading to the conversion of bioavailable nitrate into relatively non-bioavailable nitrogen gas. One of the dominant habitat types on coral reefs are carbonate sands formed from the breakdown of carbonate produced by calcifying organisms (Eyre et al., 2014). Quantifying denitrification in sandy sediments is complicated by the fact that these sediments are permeable, allowing water movement through the sediment, which can both enhance and reduce denitrification (Cook et al., 2006; Kessler et al., 2012; Sokoll et al., 2016). Reproducing these conditions while measuring denitrification is difficult; thus, models combined with a mechanistic understanding of the primary controls on denitrification offer a promising approach to quantifying denitrification in these environments (Evrard et al., 2013; Kessler et al., 2012).

For denitrification to take place under anoxic conditions, a supply of nitrate and organic matter is required. Advective flushing of permeable sediments leads to deeper oxygen penetration into these sediments, which will inhibit denitrification (Evrard et al., 2013; Precht et al., 2004). Nitrate may be supplied to the denitrification zone from either the water column or nitrification within the sediment. It has been reported that nitrification within the sediment may be a significant source of nitrate to fuel denitrification (Marchant et al., 2016; Rao et al., 2007), although modelling studies suggest that, even in the presence of nitrification, flow fields in permeable sediments may lead to little coupling between nitrification and denitrification (Kessler et al., 2013). In systems with high bottom-water nitrate concentrations, high rates of denitrification have been reported (Gao et al., 2012; Sokoll et al., 2016). Previous studies have also reported high rates of denitrification in carbonate sands (Eyre et al., 2008, 2013b), which is surprising given the low nitrate concentrations in the overlying water and their highly oligotrophic nature. One possible explanation for high rates of denitrification is the coupling of nitrification and denitrification within microniches associated with sand grains (Jahnke, 1985; Rao et al., 2007; Santos et al., 2012). In this hypothesis, mineralization of organic matter within or on sand grains consumes oxygen at a rate greater than it can diffuse into the grain, causing anoxia within the grain. This allows the bulk sediment to be simultaneously oxic, which promotes nitrification, while anoxic sites within grains allow denitrification to take place closely coupled to nitrification. A recent study has shown that incorporating intra-granular porosity into model simulations of column experiments can lead to better agreement with observations (Kessler et al., 2014) in carbonate sediments, but direct experimental evidence for this phenomenon is lacking.

Direct measurement of oxygen concentrations and denitrification rates within sediment grains is not possible with current technologies. Nevertheless, there are means by which the hypothesis that denitrification takes place within sediment grains can be tested. If denitrification is taking place within anoxic intra-granular niches, then it should be observed under oxygen concentrations above zero within flow-through reactors (FTRs) and/or stirred reactors (SRs). Another factor complicating our understanding of denitrification is the kinetics of this process relative to total sediment respiration. It has previously been shown that denitrification rates in permeable sediments are very low compared to total respiration (Bourke et al., 2017; Evrard et al., 2013) and that this pattern is consistent globally (Marchant et al., 2016) in silicate sands. There have, however, been no analogous studies on potential denitrification rates relative to respiration rates in anoxic flow-through reactors in carbonate sediments.

To address these knowledge gaps, this study had two objectives. The first was to measure potential denitrification rates in carbonate sediments and compare these to total metabolism measured in flow-through column experiments. The second was to investigate the effect of oxygen concentration on denitrification using both flow-through and stirred reactors to experimentally test the plausibility of anoxic microniches leading to enhanced rates of denitrification in carbonate sediments.

2 Methods

2.1 Flow-through reactor experiments

Sediments were collected from three sites at Heron Island (23°27′ S, 151°55′ E) in the Southern Great Barrier Reef, Australia, during 17–21 October 2015. Previously reported water column nutrient concentrations at this site are NO_3^- 0.05–0.7 µM, NH_4^+ 0.05–1.8 µM and orthophosphate 0.35–0.5 µM. The sediment organic carbon content is < 0.24 %, and benthic chlorophyll *a* has been reported as ranging from 11 to 15 mg m^{-2} (Eyre et al., 2008; Glud et al., 2008). Site 1 is located adjacent to the Heron Island research station (23°26′37″ S, 151°54′46″ E; water depth of 0.5 m at low tide), site 2 is located at Shark Bay (23°26′37″ S, 151°55′09″ E; water depth of 0.5 m low tide) and site 3 is located in the lagoon approximately 4 km east of the island (23°27′04″ S, 151°57′28″ E; water depth of 2 m at low tide). Site 3 was the most uniformly coarse and permeable site, and site 2 had the smallest median grain size and lowest permeability (Table 1). Sediments were packed into three replicate FTRs (4.6 cm diameter, 4 cm length) for each experiment, as described by Evrard et al. (2013), within 2 h of collection. Fresh unfiltered seawater was collected from in front of the research station, and the columns were percolated at a flow rate of ~ 200 mL h^{-1}. The volume of the FTRs was 66 mL (~ 33 mL porosity corrected), giving a retention time of ~ 10 min (corrected for sediment porosity). Reaction rates were calculated per volume of wet sediment. The flow velocity was 24 cm h^{-1} and was chosen as it was estimated to give a small but easily detectable change in ^{15}N–N_2. This is the upper end of those expected around ripples of 0.14–26 cm h^{-1} (Precht et al., 2004) and higher than those used by Santos et al. (2012). We deliberately did this to ensure any boundary layers at the grain surface were at a minimum, and any effect observed here could be ascribed to intra-granular porosity. Diffusive chemical gradients were manipulated by changing oxygen and nitrate concentrations within the flow-through reactors.

Two experiments were undertaken as follows. First, the effect of NO_3^- concentration on denitrification was measured as described by Evrard et al. (2013). Repacked FTRs were percolated with anoxic seawater (30 min purging with Ar), which was sequentially amended with 18, 37, 75, 150 and 300 µM^{15} NO_3^-. The oxygen concentration at the column inlets and outlets were monitored in real time using Firesting optical dissolved oxygen flow-through cells (Pyro Science), which had a detection limit of ~ 3 µM O_2 and a precision of ~ 1 %. After ~ 3 retention times (~ 30 min) at each nitrate concentration, a sample of the column effluent was collected directly into glass syringes, transferred into an exetainer and preserved with 250 µL 50 % w/v $ZnCl_2$. Second, the effect of oxygen on denitrification was measured as described by Evrard et al. (2013). Columns were percolated with unfiltered seawater amended with 150 µM^{15} NO_3^- (99 % Cambridge scientific), and the oxygen concentration was reduced incrementally in the reservoir by sporadic purging with Ar. Samples of column effluent were collected and preserved as described above.

2.2 Stirred reactor experiments

Samples for SR experiments were collected from sites 1 and 2 on 14 March 2017 and sent by overnight courier to Monash University, where they were submerged in aerated artificial seawater made from "Redcoral" nitrate- and phosphate-free sea salts at 23 °C amended with

Table 1. Sediment grain size, permeability, grain porosity and sediment oxygen consumption rate (with SD) at the three study sites. O_2 consumption and denitrification rates are from flow-through reactor experiments.

Site	Coordinates	Median grain size mm	% 1.18–3 mm	% 0.5–1.18 mm	% 0.125–0.5 mm	% < 0.124 mm	Permeability $\times 10^{-12}$ m^2	Bulk sediment porosity Vol/vol	Grain porosity Vol/vol	O_2 consumption µmol L^{-1} h^{-1}	V_{max} denitrification rate µmol L^{-1} h^{-1}
1	23°26′37″ S 151°54′46″ E	0.9	32	54	13.4	0.6	27	0.55	0.31	350 (50)	11.2 (0.3)
2	23°26′37″ S 151°55′09″ E	0.7	20	44	29	8	24	0.48	N/A	270 (80)*	2 (0.5)
3	23°27′04″ S, 151°57′28″ E	0.9	32	52	16	1	30	0.56	0.32	466 (9)	28 (2)

N/A = not analysed. * $n = 2$, range shown.

$\sim 50\,\mu M\ ^{15}NO_3^-$. SR experiments were undertaken on sieved sediments (< 2 mm) at sites 1 and 2 and the coarse fraction (2–10 mm) from site 1 in 115 mL glass vessels, capped with a rubber bung ensuring no air bubbles were present as described by Gao et al. (2010). The reactor was stirred using a magnetic stirrer bar at ~ 150 rpm, which was the speed required to re-suspend all the added sediment (15 mL). Water samples were withdrawn though a port into a syringe over a period of 3–5 h and simultaneously replaced with the same volume of artificial seawater (~ 15 mL) and preserved for nitrogen isotope analysis as described above. Oxygen was logged using a Firesting (Pyro Science) needle O_2 sensor inserted through the rubber bung.

2.3 Analytical methods

Samples for $^{28}N_2$, $^{29}N_2$ and $^{30}N_2$ analysis had a 4 mL He headspace inserted into the exetainer and were shaken for 5 min before the headspace gas was analysed on a Sercon 20–22 isotope ratio mass spectrometer, coupled to an autosampler and gas chromatography column to separate O_2 and N_2. Air was used as the calibration standard, and tests showed no false mass 30 signal compared to pure N_2 injections. The precision of the analysis of the ratios $^{29}N_2/^{28}N_2$ and $^{30}N_2/^{28}N_2$ was 0.2 and 5 %, respectively. For the analysis of 2 µmol N, this equates to an excess ^{15}N of 2.5×10^{-5} µmol for $^{29}N_2$ and 7.85×10^{-5} µmol for $^{30}N_2$. Assuming all of the N_2 production was in the form of $^{30}N_2$, this results in an equivalent detectable production rate of 0.014 nmol mL^{-1} h^{-1} in the column experiments. Rates of denitrification were calculated using the isotope pairing equations (Nielsen, 1992), and we present the total rate of denitrification ($D_{14} + D_{15}$) here. Rates of anammox were estimated based on Eq. (23) given in Risgaard Petersen et al. (2003). Sediment permeability was measured using the constant-head method (Reynolds, 2008), and sediment porosity was measured by drying a known volume of sediment saturated with freshwater. Images of the grains were taken using a Motic dissecting microscope with a 5 MP Moticam. The porosity of the sand grains was measured using mercury porosimetry at Particle and Surface Sciences Pty Ltd. Sediment grain size was measured using test sieves with mesh sizes of 2, 1.18, 0.5 and 0.125 mm.

Figure 1. Images of carbonate grains from site 3 at Heron Island. Scale bar is 1 mm.

3 Results

Sites 1 and 3 had the coarsest median grain size of 0.9 mm, whereas site 2 had a median grain size of 0.7 mm. All sites had 20–30 % sediment with grain sizes in the range of 1.18–3 mm, the sediment permeability was also similar at all sites ranging from 24 to 30×10^{-12} m^2 and the bulk porosity was similar at all sites ranging from 0.48 to 0.56 at the three sites (Table 1). Images of the sand grains showed them to be porous (Fig. 1), and this was confirmed by mercury porosimetry, which revealed the sand grains had a porosity of ~ 0.32 at sites 1 and 3, site 2 not measured (Table 1).

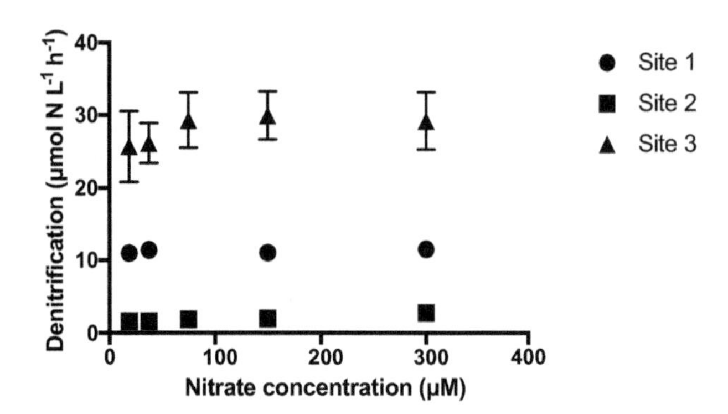

Figure 2. Denitrification rates as a function of nitrate concentration at the inlet of the columns at the three sites studied. Error bars for sites 1 and 2 are smaller than the marker symbols.

In the FTR experiments, rates of denitrification were constant above NO_3^- concentrations of 18 μM at all three study sites and were highest at site 3, which had the highest sediment oxygen consumption rates, and lowest at site 2, which had the lowest oxygen consumption rates (Fig. 2). Rates of anammox comprised < 16 % of nitrogen production (data not shown) in the anoxic flow-through columns. Plots of oxygen concentrations showed that concentrations of oxygen at the column inlets dropped in a stepwise manner when they were purged with Ar, and this was reflected at the column outlets with a delay of ~ 10 min, consistent with the theoretical column retention time (Fig. 3). Rates of denitrification were generally negligible at oxygen concentrations > 0 μM at the column outlets, except for site 1, where denitrification was observed at ~ 10 μM O_2 at the column outlet (Fig. 4).

The SR experiments showed small amounts of $^{15}N–N_2$ production once the oxygen concentration in the reactor dropped below 20 μM at site 1 (Fig. 5). At site 2, $^{15}N–N_2$ production was only observed once the oxygen concentration approached 0 μM. The coarse fraction at site 1 showed mixed results, with two of the four samples showing no clear $^{15}N–N_2$ production above 20 μM O_2 and two samples showing clear evidence of $^{15}N–N_2$ production at < 50 μM O_2.

4 Discussion

4.1 Methodological considerations

Before discussing the results in detail, we briefly consider the methods used here and potential shortcomings. Firtly, we only used one relatively fast flow rate in these FTR experiments. We chose this flow rate as we estimated this was the maximum flow rate we could use that would minimize boundary layers within the column while giving detectable production of $^{15}N–N_2$. These flow rates are in the upper range of those previously reported in sediments where porewater flow is driven by flow-topography interactions (Precht and Huettel, 2004), as we expected to be the case here. Second, it is possible that $^{15}N–N_2$ production had not reached a steady state after the manipulation of oxygen and nitrate concentrations within the FTRs. Conceptually, nitrate from the bulk porewater will diffuse into the sediment grain, where denitrification will take place, and the produced $^{15}N–N_2$ will diffuse out again before being washed out of the column. If we use a grain size of 2 mm (twice the median grain size), this means a maximum diffusion distance of ~ 1 mm to a putative denitrification zone within a sediment grain. The diffusion timescale for nitrate molecule can be calculated using Eq. (1):

$$t = L^2/2D_s, \tag{1}$$

where t is time, L is distances and D_s is the diffusion coefficient (Schulz and Zabel, 2005). For nitrate, with a diffusion coefficient of $1.7 \times 10^{-5}\ cm^2\ s^{-1}$ at 25 °C and a salinity of 35 corrected for a grain porosity of 0.3 according to Iversen and Jørgensen (1993), this gives a timescale of ~ 10 min. For nitrate to diffuse in and N_2 to diffuse out, we would therefore expect this to take a maximum of ~ 20 min, which is less than the time we waited before sampling in the column and the time interval between samples in the SRs. For the coarse fraction used in the stirred reactors, the samples taken were unlikely to represent steady state, and therefore the rates of denitrification measured under oxic conditions can be taken as a conservative minimum.

Third, we used oxygen consumption as a proxy for respiration in these sediments. It has previously been shown that ~ 50 % of oxygen consumption in sediments can be driven by the oxidation of reduced solutes (Cook et al., 2007). In this case, however, we believe that this was unlikely because we waited > 14 h before oxygen consumption measurements commenced, after which we would expect all the reduced solutes to have been either washed out or oxidized. We therefore believe that the vast majority of O_2 consumption was respiration as opposed to reduced solute oxidation. Finally, breakthrough curves are often used to quantify column retention time and dispersivity. In this instance, we did not undertake breakthrough curve measurements, as we have previously shown this column set-up to give a very distinct plug flow (Evrard et al., 2013). The offset of 10 min between the purging of oxygen in the reservoir and the response at the column outlet qualitatively confirms the theoretical retention time of the columns for these experiments.

4.2 Comparison of potential denitrification rates with previous studies

The denitrification rates measured in the present study spanned the range of flow-through reactor rates (~< 1–32 $\mu mol\ L^{-1}\ h^{-1}$) previously observed in silicate sands (Evrard et al., 2013; Kessler et al., 2012; Marchant et al., 2014; Rao et al., 2007). The availability, and composition, of organic matter is expected to be a key factor controlling potential denitrification rates (Eyre et al., 2013a; Seitzinger,

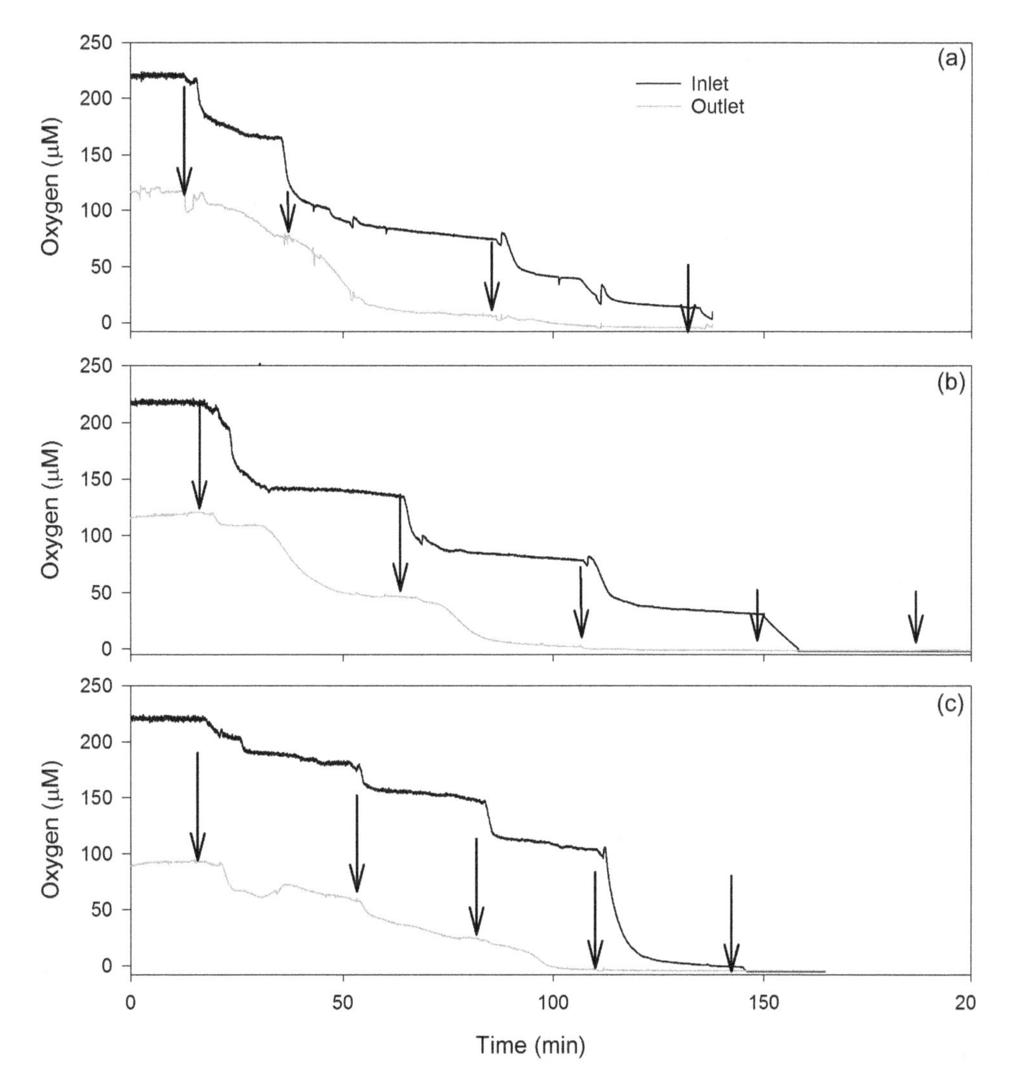

Figure 3. Example time series of oxygen concentrations at the FTR inlet and outlet at sites 1 **(a)** 2, **(b)** and 3 **(c)**. Arrows mark sampling points for $^{15}N–N_2$; note $^{15}NO_3^-$ tracer was added > 30 min before the first sampling point (i.e. before the O_2 trace commences).

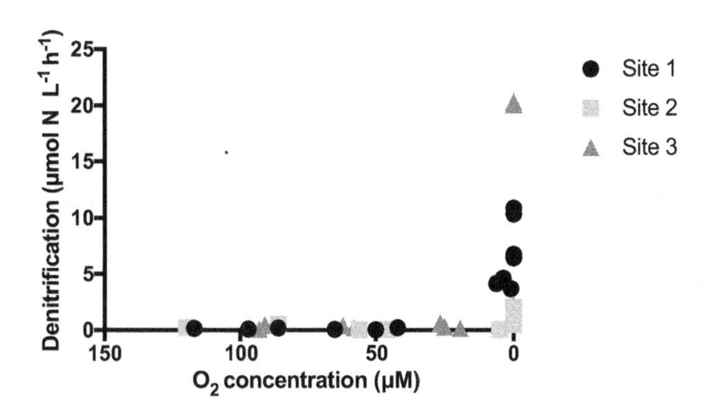

Figure 4. Denitrification rate as a function of oxygen concentration at the outlet of the columns at the three study sites.

1988), and the importance of this in permeable sediments has also been recently underscored by Marchant et al. (2016), who observed a strong relationship between potential denitrification rate and sediment oxygen consumption. If the results of the previous studies are plotted vs. sediment oxygen consumption rate, a significant relationship is observed with an r^2 of 0.92 (Fig. 6). Sites 2 and 3 in the present study seemed to deviate significantly from this relationship, as they were the only data points to lie outside the 99 % prediction interval, while site 1 sat close to the line of best fit. Omitting the study of Kessler et al. (2012) still led to sites 2 and 3 being outside the prediction intervals, with site 2 being below and site 3 above the relationship observed for silicate sands. This suggests that the slope of the relationship between denitrification and oxygen consumption rate in this study differs from previous studies. It has recently been shown that much of the metabolism in permeable sediments is dominated by

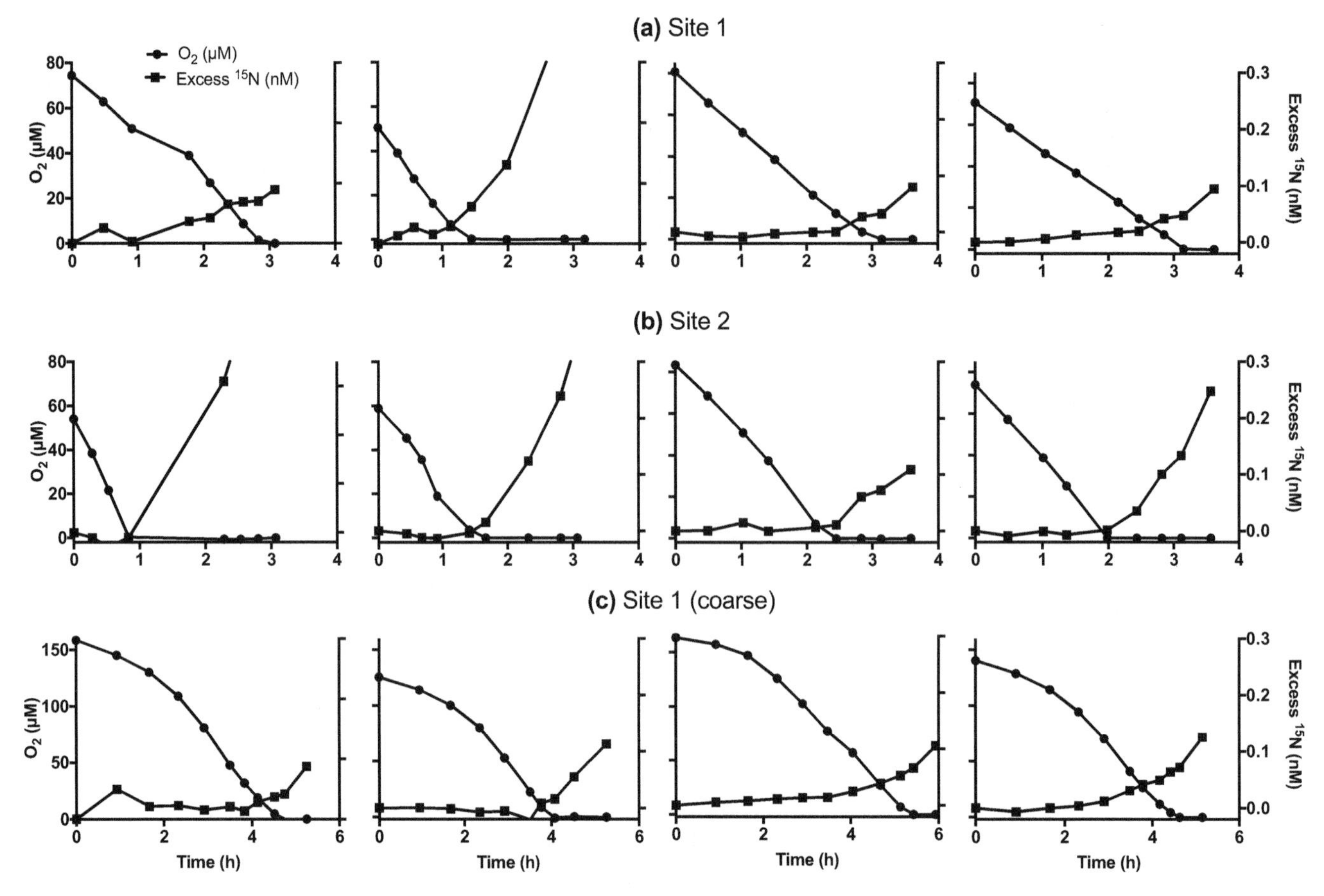

Figure 5. Time series of oxygen and excess $^{15}N–N_2$ in stirred reactor experiments on sediments taken from sites 1 **(a)** and 2 **(b)** as well as the coarse fraction of sediment (> 2 mm) collected from site 1 **(c)**. Each row shows four replicate experiments for each site.

algae, rather than bacteria (Bourke et al., 2017). Given that bacteria undertake denitrification, it is likely that there was relatively more algal respiration occurring at site 2. We speculate that site 2, which was the most sheltered and had the finest sediment, was dominated by microphytobenthos, while site 3, which had coarser sediment and higher turbulence in the outer lagoon, had a larger advective supply of phytodetritus from the water column owing to higher flushing rates (Huettel and Rusch, 2000).

4.3 Denitrification within carbonate sand grains

The experiments performed here showed denitrification was able to take place at oxygen concentrations below 20 μM at site 1 in the FTR and SR experiments and as high as 50 μM in the coarse fraction in the SR experiments (Figs. 4 and 5). It has previously been shown that nanomolar concentrations of oxygen can inhibit denitrification (Dalsgaard et al., 2014), and one possible explanation is that denitrification was taking places within anoxic niches within the grains. Theoretically, the critical radius (r) of a particle at which anoxia will occur in the centre can be calculated from Eq. (2) (Jørgensen, 1977):

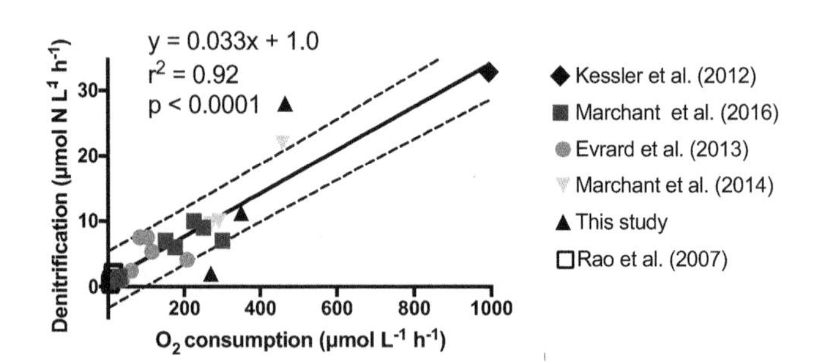

Figure 6. Denitrification rate as a function of oxygen consumption in this study and previously published studies. The solid line is the line of best fit for previous studies in silicate sands; the dashed lined are the 99 % prediction interval. Data from this study are from the flow-through reactor experiments.

$$r = (6D_sC/J)^{1/2}, \quad (2)$$

where D_s is the diffusion coefficient (corrected for tortuosity), C is the oxygen concentration at the particle surface and J is the volumetric oxygen consumption rate. At sediment respiration rates of 270–460 $\mu mol\,L^{-1}\,h^{-1}$ observed in this study, we would expect to see the centre of particles $\sim$ 1 mm in diameter become anoxic only at oxygen concentrations < 5 µM, particles at 2 mm diameter become anoxic at oxygen concentrations of $\sim$ 10–20 µM and particles > 4 mm be anoxic < 50 µM O_2. Given that $\sim$ 20–30 % of the particles fell in this size range 1–3 mm, we would expect denitrification to commence at O_2 concentrations of $\sim$ 10–20 µM if significant rates of denitrification were taking place within the particles, which is consistent with our findings for site 1. The finding that denitrification could take place in the coarse fraction at oxygen concentrations as high as 50 µM is also consistent with this, as this size class encompassed the range $\sim$ 2–10 mm.

In addition to denitrification taking places within anoxic microniches, it is also possible that denitrification under oxic conditions is occurring as has been previously reported (Gao et al., 2012). We believe this is unlikely for the following reasons. Firstly, no denitrification was detected under bulk oxic conditions in the finest sediment at site 2, suggesting organisms responsible for oxic denitrification were not active in permeable carbonate sediments. Second, oxic denitrification rates were higher in the FTR than in the SR experiments. In the SR experiments, a maximum oxic denitrification rate (O_2 > 1 µM) of 1.4 $\mu mol\,N\,L^{-1}\,h^{-1}$ (30 % of the anoxic rate) was observed at site 1, which compares to a maximum oxic denitrification rate of 4.6 $\mu mol\,N\,L^{-1}\,h^{-1}$ (45 % of the anoxic rate) at site 1 in the FTRs. In the FTRs, there will be a much thicker boundary layer limiting the diffusion of oxygen into the grains than in SRs, and hence a greater anoxic volume and hence denitrification rate than in the stirred reactors. This observation cannot easily be explained by the presence of denitrification under oxic conditions. We note that our maximum rate of denitrification under bulk oxic conditions measured in the FTR reactors (4.6 $\mu mol\,N\,L^{-1}\,h^{-1}$) is at the lower end of denitrification rates reported under oxic conditions in silicate sands of $\sim$ 6–17 $\mu mol\,N\,L^{-1}\,h^{-1}$ (Gao et al., 2010; Marchant et al., 2017), suggesting that denitrification under oxic conditions, where it occurs, has a greater enhancement effect on total denitrification than denitrification in anoxic microniches.

4.4 Implications for nitrogen cycling

Our data suggest that potential anoxic denitrification rates in tropical carbonate sands are low compared to total respiration rates as has previously been observed in silicate sands. We also found evidence to support the hypothesis that denitrification within porous sand grains was taking place. Using a simulation model of a flow field within a ripple, it has previously been shown that, in the absence of any intra-granular porosity effect, potential denitrification rates in the range of those measured here scale up to only $\sim$ 5 $\mu mol\,m^{-2}\,h^{-1}$; however under the same conditions with intra-granular porosity, rates increased by an order of magnitude to 50 $\mu mol\,m^{-2}\,h^{-1}$ owing to intense coupled nitrification denitrification within the sediment grains near the oxic–anoxic interface (Kessler et al., 2014). Previous chamber measurements for sites 1 and 2 typically show denitrification rates on the order of 60 $\mu mol\,m^{-2}\,h^{-1}$, and no significant difference was observed between the two sites (Eyre et al., 2013b). It is therefore possible that intra-granular porosity can explain some of the discrepancy between the chamber and modelled results, particularly at site 1. At site 2, however, where we saw no evidence for denitrification under oxic conditions, this is less clear. One possible explanation is that the chambers incorporate a large volume of sediment, which may include larger grains than the small subsample used in reactor experiments. We also note that for the experiments where we did observe denitrification under bulk oxic conditions the rates were highly variable, which may suggest a subset of sediment grains (shell vs. coral derived) or possibly organisms such as foraminifera (Risgaard-Petersen et al., 2006) may play a disproportionate role in denitrification. Under this scenario it is also possible that the chamber experiments at site 2 enclosed these sediment types, which may have been excluded by chance in the relatively small volume of sediment used in the reactor experiments. Another possible reason for the discrepancies between the chamber and modelled rates are artefacts associated with using chambers in permeable sediments. Model simulations have shown that there may be "wash-out" of nitrogen accumulated within porewaters which can enhance measured nitrogen fluxes (Cook et al., 2006), possibly explaining the higher chamber rates. However, strong relationships between respiration and denitrification in permeable carbonate sands measured using chambers suggest the denitrification rates are reliable (Eyre et al., 2013b).

Overall, these results suggest that denitrification may take place within anoxic sites in porous carbonate grains with porewater O_2 concentrations of up to 50 µM. The broader ecological significance of this, however, remains to be elucidated. We suggest further studies be undertaken to investigate (1) the effect of different grain types (coral vs. shell derived), grain sizes and organisms (e.g. foraminifera) on denitrification; (2) the extent of coupling between nitrification and denitrification within carbonate grains; and (3) the use of flume experiments in combination with ^{15}N tracers to experimentally test the extent of enhanced denitrification under realistic flow fields.

Author contributions. All authors contributed to the design, undertaking the experiments, data interpretation and manuscript preparation.

Competing interests. The authors declare that they have no conflict of interest.

Acknowledgements. This work was supported by the Australian Research Council grants DP150102092 and DP150101281 to Bradley David Eyre and Perran Louis Miall Cook respectively. We thank Vera Eate for analysis of $^{15}N–N_2$ and Michael Bourke for assistance in the laboratory. We thank five anonymous reviewers whose comments have helped improve this manuscript.

Edited by: Tina Treude

References

Bourke, M. F., Marriott, P. J., Glud, R. N., Hassler-Sheetal, H., Kamalanathang, M., Beardall, J., Greening, C., and Cook, P. L. M.: Metabolism in anoxic permeable sediments is dominated by eukaryotic dark fermentation, Nat. Geosci., 10, 30–35, 2017.

Cook, P. L. M., Wenzhöfer, F., Rysgaard, S., Galaktionov, O. S., Meysman, F. J. R., Eyre, B. D., Cornwell, J. C., Huettel, M., and Glud, R. N.: Quantification of denitrification in permeable sediments: insights from a two dimensional simulation analysis and experimental data, Limnol. Oceanogr.-Meth., 4, 294–307, 2006.

Cook, P. L. M., Wenzhöfer, F., Glud, R. N., Janssen, F., and Huettel, M.: Benthic solute exchange and carbon mineralisation in two subtidal sandy sediments: effect of flow, Limnol. Oceanogr., 52, 1943–1963, 2007.

Dalsgaard, T., Stewart, F. J., Thamdrup, B., De Brabandere, L., Revsbech, N. P., Ulloa, O., Canfield, D. E., and DeLong, E. F.: Oxygen at nanomolar levels reversibly suppresses process rates and gene expression in anammox and denitrification in the oxygen minimum zone off Northern Chile, mBio, 5, 01966–01914, 2014.

De'ath, G., Fabricius, K. E., Sweatman, H., and Puotinen, M.: The 27-year decline of coral cover on the Great Barrier Reef and its causes, P. Natl. Acad. Sci. USA, 109, 17995–17999, 2012.

Evrard, V., Glud, R. N., and Cook, P. L. M.: The kinetics of denitrification in permeable sediments, Biogeochemistry, 113, 563–572, 2013.

Eyre, B. D., Glud, R. N., and Pattern, N.: Coral mass spawning – a natural large-scale nutrient enrichment experiment, Limnol. Oceanogr., 53, 997–1013, 2008.

Eyre, B. D., Maher, D. T., and Squire, P.: Quantity and quality of organic matter (detritus) drives N_2 effluxes (denitrification) across seasons, benthic habitats and estuaries, Global Biogeochem. Cy., 27, 1–13, 2013a.

Eyre, B. D., Santos, I. R., and Maher, D. T.: Seasonal, daily and diel N_2 effluxes in permeable carbonate sediments, Biogeosciences, 10, 2601–2615, https://doi.org/10.5194/bg-10-2601-2013, 2013b.

Eyre, B. D., Andersson, A. J., and Cyronak, T.: Benthic coral reef calcium carbonate dissolution in an acidifying ocean, Nat. Clim. Change, 4, 969–976, 2014.

Gao, H., Schreiber, F., Collins, G., Jensen, M. M., Kostka, J. E., Lavik, G., de Beer, D., Zhou, H. Y., and Kuypers, M. M. M.: Aerobic denitrification in permeable Wadden Sea sediments, ISME J., 4, 417–426, 2010.

Gao, H., Matyka, M., Liu, B., Khalili, A., Kostka, J. E., Collins, G., Jansen, S., Holtappels, M., Jensen, M. M., Badewien, T. H., Beck, M., Grunwald, M., de Beer, D., Lavik, G., and Kuypers, M. M. M.: Intensive and extensive nitrogen loss from intertidal permeable sediments of the Wadden Sea, Limnol. Oceanogr., 57, 185–198, 2012.

Glud, R. N., Eyre, B. D., and Patten, N.: Biogeochemical responses to mass coral spawning at the Great Barrier Reef: effects on respiration and primary production, Limnol. Oceanogr., 53, 1014–1024, 2008.

Howarth, R. W. and Marino, R.: Nitrogen as the limiting nutrient for eutrophication in coastal marine ecosystems: evolving views over three decades, Limnol. Oceanogr., 51, 364–376, 2006.

Huettel, M. and Rusch, A.: Transport and degradation of phytoplankton in permeable sediment, Limnol. Oceanogr., 45, 534–549, 2000.

Iversen, N. and Jørgensen, B. B.: Diffusion coefficients of sulfate and methane in marine sediments – influence of porosity, Geochim. Cosmochim. Ac., 57, 571–578, 1993.

Jahnke, R.: A model of microenvironements in deep sea sediments – formation and effects on porewater profiles, Limnol. Oceanogr., 30, 956–965, 1985.

Jørgensen, B. B.: Bacterial sulfate reduction within reduced microniches of oxidized marine sediments, Mar. Biol., 41, 7–17, 1977.

Kessler, A. J., Glud, R. N., Cardenas, M. B., Larsen, M., Bourke, M., and Cook, P. L. M.: Quantifying denitrification in rippled permeable sands through combined flume experiments and modelling, Limnol. Oceanogr., 57, 1217–1232, 2012.

Kessler, A. J., Glud, R. N., Cardenas, M. B., and Cook, P. L. M.: Transport zonation limits coupled nitrification-denitrification in permeable sediments, Environ. Sci. Technol., 47, 13404–13411, 2013.

Kessler, A. J., Cardenas, M. B., Santos, I. R., and Cook, P. L. M.: Enhancement of denitrification in permeable carbonate sediment due to intra-granular porosity: a multi-scale modelling analysis, Geochim. Cosmochim. Ac., 141, 440–453, 2014.

Marchant, H. K., Lavik, G., Holtappels, M., and Kuypers, M. M. M.: The fate of nitrate in intertidal permeable sediments, Plos One, 9, https://doi.org/10.1371/journal.pone.0104517, 2014.

Marchant, H. K., Holtappels, M., Lavik, G., Ahmerkamp, S., Winter, C., and Kuypers, M. M. M.: Coupled nitrification-denitrification leads to extensive N loss in subtidal permeable sediments, Limnol. Oceanogr., 61, 1033–1048, 2016.

Marchant, H. K., Ahmerkamp, S., Lavik, G., Tegetmeyer, H. E., Graf, J., Klatt, J. M., Holtappels, M., Walpersdorf, E., and Kuypers, M. M. M.: Denitrifying community in coastal sediments performs aerobic and anaerobic respiration simultaneously, ISME J., 11, 1799–1812, 2017.

Nielsen, L. P.: Denitrification in sediment determined from nitrogen isotope pairing, FEMS Microbiol. Ecol., 86, 357–362, 1992.

Precht, E. and Huettel, M.: Rapid wave driven porewater exchange in a permeable coastal sediment, J. Sea. Res., 51, 93–107, 2004.

Precht, E., Franke, U., Polerecky, L., and Huettel, M.: Oxygen dynamics in permeable sediments with wave-driven porewater exchange, Limnol. Oceanogr., 49, 693–705, 2004.

Rao, A. M. F., McCarthy, M. J., Gardner, W. S., and Jahnke, R. A.: Respiration and denitrification in permeable continental shelf deposits on the South Atlantic Bight: rates of carbon and nitrogen cycling from sediment column experiments, Cont. Shelf Res., 27, 1801–1819, 2007.

Reynolds, W. D.: Saturated hydraulic properties: laboratory methods, in: Soil Sampling and Method of Analysis, edited by: Carter, M. R. and Gregorich, E. G., CRC Press, Boca Raton, Florida, 1013–1024, 2008.

Risgaard Petersen, N., Nielsen, L. P., Rysgaard, S., T, D., and Meyer, R. L.: Application of the isotope pairing technique in sediments where anammox and denitrification coexist, Limnol. Oceanogr.-Meth., 1, 63–73, 2003.

Risgaard-Petersen, N., Langezaal, A. M., Ingvardsen, S., Schmid, M. C., Jetten, M. S. M., Op den Camp, H. J. M., Derksen, J. W. M., Pina-Ochoa, E., Eriksson, S. P., Nielsen, L. P., Revsbech, N. P., Cedhagen, T., and van der Zwaan, G. J.: Evidence for complete denitrification in a benthic foraminifer, Nature, 443, 93–96, 2006.

Santos, I. R., Eyre, B. D., and Glud, R. N.: Influence of porewater advection on denitrification in carbonate sands: evidence from repacked sediment column experiments, Geochim. Cosmochim. Ac., 96, 247–258, 2012.

Schulz, H. D. and Zabel, M.: Marine Geochemistry, Springer, Berlin, 2005.

Seitzinger, S. P.: Denitrification in freshwater and coastal marine ecosystems: ecological and geochemical significance, Limnol. Oceanogr., 33, 702–724, 1988.

Sokoll, S., Lavik, G., Sommer, S., Goldhammer, T., Kuypers, M. M. M., and Holtappels, M.: Extensive nitrogen loss from permeable sediments off North-West Africa, J. Geophys. Res.-Biogeo., 121, 1144–1157, 2016.

16

Reviews and syntheses: Ice acidification, the effects of ocean acidification on sea ice microbial communities

Andrew McMinn

Institute of Marine and Antarctic Science, University of Tasmania, Hobart 7001, Tasmania, Australia

Correspondence to: Andrew McMinn (andrew.mcminn@utas.edu.au)

Abstract. Sea ice algae, like some coastal and estuarine phytoplankton, are naturally exposed to a wider range of pH and CO_2 concentrations than those in open marine seas. While climate change and ocean acidification (OA) will impact pelagic communities, their effects on sea ice microbial communities remain unclear.

Sea ice contains several distinct microbial communities, which are exposed to differing environmental conditions depending on their depth within the ice. Bottom communities mostly experience relatively benign bulk ocean properties, while interior brine and surface (infiltration) communities experience much greater extremes.

Most OA studies have examined the impacts on single sea ice algae species in culture. Although some studies examined the effects of OA alone, most examined the effects of OA and either light, nutrients or temperature. With few exceptions, increased CO_2 concentration caused either no change or an increase in growth and/or photosynthesis. In situ studies on brine and surface algae also demonstrated a wide tolerance to increased and decreased pH and showed increased growth at higher CO_2 concentrations. The short time period of most experiments (< 10 days), together with limited genetic diversity (i.e. use of only a single strain), however, has been identified as a limitation to a broader interpretation of the results.

While there have been few studies on the effects of OA on the growth of marine bacterial communities in general, impacts appear to be minimal. In sea ice also, the few reports available suggest no negative impacts on bacterial growth or community richness.

Sea ice ecosystems are ephemeral, melting and re-forming each year. Thus, for some part of each year organisms inhabiting the ice must also survive outside of the ice, either as part of the phytoplankton or as resting spores on the bottom. During these times, they will be exposed to the full range of co-stressors that pelagic organisms experience. Their ability to continue to make a major contribution to sea ice productivity will depend not only on their ability to survive in the ice but also on their ability to survive the increasing seawater temperatures, changing distribution of nutrients and declining pH forecast for the water column over the next centuries.

1 Introduction

Sea ice is widely recognized as one of the most extreme habitable environments on earth (Thomas and Dieckmann, 2003; Martin and McMinn, 2017). Organisms living within it can be exposed to temperatures of below −20 °C and salinities greater than 200 for extended periods of time (Arrigo, 2014; Thomas and Dieckmann, 2003). These environments also endure long periods of darkness and extremes in nutrient concentration, dissolved gases (O_2, CO_2) and pH (Thomas and Dieckmann, 2003; McMinn et al., 2014). In spite of these conditions, some sea ice habitats are very productive and often support dense microbial biomass (Arrigo, 2014). They also play a key supporting role in maintaining energy flow through polar marine food webs in winter and spring when sea ice covers much of the ocean surface and there is very little pelagic primary production (Arrigo and Thomas, 2004).

During the annual cycle of sea ice formation and melting, associated organisms are naturally exposed to a large range in gas concentrations, including CO_2. Gases are expelled from the forming ice in autumn and winter and become concentrated in the trapped brine (Fig. 1). Consequently, there is a relationship between the CO_2 concentration of the brine and that of the original sea water. As dissolved CO_2 concentra-

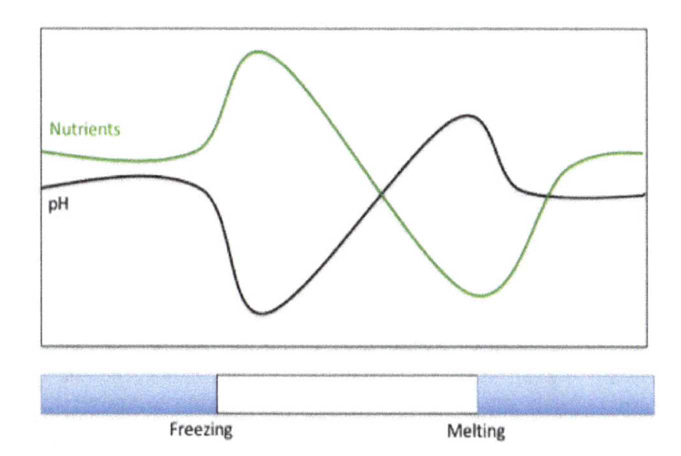

Figure 1. Generalized annual succession of pH and nutrient concentration in sea ice brine ecosystems in response to ice formation in winter and melting in spring. On the bar at the bottom of the figure, blue represents open water and white represents the presence of ice. "Nutrients" is a general trend showing the concentration of inorganic macronutrients.

tions continue to rise with ocean acidification (OA), there will also be a commensurate increase in the CO_2 concentration in the trapped brine. Macronutrients are also expelled from growing ice and concentrated in the brine in autumn but are taken up for cell growth in subsequent months (Fig. 1). This review will examine the effects of increasing CO_2 concentrations and decreasing pH on sea ice microbial communities. The review focuses on Antarctic sea ice communities as few Arctic studies have been identified.

2 Natural pH fluctuations

There are strong seasonal cycles in pH in the seasonal ice-covered waters of the Southern Ocean due to brine drainage from forming sea ice in winter and high levels of biological activity in summer (Miller et al., 2011). In Prydz Bay and the Ross Sea this has led to seasonal differences of up to 0.6 of a pH unit (Gibson and Trull, 1999; McNeil et al., 2010), which is greater than the 0.4 increase usually predicted for the end of the century (IPCC, 2014). Large seasonal differences in pH have also been observed in McMurdo Sound, Antarctica (Kapsenberg et al., 2015; Matson et al., 2014), where pH was relatively high but stable during winter and was lower and dynamic in summer. Large diurnal fluctuations in pH beneath the sea ice were also observed and it was suggested that these changes could be almost entirely explained by the biological processes of photosynthesis and respiration (Matson et al., 2014).

During the early stages of ice formation in early winter, the concentration of CO_2 within the ice is determined solely by physical processes associated with freezing and brine rejection. While low temperatures increase the solubility of CO_2, increasing brine salinity during sea ice growth leads to supersaturation (Papadimitriou et al., 2003). Change in temperature is subsequently the dominant factor determining CO_2 concentration in brine channels (Geilfus et al., 2014). However, the incorporated sea ice algae take up CO_2 for photosynthesis and, together with other components of the microbial food web such as bacteria, release it during respiration. The balance of these processes commonly produces brine CO_2 concentrations that are lower in summer and higher in winter relative to atmospheric levels (Gleitz et al., 1996; Thomas and Dieckmann, 2003). Hare et al. (2013) examined the development of pH in an outdoor experimental sea ice facility in Winnipeg, Canada. Experiments were run over several weeks and ice attained a thickness of up to 22 cm. pH profiles had a characteristic "C" profile with pH values greater than 9 at the surface and in the bottom 2 cm. Interior values were below that of the source water (8.4) with some as low as 7.1. The reason for this pattern was thought to be entirely due to abiotic factors as no algal growth was observed. The higher values at the surface and bottom reflected equilibrium with the atmosphere and underlying sea water, respectively, while those in the middle reflected CO_2 rejection during brine formation.

Changes in pH and CO_2 concentration in natural brines from the Weddell Sea in early summer have been documented by Papadimitriou et al. (2007). The in situ pH varied between 8.41 and 8.82, up more than 0.7 units from the underlying sea water. The concentration of CO_2 ranged from 3.1 to 15.9 $\mu mol\ kg^1$ and variations were consistent with biological activity (Papadimitriou et al., 2007).

As global temperatures warm, sea ice extent will inevitably decrease and the ice will form later in the season and melt sooner. As a consequence, average ice thickness will probably decrease. These large-scale changes are already clearly evident in the Arctic (Barhhart et al., 2016). Thinner ice would result in average irradiances within and beneath the ice being higher, although this would also be modified by changes in snow thickness. However, there are still large uncertainties in predictions of future changes in precipitation over sea ice and, while a warmer atmosphere might be expected to increase snowfall, there is no clear evidence of this happening so far or whether in future it will fall as snow or rain (Bracegirdle et al., 2008; Leonard and Maksym, 2011).

3 Biological communities

Sea ice is not a solid, uniform, homogenous layer but instead contains many different habitats that can be colonized by a large range of organisms. Each of these habitats is characterized by different physical and chemical conditions. Initially any organism present in the underlying water column has the potential to be trapped within the ice as it forms. The early process of ice formation (frazil ice formation) can concentrate the phytoplankton biomass many times above that of the surrounding water column (Garrison et al., 1983). All species present in the water column can be harvested, although some, such as *Nitzschia stellata*, have the ability to greatly increase the likelihood of their capture by the release of ice active substances (Raymond et al., 1994; Ugalde et al., 2014). Young ice typically contains a highly diverse microalgal community (Scott et al., 1994) but this diversity decreases with time as those species that are better adapted to this environment outcompete those that are less well adapted. The biological communities can be differentiated by where they occur in the ice: in brine channels within the ice (brine communities), on the undersurface of the ice (bottom communities) or at the snow–ice interface (surface/infiltration communities) (Horner 1985; Arrigo, 2014; Bluhm et al., 2017) Bottom communities, which are dominant in both Arctic and Antarctic coastal locations, typically inhabit a benign and equitable environment characterized by minimal changes in temperature and salinity, maximum access to nutrients but low light levels. Brine communities, which dominate in pack ice, experience extreme fluctuations in temperature and salinity, high to very low nutrient concentrations and intermediate light levels (Thomas and Dieckmann, 2003). Surface communities, which are often the dominant community in summer in pack ice, have access to the highest light but also experience the lowest temperature and nutrient levels. These communities also experience different CO_2 / pH conditions. Bottom communities are constantly in contact with the underlying water column and typically experience the bulk ocean properties. Brine and surface communities, by contrast, experience supersaturated CO_2 concentrations during early ice formation but very low levels later in the season when the CO_2 is exhausted by photosynthesis. In coastal fast-ice communities >90 % of the algal biomass and productivity is concentrated in the bottom 10 cm. However, in pack ice communities the biomass is more evenly spread, with significant biomass and productivity occurring both in brine channels and at the snow–ice interface (Meiners et al., 2012).

Major micro algal groups have been found to respond differently to CO_2 availability depending on physiological factors such as the presence of a carbon concentrating mechanism (CCM), which enables cells to use bicarbonate or the type of RuBisCo present (Eberlein et al., 2014). Thus, the response of sea ice algal communities to ocean acidification will reflect their taxonomic composition. Diatoms are often the most abundant and also the most studied group of sea ice microbes, but other eukaryotic and prokaryote groups are often important. While diatoms tend to dominate bottom communities (Garrison, 1991), dinoflagellates and other flagellate groups tend to dominate interior and surface communities (Stoecker et al., 1992; Thomson et al., 2006). There is also a diverse bacterial community (Mock and Thomas, 2005; Bowman et al., 2012; Bowman, 2013; Bluhm et al., 2017) and evidence of an active microbial loop (Martin et al., 2012). While most diatoms, including sea ice species, have been found to possess a CCM, this has not been tested on most other groups and so their response to increase in CO_2 availability is unknown.

4 Ocean acidification experiments

4.1 Sea ice algae

Studies on ocean acidification effects on sea ice algae and bacteria have taken a number of different approaches. Some have just examined the effects of increased CO_2 concentration in isolation (Coad et al., 2016; Torstensson et al., 2015), while others have looked at the combined effects of co-stressors such as temperature (Pančić et al., 2015; Torstensson, 2012a, b), light (Xu et al., 2014a; Heiden et al., 2016) or iron (Xu et al., 2014b). Most studies have been undertaken on cultures of single species in a laboratory (Mitchell and Beardall, 1996; Xu et al., 2014a; Pančić et al., 2015; Torstensson et al., 2012a, b; Young et al., 2015), although there has been one study of a brine community (Coad et al., 2016) and two in situ experiments on brine communities (McMinn et al., 2014, 2017). All but one of these studies (Torstensson et al., 2015) was conducted on a timescale of less than 30 days and most were less than 10 days (Table 1).

Studied in isolation, increased CO_2 concentration seems only to have negative effects on growth and photosynthesis at levels above $\sim 1000\,\mu$atm (Torstensson et al., 2015). When a natural diatom-dominated sea ice brine community was incubated over a CO_2 gradient from 400 to 11 300 µatm for up to 18 days, only the treatment incubated at the most extreme concentration was significantly different from the control and other treatments (Coad et al., 2016). In a further experiment, they simulated future spring conditions by allowing a surface ice community to melt into seawater with the same large CO_2 gradient. Once again, only the most extreme CO_2 concentrations induced significant reductions in growth and photosynthesis. These results are consistent with most other studies on polar marine diatoms, which show increased growth rates and photosynthetic performance at CO_2 concentrations up to at least 1000 µatm (Trimborn et al., 2013; Hoppe et al., 2013). Future increases in marine CO_2 concentration, however, are likely to be accompanied by changes in temperature, light, macro nutrient and iron concentrations (Boyd et al., 2016b) and interactions between these and other factors are likely to produce unpredictable results (Boyd and Hutchins, 2012). However, not all of these changes are likely to occur within the sea ice environment.

Studies on current Southern Ocean phytoplankton communities have identified the importance of iron and irradiance as being the key drivers determining changes in community composition and productivity (Feng et al., 2010). Others have suggested that future changes in pCO_2, temperature and stratification, which affect average irradiance and nutrient availability, will also be significant (Boyd et al., 2016a;

Table 1. Ocean acidification experiments on sea ice microorganisms. The column "+/0/−" indicates whether the organisms in the study responded positively "+", no change "0" or negatively "−" to the change in CO_2 concentration.

Species/community	Days	$T°$	CO_2 (µatm)	Site	+/0/−	Author
Navicula directa	7	0.5–4.5°	380–960	Lab	0	Torstensson et al. (2012a)
Nitzschia lecointei	14	−1.8–2.5°	390–960	Lab	−/0	Torstensson et al. (2013)
Brine community	8	−2.5°	587–6066	in situ	+	McMinn et al. (2014)
Chlamydomonas sp.	28	5°	390–1000	Lab	0	Xu et al. (2014a)
Fragilariopsis cylindrus	7	1–8°	Ph 7.1–8.0	Lab	0	Pančić et al. (2015)
Nitzschia lecointei	194	−1.8°	280–960	Lab	0/−	Torstensson et al. (2015)
Brine community	8	−5°	400–11 300	lab	+	Coad et al. (2016)
Fragilariopsis curta	4	4°	180–1000	Lab	+	Heiden et al. (2016)
Surface brine community	6	0°	45–4102	in situ	+	McMinn et al. (2017)

Constable et al., 2014). However, not all of these factors are directly relevant to ice algal communities. For instance, unlike the oceans, the temperature of sea ice habitats cannot rise above ~0 °C. Increasing sea water temperatures may well change the temporal and spatial distribution of sea ice and ice thickness but they will have only a small effect on micro algal physiology.

The combined effects of temperature and CO_2 concentration on the bipolar sea ice/phytoplankton species, *Fragilariopsis cylindrus* was examined in short incubations (7 days) by Pančić et al. (2015). Significant interactions between temperature and pH were identified, but the two factors produced opposite effects. Growth rates increased with increasing temperatures but decreased with decreasing pH, resulting in little overall change with treatment. As a result, it was concluded that *F. cylindrus* would largely be unaffected by increasing ocean acidification. These experiments were conducted over a wide range of temperature (1 to 8 °C) and pH (7.1 to 8.0). A similar study on the sea ice/benthic diatom *Navicula directa* likewise found only minor responses to increased CO_2 with no synergistic effects with temperature, although a smaller range of temperature (0.5–4.5 °C) and pH (7.9–8.2) was used (Torstensson et al., 2012a). These authors emphasized the need to examine multiple strains of a species and over longer incubation times before conclusions could be reached on its long-term response. It should be noted that, while both these studies used sea ice algae taxa, the cells were exposed to temperatures well above 0 °C, i.e. temperatures that could never be experienced by cells living within sea ice. Although the maximum temperature of sea ice is fixed, the average temperature is still likely to rise. This could result in more intensive heterotrophic processes, via increased grazing rates and nutrient regeneration (Melnikov, 2009). However, it is probable that these processes, which already occur in spring and summer, will simply develop a little earlier.

Future changes in iron supply have been identified as likely to have a significant but unquantified effect on much of the world's phytoplankton (Shi et al., 2010), but the effects on sea ice communities are unknown: to date there have been no studies on the combined effects of increased CO_2– and iron limitation on sea ice algae. A study of Southern Ocean phytoplankton, which contains many of the same species as sea ice communities, under co-limitation by CO_2, iron and other factors found that competition was likely to induce taxonomic changes in community composition, favouring small diatoms (Xu et al., 2014b). Algae growing at the ice water interface (bottom communities), like the phytoplankton, experience iron-limiting conditions and show evidence of chronic iron stress (Pankowski and McMinn, 2008, 2009). However, the iron concentration within brine channels is one to two orders of magnitude greater than in the underlying seawater (Lannuzel et al., 2011) and sea ice brine algae have been shown not to be iron stressed at all (Pankowski and McMinn, 2008, 2009). Under these circumstances, iron cannot be considered a co-stressor.

The combined effect of CO_2 concentration and increased light on the ice diatom, *Fragilariopsis curta* was examined by Heiden et al. (2016). In this study growth and photophysiology were not stimulated at relevant light (20 µmol photons $m^{-2}\,s^{-1}$) and OA-relevant elevated pCO_2 (1000 µatm). These authors also showed that there were large variations in species-specific responses. The growth and photosynthetic response of the chlorophyte *Chlamydomonas*, isolated from Antarctic sea ice, to ocean acidification and photoperiod, was examined by Xu et al. (2014a) over the course of 28 days. While *Chlamydomonas* is not a common ice algal taxon, it nonetheless represents an organism adapted to living in the ice. Their data showed that CO_2 concentration had a minimal effect on the response to differing photoperiods.

There have been few in situ experiments with natural sea ice communities. McMinn et al. (2014, 2017) incubated dinoflagellate-dominated brine algal communities in situ in McMurdo Sound, Antarctica. The 2014 study was conducted in spring, while the 2017 study was conducted in late summer. They used sack holes to extract the brine (McMinn et al., 2009), adjusted the CO_2 concentrations of the treatments and then returned them to the same depth within the ice at which they were collected. Incubations were relatively short, up to 6 days. In both studies growth and photosynthesis were only affected when the pH fell below approximately 7.5. In a companion set of experiments in both studies, the carbon system was manipulated to hold the pH constant at ~ 8.0 while providing a gradient in CO_2 concentrations up to 2700 µatm. Growth increased by approximately 20 % and remained constant at even the highest concentrations. This 20 % increase in growth with increased CO_2 supply is of a similar scale to those estimated for a variety of temperate phytoplankton taxa on a similar CO_2 gradient (Rost et al., 2003; Riebesell, 2004).

The concentration of dissolved CO_2 in seawater is usually considered insufficient to maintain maximum growth rates in phytoplankton. As a consequence, most species have developed strategies to alleviate this stress, collectively referred to as carbon concentrating mechanisms (CCMs) (Raven et al., 2011). The presence of a CCM allows most marine microalgae, and diatoms in particular, to concentrate inorganic carbon as CO_2 and/or HCO_3 (Giordano et al., 2005) and thus alleviate much of this stress. Ocean acidification increases the natural CO_2 concentration and so potentially favours those species without a CCM; however, it will also reduce the need to invest in this energetically demanding function for species with CCMs (Beardall and Raven, 2004). Inorganic carbon uptake by the sea ice diatom *Nitzschia stellata* was examined by Mitchell and Beardall (1996). They demonstrated that this taxon was capable of carbon-dependent photosynthesis at rates greater than the CO_2 supply rate to the cell and that much of the carbonic anhydrase activity was associated with the cell surface; these characteristics demonstrated the presence of an active CCM. Other studies (Tortell et al., 2013; Gibson et al., 1999; McMinn et al., 1999) have also found evidence of CCM activity from either carbon isotope analysis or carbonic anhydrase activity. So far, all sea ice species examined have shown evidence of an active CCM. However, little is still known about how the presence or up/down regulation of CCMs will respond to future changes in CO_2 concentration (Raven et al., 2011).

While most sea ice algal OA research has focused on Antarctic communities, those from the Arctic would be expected to respond in a similar way as they share similar taxonomic compositions, i.e. dominated by diatoms, and have similar temperature, salinity and irradiance profiles (Horner, 1985). There are, however, some structural differences in sea ice ecosystems between the two regions, resulting in brine algal communities being virtually absent from the Arctic and the vast bulk of the algal biomass being located on the bottom of the ice (Horner, 1985; Thomas and Dieckmann, 2003). As in the Antarctic, bottom communities are generally less exposed to extreme fluctuations in temperature, salinity, gases and nutrients. Pančić et al. (2015) examined the response of the bipolar sea ice diatom species, *F. cylindrus*, isolated from the Greenland Sea, and concluded that, as in the Antarctic, OA would have little effect on the growth of this species.

4.2 Bacteria

Sea ice contains an abundant and diverse psychrophilic bacterial community (Mock and Thomas, 2005; Bowman et al., 2012; Bowman, 2013). Ocean acidification effects on marine bacterial communities in general appear to be minimal (Oliver et al., 2014; Lin et al., 2017). They are able to mitigate the effect of decreased pH by enhancing the expression of genes encoding proton pumps, such as respiration complexes, proteorhodopsin and membrane transporters (Bunse et al., 2016). Bacterial growth rates usually appear to be closely coupled with phytoplankton growth (Engel et al., 2013). Likewise, in sea ice studies, bacterial growth rates have been found to increase with increasing CO_2 concentration, probably reflecting an increase in sea ice algae and an increased rate of DOC production (Torstensson et al., 2015). A pH range of 8.203–9.041 has also been found to have no effect on the community richness or diversity of Antarctic fast-ice bacteria (Torstensson et al., 2013).

5 Discussion and summary

Working with sea ice algal communities is implicitly more difficult than working with pelagic phytoplankton communities because, like benthic microbial ecosystems, they live in biofilms that have steep and highly structured vertical chemical and physical gradients. Most studies on sea ice algae physiology, including responses to pH and CO_2, have removed the cells from their ice substrate and suspended them in water. This process both exposes cells to osmotic shock and also destroys the highly structured vertical gradients. In situ studies, where the communities are not handled or moved but are more technically challenging, are likely to produce the most meaningful results. In situ methods have been used in studies of sea ice algal photosynthesis and photophysiology for several years (McMinn et al., 2000, 2012; Kühl et al., 2001). An innovative in situ methodology for studying OA effects of on sea ice bottom communities has also been recently published by Barr et al. (2017), although no results from this study are yet forthcoming. This in situ methodological approach is similar to that used in Free Ocean CO_2 Enrichment system (FOCE) on Antarctic benthic communities (Stark et al., 2017).

While studying sea ice bottom communities by in situ methods is achievable, studying brine communities in situ presents additional problems. These communities, which are entombed in brine channels and pockets, are often completely isolated and invisible. Currently used methods, such as variable fluorescence and microsensors, are unable to penetrate the ice matrix to measure biological responses. Even the in situ chemistry of the brine chambers cannot be measured. Studies by McMinn et al. (2012, 2014) extracted the brine algae from the ice by drainage into sack holes, manipulated the carbon chemistry and then reinserted the algal samples back into the ice. While not truly in situ, the algae in these studies were at least exposed to natural irradiances and temperatures. Future studies on the effects of OA on sea ice algal communities will also need to focus on using in situ approaches and more accurately reproducing natural conditions.

Most studies on the effects of OA on ice algae have used only short incubation times, i.e. less than 10 days. In the first long-term experiment (194 days), Torstensson et al. (2015) showed that significant impacts were not detected until after 147 days. Interestingly, however, most sea ice environments are both ephemeral and constantly changing. As pH and CO_2 concentrations within the ice are determined by temperature, natural sea ice communities are very unlikely to be exposed to a constant pH environment over this length of time. While longer experiments would seem to give more reliable responses, there is little point in extending incubation times beyond the periodicity of natural change. Future studies could include these fluctuations in their experimental design.

It is well known that the physiological responses of diatoms in culture changes with time (Anderson, 2005). Some of the culture experiments discussed herein used strains of diatoms that had been isolated decades earlier (Xu et al., 2014; Heiden et al., 2016). Therefore, the extreme age of the culture used needs to be taken into account when interpreting the physiological response. Where possible, culture experiments should utilize freshly isolated strains to optimize meaningful physiological responses.

Brine algae and, to a lesser degree, bottom algae are naturally exposed to a large range of pH and CO_2 concentrations during seasonal ice formation and melting (Hare et al., 2013), and this range is typically greater than predicted changes resulting from OA. As a result, it is not surprising that ice algae appear to be well adapted to changes in carbon chemistry and show only minimal responses to these changes. Likewise, many coastal phytoplankton communities, which are also exposed to large diurnal and seasonal changes in pH and CO_2 concentrations, show either a positive response or no response to OA (Grear et al., 2017). Larger responses have been detected in open ocean areas where natural fluctuations are much lower (Li et al., 2016).

Sea ice ecosystems are ephemeral and in most relevant locations they melt and re-form each year. Thus, for some part of each year organisms inhabiting the ice must also survive outside of the ice, either as part of the phytoplankton or as resting spores on the bottom. During these times, they will be exposed to the full range of co-stressors that pelagic organisms experience (Xu et al., 2014c). Some sea ice taxa go into dormancy during these periods, e.g. *Polarella glacialis* (Thompson et al., 1996), while others, such as *Fragilariopsis curta* and *Fragilariosis cylindrus*, make a major contribution to phytoplankton biomass. Most sea ice microorganisms do not appear to be badly impacted by moderate increases in CO_2. However, their ability to continue to make a major contribution to sea ice productivity will depend not only on their ability to survive in the ice but also on their ability to survive the increasing seawater temperatures, changing distribution of nutrients and declining pH forecast for the water column over the next centuries (Boyd and Hutchins, 2012).

Competing interests. The author declares that he has no conflict of interest.

Special issue statement. This article is part of the special issue "The Ocean in a High-CO_2 World IV". It is a result of the 4th International Symposium on the Ocean in a High-CO_2 World, Hobart, Australia, 3–6 May 2016.

Acknowledgements. Australian Antarctic Science Grants programme and Antarctic Gateway Partnership are acknowledged for their continuous financial support for sea ice research in general and for this review in particular.

Edited by: Elizabeth H. Shadwick

References

Anderson, R. A.: Algal culturing techniques, Elsevier Academic Press, Burlington MA, USA, 2005.

Arrigo, K. R.: Sea Ice Ecosystems, Ann. Rev. Mar. Sci., 6, 439–467, 2014

Arrigo, K. R. and Thomas, D. N.: Large scale importance of sea ice biology in the Southern Ocean, Antarct. Sci., 16, 471–486, 2004.

Barhhart, K. R., Miller, C. R., Overeem, I., and Kay, J. E.: Mapping the future expansion of Arctic open water, Nature Climate Change, 6, 280–285, 2016.

Barr, N. C., Lohrer, A. M., and Cummings, V. J.: An *in situ* incubation method for measuring the productivity and responses of under-ice algae to ocean acidification and warming in polar marine habitats, Limnol. Oceanogr.-Meth., 15, 264–275, https://doi.org/10.1002/lom3.10154, 2017.

Beardall, J. and Raven J. A.: The potential effects of global climate change on microalgal photosynthesis, growth and ecology, Phycologia, 43, 26–40, 2004.

Bluhm, B. A., Hop, H., Melnikov, I. A., Poulin, M., Vihtakari, M., Collins, M. R., Gradinger, R., Juul-Pedersen, T., and von Quillfeldt, C.: Sea ice biota, State of the Arctic Report, 32–60, 2013.

Bowman, J. P.: Sea ice microbial communities, in: The Prokaryotes, Prokaryotic communities and ecophysiology, edited by: Rosenberg, E., DeLong, E. F., and Lory, S., Stackbrandt Eand Thompson F, Springer-Verlag, Berlin, 139–161, 2013.

Bowman, J. S., Rasmussen, S., Blom, N., Deming, J. W., Rysgaard, S., and Sicheritz-Ponten, T.: Microbial community structure of

Arctic multiyear sea ice and surface seawater by 454 sequencing of the 16S RNA gene, ISME J., 6, 11–20, 2012.

Boyd, P. W., Cornwall, C. E., Davidson, A., Doney, S. C., Fourquez F., Hurd, C. L., and McMinn, A.: Environmental variability, heterogeneity, ocean biota, and climate change, Glob. Change Biol., 22, 2633–2650, 2016a.

Boyd, P. W., Dillingham, P. W., McGraw, C. M., Armstrong, E. A., Cornwall, C. E., Feng, C. L., Hurd, C. L., Gault-Ringold, M., Roleda, M. Y., Timmins-Schiffman, E., and Nunn, B. L.: Physiological responses of a Southern Ocean diatom to complex future ocean conditions, Nature Climate Change, 6, 207–216, 2016b.

Boyd, P. W. and Hutchins, D. A.: Understanding the responses of ocean biota to a complex matrix of cumulative anthropogenic change, Mar. Ecol.-Prog. Ser., 470, 125–135, 2012.

Bracegirdle, T. J., Connolley, W. M., and Turner, J.: Antarctic climate change over the twenty first century, J. Geophys. Res., 113, D03103, https://doi.org/10.1029/2007JD008933, 2008.

Bunse, C., Lundin, D., Karlsson, C. M. G., Vila-Costa, M., Palovaara, J., Akram, G., Svensson, L., Holmfeldt, K., González, J. M., Calvo, E., Pelejero, C., Marrasé, C., Dopson, M., Gasol, J. M., and Pinhassi, J.: Response of marine bacterioplankton pH homeostasis gene expression to elevated CO_2, Nature Climate Change, 5, 483–491, 2016.

Coad, T., McMinn, A., Nomura, D., and Martin, A.: Effect of elevated CO_2 concentration on the photophysiology and growth of microalgae in Antarctic pack ice algal communities, Deep-Sea Res. Pt. II, 131, 160–169, 2016.

Constable, A. J., Melbourne-Thomas, J., Corney, S. P., Arrigo, K. R., Barbraud, C., Barnes, D. K. A., Bindoff, N. L., Boyd, P. W., Brandt, A., Costa, D. P., Davidson, A. T., Ducklow, H. W., Emmerson, L., Fukuchi, M., Gutt, J., Hindell, M. A., Hofmann, E. E., Hosie, G., Iida, H., Jacob, S., Johnston, N. M., Kawaguchi, S., Kokubun, N.,Koubbi, P., Lea, M. A., Makhado, A., Massom, R., Meiners, K., Meredith, M. P., Murphy, E. J., Nicol, S., Reid, K., Richerson, K., Riddle, M. J., Rintoul, S., Smith Jr., W. O., Southwell, C., Stark, J. S., Sumner, M., Swadling, K. M., Takahashi, K. T., Trathan, P. N., Welsford, D. C., Weimerskirch, H., Westwood, K. J., Wienecke, B. C., Wolf-Gladrow, D. C., Wright, S. W., Xavier, J. C., and Ziegler, P.: Climate change and Southern Ocean ecosystems I: how changes in physical habitats directly affect marine biota, Glob. Change Biol. 20, 3004–3025, 2014.

Eberlein, T., van de Waal, D. B., and Rost, B.: Differential effects of ocean acidification on carbon acquisition in two bloom-forming dinoflagellate species, Physiol. Plant, 151, 468–479, 2014.

Engel, A., Borchard, C., Piontek, J., Schulz, K. G., Riebesell, U., and Bellerby, R.: CO_2 increases ^{14}C primary production in an Arctic plankton community, Biogeosciences, 10, 1291–1308, https://doi.org/10.5194/bg-10-1291-2013, 2013.

Feng, Y., Hare, C. E., Rose, J. M., Handy, S. M., Ditullio, G. R., Lee, P. A., Smith, W. O., Peloquin, J., Tozzi, S., and Sun, J.: Interactive effects of iron, irradiance and CO_2 on Ross Sea phytoplankton, Deep-Sea Res. Pt. I, 57, 368–383, 2010.

Garrison, D. L.: Antarctic sea ice biota, Am. Zool., 31, 17–37, 1991.

Garrison, D. L., Ackley, S. F., and Buck, K. R.: A physical mechanism for establishing algal populations in frazil ice, Nature, 306, 363–365, 1983.

Geilfus, N.-X., Tison, J.-L., Ackley, S. F., Galley, R. J., Rysgaard, S., Miller, L. A., and Delille, B.: Sea ice pCO_2 dynamics and air–ice CO_2 fluxes during the Sea Ice Mass Balance in the Antarctic (SIMBA) experiment – Bellingshausen Sea, Antarctica, The Cryosphere, 8, 2395–2407, doi:10.5194/tc-8-2395-2014, 2014.

Gibson, J. A. and Trull, T. W.: Annual cycle of fCO_2 under sea-ice and in open water in Prydz Bay, East Antarctica, Mar. Chem., 66, 187–200, 1991.

Giordano, M., Beardall, J., and Raven, J. A.: CO_2 concentrating mechanisms in algae: mechanisms, environmental modulation, and evolution, Ann. Rev. Plant Biol., 56, 99–131, 2005.

Gleitz, M., Kukert, H., Riebesell, U., and Dieckmann, G. S.: Carbon acquisition and growth of Antarctic sea ice diatoms in closed bottle incubations, Mar. Ecol.-Prog. Ser., 135, 169–177, 1996.

Grear, J. S., Rynearson, T. A., Montalbano, A. L., Govenar, B., Menden-Deuer, S.: pCO_2 effects on species composition and growth of an estuarine phytoplankton community, Estuar. Coast. Shelf Sci., 190, 40–49, 2017.

Hare, A. A., Wang, F., Barber, D., Geilfus, N-X., Galley, R. J., and Rysgaard, S.: pH evolution in sea ice grown at an outdoor experimental facility, Mar. Chem., 154, 46–54, 2013.

Heiden, J. P., Bischof, K., and Trimborn, S.: Light intensity modulates the response of two Antarctic diatom species to ocean acidification, Front. Mar. Sci., 3, 260, https://doi.org/10.3389/fmars.2016.00260, 2016.

Hoppe, C. J. M., Hassler, C. S., Payne, C. D., Tortell, P. D., Rost, B., and Trimborn, S.: Iron Limitation Modulates Ocean Acidification Effects on Southern Ocean Phytoplankton Communities, PLoS One, 8, e79890. https://doi.org/10.1371/journal.pone.0079890, 2013.

Horner, R. A.: Sea ice biota, CRC Press, Boca Raton, 1–215, 1985.

IPCC: Climate Change 2014: Synthesis Report, Contribution of Working Groups I, II and III to the Fifth Assessment Report of the Intergovernmental Panel on Climate Change, Core Writing Team, edited by: Pachauri, R. K. and Meyer, L. A., IPCC, Geneva, Switzerland, 151 pp., 2014.

Kapsenberg, L., Kelley, A. L., Shaw, E. C., Martz, T. R., and Hofmann, G. E. Near-shore Antarctic pH variability has implications for the design of ocean acidification experiments, Sci. Rep., 5, 9638, https://doi.org/10.1038/srep09638, 2015.

Kühl, M., Glud, R. N., Borum, J., Roberts, R., and Rysgaard, S.: Photosynthetic performance of surface-associated algae below sea ice as measured with a pulse-amplitude-modulated (PAM) fluorometer and O_2 microsensors, Mar. Ecol.-Prog. Ser., 233, 1–14, 2001.

Lannuzel, D., Bowie, A. R., van der Merwe, P. C., Townsend, A. T., and Schoemann, V.: Distribution of dissolved and particulate metals in Antarctic sea ice, Mar. Chem., 124, 134–146, 2011.

Leonard, K. C. and Maksym, T.: The importance of wind-blown snow redistribution to snow accumulation on Bellingshausen Sea ice, Ann. Glaciol., 52, 271–278, 2011.

Li, F., Wu, Y., Hutchins, D. A., Fu, F., and Gao, K.: Physiological responses of coastal and oceanic diatoms to diurnal fluctuations in seawater carbonate chemistry under two CO_2 concentrations, Biogeosciences, 13, 6247–6259, https://doi.org/10.5194/bg-13-6247-2016, 2016.

Lin, X., Huang, R., Li, Y., Wu, Y., Hutchins, D. A., Dai, M., and Gao, K.: Insignificant effects of elevated CO_2 on bacterioplankton community in a eutrophic coastal mesocosm experiment, Biogeosciences Discuss., https://doi.org/10.5194/bg-2017-10, in review, 2017.

Martin, A. and McMinn, A.: Sea ice, extremophiles and life in extra-terrestrial ocean worlds, Int. J. Astrobiol., accepted, https://doi.org/10.1017/S1473550416000483, 2017.

Martin, A., McMinn, A., Davey, S. K., Anderson, J. M., Miller, H. C., Hall, J. A., and Ryan, K. G.: Preliminary evidence for the microbial loop in Antarctic sea ice using microcosm simulations, Antarct. Sci., 24, 547–553, 2012.

Matson, G., Washburn, L., Martz, T. R., and Hofmann, G. E.: Abiotic versus biotic drivers of ocean pH variation under fast sea ice in McMurdo Sound, Antarctica, PLoS ONE, 9, e107239, https://doi.org/10.1371/journal.pone.0107239, 2014.

McMinn, A., Skerratt, J., Trull, T., Ashworth, C., and Lizotte, M.: Nutrient stress gradient in the bottom 5 cm of fast ice, McMurdo Sound, Antarctica, Polar Biol., 21, 220–227, 1999.

McMinn, A., Ashworth, C., and Ryan, K.: *In situ* net primary productivity of an Antarctic fast ice bottom algal community, Aquat. Microb. Ecol., 21, 177–185, 2000.

McMinn, A., Gradinger, R., and Nomura, D.: 3.8 Biochemical properties of Sea ice, in: Handbook of Field techniques in sea ice research, edited by: Eicken, H., University of Alaska Press, USA, 2009.

McMinn, A., Ashworth, C., Bhagooli, R., Martin, A., Ralph, P., Ryan, K., and Salleh, S.: Antarctic coastal microalgal primary production and photosynthesis, Mar. Biol., 159, 2827–2837, 2012.

McMinn, A., Muller, M., Martin, A., and Ryan, K. G.: The response of Antarctic sea ice algae to changes in pH and CO_2, PloS One, 9, e86984, https://doi.org/10.1371/journal.pone.0086984, 2014.

McMinn, A., Muller, M., Martin, A., and Ryan, K. G.: Effects of changing pH and CO_2 concentration on a late summer surface sea ice community, Mar. Biol., 164, 87, 2017.

McNeil, B. I., Tagliabue, A., and Sweeney, C.: A multi-decadal delay in the onset of corrosive "acidified" waters in the Ross Sea of Antarctica due to strong air-sea CO_2 disequilibrium, Geophys. Res. Lett., 37, L19607, https://doi.org/10.1029/2010gl044597, 2010.

Meiners, K. M., Vancoppenolle, M., Shanassekos, S., Dieckmann, D. S., Thomas, D. N., Tison, J.-L., Arrigo, K. R., Garrison, D., McMinn, A., Lannuzel, D., van der Merwe, P., Swadling, K. M., Smith Jr., W. O., Melnikov, I., and Raymond, B.: Chlorophyll *a* in Antarctic sea ice from historical ice core data, Geophys. Res. Lett., 39, 21602–21602, 2012.

Melnikov, I. A.: Recent sea ice ecosystem in the Arctic Ocean: a review, in: Influence of climate change on the changing Arctic and sub-Arctic conditions, edited by: Nihoul, J. C. J. and Kostiano, A. G., NATO Science for Peace and Security Series C-Environmental Security, 57–71, 2009.

Miller, L. A., Papakyriakou, T. N., Collins, E. R., Deming, J. W., Ehn, J. K., Macdonald, R. W., Mucci, A., Owens, O., Raudsepp, M., and Sutherland, N.: Carbon dynamics in sea ice: A winter flux time series, J. Geophys. Res., 116, C02028, https://doi.org/10.1029/2009JC006058, 2011.

Mitchell, C. and Beardall, J.: Inorganic carbon uptake by an Antarctic sea-ice diatom, *Nitzschia frigida*, Polar Biol., 16, 95–99, 1996.

Mock, T. and Thomas, D. N.: Minireview: Recent advances in sea ice microbiology, Environ. Microbiol., 7, 605–619, 2005.

Oliver, A. E., Newbold, L. K., Whiteley, A. S., and van der Gast, C. J.: Marine bacterial communities are resistant to elevated carbon dioxide levels, Environ. Microbiol. Rep., 6, 574–582, 2014.

Pančić, M., Hansen, P. J., Tammilehto, A., and Lundholm, N.: Resilience to temperature and pH changes in a future climate change scenario in six strains of the polar diatom Fragilariopsis cylindrus, Biogeosciences, 12, 4235–4244, https://doi.org/10.5194/bg-12-4235-2015, 2015.

Pankowski, A. and McMinn, A.: Ferredoxin and flavodoxin in eastern Antarctica pack ice, Polar Biol., 31, 1153–1165, 2008.

Pankowski, A. and McMinn, A.: Iron availability regulates growth, photosynthesis and production of ferredoxin and flavodoxin in Antarctic sea ice diatoms, Aquat. Biol., 4, 273–288, 2009.

Papadimitriou, S., Kennedy, H., Katner, G., Dieckmann, G. S., and Thomas, D. N.: Experimental evidence for carbonate precipitation and CO_2 degassing during sea ice formation, Geochem. Cosmochem. Ac., 68, 1749–1761, 2003.

Papadimitriou, S., Thomas, D. N., Kennedy, H., Haas, C., Kuosa, H., Krell, A., and Dieckmann, G. S.: Biogeochemical composition of natural sea ice brines from the Weddell Sea during early austral summer, Limnol. Oceanogr., 52, 1809–1823, 2007.

Raven, J. A., Giordano, M., Beardall, J., and Maberly, S.: Algal and aquatic plant carbon concentrating mechanisms in relation to environmental change, Photosynth. Res., 109, 281–296, 2011.

Raymond, I. A., Sullivan, C. W., and Devries, A. L.: Release of an ice-active substance by Antarctic sea-ice diatoms, Polar Biol., 14, 71–77, 1994.

Riebesell, U.: Effects of CO_2 enrichment on marine phytoplankton, J. Oceanogr., 60, 719–729, 2004.

Rost, B., Riebesell, U., Burkhardt, S., and Sultemeyer, D.: Carbon acquisition of bloom forming marine phytoplankton, Limnol. Oceanogr., 48, 55–67, 2003.

Scott, P., McMinn, A., and Hosie, G.: Physical parameters influencing diatom community structure in eastern Antarctic sea ice, Polar Biol., 14, 507–517, 1994.

Shi, D., Xu, Y., Hopkinson, F., and Morel, M. M.: Effect of ocean acidification on iron availability to marine phytoplankton, Science, 327, 676–679, 2010.

Stark, J., Roden, N., Johnstone, G., Milnes, M., Black, J., Whiteside, S., Kirkwood, B., Newbery, K., Stark, S., van Ooijen, E., Tilbrook, B., Peltzer, E., and Roberts, D.: Manipulating seawater pH under sea ice: carbonate chemistry of an *in situ* free-ocean CO_2 enrichment experiment in Antarctica (antFOCE), Nature scientific reports, in press, 2017.

Stoecker, D. K., Buck, K. R., and Putt, M.: Changes in the sea-ice brine community during the spring-summer transition, McMurdo Sound, Antarctica, 1. Photosynthetic protists, Mar. Ecol.-Prog. Ser., 84, 265–278, 1992.

Thomas, D. N. and Dieckmann, G. S.: Sea Ice, an introduction to its physics, chemistry, biology and geology, Blackwell, Oxford, UK, 2003.

Thomson, P. G., McMinn, A., Kiessling, I., Watson, M., and Goldsworthy, P. M.: Composition and succession of dinoflagellates and Chrysophytes in the upper fast ice of Davis Station, East Antarctica, Polar Biol., 29, 337–345, 2006.

Torstensson, A., Chierici, M., and Wulff, A.: The influence of increased temperature and carbon dioxide levels on the benthic/sea ice diatom *Navicula directa*, Polar Biol., 35, 205–214, 2012a.

Torstensson, A., Dinasquet, J., Chierici, M., Fransson, A., Riemann, L., and Wulff, A.: Physicochemical control of bacterial and protest community composition and diversity in Antarctic sea ice, Environ. Microbiol., 17, 3869–3881, 2012b.

Torstensson, A., Hedblom, M., Andersson, J., Andersson, M. X., and Wulff, A.: Synergism between elevated pCO_2 and temperature on the Antarctic sea ice diatom Nitzschia lecointei, Biogeosciences, 10, 6391–6401, https://doi.org/10.5194/bg-10-6391-2013, 2013.

Torstensson, A., Hedblom, M., Bjork, M. M., Chierici, C., and Wulff, A.: Long-term acclimation to elevated pCO_2 alters carbon metabolism and reduces growth in the Antarctic diatom *Nitzschia lecointei*, Proc. R. Soc. B, 282, 1513, https://doi.org/10.1098/rspb.2015.1513, 2015.

Tortell, P. D., Mills, M. M., Payne, C. D., Maldonado, M. T., Chierici, M., Fransson, A., Alderkamp, A. C., and Arrigo, K. R.: Inorganic C utilization and C isotope fractionation by pelagic and sea ice algal assemblages along the Antarctic continental shelf, Mar. Ecol.-Prog. Ser., 483, 47–66, 2013.

Trimborn, S., Breneis, T., Sweet, E., and Rost, B.: Sensitivity of Antarctic phytoplankton species to ocean acidification: Growth, carbon acquisition, and species interaction, Limnol. Oceanogr., 58, 997–1007, 2013.

Ugalde, S. C., Martin, A., Meiners, K. M., McMinn, A., and Ryan, K.: Extracellular organic carbon dynamics during an algal bloom in Antarctic fast ice, Aquat. Micobiol. Ecol., 131, 123–139, 2014.

Xu, D., Wang, Y., Fan, X., Wang, D., Ye, N., Zhang, X., Mou, S., Guan, Z., and Zhuang, Z.: Long-term experiment on physiological responses to synergetic effects of ocean acidification and photoperiod in the Antarctic sea ice algae *Chlamydomonas* sp. ICE-L, Environ. Sci. Technol., 48, 7738–7746, 2014a.

Xu, K., Fu, F., and Hutchins, D. A.: Comparative responses of two dominant Antarctic phytoplankton taxa to interactions between ocean acidification, warming, irradiance, and iron availability, Limnol. Oceanogr., 59, 1919–1931, 2014b.

Xu, J., Gao, K., Li, Y., and Hutchins, D. A.: Physiological and biochemical responses of diatoms to projected ocean temperatures, Mar. Ecol.-Prog. Ser., 515, 73–81, 2014c.

Young, J. N., Goldman, J. A. L., Kranz, S. A., Tortell, P. D., and Morell, F. M. M.: Slow carboxylation of Rubisco constrains the rate of carbon fixation during Antarctic phytoplankton blooms, New Phytol., 205, 172–181, 2015.

17

Effects of ultraviolet radiation on photosynthetic performance and N_2 fixation in *Trichodesmium erythraeum* IMS 101

Xiaoni Cai[1,2], David A. Hutchins[2], Feixue Fu[2], and Kunshan Gao[1]

[1]State Key Laboratory of Marine Environmental Science, Xiamen University, Xiamen, Fujian, 361102, China
[2]Department of Biological Sciences, University of Southern California, 3616 Trousdale Parkway, Los Angeles, California, 90089, USA

Correspondence to: Kunshan Gao (ksgao@xmu.edu.cn)

Abstract. Biological effects of ultraviolet radiation (UVR; 280–400 nm) on marine primary producers are of general concern, as oceanic carbon fixers that contribute to the marine biological CO_2 pump are being exposed to increasing UV irradiance due to global change and ozone depletion. We investigated the effects of UV-B (280–320 nm) and UV-A (320–400 nm) on the biogeochemically critical filamentous marine N_2-fixing cyanobacterium *Trichodesmium* (strain IMS101) using a solar simulator as well as under natural solar radiation. Short exposure to UV-B, UV-A, or integrated total UVR significantly reduced the effective quantum yield of photosystem II (PSII) and photosynthetic carbon and N_2 fixation rates. Cells acclimated to low light were more sensitive to UV exposure compared to high-light-grown ones, which had more UV-absorbing compounds, most likely mycosporine-like amino acids (MAAs). After acclimation under natural sunlight, the specific growth rate was lower (by up to 44 %), MAA content was higher, and average trichome length was shorter (by up to 22 %) in the full spectrum of solar radiation with UVR, than under a photosynthetically active radiation (PAR) alone treatment (400–700 nm). These results suggest that prior shipboard experiments in UV-opaque containers may have substantially overestimated in situ nitrogen fixation rates by *Trichodesmium*, and that natural and anthropogenic elevation of UV radiation intensity could significantly inhibit this vital source of new nitrogen to the current and future oligotrophic oceans.

1 Introduction

Global warming is inducing shoaling of the upper mixed layer and enhancing a more frequent stratification of the surface layer, thus exposing phytoplankton cells which live in the upper mixed layer to higher depth-integrated irradiance including UV radiation (Häder and Gao, 2015). The increased levels of UV radiation have generated concern about their negative effects on aquatic living organisms, particularly phytoplankton, which require light for energy and biomass production.

Cyanobacteria are the largest and most widely distributed group of photosynthetic prokaryotes on the Earth, and they contribute markedly to global CO_2 and N_2 fixation (Sohm et al., 2011). Fossil evidence suggests that cyanobacteria first appeared during the Precambrian era (2.8 to 3.5×10^9 years ago) when the atmospheric ozone shield was absent (Sinha and Häder, 2008). Cyanobacteria have thus often been presumed to have evolved under more elevated UV radiation conditions than any other photosynthetic organisms, possibly making them better equipped to handle UV radiation.

Nevertheless, a number of studies have shown that UV-B impairs not only the DNA, pigmentation, and protein structures of cyanobacteria but also several key metabolic activities, including growth, survival, buoyancy, nitrogen metabolism, CO_2 uptake, and ribulose 1,5-bisphosphate carboxylase activity (Rastogi et al., 2014). To deal with UV stress cyanobacteria have evolved a number of defense strategies, including migration to escape from UV radiation, efficient DNA repair mechanisms, programmed cell death, the production of antioxidants, and the biosynthesis of UV-absorbing compounds, such as mycosporine-like amino acids

(MAAs) and scytonemin (Rastogi et al., 2014; Häder et al., 2015).

The non-heterocystous cyanobacterium *Trichodesmium* plays a critical role in the marine nitrogen cycle, as it is one of the major contributors to oceanic nitrogen fixation (Capone et al., 1997) and furthermore is an important primary producer in the tropical and subtropical oligotrophic oceans (Carpenter et al., 2004). This global importance of *Trichodesmium* has motivated numerous studies regarding the physiological responses of *Trichodesmium* to environmental factors, including visible light, phosphorus, iron, temperature, and CO_2 (Kranz et al., 2010; Shi et al., 2012; Fu et al., 2014; Spungin et al., 2014; Hutchins et al., 2015). However, to the best of our knowledge, there have been no reports on how UV exposure may affect *Trichodesmium*.

Trichodesmium spp. have a cosmopolitan distribution throughout much of the oligotrophic tropical and subtropical oceans, where there is a high penetration of solar UV-A and UV-B radiation (Carpenter et al., 2004). It also frequently forms extensive surface blooms (Westberry and Siege, 2006), where it is presumably exposed to very high levels of UV radiation. Moreover, in the ocean, *Trichodesmium* populations may experience continuously changing irradiance intensities as a result of vertical mixing. Cells photoacclimated to reduced irradiance at lower depths might be subject to solar UV radiation (UVR) damage when they are vertically delivered close to the sea surface due to mixing. Therefore, this unique cyanobacterium may have developed defensive mechanisms to overcome harmful effects of frequent exposures to intense UV radiation. Understanding how its N_2 fixation and photosynthesis respond to UV irradiance will thus further our knowledge of its ecological and biogeochemical roles in the ocean.

When estimating N_2 fixation using incubation experiments in the field, marine scientists have typically excluded UV radiation by using incubation bottles made of UV-opaque materials like polycarbonate (Capone et al., 1998; Olson et al., 2015). Thus, it seems possible that most shipboard measurements of *Trichodesmium* N_2 fixation rates could be overestimates of actual rates under natural UV exposure conditions in the surface ocean. Our experiments were specifically designed because of the importance of *Trichodesmium* in the input of carbon and nitrogen on oligotrophic oceans, and the lack of studies about the impact of enhanced UVR on the C and N fixation. In this study, *Trichodesmium* was exposed to spectrally realistic irradiances of UVR in laboratory experiments to examine the short-term effects of UVR on photosynthesis and N_2 fixation. In addition, *Trichodesmium* was grown under natural solar irradiance outdoors in order to assess UV impacts on longer timescales, and to test for induction of protective mechanisms to ameliorate chronic UV exposure effects.

2 Materials and methods

2.1 Experimental design

The experiments to evaluate how UVR affects photosynthesis and N_2 fixation of *Trichodesmium* were carried out in indoor and outdoor environments as follows, with the study divided into two parts: (1) a short-term experiment under a solar simulator (refer to Fig. S1 in the Supplement for the spectrum) to examine the responses of *Trichodesmium erythraeum* IMS 101 to a range of acute UV radiation exposures, and (2) a long-term UV experiment under natural sunlight to examine acclimated growth and physiology of *Trichodesmium* IMS 101. The first set of experiments was intended to mimic intense but transitory UV exposures, as might occur sporadically during vertical mixing, while the second set was intended to give insights into responses during extended near-surface UV exposures, such as during a surface bloom event.

2.2 Short-term UV experiment

Trichodesmium erythraeum IMS101 strain was isolated from the North Atlantic Ocean (Prufert-Bebout et al., 1993) and maintained in laboratory stock cultures in exponential growth phase in autoclaved artificial seawater enriched with nitrogen-free YBCII medium (Chen et al., 1996). For the short-term UV experiment, the cells were grown under low light (LL; 70 µmol photons $m^{-2}\,s^{-1}$) and hight light (HL; 400 µmol photons $m^{-2}\,s^{-1}$) (12 : 12 light : dark) photosynthetically active radiation (PAR) for at least 50 generations (about 180 days) prior to the UV experiments. These two light levels represent growth sub-saturating and super-saturating levels for *Trichodesmium* (Cai et al., 2015). Cultures were grown in triplicate using a dilute semi-continuous culture method, with medium renewed every 4–5 days at 25 °C. The cell concentration was maintained at $< 5 \times 10^4$ cell mL^{-1}.

To determine the short-term responses of *Trichodesmium* IMS101 to UV radiation, subcultures of *Trichodesmium* IMS101 were dispensed at a final cell density of 2–4×10^4 cells mL^{-1} into containers that allow transmission of all or part of the UV spectrum, including 35 mL quartz tubes (for measurements of carbon fixation or measurements of fluorescence parameters), 100 mL quartz tubes (for pigment measurements), or 13 mL gas-tight borosilicate glass vials (for N_2 fixation measurements). Three triplicated radiation treatments were implemented: (1) PAB (PAR + UV-A + UV-B) treatment, using tubes covered with Ultraphan film 295 (Digefra, Munich, Germany), thus receiving irradiances > 295 nm; (2) PA (PAR + UV-A) treatment, using tubes covered with Folex 320 film (Montagefolie, Folex, Dreieich, Germany), and receiving irradiances > 320 nm; and (3) P treatment – tubes covered with Ultraphan film 395 (UV Opak, Digefra), with samples receiving irradiances above 395 nm, representing PAR (400–700 nm). Since the transmission spectrum of the borosilicate glass was similar to that of Ultraphan film 295, the borosilicate glass vials for N_2 fixation measurements of PAB treatment were uncovered. Transmission spectra of these tubes (quartz and borosilicate) and the various cut-off foils used in this study are shown in Fig. S1.

The experimental tubes were placed under a solar simulator (Sol 1200W; Dr. Hönle, Martinsried, Germany) at a distance of 110 cm from the lamp, and maintained in a circulating water bath for temperature control (25 °C) (CTP-3000, Eyela, Japan). Irradiance intensities were measured with a LI-COR 2π PAR sensor (PMA2100, Solar Light, USA) that has channels for PAR (400–700 nm), UV-A (320–400 nm), and UV-B (280–320 nm). Measured values at the 110 cm distance were 87 W m^{-2} (PAR, ca. 400 µmol photons m^{-2} s^{-1}), 28 W m^{-2} (UV-A), and 1 W m^{-2} (UV-B). For the fluorescence measurements, samples were exposed under a solar simulator for 60 min and measurements of fluorescence parameters were performed during the exposure (see below). Due to analytical sensitivity issues, for the carbon and N_2 incorporation measurements, the exposure duration was 2 h, and for the measurements of UVAC (UV-absorbing compounds) contents, the exposure time was 10 h.

2.3 Long-term UV experiment

To assess the long-term effects of solar ultraviolet radiation on *Trichodesmium* IMS101, an outdoor experiment was carried out during the winter (1 to 26 January 2014) in subtropical Xiamen, China. Cell cultures of 300–400 mL were grown in 500 mL quartz vessels exposed to 100 % daytime natural solar irradiance (surface ocean irradiance) (daytime PAR average of ~ 120 W m^{-2}, highest PAR at noon ~ 300 W m^{-2}). All of the quartz vessels were placed in a shallow water bath at 25 °C using a temperature control system (CTP-3000, Eyela, Japan). Two triplicated radiation treatments were implemented: (1) treatment P – PAR alone (400–700 nm), tubes covered with Ultraphan film 395 (UV Opak, Digefra); (2) treatment PAB – PAR + UV-A + UV-B (295–700 nm), unwrapped quartz tubes. Incident solar radiation was continuously monitored with a broadband Eldonet filter radiometer (Eldonet XP, Real Time Computer, Möhrendorf, Germany) that was placed near the water bath. Daily doses of solar PAR, UV-A and UV-B during the experiments are shown in Fig. S2. The photoperiod during the outdoor incubation was 11 : 13 light : dark (light period from 07:00 to 18:00 local time). Cells were maintained in exponential growth phase (cell density $< 5 \times 10^4$), with dilutions (after sunset) every 4 days. All parameters were measured after acclimation under P or PAB radiation for a week.

In order to evaluate adaptation responses of *Trichodesmium* to natural solar irradiance, all parameters were obtained after one week acclimation outdoor. Specific growth rate (μ, d^{-1}) of *Trichodesmium* IMS101 was determined based on the change in cell concentrations over 4 days during the 8th–11th and 12th–15th day using microscopic counts (Cai et al., 2015); the corresponding total dose from day 8 to day 11 and from day 12 to day 15 were 17.03 and 18.51 MJ m^{-2}, respectively. Chl *a* content was measured at the 11th, 15th, and 19th day, and Chl *a*-specific absorption spectrum was measured at the 18th day. Carbon and N_2 fixation rate were measured at 11:00–13:00 on the 18th day; the diel solar irradiance record on that day is given in Fig. S3. In order to separate the respective effects of UV-A and UV-B on carbon and N_2 fixation, a shift experiment was carried out: subcultures from either P or PAB treatments were transferred into another P (PAR), PA (PAR + UV-A), or PAB (PAR + UV-A + UV-B) treatment, which were marked as P′, PA′ and PAB′ treatments, respectively (namely P-grown cells divided into P′, PA′, and PAB′ treatments; PAB-grown cells also divided into P′, PA′, and PAB′ treatments). For carbon and N_2 fixation measurements, 35 mL quartz tubes and 13 mL gas-tight borosilicate glass vials were used, respectively, as described below. Triplicate samples were used for each radiation treatment for carbon and N_2 fixation, and the incubations were performed under 100 % solar irradiance for 2 h.

3 Measurements and analyses

3.1 Effective photochemical quantum yield

Effective photochemical quantum yield (F'_V/F'_M) is generally considered to be light quantum use efficiency. We use this parameter to indicate photosystem II activity. During the exposure under the solar simulator in the short-term experiment, small aliquots of cultures (2 mL) were withdrawn at time intervals of 3–10 min and immediately measured (without any dark adaptation) using a pulse-amplitude-modulated (PAM) fluorometer (Xe-PAM, Walz, Germany). The quantum yield of PSII (F'_V/F'_M) was determined by measuring the instant maximum fluorescence (F'_M) and the steady-state fluorescence (F_t) under the actinic light. The maximum fluorescence (F'_M) was determined using a saturating light pulse (4000 µmol photons m^{-2} s^{-1} in 0.8 s) with the actinic light level set at 400 µmol photons m^{-2} s^{-1}, similar to the PAR level during the solar simulator exposure. The quantum yield was calculated as $F'_V/F'_M = (F'_M - F_t)/F'_M$ (Genty et al., 1989).

3.2 Chlorophyll-specific absorption spectra and UV-absorbing compounds (UVACs)

Chl *a*-specific absorption spectra were measured on the 18th day, after consecutive sunny days. Cellular absorption spectra were measured using the "quantitative filter technique" (Kiefer and SooHoo, 1982; Mitchell, 1990). The cells were filtered onto GF/F glass fiber filters and scanned from 300 to 800 nm using a 1 nm slit in a spectrophotometer equipped with an integrating sphere to collect all the transmitted or forward-scattered light (i.e., light diffused by the filter and the quartz diffusing plate). Filters soaked in culture medium were used as blanks. Chlorophyll-specific absorption cross-sections (a^*) were calculated according to Cleveland and Weidemann (1993) and Anning et al. (2000). Content of Chl *a* and UV-absorbing compounds (UVACs) were measured by filtering the samples onto GF/F filters and subsequently extracted in 4 mL of 100 % methanol overnight in darkness at 4 °C. The absorption of the supernatant was measured by a scanning spectrophotometer (Beckman Coulter Inc., Fullerton, CA, USA). The concentration of Chl *a* was calculated according to Ritchie (2006). The main absorption values for UV-absorbing compounds ranged between wavelengths of 310 and 360 nm, and the peak absorption value at 332 nm was used to estimate total absorptivity of UVACs according to Dunlap et al. (1995). The absorptivity of UVACs was finally normalized to the Chl *a* content ($\mu g\,(\mu g\,\mathrm{Chl}\,a)^{-1}$).

Trichodesmium IMS101 UVACs content was compared to that of three other marine phytoplankton species, including *Chlorella* sp., *Phaeodactylum tricornutum*, and *Synechococcus* WH7803, representing a green alga, a diatom, and a unicellular cyanobacterium, respectively. All cultures were maintained under the same conditions (25° C, 150 µmol photons $m^{-2}\,s^{-1}$) for several days prior to pigment extraction. The absorption spectra were measured using the same method in *Trichodesmium* by filtering the samples on GF/F filters, which were subsequently extracted in 4 mL of 100 % methanol overnight at 4 °C. The absorption spectra of the supernatant were scanned from 250 to 800 nm in a spectrophotometer (Beckman Coulter Inc., Fullerton, CA, USA). The optical density (OD) values were then normalized to OD (662 nm) at Chl *a* peak.

3.3 Carbon fixation rate

Carbon fixation rates of both short- and long-term experiments were measured using the ^{14}C method. Samples of 20 mL were placed in 35 mL quartz tubes and inoculated with 5 µCi (0.185 MBq) of labeled sodium bicarbonate (ICN Radiochemicals) and were then maintained under the corresponding radiation treatments for 2 h. After incubation, the cells were filtered onto Whatman GF/F filters (Φ25 mm) and stored at −20 °C until analysis. To determine the radioactivity, the filters were thawed and then exposed to HCl fumes overnight and dried at 60 °C for 4 h before being placed in scintillation cocktail (Hisafe 3, Perkin-Elmer, Shelton, CT, USA) and measured with a scintillation counter (Tri-Carb 2800TR, Perkin-Elmer, Shelton, CT, USA) as previously described (Cai et al., 2015).

3.4 N_2 fixation rate

Rates of N_2 fixation for both short- and long-term experiments were measured in parallel with the carbon fixation measurements using the acetylene reduction assay (ARA) (Capone, 1993). Samples of 5 mL subcultures were placed in 13 mL gas-tight borosilicate vials (described above), and 1 mL of acetylene was injected into the headspace before incubating for 2 h under the corresponding radiation treatment conditions. A 500 µL headspace sample was then analyzed in a gas chromatograph equipped with a flame-ionization detector and quantified relative to an ethylene standard. The ethylene produced was calculated using the Bunsen gas solubility coefficients according to Breitbarth et al. (2004) and an ethylene production to N_2 fixation conversion factor of 4 was used to derive N_2 fixation rates, which were then normalized to cell number.

4 Data analysis

The inhibition of ΦPSII, carbon fixation, and N_2 fixation due to UVR, UV-A, or UV-B was calculated as

$$\text{UVR-induced inhibition} = (I_P - I_{PAB})/I_P \times 100\,\%$$
$$\text{UV-A-induced inhibition} = (I_P - I_{PA})/I_P \times 100\,\%$$
$$\text{UV-B-induced inhibition} = \mathrm{UVR_{inh}} - \mathrm{UVA_{inh}},$$

where I_P, I_{PA}, and I_{PAB} indicate the values of carbon fixation or N_2 fixation in the P, PA, and PAB treatments, respectively. Repair (r) and damage (k) rates during the 60 min exposure period in the presence of UV were calculated using the Kok model (Heraud and Beardall, 2000):

$$P/P_{\mathrm{initial}} = r/(r+k) + k/(r+k) \times \exp(-(r+k) \times t),$$

where P_{initial} and P were the yield values at the beginning and at exposure time t. Three replicates for culture conditions or each radiation condition was used in all experiments, and the data are plotted as mean and standard deviation values. Two-way ANOVA tests were used to determine the interaction between acclimatization conditions and UVR at a significance level of $p = 0.05$.

5 Results

For the short-term UV experiment, the effects of acute UVR exposure on cells grown under LL and HL conditions are shown in Fig. 1. For the cells grown under LL condition, the F'_V/F'_M declined sharply within 10 min after first exposure in all radiation treatments and then leveled off. F'_V/F'_M decreased less in the samples receiving PAR alone (to 43 % of the initial value) than those additionally receiving UV-A (to 30 % of the initial value) or UV-A + UV-B (to 24 % of the initial value) (Fig. 1a). The F'_V/F'_M value of PA

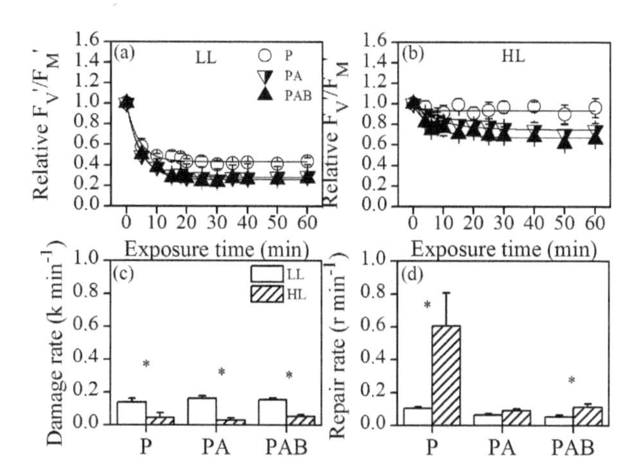

Figure 1. Changes of effective quantum yield (F'_V/F'_M) of *Trichodesmium* IMS101 grown under **(a)** LL and **(b)** HL conditions while exposed to PAR (P), PAR + UVA (PA), and PAR + UVA + UVB (PAB) under a solar simulator for 60 min. PSII damage (**c**; k, in min^{-1}) and repair rates (**d**; r, in min^{-1}) of LL- and HL-grown cells were derived from the yield decline curve in the upper panels. Asterisks above the histogram bars indicate significant differences between LL- and HL-grown cells. Values are the mean ± SD for triplicate incubations.

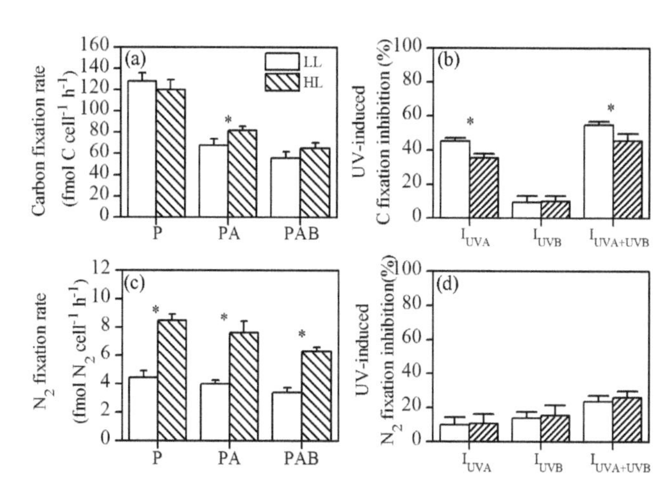

Figure 2. Photosynthetic carbon fixation rate (**a**; $fmol\,C\,cell^{-1}\,h^{-1}$) and UV-induced C fixation inhibition **(b)**, N_2 fixation rate (**c**; $fmol\,N_2\,cell^{-1}\,h^{-1}$), and corresponding UV-induced N_2 fixation inhibition **(d)** of *Trichodesmium* IMS101 grown under LL and HL conditions. Asterisks above the histogram bars indicate significant differences between LL- and HL-grown cells. Values are the mean ± SD for triplicate incubations.

and PAB treatments were significantly lower compared to the PAR treatment ($p = 0.03$ and $p < 0.01$, respectively). F'_V/F'_M of HL-grown cells declined less and more slowly compared to the LL-grown cells. The F'_V/F'_M of HL cells under PAR alone remained more or less constant during the exposure, since the PAR level was similar to the growth level of HL (400 μmol photons $m^{-2}\,s^{-1}$). In contrast, the F'_V/F'_M decreased to 75 and 65 % of its initial value for the PA and PAB treatment, respectively, and were significantly lower than the P treatment ($p < 0.01$) (Fig. 1b).

The damage and repair rates of the PSII reaction center estimated from the exponential decay in the effective quantum yield showed higher damage and lower repair rates in the LL-grown cells than in the HL-grown ones (Fig. 1c, d). The PSII damage rates (k, min^{-1}) of LL-grown cells were 0.14, 0.16 and 0.15 min^{-1} in the P, PA, and PAB treatments, respectively, about 2 times faster than in the cells grown under HL conditions (Fig. 1c). The PSII repair rates (r, min^{-1}) of LL-grown cells were 0.1, 0.06, and 0.05 min^{-1} in the P, PA, and PAB treatments, which were 83 % ($p < 0.01$), 33 % ($p < 0.01$), and 54 % ($p < 0.01$) lower than in HL-grown cells, respectively (Fig. 1d). The damage rate was not significantly different among P, PA, and PAB treatments within either of the LL- and HL-grown treatments ($p > 0.05$), but the repair rate was much higher in the P treatment without UV than in PA or PAB treatments in the HL-grown cells ($p < 0.01$).

The photosynthetic carbon fixation and N_2 fixation rates during the UV exposure are shown in Fig. 2. The HL-grown cells had 17 % higher photosynthetic carbon fixation rates than the LL-grown ones under the PA treatment ($p < 0.01$); however, the LL- and HL-grown cells did not show significant differences in carbon fixation rates under the P and PAB treatments ($p = 0.29$, and $p = 0.06$). In the presence of UV radiation, carbon fixation was significantly inhibited in both LL and HL-grown cells (Fig. 2a). Carbon fixation inhibition induced by UV-A was about 35–45 %, much larger than that induced by UV-B, which caused only about a 10 % inhibition of carbon fixation ($p < 0.01$). The UV-A exposed carbon fixation rate was significantly higher in the LL-grown cells than in HL-grown cells ($p < 0.01$), while UV-B did not cause a significant difference in inhibition between the HL- and LL-grown cells ($p = 0.88$) (Fig. 2b). N_2 fixation rates were about 2-fold higher in HL-grown cells in all radiation treatments (Fig. 2c, $p < 0.01$), but the UV-induced N_2 fixation inhibition showed no significant differences between the LL- and HL-grown cells regardless of UV-A or UV-B exposures (Fig. 2d, $p = 0.80$, 0.62, 0.39 for UVA-, UVB-, and UVR-induced inhibition, respectively).

Compared to other phytoplankton under the same growth conditions, *Trichodesmium* IMS101 had much higher absorbance in the UV region (300–400 nm) (Fig. 3a). In this study, the absorbance at 332 nm of HL-grown cells was about 2-fold higher compared to LL-grown ones (Fig. 3b). However, the cellular Chl *a* content (data not shown) and UVACs contents of both LL- and HL-grown cells did not present differences between radiation treatments after exposure to UV for 10 h (Fig. 3c).

For the long-term UV experiment, after being acclimated under full natural solar radiation for 7 days, the specific growth rates of cells grown under the PAB treatment were 0.15 ± 0.01 and 0.14 ± 0.06 during the 8th–11th day and

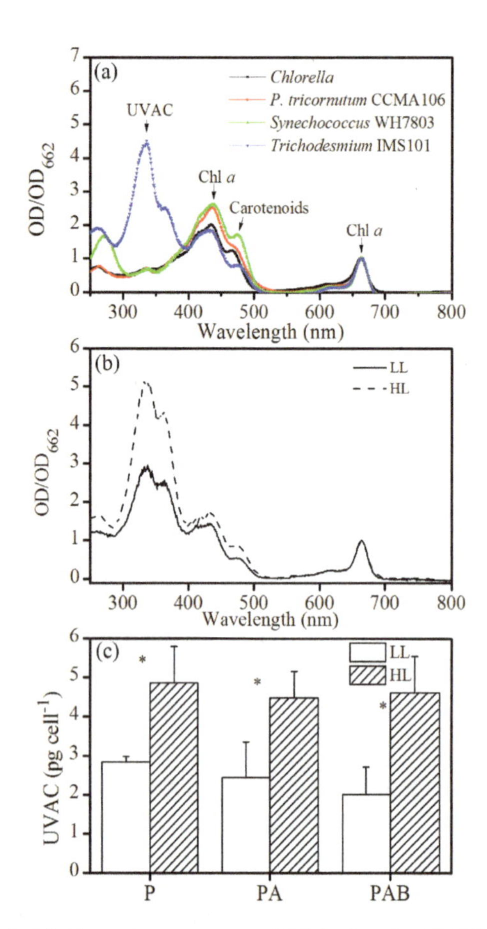

Figure 3. (a) Absorption spectrum of *Trichodesmium* IMS101 compared to other phytoplankton. Pigments were extract by 100 % methanol. OD value normalized to OD_{662} (Chl *a*). **(b)** Absorption spectrum of the *Trichodesmium* IMS101 grown under LL and HL conditions, with OD value normalized to OD_{662} (Chl *a*). **(c)** Cellular contents of UVACs of *Trichodesmium* IMS101 grown under LL and HL conditions after exposure to PAR (P), PAR + UVA (PA), and PAR + UVA + UVB (PAB) under a solar simulator for 10 h. Asterisks above the histogram bars indicate significant differences between LL- and HL-grown cells. Values are the mean ± SD for triplicate incubations.

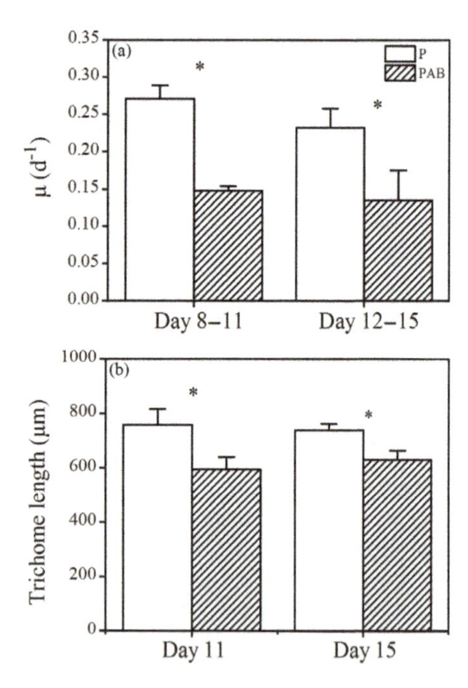

Figure 4. (a) Specific growth rate (measured during 8th–11th and 12th–15th days) of *Trichodesmium* IMS101 grown under solar PAR (P) and PAR + UVA + UVB (PAB). Corresponding total solar doses from day 8 to day 11 and from day 12 to day 15 were 17.03 and 18.51 MJ, respectively. **(b)** Trichome length (measured on the 11th and 15th day) of *Trichodesmium* IMS101 grown under solar PAR (P) and PAR + UVA + UVB (PAB). The asterisks indicate significant differences between radiation treatments. Values are the mean ± SD for triplicate cultures.

12th–15th day periods, respectively. These growth rates were significantly lower by 44 and 39 % compared to cells grown under the P treatment, respectively (Fig. 4a, $p = 0.014$ and $p = 0.03$). The mean trichome lengths of P treatment cells on the 11th and 15th day were 758 ± 56 and 726 ± 19 μm, while addition of UVR significantly reduced the trichome length by 22 % (day 11, $p = 0.02$)and 11 % (day 15, $p = 0.02$).

Analysis of the Chl *a*-specific absorption spectra, $a^*(\lambda)$, demonstrated that UVR had a major effect on the absorbance of UV regions and phycobilisomes (Fig. 5). The optical absorption spectra revealed a series of peaks in the UV and visible wavelengths corresponding to the absorption peaks of UVACs at 332 nm, Chl *a* at 437 and 664 nm, phycourobilin (PUB) at 495 nm, phycoerythrobilin (PEB) at 545 nm, phycoerythrocyanin (PEC) at 569 nm, and phycocyanin (PC) at 627 nm. In the UV region, the $a^*(\lambda)$ value was higher in the PAB treatment cultures than in the P treatment cultures (Fig. 5). The UVR treatments did not show clear effects on Chl *a* content compared to acclimation to P alone measured on different days (Fig. S3). However, the ratio of UVACs to Chl *a* was increased by 41 % in the PAB compared to the P treatment ($p < 0.01$).

The cells grown in the long-term P and PAB treatments showed different responses for carbon and N_2 fixation after being transferred to short-term P′, PA′, and PAB′ radiation treatments at noon on the 18th day (Fig. 6). P- and PAB-acclimated cells did not show significant differences in carbon fixation among all short-term P′, PA′ and PAB′ treatments (Fig. 6a, $p = 0.17$, $p = 0.22$, $p = 0.51$, respectively), nor in the UV-induced inhibition of carbon fixation (Fig. 6b, $p > 0.05$). However, inhibition induced by UV-A at short exposures was about 58 % in both P and PAB treatments and significantly higher than inhibition induced by UV-B radiation (Fig. 6b, $p < 0.01$).

N_2 fixation rates of P-acclimated cells were significantly higher than PAB-acclimated cells in all P′, PA′, and PAB′ treatments (Fig. 6c, $p < 0.01$). The N_2 fixation inhibition induced by UV-A of PAB-acclimated cells was 49 %, significantly higher by 47 % than that of P-

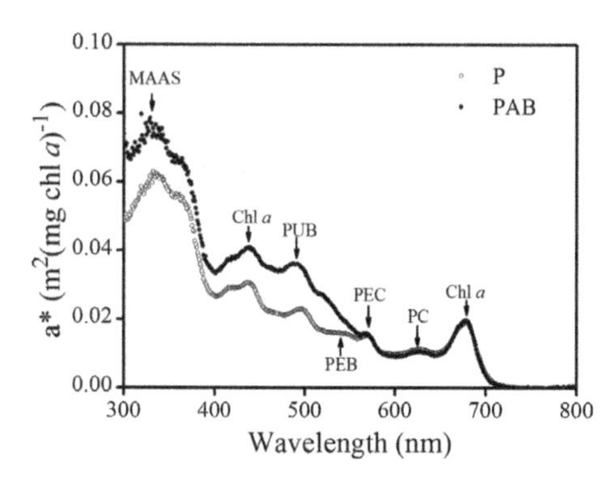

Figure 5. Chl *a*-specific absorption spectrum (a^*) of *Trichodesmium* IMS101 grown under solar PAR (P) and PAR + UVA + UVB (PAB). The measurements were taken on the 18th day. The absorption peaks of MAAs (330 nm), PUB (495 nm), PEB (545 nm), PEC (569 nm), PC (625 nm), and Chl *a* (438 and 664 nm) are indicated.

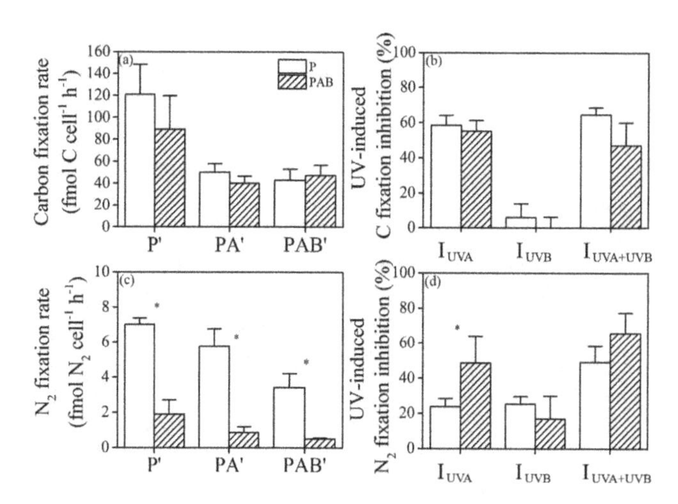

Figure 6. Photosynthetic carbon fixation rate (**a**; fmol C cell^{-1} h^{-1}) and UV-induced C fixation inhibition **(b)**, N_2 fixation rate (**c**; fmol N_2 cell^{-1} h^{-1}), and corresponding UV-induced N_2 fixation inhibition **(d)** of *Trichodesmium* IMS101 grown under solar PAR (P) and PAR + UVA + UVB (PAB) transferred to other P′, PA′ and PAB′ treatments. The measurement was taken on the 18th day at 11:00–13:00. Asterisks above the histogram bars indicate significant differences between P and PAB treatments. Values are the mean ± SD for triplicate incubations.

acclimated cells ($p = 0.03$), while there was no significant difference in UVB-induced N_2 fixation inhibition between P- and PAB-acclimated cells (Fig. 6d, $p = 0.62$). The carbon fixation rates measured under P (P treated cells to P′) and PAB (PAB treated cells to PAB′) conditions were 89.2 and 47.1 fmol C cell^{-1} h^{-1}, respectively, while N_2 fixation rates measured under those conditions were 1.9 and 0.5 fmol N_2 cell^{-1} h^{-1}. UVR exposure lowered estimates of carbon and N_2 fixation rates by 47 and 65 %, respectively.

6 Discussion

Our study shows that growth, photochemistry, photosynthesis, and N_2 fixation in *Trichodesmium* sp. are all significantly inhibited by UVR, including both UV-A and UV-B. These effects occur in both short-term, acute exposures and after extended exposures during acclimated growth. These results are ecologically relevant, since this cyanobacterium is routinely exposed to elevated solar irradiances in its tropical habitat either transiently, during vertical mixing, or over longer periods during surface blooms. *Trichodesmium* provides a biogeochemically critical source of new N to open-ocean food webs, so significant UV inhibition of its growth and N_2 fixation rates could have major consequences for ocean biology and carbon cycling.

Short exposure to UVR causes a significant decline in the quantum yield of photosystem II (PSII) fluorescence of *Trichodesmium*, which is consistent with damage to critical PSII proteins such as D1 in a brackish water cyanobacterium *Arthrospira (Spirulina) platensis* (Wu et al., 2011). UV-induced degradation of D1 proteins results in inactivation of PSII, leading to reduction in photosynthetic activity (Campbell et al., 1998). In addition, studies of various microbial mats have shown that RuBisco activity and supply of ATP and NADPH are inhibited under UV exposure, which might also lead to the reduction in photosynthetic carbon fixation (Cockell and Rothschild, 1999; Sinha et al., 1996, 1997).

Exposure to UVR had an impact on nitrogenase activity in *Trichodesmium*, since both the short- and the long-term UV exposure led to significant reduction in N_2 fixation of up to 30 % (short-term) or ∼ 60 % (long-term) (Figs. 2d and 6d). Studies on the freshwater cyanobacterium *Anabaena* sp. (subg. *Dolichospermum*) have shown a 57 % decline in N_2 fixation rate after 30 min of exposure to UVR of 3.65 W (Lesser, 2007). Some rice-field cyanobacteria completely lost N_2 fixation activity after 25–40 min of exposure to UV-B from a 2.5 W source (Kumar et al., 2003). In our results, long-term exposure to UV led to higher inhibition of N_2 fixation, implying that accumulated damage to the key N_2-fixing enzyme, nitrogenase, could have occurred during the growth period under solar radiation in the presence of UVR.

Compared to N_2 fixation, UVR induced an even higher degree of inhibition of carbon fixation. The carbon fixation rate decreased by 50 % in the presence of UVR. UV-A induced higher inhibition than UV-B, indicating that although UV-B photons (295–320 nm) are in general more energetic and damaging than UV-A (320–400 nm), the greater fluxes of UV-A caused more inhibition of carbon fixation, which was consistent with other studies of spectral dependence of UV effects (Cullen and Neale, 1994; Neale, 2000). This finding is ecologically significant, since UV-A penetrates much deeper into clear open ocean and coastal seawater than does UV-B.

Compared to low-light-grown cells, the high-light-grown ones were more resistant to UVR, which was reflected in the lower PSII damage rate and faster recovery rate in the presence of UVR, as well as the significantly lower levels of carbon fixation inhibition caused by UV-A and/or UV-B. Such a reduced sensitivity to UVR coincided well with a significant increase in UV-absorbing compounds in the HL-grown cells compared to the LL-grown ones. Similar dependence of photosynthetic sensitivity to UV inhibition on growth light levels has been reported in other species of phytoplankton (Litchman and Neale, 2005; Sobrino and Neale, 2007). A red-tide dinoflagellate *Gymnodinium sanguineum* Hirasaka accumulates 14-fold higher MAAs in high-light-grown cells (76 $W m^{-2}$) than in low-light-grown ones (15 $W m^{-2}$) and the former ones have lower sensitivity to UVR at wavelengths strongly absorbed by the MAAs (Neale et al., 1998). The sensitivity of PSII quantum yield to UV exposure in *Synechococcus* WH7803 was also less in high-light-grown versus low-light-grown cells (Garczarek et al., 2008). In addition, it has been observed that phytoplankton from turbid waters or acclimated to low-light conditions are more sensitive to UVR than those from clear waters (Villafañe et al., 2004; Litchman and Neale, 2005; Helbing et al., 2015). These observations suggest that *Trichodesmium* spp. may acclimate to growth in the upper mixed layer by producing UV-absorbing compounds, making them more tolerant of UVR than cells living at deeper depths.

Although UVR can clearly cause damage to PSII and inhibit physiological processes in *Trichodesmium* sp., this cyanobacterium has evolved protective biochemical mechanisms to deal with UVR in their natural high-UV habitat. One important class of UV-absorbing substances consists of MAAs and scytonemin. These compounds strongly absorb in the UV-A and/or UV-B region of the spectrum and dissipate their energy as heat without forming reactive oxygen species, protecting the cells from UV and from photooxidative stress (Banaszak, 2003). The MAAs, which have strong UV-absorption maxima between 310 and 362 nm (Sinha and Häder, 2008) as identified by high-performance liquid chromatography in other studies, consist of a group of small, water-soluble compounds, including asterina-332 ($\lambda max = 332$) and shinorine ($\lambda max = 334$), which are the most abundant, as well as mycosporine-glycine ($\lambda max = 310$), porphyra-334 ($\lambda max = 334$), and palythene ($\lambda max = 360$) (Shick and Dunlap, 2002; Subramaniam et al., 1999). As was found previously in *Trichodesmium* spp., high absorbance in the UV region is mainly due to the presence of MAAs, with absorbance maxima between 310 and 362 nm (Sinha and Häder, 2008).

Our investigation strongly suggests that *Trichodesmium* is able to synthesize MAAs ($\lambda max \sim 330$ and 360 nm) in response to elevated PAR and UVR. Synthesis of MAAs has been reported to be stimulated by high PAR and UVR in other phytoplankton (Karsten et al., 1998; Vernet and Whitehead, 1996; Sinha et al., 2001). Our high-light-grown cells were more tolerant of UVR, likely at least partly due to their ability to synthesize double the amount of MAAs in comparison to low-light-grown ones (Fig. 3b). It has been showed that accumulation of MAAs may represent a natural defensive system against exposure to biologically harmful UVR (Karsten et al., 1998) and cells with high concentrations of MAAs are more resistant to UVR than cells with small amounts of these compounds (Garcia-Pichel and Castenholz, 1993). In fact, MAA concentrations varying between 0.9 and 8.4 µg mg (dry weight)$^{-1}$ have been measured in cyanobacterial isolates (Garcia-Pichel and Castenholz, 1993), and ratios of MAAs to Chl *a* in the range of 0.04 to 0.19 have been reported in cyanobacterial mats (Quesada et al., 1999). In our study, we found that *Trichodesmium* contained a much higher concentration of MAAs (the highest value in HL-grown cells is 5 pg cell^{-1}) and that the ratio of these compounds to Chl *a* was 5, consistent with previous reports in regard to *Trichodesmium* (Subramaniam et al., 1999), which is much higher than in other phytoplankton. This acclimatization capacity depending on intensity and spectral quality of radiation could be a major reason for the ability of *Trichodesmium* to grow and form extensive surface blooms under strong irradiation in the oligotrophic oceans.

In our study, no significant changes in the amount of MAAs were observed after 10 h of exposure to UVR under the solar simulator. In contrast, a significant increase of 23 % in the concentration of MAAs was observed in cells treated with the full solar spectrum compared to PAR-treated ones grown outdoors after consecutive sunny days (on the 18th). It seems that the synthesis of MAAs takes a relatively long time. Other studies have shown the time required for induction of MAAs in other cyanobacteria is dependent on UV doses and species and shows a circadian rhythm (Sinha et al., 2001, 2003).

Long-term exposure to high solar UVR significantly not only reduced *Trichodesmium*'s growth rate (by 37–44 %) but also significantly shortened its average trichome length (less cell per filament) (Fig. 4). The decreased growth rates correlated with decreased trichome length are consistent with our previous studies under different light levels without UVR (Cai et al., 2015). It has been reported that enhanced UVR is one of the environmental factors that not only inhibits the growth of cyanobacteria but also changes their morphology (Rastogi et al., 2014). Natural solar UVR can suppress formation of heterocysts and shorten the filament length of *Anabaena* sp. PCC7120, because UVR may affect calcium signaling then the expression of the key genes responsible for cell differentiation (Gao et al., 2007). Natural levels of solar UVR in southern China were also found to break the filaments and alter the spiral structure of *Arthrospira* (*Spirulina*) *platensis*, with a compressed helix that lessens UV exposures for the cells (Wu et al., 2005). Cells in the trichomes of the estuarine cyanobacterium *Lyngbya aestuarii* coil and then form small bundles in response to UV-B irradiation (Rath and Adhikari, 2007). However, the shortened trichomes of

Trichodesmium in this work may be a result of UV-inhibited growth rather than a responsive strategy against UV.

Carbon fixation in the long-term experiment showed similar patterns with the short-term UV experiment, demonstrating that UV-A played a larger role in inhibiting carbon fixation than UV-B. Since the ratio of UV-B to UV-A is lower in natural solar light (1 : 50) than under our artificial UVR (1 : 28), the inhibitory effects of UV-B were smaller compared to UV-A in the cultures under sunlight. Carbon fixation and N_2 fixation rates measured outdoors indicated that UV-induced carbon fixation inhibition recovers quickly following transfer to PAR conditions, while the UV-induced N_2 fixation inhibition does not (Fig. 6a, c). Factors that might be responsible include lower turnover rate of nitrogenase than that of RuBisco, more UV-induced damage to nitrogenase with lower efficiency of repair (Kumar et al., 2003), and indirect harm caused by reactive oxygen species induced by UV (Singh et al., 2014).

The UV effects in our study were measured under conditions that minimized self-shading, namely during growth as single filaments. However, in its natural habitat *Trichodesmium* often grows in a colonial form, with packages of many cells held together by an extracellular sheath (Capone et al., 1998). In such colonial growth forms, the effective cellular path lengths for UVR are likely greatly increased, thereby amplifying the overall sunscreen factor for the colony. *Trichodesmium* spp. might use this colony strategy to protect themselves from natural UV damage in the ocean.

Our investigation shows that this cyanobacterium appears to have evolved the ability to produce exceptionally high levels of UV protective compounds, likely MAAs. However, even this protective mechanism is insufficient to prevent substantial inhibition of nitrogen and carbon fixation in the high-irradiance environment where this genus lives. *Trichodesmium* spp. are distributed in the upper layers of the euphotic zone in oligotrophic waters, and their population densities are generally greatest at relatively shallow depths (20 to 40 m) in the upper water column (Capone et al., 1997). It seems likely that UV inhibition therefore significantly reduces the amount of critical new nitrogen supplied by *Trichodesmium* to the N-limited oligotrophic gyre ecosystems, a possibility that has not been generally considered in regional or global models of the marine nitrogen cycle. On the other hand, the UV-absorbing compounds (most likely MAAs) are expensive to make in terms of nitrogen in particular (Singh et al., 2008). Decreased nitrogen supplied may increase sensitivity of phytoplankton assemblages to UV further (Litchman et al., 2002), thus potentially creating a positive feedback between N limitation and the UV sensitivity.

Trichodesmium can form dense, extensive blooms in the surface oceans, and a frequently cited estimate of global nitrogen fixation rates by *Trichodesmium* blooms is $\sim 42\,\mathrm{Tg\,N\,yr^{-1}}$ (Westberry et al., 2006). Previous biogeochemical models of global N_2 fixation have emphasized controls by many environmental factors, including solar PAR, temperature, wind speed, and nutrient concentrations (Luo et al., 2014), but have largely neglected the effects of UVR. When estimating N_2 fixation using incubation experiments in the field, however, marine scientists have typically excluded UVR by using incubation bottles made of UV-opaque materials like polycarbonate (Olson et al., 2015). Our results suggest that under solar radiation at the surface ocean, including realistic levels of UVR inhibition lowers estimates of carbon fixation and N_2 fixation by around 47 and 65 %, respectively (Fig. 6).

Thus, it seems likely that shipboard measurements and possibly current model projections of *Trichodesmium* N_2 fixation and primary production rates that do not take into account UV inhibition could be substantial overestimates. However, our study was only carried out under full solar radiation, simulating sea surface conditions, so further studies are needed to investigate depth-integrated UV inhibition. Moreover, the response to UVR may be taxon-specific. For example, unicellular N_2-fixing cyanobacteria such as the genus *Crocosphaera*, with smaller cell size and thus greater light permeability, may be more vulnerable to UVR than *Trichodesmium* (Wu et al., 2015). In the future, as enhanced stratification and decreasing mixed layer depth expose cells to relatively higher UV levels, differential sensitivities to UVR may result in changes in diazotroph community composition. Such UV-mediated assemblage shifts could have potentially major consequences for marine productivity, and for the global biogeochemical cycles of nitrogen and carbon. Future research would be necessary to confirm and/or deepen the consequences of UV effects in carbon and nitrogen cycle in the ocean.

Competing interests. The authors declare that they have no conflict of interest.

Acknowledgements. This study was supported by grants from the national key R&D program (2016YFA0601400), the National Natural Science Foundation (41430967, 41720104005), and the joint project of the National Natural Science Foundation of China and Shandong province (no. U1606404) to Kunshan Gao, and by US National Science Foundation grants OCE 1260490 and OCE 1538525 to Feixue Fu and David A. Hutchins. Feixue Fu's visit to Xiamen was supported by MEL's visiting scientists programs. The authors acknowledge financial support from the China Scholarship Council during Xiaoni Cai's visit to the University of Southern California. The authors would like to thank Nana Liu and Xiangqi Yi from Xiamen University for their kind assistance during the experiments.

Edited by: Gerhard Herndl

References

Anning, T., Maclntyre, H. L., Sammes, S. M. P. a. P. J., Gibb, S., and Geider, R. J.: Photoacclimation in the marine diatom *Skeletonema costatum*, Limnol. Oceanogr., 45, 1807–1817, 2000.

Banaszak, A. T.: Photoprotective physiological and biochemical responses of aquatic organisms, in: UV effects in aquatic organisms and ecosystems, edited by: Helbling, E. W. and Zagarese, H. E., Comprehensive Series in Photosciences, Royal Society of Chemistry, Cambridge, UK, 329–356, 2003.

Breitbarth, E., Mills, M. M., Friedrichs, G., and LaRoche, J.: The Bunsen gas solubility coefficient of ethylene as a function of temperature and salinity and its importance for nitrogen fixation assays, Limnol. Oceanogr.-Meth., 2, 282–288, 2004.

Cai, X., Gao, K., Fu, F., Campbell, D., Beardall, J., and Hutchins, D.: Electron transport kinetics in the diazotrophic cyanobacterium *Trichodesmium* spp. grown across a range of light levels, Photosyn. Res., 124, 45–56, https://doi.org/10.1007/s11120-015-0081-5, 2015.

Campbell, D., Eriksson, M. J., Oquist, G., Gustafsson, P., and Clarke, A. K.: The cyanobacterium *Synechococcus* resists UV-B by exchanging photosystem II reaction-center D1 proteins, P. Natl. Acad. Sci. USA, 95, 364–369, 1998.

Capone, D.: Determination of nitrogenase activity in aquatic samples using the acetylene reduction procedure, in: Handbook of methods in aquatic microbial ecology, edited by: Kemp, P. F., Sherr, B. F., Sherr, E. B., and Cole, J. J., Lewis Publishers, Boca Raton, Fla, 621–631, 1993.

Capone, D., Zehr, J., Paerl, H., and Bergman, B.: *Trichodesmium*, a globally significant marine cyanobacterium, Science, 276, 1221–1227, 1997.

Capone, D. G., Subramaniaml, A., Joseph, P., Carpenters, E. J., Johansen, M., and Ronald, L.: An extensive bloom of the N_2-fixing cyanobacterium *Trichodesmium erythraeum* in the central Arabian Sea, Mar. Ecol.-Prog. Ser., 172, 281–292, 1998.

Carpenter, E. J., Subramaniam, A., and Capone, D. G.: Biomass and primary productivity of the cyanobacterium *Trichodesmium* spp. in the tropical N Atlantic ocean, Deep-Sea Res. Pt I, 51, 173–203, https://doi.org/10.1016/j.dsr.2003.10.006, 2004.

Chen, Y. B., Zehr, J. P., and Mellon, M.: Growth and nitrogen fixation of the diazotrophic filamentous nonheterocystous cyanobacterium *Trichodesmium* sp. IMS101 in defined media: evidence for a circadian rhythm, J. Phycol., 32, 916–923, 1996.

Cleveland, J. S. and Weidemann, A. D.: Quantifying Absorption by Aquatic Particles: A Multiple Scattering Correction for Glass-Fiber, Limnol. Oceanogr., 38, 1321–1327, 1993.

Cockell, C. S. and Rothschild, L. J.: The Effects of UV Radiation A and B on Diurnal Variation in Photosynthesis in Three Taxonomically and Ecologically Diverse Microbial Mats, Photochem. Photobiol., 69, 203–210, https://doi.org/10.1111/j.1751-1097.1999.tb03274.x, 1999.

Cullen, J. J. and Neale, P. J.: Ultraviolet radiation, ozone depletion, and marine photosynthesis, Photosyn. Res., 39, 303–320, https://doi.org/10.1007/bf00014589, 1994.

Dunlap, W., Rae, G., Helbling, E., Villafañe, V., and Holm-Hansen, O.: Ultraviolet-absorbing compounds in natural assemblages of Antarctic phytoplankton, Antarct. J. US, 30, 323–326, 1995.

Fu, F.-X., Yu, E., Garcia, N. S., Gale, J., Luo, Y., Webb, E. A., and Hutchins, D. A.: Differing responses of marine N_2 fixers to warming and consequences for future diazotroph community structure, Aquat. Microb. Ecol., 72, 33–46, 2014.

Gao, K., Yu, H., and Brown, M. T.: Solar PAR and UV radiation affects the physiology and morphology of the cyanobacterium *Anabaena* sp. PCC 7120, J. Photochem. Photobiol. B., 89, 117–124, https://doi.org/10.1016/j.jphotobiol.2007.09.006, 2007.

Garcia-Pichel, F. and Castenholz, W. R.: Occurrence of UV-absorbingmycosporine-like compounds among cyanobacterial isolates and estimationof their screening capacity, Appl. Environ. Microb., 59, 163–169, 1993.

Garczarek, L., Dufresne, A., Blot, N., Cockshutt, A. M., Peyrat, A., Campbell, D. A., Joubin, L., and Six, C.: Function and evolution of the psbA gene family in marine *Synechococcus*: *Synechococcus* sp. WH7803 as a case study, ISME J., 2, 937–953, 2008.

Genty, B., Briantais, J.-M., and Baker, N. R.: The relationship between the quantum yield of photosynthetic electron transport and quenching of chlorophyll fluorescence, Biochimica et Biophysica Acta (BBA) – General Subjects, 990, 87–92, https://doi.org/10.1016/S0304-4165(89)80016-9, 1989.

Häder, D.-P. and Gao, K.: Interactions of anthropogenic stress factors on marine phytoplankton, Frontiers in Environmental Science, 3, 1–14, 2015.

Häder, D. P., Williamson, C. E., Wangberg, S. A., Rautio, M., Rose, K. C., Gao, K., Helbling, E. W., Sinha, R. P., and Worrest, R.: Effects of UV radiation on aquatic ecosystems and interactions with other environmental factors, Photochem. Photobiol. Sci., 14, 108–126, https://doi.org/10.1039/c4pp90035a, 2015.

Helbling, W., Banaszak, A. T., and Villafañe, V. E.: Global change feed-back inhibits cyanobacterial photosynthesis, Sci. Rep., 5, 14514–14514, 2015.

Heraud, P. and Beardall, J.: Changes in chlorophyll fluorescence during exposure of Dunaliella tertiolecta to UV radiation indicate a dynamic interaction between damage and repair processes, Photosyn. Res., 63, 123–134, https://doi.org/10.1023/a:1006319802047, 2000.

Hutchins, D. A., Walworth, N. G., Webb, E. A., Saito, M. A., Moran, D., McIlvin, M. R., Gale, J., and Fu, F.-X.: Irreversibly increased nitrogen fixation in *Trichodesmium* experimentally adapted to elevated carbon dioxide, Nat. Commun., 6, 8155, https://doi.org/10.1038/ncomms9155, 2015.

Karsten, U., Sawall, T., and Wiencke, C.: A survey of the distribution of UV-absorbing substances in tropical macroalgae, Phycol. Res., 46, 271–279, https://doi.org/10.1046/j.1440-1835.1998.00144.x, 1998.

Kiefer, D. A. and SooHoo, J. B.: Spectral absorption by marine particles of coastal waters of Baja California, Limnol. Oceanogr., 27, 492–499, 1982.

Kranz, S. A., Levitan, O., Richter, K. U., Prasil, O., Berman-Frank, I., and Rost, B.: Combined effects of CO_2 and light on the N_2-fixing cyanobacterium *Trichodesmium* IMS101: physiological responses, Plant Physiol., 154, 334–345, https://doi.org/10.1104/pp.110.159145, 2010.

Kumar, A., Tyagi, M. B., Jha, P. N., Srinivas, G., and Singh, A.: Inactivation of cyanobacterial nitrogenase after exposure to Ultraviolet-B radiation, Curr. Microbiol., 46, 380–384, https://doi.org/10.1007/s00284-001-3894-8, 2003.

Lesser, M. P.: Effects of ultraviolet radiation on productivity and nitrogen fixation in the Cyanobacterium, *Anabaena* sp. (Newton's strain), Hydrobiologia, 598, 1–9, https://doi.org/10.1007/s10750-007-9126-x, 2007.

Litchman, E., Neale, P. J., and Banaszak, A. T.: Increased sensitivity to ultraviolet radiation in nitrogen-limited dinoflagellates: Photoprotection and repair, Limnol. Oceanogr., 47, 86–94, 2002.

Litchman, E. and Neale, P. J.: UV effects on photosynthesis, growth and acclimation of an estuarine diatom and cryptomonad, Mar. Ecol.-Prog. Ser., 300, 53–62, 2005.

Luo, Y.-W., Lima, I. D., Karl, D. M., Deutsch, C. A., and Doney, S. C.: Data-based assessment of environmental controls on global marine nitrogen fixation, Biogeosciences, 11, 691–708, https://doi.org/10.5194/bg-11-691-2014, 2014.

Mitchell, B. G.: Algorithms for determining the absorption coefficient for aquatic particulates using the quantitative filter technique, Orlando'90, 137–148, 1990.

Neale, P. J., Banaszak, A. T., and Jarriel, C. R.: Ultraviolet sunscreens in *Gymnodinium sanguineum* (Dinophyceae): mycosporine-like amino acids protect against inhibition of photosynthesis, J. Phycol., 34, 928–938, 1998.

Neale, P. J.: Spectral weighting functions for quantifying the effects of ultraviolet radiation in marine ecosystems, edited by: De Mora, S. J, Demers, S., and Vernet, M., Effects of UV Radiation on the Marine Environment, Cambridge University Press, Cambridge, UK, 73–100, 2000.

Olson, E. M., McGillicuddy, D. J., Dyhrman, S. T., Waterbury, J. B., Davis, C. S., and Solow, A. R.: The depth-distribution of nitrogen fixation by *Trichodesmium* spp. colonies in the tropical–subtropical North Atlantic, Deep-Sea Res. Pt. I, 104, 72–91, https://doi.org/10.1016/j.dsr.2015.06.012, 2015.

Prufert-Bebout, L., Paerl, H. W., and Lassen, C.: Growth, nitrogen fixation, and spectral attenuation in cultivated *Trichodesmium* species, Appl. Environ. Microb., 59, 1367–1375, 1993.

Quesada, A., Vincent, W. F., and Lean, D. R. S.: Community and pigment structure of Arctic cyanobacterial assemblages: the occurrence and distribution of UV-absorbing compounds, FEMS Microbiol. Ecol., 28, 315–323, https://doi.org/10.1111/j.1574-6941.1999.tb00586.x, 1999.

Rastogi, R. P., Sinha, R. P., Moh, S. H., Lee, T. K., Kottuparambil, S., Kim, Y. J., Rhee, J. S., Choi, E. M., Brown, M. T., Hader, D. P., and Han, T.: Ultraviolet radiation and cyanobacteria, J. Photoch. Photobio. B, 141, 154–169, https://doi.org/10.1016/j.jphotobiol.2014.09.020, 2014.

Rath, J. and Adhikary, S. P.: Response of the estuarine cyanobacterium *Lyngbya aestuarii* to UV-B radiation, J. Appl. Phycol., 19, 529–536, 2007.

Ritchie, R. J.: Consistent sets of spectrophotometric chlorophyll equations for acetone, methanol and ethanol solvents, Photosyn. Res., 89, 27–41, https://doi.org/10.1007/s11120-006-9065-9, 2006.

Shi, D., Kranz, S. A., Kim, J. M., and Morel, F. M. M.: Ocean acidification slows nitrogen fixation and growth in the dominant diazotroph *Trichodesmium* under low-iron conditions, P. Natl. Acad. Sci. USA, 109, E3094–E3100, 2012.

Shick, J. M. and Dunlap, W. C.: Mycosporine-like amino acids and related Gadusols: biosynthesis, acumulation, and UV-protective functions in aquatic organisms, Annu. Rev. Physiol., 64, 223–262, https://doi.org/10.1146/annurev.physiol.64.081501.155802, 2002.

Singh, S. P., Kumari, S., Rastogi, R. P., Singh, K. L., and Sinha, R. P.: Mycosporine-like amino acids (MAAs): chemical structure, biosynthesis and significance as UV-absorbing/screening compounds, Indian J. Exp. Biol., 46, 7–17, 2008.

Singh, S. P., Rastogi, R. P., Hader, D. P., and Sinha, R. P.: Temporal dynamics of ROS biogenesis under simulated solar radiation in the cyanobacterium *Anabaena variabilis* PCC 7937, Protoplasma, 251, 1223–1230, https://doi.org/10.1007/s00709-014-0630-3, 2014.

Sinha, R. P. and Häder, D.-P.: UV-protectants in cyanobacteria, Plant Sci., 174, 278–289, https://doi.org/10.1016/j.plantsci.2007.12.004, 2008.

Sinha, R. P., Singh, N., Kumar, A., Kumar, H. D., Häder, M., and Häder, D. P.: Effects of UV irradiation on certain physiological and biochemical processes in cyanobacteria, J. Photoch. Photobio. B, 32, 107–113, https://doi.org/10.1016/1011-1344(95)07205-5, 1996.

Sinha, R. P., Singh, N., Kumar, A., Kumar, H. D., and Häder, D.-P.: Impacts of ultraviolet-B irradiation on nitrogen-fixing cyanobacteria of rice paddy fields, J. Plant. Physiol., 150, 188–193, https://doi.org/10.1016/S0176-1617(97)80201-5, 1997.

Sinha, R. P., Klisch, M., Walter Helbling, E., and Häder, D.-P.: Induction of mycosporine-like amino acids (MAAs) in cyanobacteria by solar ultraviolet-B radiation, J. Photoch. Photobio. B, 60, 129–135, https://doi.org/10.1016/S1011-1344(01)00137-3, 2001.

Sinha, R. P., Ambasht, N. K., Sinha, J. P., Klisch, M., and Häder, D.-P.: UV-B-induced synthesis of mycosporine-like amino acids in three strains of *Nodularia* (cyanobacteria), J. Photoch. Photobio. B, 71, 51–58, https://doi.org/10.1016/j.jphotobiol.2003.07.003, 2003.

Sobrino, C. and Neale, P. J.: Short-term and long-term effects of temperature on photosynthesis in the diatom *Thalassiosira Pseudonana* under UVR exposures, J. Phycol., 43, 426–436, https://doi.org/10.1111/j.1529-8817.2007.00344.x, 2007.

Sohm, J. A., Webb, E. A., and Capone, D. G.: Emerging patterns of marine nitrogen fixation, Nat. Rev. Microbiol., 9, 499–508, https://doi.org/10.1038/nrmicro2594, 2011.

Spungin, D., Berman-Frank, I., and Levitan, O.: *Trichodesmium*'s strategies to alleviate P-limitation in the future acidified oceans, Environ. Microbiol., 16, 1935–1947, https://doi.org/10.1111/1462-2920.12424, 2014.

Subramaniam, A., Carpenter, E. J., Karentz, D., and Falkowski, P. G.: Bio-optical properties of the marine diazotrophic cyanobacteria *Trichodesmium* spp. I. Absorption and photosynthetic action spectra, Limnol. Oceanogr., 44, 608–617, 1999.

Vernet, M. and Whitehead, K.: Release of ultraviolet-absorbing compounds by the red-tide dinoflagellate *Lingulodinium polyedra*, Mar. Biol., 127, 35–44, https://doi.org/10.1007/bf00993641, 1996.

Villafañe, V. E., Barbieri, E. S., and Helbling, E. W.: Annual patterns of ultraviolet radiation effects on temperate marine phytoplankton off Patagonia, Argentina, J. Plankton. Res., 26, 167–174, https://doi.org/10.1093/plankt/fbh011, 2004.

Westberry, T. K. and Siegel, D. A.: Spatial and temporal distribution of *Trichodesmium* blooms in the world's oceans, Global Biogeochem. Cy., 20, GB4016, https://doi.org/10.1029/2005gb002673, 2006.

Wu, H., Gao, K., Villafane, V. E., Watanabe, T., and Helbling, E. W.: Effects of solar UV radiation on morphology and photosynthesis of filamentous cyanobacterium *Arthrospira platensis*, Appl. Environ. Microb., 71, 5004–5013, https://doi.org/10.1128/AEM.71.9.5004-5013.2005, 2005.

Wu, H., Abasova, L., Cheregi, O., Deák, Z., Gao, K., and Vass, I.: D1 protein turnover is involved in protection of Photosystem II against UV-B induced damage in the cyanobacterium *Arthrospira (Spirulina) platensis*, J. Photoch. Photobio. B, 104, 320–325, https://doi.org/10.1016/j.jphotobiol.2011.01.004, 2011.

Wu, Y., Li, Z., Du, W., and Gao, K.: Physiological response of marine centric diatoms to ultraviolet radiation, with special reference to cell size, J. Photoch. Photobio. B, 153, 1–6, https://doi.org/10.1016/j.jphotobiol.2015.08.035, 2015.

Permissions

All chapters in this book were first published in BIOGEOSCIENCES, by Copernicus Publications; hereby published with permission under the Creative Commons Attribution License or equivalent. Every chapter published in this book has been scrutinized by our experts. Their significance has been extensively debated. The topics covered herein carry significant findings which will fuel the growth of the discipline. They may even be implemented as practical applications or may be referred to as a beginning point for another development.

The contributors of this book come from diverse backgrounds, making this book a truly international effort. This book will bring forth new frontiers with its revolutionizing research information and detailed analysis of the nascent developments around the world.

We would like to thank all the contributing authors for lending their expertise to make the book truly unique. They have played a crucial role in the development of this book. Without their invaluable contributions this book wouldn't have been possible. They have made vital efforts to compile up to date information on the varied aspects of this subject to make this book a valuable addition to the collection of many professionals and students.

This book was conceptualized with the vision of imparting up-to-date information and advanced data in this field. To ensure the same, a matchless editorial board was set up. Every individual on the board went through rigorous rounds of assessment to prove their worth. After which they invested a large part of their time researching and compiling the most relevant data for our readers.

The editorial board has been involved in producing this book since its inception. They have spent rigorous hours researching and exploring the diverse topics which have resulted in the successful publishing of this book. They have passed on their knowledge of decades through this book. To expedite this challenging task, the publisher supported the team at every step. A small team of assistant editors was also appointed to further simplify the editing procedure and attain best results for the readers.

Apart from the editorial board, the designing team has also invested a significant amount of their time in understanding the subject and creating the most relevant covers. They scrutinized every image to scout for the most suitable representation of the subject and create an appropriate cover for the book.

The publishing team has been an ardent support to the editorial, designing and production team. Their endless efforts to recruit the best for this project, has resulted in the accomplishment of this book. They are a veteran in the field of academics and their pool of knowledge is as vast as their experience in printing. Their expertise and guidance has proved useful at every step. Their uncompromising quality standards have made this book an exceptional effort. Their encouragement from time to time has been an inspiration for everyone.

The publisher and the editorial board hope that this book will prove to be a valuable piece of knowledge for researchers, students, practitioners and scholars across the globe.

List of Contributors

Sang H. Lee, Bo Kyung Kim, Yu Jeong Lim, HuiTae Joo, Jae Joong Kang and Dabin Lee
Department of Oceanography, Pusan National University, Geumjeong-gu, Busan 609-735, South Korea

Jisoo Park, Sun-Yong Ha and Sang Hoon Lee
Korea Polar Research Institute, Incheon 406-840, South Korea

Yujie Bai, Yun Tian, Jingyong Ma, Wei Feng and Bin Wu
Yanchi Research Station, School of Soil and Water Conservation, Beijing Forestry University, Beijing 100083, China

Duo Qian
Beijing Vocational College of Agriculture, Beijing 102442, China

Charles P.-A. Bourque
Faculty of Forestry and Environmental Management, 28 Dineen Drive, University of New Brunswick, New Brunswick, E3B5A3, Canada

Heli Peltola
Faculty of Science and Forestry, School of Forest Sciences, University of Eastern Finland, Joensuu, 80101, Finland

Tianshan Zha and Xin Jia
Yanchi Research Station, School of Soil and Water Conservation, Beijing Forestry University, Beijing 100083, China
Key Laboratory of State Forestry Administration on Soil and Water Conservation, Beijing Forestry University, Beijing, China

Yan Liu and Wenhua Xiang
Faculty of Life Science and Technology, Central South University of Forestry and Technology, Changsha 410004, Hunan, China

Xiaoyong Chen
National Engineering Laboratory for Applied Technology of Forestry and Ecology in South China, Central South University of Forestry and Technology, Changsha 410004, Hunan, China
Division of Science, College of Arts and Sciences, Governors State University, University Park, Illinois 60484, USA

Pifeng Lei and Wende Yan
Faculty of Life Science and Technology, Central South University of Forestry and Technology, Changsha 410004, Hunan, China
National Engineering Laboratory for Applied Technology of Forestry and Ecology in South China, Central South University

Keith F. Lewin, Andrew M. McMahon, Kim S. Ely, Shawn P. Serbin and Alistair Rogers
Environmental and Climate Sciences Department, Brookhaven National Laboratory, Upton, NY 11973, USA

S. Nemiah Ladd
Department of Surface Waters - Research and Management, Eawag, Swiss Federal Institute of Aquatic Science and Technology, 6047 Kastanienbaum, Switzerland
Department of Earth Sciences, ETH Zürich, 8092 Zürich, Switzerland

Nathalie Dubois
Department of Earth Sciences, ETH Zürich, 8092 Zürich, Switzerland
Department of Surface Waters - Research and Management, Eawag, Swiss Federal Institute of Aquatic Science and Technology, 8600 Dübendorf, Switzerland

Carsten J. Schubert
Department of Surface Waters - Research and Management, Eawag, Swiss Federal Institute of Aquatic Science and Technology, 6047 Kastanienbaum, Switzerland
Institute of Biogeochemistry and Pollutant Dynamics, ETH Zürich, 8092 Zürich, Switzerland

Yulong Zhang
College of Land and Environment, Shenyang Agricultural University, Shenyang, 110161, Liaoning, China

Zhanxiang Sun, Jiaming Zheng, Wei Bai, Liangshan Feng, Chen Feng, Zhe Zhang, Ning Yang and Yang Liu
Tillage and Cultivation Research Institute, Liaoning Academy of Agricultural Sciences, Shenyang, 110161, Liaoning, China

Yue Zhang and Lizhen Zhang
College of Resources and Environmental Sciences, China Agricultural University, Beijing, 100193, China

Jochem B. Evers
Wageningen University, Centre for Crop Systems Analysis (CSA), Droevendaalsesteeg 1, 6708 PB Wageningen, the Netherlands

Qian Cai
College of Land and Environment, Shenyang Agricultural University, Shenyang, 110161, Liaoning, China
Tillage and Cultivation Research Institute, Liaoning Academy of Agricultural Sciences, Shenyang, 110161, Liaoning, China

Jianzhong Su, Minhan Dai, Lifang Wang, Xianghui Guo, Huade Zhao and Fengling Yu
State Key Laboratory of Marine Environmental Science, Xiamen University, Xiamen, China

Jianping Gan
Department of Mathematics and Division of Environment, Hong Kong University of Science and Technology, Kowloon, Hong Kong SAR, China

Biyan He
State Key Laboratory of Marine Environmental Science, Xiamen University, Xiamen, China
College of Food and Biological Engineering, Jimei University, Xiamen, China

Jian Sun and Ning Zong
Key Laboratory of Ecosystem Network Observation and Modeling, Institute of Geographic Sciences and Natural Resources Research, Chinese Academy of Sciences, Datun Road, Beijing 100101, China

Qingping Zhou
Institute of Qinghai–Tibetan Plateau, Southwest Minzu University, Chengdu 610041, China

Shuli Niu
Key Laboratory of Ecosystem Network Observation and Modeling, Institute of Geographic Sciences and Natural Resources Research, Chinese Academy of Sciences, Datun Road, Beijing 100101, China
Department of Resources and Environment, University of Chinese Academy of Sciences, Beijing 100049, China

Linghao Li
State Key Laboratory of Vegetation and Environmental Change, Institute of Botany, Chinese Academy of Sciences, Xiangshan, Beijing 100093, China

Bing Song
Key Laboratory of Ecosystem Network Observation and Modeling, Institute of Geographic Sciences and Natural Resources Research, Chinese Academy of Sciences, Datun Road, Beijing 100101, China
State Key Laboratory of Vegetation and Environmental Change, Institute of Botany, Chinese Academy of Sciences, Xiangshan, Beijing 100093, China

Johanne H. Rydsaa, Frode Stordal and Lena M. Tallaksen
Department of Geosciences, University of Oslo, Oslo, Norway

Anders Bryn
Natural History Museum, University of Oslo, Oslo, Norway

Federico Baltar
Department of Marine Science, University of Otago, Dunedin, New Zealand
Niwa/University of Otago Research Centre for Oceanography, Dunedin, New Zealand

Blair Thomson, Christopher David Hepburn and Miles Lamare
Department of Marine Science, University of Otago, Dunedin, New Zealand

Katharine J. Crawfurd
Nioz Royal Netherlands Institute for Sea Research, Department of Marine Microbiology and Biogeochemistry and Utrecht University, 1790 AB Den Burg, Texel, the Netherlands

Santiago Alvarez-Fernandez
Alfred-Wegener-Institut Helmholtz-Zentrum für Polar- und Meeresforschung, Biologische Anstalt Helgoland, 27498 Helgoland, Germany

Kristina D. A. Mojica
Department of Botany and Plant Pathology, Cordley Hall 2082, Oregon State University, Corvallis, Oregon 97331-29052, USA

Ulf Riebesell
GEOMAR Helmholtz Centre for Ocean Research Kiel, Biological Oceanography, Düsternbrooker Weg 20, 24105 Kiel, Germany

Corina P. D. Brussaard
Nioz Royal Netherlands Institute for Sea Research, Department of Marine Microbiology and Biogeochemistry and Utrecht University, 1790 AB Den Burg, Texel, the Netherlands

Aquatic Microbiology, Institute for Biodiversity and Ecosystem Dynamics, University of Amsterdam, 1090 GE Amsterdam, the Netherlands

Perran L. M. Cook
Water Studies Centre, Monash University, Clayton 3800, Australia

Andrew McCowan
Water Technology Pty Ltd, 15 Business Park Drive, Notting Hill 3168, Australia

Yafei Zhu
Water Studies Centre, Monash University, Clayton 3800, Australia
Water Technology Pty Ltd, 15 Business Park Drive, Notting Hill 3168, Australia

Dandan Duan and Dainan Zhang
State Key Laboratory of Organic Geochemistry, Guangzhou Institute of Geochemistry, Chinese Academy of Sciences, Guangzhou, Guangdong 510640, China
University of Chinese Academy of Sciences, Beijing 100049, China

Jingfu Wang and Jing'an Chen
State Key Laboratory of Environmental Geochemistry, Institute of Geochemistry, Chinese Academy of Sciences, Guiyang 55002, China

Yu Yang and Yong Ran
State Key Laboratory of Organic Geochemistry, Guangzhou Institute of Geochemistry, Chinese Academy of Sciences, Guangzhou, Guangdong 510640, China

Sandy J. Thomalla and Thato N. Mtshali
Southern Ocean Carbon and Climate Observatory, Natural Resources and Environment, CSIR, Rosebank, Cape Town 7700, South Africa

Hazel Little
Department of Oceanography, University of Cape Town, Rondebosch, Cape Town 7701, South Africa

Thomas J. Ryan-Keog
Southern Ocean Carbon and Climate Observatory, Natural Resources and Environment, CSIR, Rosebank, Cape Town 7700, South Africa
Department of Oceanography, University of Cape Town, Rondebosch, Cape Town 7701, South Africa

Perran Louis Miall Cook and Adam John Kessler
Water Studies Centre, School of Chemistry, Monash University, Clayton, Australia

Bradley David Eyre
Centre for Coastal Biogeochemistry, Southern Cross University, Lismore, Australia

Andrew McMinn
Institute of Marine and Antarctic Science, University of Tasmania, Hobart 7001, Tasmania, Australia

Kunshan Gao
State Key Laboratory of Marine Environmental Science, Xiamen University, Xiamen, Fujian, 361102, China

David A. Hutchins and Feixue Fu
Department of Biological Sciences, University of Southern California, 3616 Trousdale Parkway, Los Angeles, California, 90089, USA

Xiaoni Cai
State Key Laboratory of Marine Environmental Science, Xiamen University, Xiamen, Fujian, 361102, China
Department of Biological Sciences, University of Southern California, 3616 Trousdale Parkway, Los Angeles, California, 90089, USA

Index

CPSIA information can be obtained
at www.ICGtesting.com
Printed in the USA
BVHW011429100619
550611BV00002B/234/P